International Series in Operations Research & Management Science

Volume 216

More information about this series at http://www.springer.com/series/6161

Michael C. Fu
Editor

Handbook of Simulation Optimization

Springer

Editor
Michael C. Fu
University of Maryland
College Park, MD, USA

ISSN 0884-8289 ISSN 2214-7934 (electronic)
ISBN 978-1-4939-1383-1 ISBN 978-1-4939-1384-8 (eBook)
DOI 10.1007/978-1-4939-1384-8
Springer New York Heidelberg Dordrecht London

Library of Congress Control Number: 2014950046

Printed on acid-free paper

Springer is part of Springer Science+Business Media (www.springer.com)

Preface

Arguably, the two most powerful operations research/management science (OR/MS) techniques are simulation and optimization. Simulation in this book will refer to stochastic simulation, whereby there is randomness in the system, also known as Monte Carlo simulation. Optimization dates back many centuries and is generally considered the older of the two siblings. Both approaches were propelled forward by the advent of the digital computer over half a century ago, leading up to the present golden age when both routinely address complex large-scale real-world problems and both are implemented in a large variety of computer software packages. However, combining the two techniques is a more recent development, and software effectively integrating the two is relatively limited; thus, simulation optimization remains an exciting and fertile area of research. The purpose of the handbook is to provide an overview of the state of the art of simulation optimization, comprising a survey of the most well-established approaches and a sampling of recent research advances in theory/methodology.

The single volume should serve as a reference for those already in the field and as a means for those new to the field for understanding and applying the main approaches to problems of interest. The intended audience includes researchers, practitioners, and graduate students in the business/engineering fields of operations research, management science, operations management, and stochastic control, as well as in economics/finance and computer science.

In addition to sincerely thanking the authors for their contributions, which have resulted in a high-quality volume, I wish to gratefully acknowledge the support and encouragement of Fred Hillier, the Series Editor, whose persistence ensured that I would not give up on this project when other commitments appeared overwhelming. Last but certainly not least, I'd like to thank Marie Chau for her superb editorial assistance.

College Park, MD, USA
February 2014

Michael C. Fu

Contents

Contributors

Sigrún Andradóttir Georgia Institute of Technology, Atlanta, GA, USA

Güzin Bayraksan The Ohio State University, Columbus, OH, USA

Marie Chau University of Maryland, College Park, MD, USA

Chun-Hung Chen George Mason University, Fairfax, VA, USA

Stephen E. Chick INSEAD, Fountainbelau, France

Michael C. Fu University of Maryland, College Park, MD, USA

Saeed Ghadimi University of Florida, Gainesville, FL, USA

Abhijit Gosavi Missouri University of Science and Technology, Rolla, MO, USA

Shane G. Henderson Cornell University, Ithaca, NY, USA

Tito Homem-de-Mello Universidad Adolfo Ibañez, Santiago, Chile

L. Jeff Hong Hong Kong University of Science and Technology, Clear Water Bay, Hong Kong

Jiaqiao Hu Stony Brook University, Stony Brook, NY, USA

Sujin Kim National University of Singapore, Singapore, Singapore

Jack P.C. Kleijnen Tilburg University, Tilburg, The Netherlands

Guanghui Lan University of Florida, Gainesville, FL, USA

Loo Hay Lee National University of Singapore, Singapore, Singapore

Barry L. Nelson Northwestern University, Evanston, IL, USA

Raghu Pasupathy Purdue University, West Lafayette, IN, USA

Nugroho A. Pujowidianto Hewlett-Packard Singapore, Singapore

Jie Xu George Mason University, Fairfax, VA, USA

Zelda B. Zabinsky University of Washington, Seattle, WA, USA

Selected Abbreviations and Notation

a.s.	almost surely
c.d.f.	cumulative distribution function
CLT	central limit theorem
p.d.f.	probability density function
p.m.f.	probability mass function
e.g.	for example (exempli gratia)
i.e.	that is (id est)
i.i.d.	independent and identically distributed
s.t.	subject to, or such that
w.p.	with probability
cf.	compare, or see (confer)
$\mathsf{E}[\cdot]$	expectation
$\mathsf{P}(\cdot)$	probability
$\mathrm{Var}[\cdot]$	variance
$\mathrm{Cov}(\cdot)$	covariance
$f(\cdot)$	objective function (except in Chaps. 2, 5, and 12, where it is a p.d.f.)
$g(\cdot)$	gradient estimate or p.d.f. (objective function in Chaps. 2 and 9)
x or $\mathbf{x}$	decision variables (vector)
$x^\top$ or $A^\top$	transpose of vector x or matrix A
$Y(x,\xi)$	output random variable
$\overline{Y}$	mean or sample average of Y
$\hat{f}$	estimate of f
ξ	randomness
θ, $J(\theta)$	parameter, performance measure (Chap. 5 only)
Θ	constrained parameter set
$\lvert x\rvert$, $\lvert\Theta\rvert$	absolute value (or modulus) of real-valued x, size of set Θ
$\lVert\cdot\rVert$	norm, e.g., Euclidean or L^2
$\mathbf{1}\{\cdot\}$	indicator function of the set $\{\cdot\}$
∇	gradient operator
$\equiv$ or $:=$	equal by definition

$\overset{d}{=}$	equal in distribution
$\overset{d}{\to}$	convergence in distribution
$\overset{p}{\to}$	convergence in probability
$\Longrightarrow$	implies
$\Longleftrightarrow$	if and only if
$\mathcal{N}(\mu,\sigma^2)$	normal distribution with mean μ and variance σ^2
$\mathcal{N}(\mu,\Sigma)$	multivariate normal distribution with mean (vector) μ and covariance matrix Σ
$U(a,b)$	(continuous) uniform distribution on $[a,b]$
$(x)^+$	$\max\{x,0\}$
$(x)^-$	$\max\{-x,0\}$
$\lceil x \rceil$	smallest integer greater than or equal to x (ceiling function)
$f(h) \in O(h)$	$\lim_{h\to 0}\sup f(h)/h < \infty$ or $\lim_{h\to\infty}\sup f(h)/h < \infty$
$f(h) \in o(h)$	$\lim_{h\to 0} f(h)/h = 0$ or $\lim_{h\to\infty} f(h)/h = 0$

Chapter 1
Overview of the Handbook

Michael C. Fu

Abstract This chapter provides a brief introduction to the handbook, including an overview of the contents of the other chapters.

Consider the optimization problem

$$\min_{x \in \Theta} f(x), \tag{1.1}$$

where f is the objective function, x represents the decision variables, and Θ is the feasible region or constraint set. All of the chapters with the exception of the final one basically address this problem in some form. As in the deterministic optimization domain, one dichotomy is whether the decision variables are discrete (ordered or unordered, finite or infinite) or continuous, or a mixture of the two. Most real-world simulation optimization problems have multiple objectives. Although this handbook focuses on the case of a single objective function, this (generally nonlinear) function could be understood as having already subsumed competing objectives through an appropriate combination or having moved all other objectives to the constraints in the constrained optimization setting (discussed in many of the chapters, but see especially Chap. 9, which focuses on stochastic constraints). Other approaches to multi-objective optimization such as goal programming and Pareto-optimal solutions are discussed briefly in Chaps. 10 and 11.

The term "simulation optimization" in this handbook generally refers to optimization in the setting where the objective function f cannot be computed exactly, i.e., it is estimated with some noise. More generally, such a setting is also known as "stochastic optimization," but simulation is used in its place here to emphasize the specific context of using (stochastic or Monte Carlo) simulation for the estimation. Thus, f is not directly available; rather, realizations of random variables (simulation replication outputs) are observed. For simplicity, assume that these observations can be summarized by a single random variable, call it $Y(x, \xi)$, where ξ represents the randomness, e.g., a stream of underlying random numbers in a stochastic simulation.

M.C. Fu (✉)
University of Maryland, College Park, MD, USA
e-mail: mfu@umd.edu

M.C. Fu (ed.), *Handbook of Simulation Optimization*, International Series in Operations Research & Management Science 216, DOI 10.1007/978-1-4939-1384-8_1

For most of the handbook, the objective function is the *expectation* of this output random variable (response), i.e.,

$$f(x) = \mathsf{E}[Y(x,\xi)],$$

but more generally it could be some other performance measure such as a quantile, i.e.,

$$f(x) = q_\alpha(x),$$

where the α-quantile q_α, $0 < \alpha < 1$, of Y is defined by

$$q_\alpha(x) = \sup\{y : \mathsf{P}(Y(x,\xi) \le y) \le \alpha\},$$

or equivalently $\alpha = \mathsf{P}(Y(x,\xi) \le q_\alpha(x))$ when Y is a continuous random variable.

However, in the literature, simulation optimization is also used sometimes to refer to the search process itself, i.e., utilizing randomization in guiding the search. These types of algorithms are the focus of Chaps. 11 and 12. And in some cases, simulation (or stochastic) optimization references both sources of stochasticity; hence, the book by Spall [32] is titled "Stochastic Search and Optimization" to cover both. Another term used is "optimization via simulation" as in the next chapter and in the early review paper on the topic [7].

To overview the contents, the handbook can be roughly categorized as follows. Discrete optimization is treated exclusively in Chaps. 2 and 3, whereas the rest of the handbook focuses mainly on continuous optimization, although the discussion of random search in Chap. 10 explicitly starts with the discrete setting. Specifically, Chap. 2 treats the discrete setting along the entire spectrum of possibilities: when the number of alternatives is relatively small, when the number of alternatives is very large but finite, and when the number of alternatives is (countably) infinite. In the first case, the statistical ranking & selection framework is followed, and several practical algorithms are presented using the probability of correct selection criterion. In the second case, the approach of ordinal optimization—the idea that ordering converges (exponentially) fast versus the canonical Monte Carlo inverse square root rate for estimation—is discussed. In the third case, random search methods tailored to the simulation setting are described, both globally and locally convergent algorithms. Chapter 3 also addresses the problem of selecting the best for a relatively small set of alternatives, but using an optimization framework on the simulation budget allocation rather than from a statistical perspective. The two main approaches covered are optimal computing budget allocation (OCBA) and expected value of information (EVI).

The continuous setting includes the most commonly used simulation optimization techniques: response surface methodology (Chap. 4), stochastic approximation (Chaps. 5–7), sample average approximation (Chaps. 8 and 9), and random search methods (Chaps. 10–12). Coincidentally, response surface methodology (RSM) and stochastic approximation (SA) both trace their roots back to 1951, launched by the seminal papers of Box and Wilson [3] and Robbins and Monro [27], respectively. Needless to say, neither of the two settings considered simulation optimization, since simulation itself was but in its infancy (the Monte Carlo method having just been introduced in 1949 [22]), but one could argue that at present simulation optimization is the primary driver of both fields. Chapter 4 summarizes the RSM approach and presents some recent advances, including a constrained optimization and robust optimization variants. Stochastic gradient estimation methods are presented in Chap. 5, setting the stage for the subsequent two chapters on SA methods. Chapter 5 also includes some recent developments on using gradients in new ways, for both SA and RSM. An overview of SA methods is provided in Chap. 6, covering the classical Robbins–Monro [27] and Kiefer–Wolfowitz [18] algorithms, as well as succeeding algorithms such as simultaneous perturbation stochastic approximation [31], which is particularly applicable for high-dimensional simulation optimization problems. Classical asymptotic convergence properties are summarized, and practical implementation challenges that arise in applying an SA algorithm in simulation optimization, such as those having to do with the step size (gain) sequences and the projection operator, are discussed in detail. Chapter 7 presents some recently developed SA algorithms that have been successfully applied to (convex) stochastic programming problems, and summarizes some theoretical properties of these algorithms. Letting x^* denote an optimal solution for the optimization problem (1.1), the theoretical results establish *finite-time* bounds on the distance from the optimal value $f(x^*)$, in contrast to classical results that focus on asymptotic convergence properties of the estimated optimum x^*.

Chapter 8 provides a guide to sample average approximation (SAA), with a focus on characterizing when it might be a fruitful approach. A novel connection to infinitesimal perturbation analysis (IPA), one of the stochastic gradient estimation techniques discussed in Chap. 5, is made. Theoretical properties of SAA are reviewed, and the approach is contrasted with SA for simulation optimization. Note that in principle, SAA can also be applied to mixed discrete-continuous optimization problems, although the focus of Chap. 8 is on cases where there is special structure such as continuity and differentiability for continuous optimization problems. For the SAA setting, recent advances in stochastic constraints and variance reduction techniques are treated in Chap. 9. However, as mentioned earlier, stochastic constraints arise in the general simulation optimization setting, so these results can also be viewed as beneficial from the broader simulation optimization perspective.

The next three chapters treat random search, which is also covered in Chap. 2 for the discrete setting. Chapter 10 explicitly treats the structure of random search for the setting where simulation is used to estimate the objective function, whereas the focus of the random search methods of Chaps. 11 and 12 is on the search rather than the estimation. Chapter 10 begins with a review of algorithms developed for the discrete deterministic setting (primarily simulated annealing) before moving to the stochastic simulation setting. The framework specifies the generation of a population of potential candidates at each iteration. Chapter 11 analyzes the theoretical convergence properties of various random search algorithms in the deterministic setting, focusing on iterative algorithms that consider one candidate at each iteration. Implementation issues are addressed using the well-known hit-and-run sampling algorithm for generating from a probability distribution. Along the same vein, Chap. 12 treats a relatively newer class of random search algorithms called model-based methods, which work directly with a probability distribution to generate a population of candidate solutions, and then update the probability distribution based on the sampled candidates' performance. This class of methods includes the cross-entropy method [28] and model reference adaptive search [17].

The final chapter covers the dynamic stochastic setting of Markov decision processes, where simulation-based approaches have been used successfully to solve large-scale problems [4, 5]. These techniques have become to be known as approximate dynamic programming [9, 26] in the operations research/management science community and reinforcement learning [15] in the computer science community.

Each of the chapters was written independently, with editorial coordination only on topical coverage and notational usage (see table on next page). Generally, the chapters can be read independently of each other, with a few possible exceptions being Chaps. 6 & 7 and Chaps. 8 & 9, as the first in each pair provides both background and context that enhances understanding of the second; moreover, both Chaps. 6 and 8 consider procedures that make use of stochastic gradient techniques, the topic of Chap. 5. As might be discerned by the earlier summary descriptions of the chapters, there is clearly some overlap of topical coverage, e.g., random search is covered from various angles in four different chapters.

This handbook focuses exclusively on methodology. An abundant source of simulation optimization applications, covering diverse areas including manufacturing, supply chain management, transportation, health care, finance/risk, and the military, can be found in the annual Winter Simulation Conference (WSC) proceedings, freely available on the World Wide Web at

http://informs-sim.org or http://wintersim.org.

As of the writing of this handbook, simulation optimization has become a regular full track at WSC, which is the premier conference in the field of stochastic discrete-event simulation (and also includes hybrid and agent-based simulations). One might ask whether or not there has been any widespread comparisons between different approaches. Some papers that have reported comparisons between SA and SAA include the following: an (s,S) inventory control problem [11]; a multidimensional newsvendor problem [19]; three stochastic programming problems [23]: a stochastic utility problem, a stochastic max-flow problem, and a network planning problem with random demand. The Simulation Optimization Library at

http://www.simopt.org

is a testbed of simulation optimization problems; see the Web site and [24, 25] for details.

Around the turn of the millennium, commercial software for simulation optimization started becoming prevalent. As discussed in [8], the majority of these are algorithms based on ideas from deterministic optimization, predominantly robust metaheuristics, offered as add-ons to existing simulation software packages. The last section of the next chapter (Sect. 2.8) includes discussion on using these commercial packages more effectively.

Listed in the following table is the specific notation used for the simulation optimization settings in each of the chapters, with the exception of the very last chapter (Chap. 13), which explicitly treats multi-stage problems where the solution is a policy rather than a set of decision variables, so the notation is necessarily quite different. The indicated "feasible region (constraint set)" column is a misnomer in some settings, as it is only the initial region specified on the input decision variables and does not include further constraints imposed on some estimated outputs, e.g., in Chaps. 4, 8, and 9.

chapter	decision variable	objective function	output random variable	feasible region (constraint set)
1	x	$f(x)$	$Y(x,\xi)$	Θ
2	$\mathbf{x}$	$g(\mathbf{x})$	$Y(\mathbf{x},\xi)$	$\Theta \subseteq \mathbb{Z}^d$
3	i	w_i	x_{ij}	$\{1,\ldots,k\}$
4	$\mathbf{z}$	$\mathsf{E}[w_0 \vert \mathbf{z}]$	w_0	$\mathbb{R}^k$
5	x or θ	$f(x)$ or $J(\theta)$	$Y(x,\xi)$, $Y(\theta,\xi)$	$\Theta \subseteq \mathbb{R}^d$
6	x	$f(x)$	$Y(x,\xi)$	$\Theta \subseteq \mathbb{R}^d$
7	x	$f(x)$	$Y(x,\xi)$	$\Theta \subseteq \mathbb{R}^n$
8	x	$f(x)$	$Y(x,\xi)$	$\Theta \subseteq \mathbb{R}^d$
9	x	$f(x) \equiv g_0(x)$	$Y(x,\xi) \equiv G_0(x,\xi)$	$\Theta \subseteq \mathbb{R}^{d_x}$
10	x	$f(x)$	$Y(x,\xi)$	$\Theta \subseteq \mathbb{R}^d$
11	x	$f(x)$	$Y(x,\xi)$	$\Theta \subseteq \mathbb{R}^n$
12	x	$h(x)$	$H(x,\xi)$	$\Theta \subseteq \mathbb{R}^d$

The use of f as an objective function is common in deterministic optimization; for that reason it is adopted for the most part in this handbook. However, in the stochastic world, f is commonly used for probability density functions, which are needed in describing certain stochastic gradient estimation techniques in Chap. 5 and model-based search methods in Chaps. 2 and 12. Furthermore, the use of f is not commonly found in statistical models, e.g., regression; hence, it is absent from the chapter on response surface methodology approach (Chap. 4).

This handbook presupposes that the reader is familiar with the basics of stochastic simulation. If this is not the case, the following resources are recommended to provide background and fill any gaps in prerequisites:

- The entry, "Simulation of Stochastic Discrete-Event Systems" [10] in the 3rd edition of the *Encyclopedia of Operations Research and Management Science* [12].[1]

---

[1]If you are at a university or other research institution, your library might have online access to the Encyclopedia through the publisher Springer.

- The Introductory Tutorials found in the annual Winter Simulation Conference proceedings alluded to earlier.
- The simulation handbooks [1, 16], and textbooks [2, 21].

Among the topics not explicitly treated in this volume are approaches based on deterministic methods that fall under the umbrella of metaheuristics, which include genetic algorithms and tabu search, as well as direct search algorithms such as coordinate search, pattern search (Hooke–Jeeves), and the Nelder–Mead simplex algorithm. Although there is no dedicated chapter for metaheuristics, algorithms similar to simulated annealing are treated in the random search chapters. The reader is again referred to the Encyclopedia entries, "Heuristics" [20] and "Metaheuristics" [30], as well as to the edited handbooks [13, 14]. Stochastic programming is also not treated explicitly (although certainly alluded to in Chaps. 7–9), but there is again a nice Encyclopedia entry on this topic [29], where simulation has begun to play a larger role in the field. The book [6] also treats ordinal optimization and perturbation analysis, topics covered in this handbook in Chaps. 2 and 5, respectively.

Acknowledgements This work was supported in part by the National Science Foundation under Grants CMMI-0856256 and ECCS-0901543, and by the Air Force Office of Scientific Research under Grant FA9550-10-10340.

References

1. J. Banks, editor. *Handbook of Simulation: Principles, Methodology, Advances, Applications, and Practice*. John Wiley & Sons, New York, 1998.
2. J. Banks, J. S. Carson, B. L. Nelson, and D. M. Nicol. *Discrete-Event System Simulation*. Prentice-Hall, Upper Saddle River, NJ, 5th edition, 2009.
3. G. E. P. Box and K. B. Wilson. On the experimental attainment of optimum conditions. *Journal of the Royal Statistical Society. Series B (Methodological)*, 13(1):1–45, 1951.
4. H. S. Chang, M. C. Fu, J. Hu, and S. I. Marcus. *Simulation-based Algorithms for Markov Decision Processes*. Springer, New York, 2007.
5. H. S. Chang, J. Hu, M. C. Fu, and S. I. Marcus. *Simulation-Based Algorithms for Markov Decision Processes*. Springer, New York, 2nd edition, 2013.
6. C.-H. Chen, Q.-S. Jia, and L. H. Lee, editors. *Stochastic Simulation Optimization for Discrete Event Systems: Perturbation Analysis, Ordinal Optimization, and Beyond*. World Scientific, Singapore, 2013.
7. M. C. Fu. Sample path derivatives for (s,S) inventory systems. *Operations Research*, 42(2):351–364, 1994.
8. M. C. Fu. Optimization for simulation: Theory vs. practice (Feature Article). *INFORMS Journal on Computing*, 14(3):192–215, 2002.
9. M. C. Fu. Approximate dynamic programming. In S. I. Gass and M. C. Fu, editors, *Encyclopedia of Operations Research and Management Science*, pages 73–77. Springer, New York, 3rd edition, 2013.
10. M. C. Fu and D. Gross. Simulation of stochastic discrete-event systems. In S. I. Gass and M. C. Fu, editors, *Encyclopedia of Operations Research and Management Science*, pages 1410–1418. Springer, New York, 3rd edition, 2013.
11. M. C. Fu and K. J. Healy. Techniques for simulation optimization: An experimental study on an (s,S) inventory system. *IIE Transactions*, 29(3):191–199, 1997.

12. S. I. Gass and M. C. Fu, editors. *Encyclopedia of Operations Research and Management Science*. Springer, New York, 3rd edition, 2013.
13. M. Gendreau and J.-Y. Potvin, editors. *Handbook of Metaheuristics*. Springer, New York, 2nd edition, 2010.
14. F. W. Glover and G. A. Kochenberger, editors. *Handbook of Metaheuristics*. Springer, New York, 2003.
15. A. Gosavi. *Simulation-Based Optimization: Parametric Optimization Techniques and Reinforcement Learning*. Kluwer, Boston, MA, 2003.
16. S. G. Henderson and B. L. Nelson, editors. *Handbooks in Operations Research and Management Science: Simulation*. North-Holland/Elsevier, Amsterdam, 2006.
17. J. Hu, M. C. Fu, and S. I. Marcus. A model reference adaptive search method for global optimization. *Operations Research*, 55(3):549–568, 2007.
18. J. Kiefer and J. Wolfowitz. Stochastic estimation of the maximum of a regression function. *Annals of Mathematical Statistics*, 23:462–266, 1952.
19. S. Kim. Gradient-based simulation optimization. In L. F. Perrone, F. P. Wieland, J. Liu, B. G. Lawson, D. M. Nicol, and R. M. Fujimoto, editors, *Proceedings of the 2006 Winter Simulation Conference*, pages 159–167. IEEE, Piscataway, NJ, 2006.
20. M. Laguna and R. Marti. Heuristics. In S. I. Gass and M. C. Fu, editors, *Encyclopedia of Operations Research and Management Science*, pages 695–703. Springer, New York, 3rd edition, 2013.
21. A. M. Law and W. D. Kelton. *Simulation Modeling and Analysis*. McGraw-Hill, New York, 3rd edition, 2000.
22. N. Metropolis and S. Ulam. The Monte Carlo method. *Journal of the American Statistical Association*, 44(247):335–341, September 1949.
23. A. Nemirovski, A. Judisky, G. Lan, and A. Shapiro. Robust stochastic approximation approach to stochastic programming. *SIAM Journal on Optimization*, 19(4):1574–1609, 2009.
24. R. Pasupathy and S. G. Henderson. A testbed of simulation-optimization problems. In L. F. Perrone, F. P. Wieland, J. Liu, B. G. Lawson, D. M. Nicol, and R. M. Fujimoto, editors, *Proceedings of the 2006 Winter Simulation Conference*, pages 255–263. IEEE, Piscataway, NJ, 2006.
25. R. Pasupathy and S. G. Henderson. Simopt: A library of simulation optimization problems. In S. Jain, R. R. Creasey, J. Himmelspach, K. P. White, and M. Fu, editors, *Proceedings of the 2011 Winter Simulation Conference*, pages 4080–4090. IEEE, Piscataway, NJ, 2011.
26. W. B. Powell. *Approximate Dynamic Programming: Solving the Curses of Dimensionality*. Wiley, New York, 2nd edition, 2011.
27. H. Robbins and S. Monro. A stochastic approximation method. *Annals of Mathematical Statistics*, 22:400–407, 1951.
28. R. Y. Rubinstein and D. P. Kroese. *The Cross-Entropy Method: A Unified Approach to Combinatorial Optimization, Monte-Carlo Simulation, and Machine Learning*. Springer, 2004.
29. S. Sen. Stochastic programming. In S. I. Gass and M. C. Fu, editors, *Encyclopedia of Operations Research and Management Science*, pages 1486–1496. Springer, New York, 3rd edition, 2013.
30. K. Sörenson and F. W. Glover. Metaheuristics. In S. I. Gass and M. C. Fu, editors, *Encyclopedia of Operations Research and Management Science*, pages 960–970. Springer, New York, 3rd edition, 2013.
31. J. C. Spall. Multivariate stochastic approximation using simultaneous perturbation gradient approximation. *IEEE Transactions on Automatic Control*, 37:332–341, 1992.
32. J. C. Spall. *Introduction to Stochastic Search and Optimization*. Wiley-Interscience, New York, 2003.

Chapter 2
Discrete Optimization via Simulation

L. Jeff Hong, Barry L. Nelson, and Jie Xu

Abstract This chapter describes tools and techniques that are useful for optimization via simulation—maximizing or minimizing the expected value of a performance measure of a stochastic simulation—when the decision variables are discrete. Ranking and selection, globally and locally convergent random search and ordinal optimization are covered, along with a collection of "enhancements" that may be applied to many different discrete optimization via simulation algorithms. We also provide strategies for using commercial solvers.

2.1 Introduction

In this chapter we cover optimization via simulation (OvS) problems that can be represented as

$$\min g(\mathbf{x}), \quad \mathbf{x} \in \Theta, \tag{2.1}$$

where $g(\mathbf{x}) = \mathsf{E}\left[Y(\mathbf{x},\xi)\right]$. There is a single objective $g(\mathbf{x})$, which is representable as the expected value of a random variable $Y(\mathbf{x},\xi)$, where ξ represents the randomness, e.g., the random numbers in a simulation. The distribution of $Y(\mathbf{x},\xi)$ is an unknown function of the vector of decision variables $\mathbf{x}$, but realizations of $Y(\mathbf{x},\xi)$ can be observed through simulation experiments. If the problem is more naturally thought of as maximization then we can always formulate an equivalent minimization version. Henceforth we drop the argument ξ for notational convenience.

L.J. Hong
City University of Hong Kong, Kowloon Tong, Hong Kong
e-mail: jeffhong@cityu.edu.hk

B.L. Nelson
Northwestern University, Evanston, IL, USA
e-mail: nelsonb@northwestern.edu

J. Xu (✉)
George Mason University, Fairfax, VA, USA
e-mail: jxu13@gmu.edu

M.C. Fu (ed.), *Handbook of Simulation Optimization*, International Series in Operations Research & Management Science 216, DOI 10.1007/978-1-4939-1384-8_2

Our focus is on two related cases: Either $\mathbf{x}$ is an index in the set $\Theta = \{1,2,\ldots,k\}$ of feasible solutions, which may be categorical (not ordered in any way); or $\mathbf{x}$ is a vector of d integer-ordered decision variables in a feasible region $\Phi \subset \mathbb{R}^d$, possibly defined by a set of deterministic constraints. In this case we assume that Φ is compact and convex. Therefore, $\Theta = \Phi \cap \mathbb{Z}^d$ is a finite set, where $\mathbb{Z}^d$ denotes all d-dimensional vectors with integer components, and Problem (2.1) has only a finite number of feasible solutions. We refer to this class of problems as Discrete Optimization via Simulation (DOvS) problems.

The solution methods we describe assume that $\mathrm{Var}[Y(\mathbf{x})] < \infty$ for all $\mathbf{x} \in \Theta$, and that we have an estimator $\hat{g}(\mathbf{x})$ that converges with probability 1 (w.p.1) to $g(\mathbf{x})$ as we expend more and more simulation effort on solution $\mathbf{x}$. If that estimator is a sample mean, then we use the notation $\overline{Y}(\mathbf{x})$. Often, but not always, we can simulate independent and identically distributed (i.i.d.) replications, $Y_1(\mathbf{x}), Y_2(\mathbf{x}), \ldots$ at any $\mathbf{x}$.

The following examples, which are based on Nelson [48], illustrate the types of problems we consider and the issues that arise.

2.1.1 Designing a Highly Reliable System

A system works only if all of its subsystems work; the subsystems consist of components that have their own time-to-failure and repair-time distributions. The objective is to decide how many and what redundant components to use to minimize steady-state system unavailability given budget constraints. The budget is relatively tight, so altogether there are only 152 feasible configurations. Let $x \in \{1,2,\ldots,152\}$ index the configurations.

In this problem there are a small number of feasible solutions, and they can be treated as categorical. This is a choice, however, because we could also define $\mathbf{x} = (x_1, x_2, \ldots, x_d)$ where x_i is the number of redundant components of type i to include. The state of art for solving DOvS problems makes it advantageous to treat the problem as categorical if the number of solutions is relatively small because there are highly efficient solution methods that apply when it is possible to simulate all feasible solutions, at least a little. This is analogous to deterministic integer programming (IP) problems in which it is possible to exhaust the feasible region, making it pointless to apply a high-powered IP algorithm. However, unlike a deterministic IP, a single evaluation of the objective function in DOvS is (typically) not sufficient because $Y(\mathbf{x})$ is only an estimator of $g(\mathbf{x})$ that is subject to sampling error, i.e., $Y(\mathbf{x}) \neq g(\mathbf{x})$ w.p.1, even though $\mathrm{E}[Y(\mathbf{x})] = g(\mathbf{x})$. Sampling error is reduced by expending additional simulation effort at solution $\mathbf{x}$, and doing so (usually adaptively) is the central feature of solution methods.

In this problem $g(\mathbf{x})$ represents steady-state (long-run average as time goes to infinity) system unavailability, which implies that the estimator of $g(\mathbf{x})$ can be defined in one of two ways:

Extended replication average:

$$\hat{Y}(\mathbf{x}) = \frac{1}{T}\int_0^T A(t)\,dt \tag{2.2}$$

where $A(t) = 1$ if the system is unavailable, and 0 otherwise. Sampling error is controlled by increasing the run length T.

Additional replication average:

$$\overline{Y}(\mathbf{x}) = \frac{1}{n}\sum_{i=1}^{n}\left(\frac{1}{T_e - T_b}\int_{T_b}^{T_e} A_i(t)\,dt\right) = \frac{1}{n}\sum_{i=1}^{n}\hat{Y}_i(\mathbf{x}) \tag{2.3}$$

where $T_e > T_b$ are fixed times, and $A_i(t)$ is the sample path from the ith replication. Sampling error is controlled by increasing the number of i.i.d. replications n.

In Sect. 2.3 we describe ranking and selection methods for such problems.

2.1.2 Flow-Line Throughput

A three-stage flow line has finite buffer storage space in front of stations 2 and 3 (the number of spaces being denoted by x_4 and x_5) and an infinite number of jobs in front of station 1. There is a single server at each station, and the service-time distribution at station i has service rate $x_i, i = 1,2,3$. If the buffer of station i is full, then station $i-1$ is blocked and a finished job cannot be released from station $i-1$. The total buffer space and the service rates are limited by constraints on space and cost. The objective is to find a buffer allocation and service rates such that the expected throughput over a 1-year planning horizon is maximized. The deterministic constraints are $x_1+x_2+x_3 \leq 20, x_4+x_5 = 20, 1 \leq x_i \leq 20$ and $x_i \in \mathbb{Z}^+$ for $i = 1,2,\ldots,5$, implying 21,660 feasible solutions. This example is adapted from [8, 58].

Here the number of feasible solutions is probably larger than can be exhausted, so some sort of search is required. In Sects. 2.5–2.6 we describe methods to solve such problems based on adaptive random search.

Notice in this problem that the fixed 1-year planning horizon means that sampling error is reduced only by increasing the number of replications—it makes no sense to extend a replication beyond 1 year because the performance measure of interest is defined with respect to 1 year. Thus, the natural estimator is

$$\overline{Y}(\mathbf{x}) = \frac{1}{n}\sum_{j=1}^{n} Y_j(\mathbf{x})$$

where $Y_j(\mathbf{x})$ is the throughput observed in 1 year of production on replication j for solution $\mathbf{x}$.

2.1.3 Inventory Management with Dynamic Customer Substitution

A retailer faces a one-shot inventory stocking decision for six product variants at the beginning of the selling season so as to maximize the expected value of profit. No inventory replenishment can occur, and there is no salvage value for the products. Each consumer selects the available product with the highest utility for them, which may be a no-purchase option. The number of customers is Poisson, and the customer's choice behavior is modeled by a multinomial logit model. Pricing is an exogenous decision. Let $x_1, x_2, \ldots, x_6$ denote the number of each variant the retailer chooses to stock. This example is adapted from [45].

In theory, there are a countably infinite number of feasible solutions since there is no fixed upper bound on the quantity of variant i that the retailer stocks, x_i. In practice, clearly there is a level beyond which it makes no sense to stock. These level bounds can be determined by factors such as store capacity and maximal demand possible. Then, the problem can be transformed into a DOvS problem.

When solving deterministic IPs, branch-and-bound methods that relax integrality constraints are often used, which is possible because the objective function can be evaluated at non-integer values. When solving DOvS problems, however, the simulation may not make sense at fractional values of $\mathbf{x}$. For instance, it is not clear how to simulate stocking 112.3 blue shirts. As opposed to having a known linear, quadratic or even convex objective function, the function $g(\mathbf{x})$ is implied by the simulation model and little structural information about it is available. And, of course, $g(\mathbf{x})$ can only be estimated.

2.1.4 Themes

The focus of this chapter is on methods that lead to rigorously provable performance, as defined in Sect. 2.2 below (however, in Sect. 2.8 we give practically useful tips for using commercial OvS solvers based on metaheuristics). We assume little or no structural information about $g(\mathbf{x})$; as a result, the balance between expending effort on search—looking for better feasible solutions—and estimation—estimating the true objective function value of solutions we have investigated—is a core issue for DOvS algorithms.

There are three fundamental types of errors that occur in DOvS problems; the latter two occur even when the number of feasible solutions is small enough that all of them can be simulated.

1. *The optimal solution is never simulated.* This is a reality in many difficult nonlinear IPs when the feasible solutions cannot be exhausted, and DOvS is no different.

2. *The best solution that was simulated is not selected.* Sampling variability means that the best solution we simulated may not have the best estimated objective function value.
3. *We do not have a good estimate of the objective function value of the solution we do select.* When minimizing a stochastic response by selecting the solution with the smallest estimated value, there is a natural bias toward solutions whose estimated performance is better (lower) than its true expected value.

How these issues are addressed in different problem classes is the subject of the remainder of this chapter.

2.2 Optimality Conditions

Let $\Theta^* = \text{argmin}\{g(\mathbf{x}) : \mathbf{x} \in \Theta\}$ be the set of optimal solutions of Problem (2.1). Because $\Theta = \Phi \cap \mathbb{Z}^d$ and $\Phi \subset \mathbb{R}^d$ is a compact set, Θ is a finite set, i.e., $|\Theta| < \infty$. Therefore, Θ^* is guaranteed to be nonempty. Furthermore, the finiteness of Θ also implies that there exists a positive constant $\delta > 0$ such that

$$g^* \leq g(\mathbf{y}) - \delta \quad \text{forall } \mathbf{y} \in \Theta \setminus \Theta^*, \tag{2.4}$$

where $g^* = \min_{\mathbf{x} \in \Theta} g(\mathbf{x})$ is the optimal objective value. Note that Eq. (2.4) implies that optimal solutions are at least δ better than other feasible solutions.

To solve optimization problems we often need optimality conditions which (a) assure the correctness of algorithms and (b) help in designing implementable stopping rules. To illustrate the usefulness of optimality conditions, we consider the following two examples.

- In unconstrained nonlinear optimization, convergence to a stationary point whose gradient is zero is a widely used optimality condition. Many algorithms, including the steepest descent algorithm and Netwon's method, are proved to satisfy this condition [50]. In practice, all these algorithms typically stop short of convergence. But they often stop when they find a solution whose gradient is sufficiently close to zero.
- In integer linear programming, algorithms often keep track of an upper bound and a lower bound. A commonly used optimality condition is that the gap between the two bounds goes to zero. Many algorithms, including the branch-and-bound and branch-and-cut algorithms, are proved to satisfy this condition. Then in practical implementation, these algorithms often stop when the gap is small enough.

Although the set of optimal solutions Θ^* is clearly defined for DOvS problems, defining optimality conditions for DOvS algorithms is not easy for the following reasons:

1. The objective function $g(\mathbf{x})$ cannot be calculated exactly; instead, it can only be estimated by $\overline{Y}(\mathbf{x})$. This estimation noise generally makes it impossible to rank

solutions with 100 % confidence. Therefore, DOvS algorithms cannot guarantee in general to find an optimal solution with a finite amount of computational effort.
2. Typically $g(\mathbf{x})$ and $Y(\mathbf{x})$ are unknown functions that are embedded in simulation models. We do not have the structural results on $g(\mathbf{x})$ or $Y(\mathbf{x})$ that can be used to screen out (often a large number of) inferior solutions as, for instance, in branch-and-cut algorithms for integer linear programs. Therefore, to find an optimal solution of Problem (2.1), one has to evaluate all feasible solutions.
3. Although Θ is a finite set, it often has a large number of feasible solutions as in the flow-line and inventory-management examples reported in Sects. 2.1.2 and 2.1.3, respectively. Complete enumeration of all solutions may not be possible for many practical problems.

Despite these difficulties, researchers have established various optimality conditions for DOvS problems that are either theoretically convenient or practically useful. In this section we will introduce these conditions.

When Θ is small, i.e., Θ has less than a few hundreds of solutions, we may be able to simulate all solutions and select the best among them. This is known as ranking and selection (R&S). Because of the randomness in the simulation outputs, one cannot guarantee to select the best solution with 100 % confidence. Then, a practical approach is to analyze *the probability of correct selection* (PCS), i.e., $\mathsf{P}(\mathbf{x}^* \in \Theta^*)$ where $\mathbf{x}^*$ denotes the selected best solution. A commonly used optimality condition is to require R&S algorithms to achieve a predetermined PCS, i.e., $\mathsf{P}(\mathbf{x}^* \in \Theta^*) \geq 1-\alpha$. In Sect. 2.3 we will introduce such algorithms.

When Θ is large, simulating all solutions in Θ becomes practically impossible. One may soften the goal of finding an optimal solution to finding a good enough solution, where a "good enough" solution may be defined as one of the top t solutions in Θ. Suppose that one has the computational budget to evaluate n solutions in Θ. Then, it is important to know the probability that at least one of these n solutions is a top t solution, which is known as the *alignment probability* (AP). Let $T \subset \Theta$ denote the set of top t solutions, and let S denote the set of the chosen n solutions. Then, the alignment probability is defined as $\mathsf{P}(|T \cap S| \geq 1)$. A commonly used optimality condition is to require algorithms to achieve a predetermined level of AP, i.e., $\mathsf{P}(|T \cap S| \geq 1) \geq 1-\alpha$. This optimality condition is used in ordinal optimization, which will be introduced in Sect. 2.4.

Another commonly used optimality condition when Θ is large is global convergence as the amount of computational effort goes to infinity. Let $\mathbf{x}_m^*$ denote the solution that the algorithm would report as optimal if stopped at the end of iteration m and $g_m^* = g(\mathbf{x}_m^*)$. Then, the algorithm is globally convergent in probability if $g_m^* \to g^*$ in probability and globally convergent w.p.1 if $g_m^* \to g^*$ w.p.1. By Eq. (2.4), these two convergence criterion are equivalent to

$$\lim_{m\to\infty} \mathsf{P}(\mathbf{x}_m^\star \in \Theta^\star) = 1,$$

$$\mathsf{P}\left(\lim_{m\to\infty} \mathbf{1}\{\mathbf{x}_m^\star \in \Theta^\star\} = 1\right) = 1,$$

respectively, where $\mathbf{1}\{\cdot\}$ is the indicator function. As all algorithms stop in a finite amount of time, one may wonder why convergence properties are desirable. Andradóttir [2] answers this question:

> Although this [convergence] performance guarantee does not assure that the algorithm will return a "good" estimated optimal solution (because additional computational effort may be required), it is certainly a reassuring property to have. From a different perspective, it is worrisome to use a simulation optimization algorithm in practice that is not known to converge even if an infinite amount of computational effort is expended!

Many globally convergent algorithms have been proposed to solve DOvS problems and we will introduce them in Sect. 2.5.

As pointed earlier in this section, optimality conditions assure the correctness of algorithms and help in designing implementable stopping rules. Global convergence achieves the first goal by reassuring the algorithm eventually finds an optimal solution. However, to achieve global convergence when there is no special structure, algorithms have to evaluate all solutions in Θ in the limit, and it is not clear how to relax this requirement for some implementable stopping rules. To resolve this problem, local convergence has been proposed. Let $N(\mathbf{x}) \subset \Theta$ denote the local neighborhood of any solution $\mathbf{x} \in \Theta$, and let $\mathbf{x}$ be a locally optimal solution if $g(\mathbf{x}) \leq g(\mathbf{y})$ for all $\mathbf{y} \in N(\mathbf{x})$. Notice that the definition of local optimality depends on the definition of the local neighborhood and different local neighborhoods may result in different local optimal solutions. Let $\mathscr{L}$ denote the set of locally optimal solutions for the DOvS problem. Then, similar to the definition of global convergence, we may define local convergence in probability and local convergence w.p.1 as

$$\lim_{m\to\infty} \mathsf{P}(\mathbf{x}_m^\star \in \mathscr{L}) = 1,$$

$$\mathsf{P}\left(\lim_{m\to\infty} \mathbf{1}\{\mathbf{x}_m^\star \in \mathscr{L}\} = 1\right) = 1,$$

respectively. To converge to a locally optimal solution, algorithms do not need to evaluate all feasible solutions. Furthermore, because the local neighborhood is typically a small set, one can statistically test the local optimality of any solution $\mathbf{x}$, i.e.,

$$\mathrm{H}_0:\ g(\mathbf{x}) \leq \min_{\mathbf{y}\in N(\mathbf{x})} g(\mathbf{y}) \quad \text{vs} \quad \mathrm{H}_1:\ g(\mathbf{x}) > \min_{\mathbf{y}\in N(\mathbf{x})} g(\mathbf{y}),$$

and control the type I and type II errors of the test. This provides an implementable stopping rule for algorithms in a finite amount of time. In Sect. 2.6 we will introduce some locally convergent algorithms.

Other than the algorithms that are built around convergence or correct-selection guarantees, there are also many algorithms that are based on heuristics for deterministic optimization algorithms, such as genetic algorithms and tabu search. These algorithms work well for difficult deterministic integer programs, and they are somewhat tolerant of sampling variabilities. However, they typically do not satisfy any optimality conditions for DOvS problems and may be misled by sampling variabilities. These algorithms are typically used in commercial Solvers and we offer some suggestions on how to use them effectively and efficiently in Sect. 2.8.

2.3 Ranking and Selection

Methods in the category of ranking and selection (R&S) apply to problems with a relatively small number of feasible solutions, such as designing the highly reliable system described in Sect. 2.1.1. The R&S procedures that have seen broad use in optimization via simulation treat the feasible solutions as categorical, meaning that they make no attempt to exploit relationships among the solutions (other than that their similarity may make the use of common random numbers effective). The definition of "relatively small" depends to some extent on how much time it takes to simulate an alternative, since R&S procedures simulate them all, but problems with up to 1,000 feasible solutions have been solved in this way. Recently, Luo et al. [44] implemented R&S procedures on a parallel computing environment with tens to hundreds of parallel processors, and solved practical DOvS problems with more than 20,000 feasible solutions.

Our focus will be on the indifference-zone formulation of optimality, as described in Sect. 2.2, but with some discussion of a Bayesian formulation. Suppose that there are $k \geq 2$ solutions in Θ, denoted as $\mathbf{x}_1, \mathbf{x}_2, \ldots, \mathbf{x}_k$. We let $Y_j(\mathbf{x}_i)$ denote the jth observation from simulating solution $\mathbf{x}_i$. Many R&S methods assume that $Y_j(\mathbf{x}_i) \sim \mathscr{N}(g(\mathbf{x}_i), \sigma_i^2)$, where $g(\mathbf{x}_i)$ is unknown and the σ_i^2 are typically unknown and unequal. To simplify notation, we let $g(\mathbf{x}_1) \leq g(\mathbf{x}_2) \leq \cdots \leq g(\mathbf{x}_k)$ and the goal of a R&S procedure is to select solution $\mathbf{x}_1$ whose identity is unknown. Under the indifference-zone formulation, the best solution $\mathbf{x}_1$ will be selected with a probability at least $1 - \alpha$ as long as the difference between the objective values of the best and second-best solutions is at least $\delta > 0$. If there are a set of solutions whose objective values are within δ of the best solution, then all solutions in that set are acceptable.

R&S procedures were developed in the 1950s for statistical selection problems such as choosing the best treatment for a medical condition. In such contexts small numbers of alternatives with relatively equal variances (maybe even "known" variances from similar experiments) were common. Researchers have creatively built on this foundation to address problems that are of particular importance in computer simulation: unknown and unequal variances; larger numbers of feasible solutions; induced correlation across solutions due to the use of common random numbers; autocorrelation within a replication of a solution in steady-state simulation; and non-normal output data. Despite the extensive literature and the many variations that have been proposed, the foundations for most procedures can be found in three very old procedures described below.

Bechhofer's procedure [4] is one of the earliest and simplest indifference-zone selection procedures. It assumes that $\sigma_1^2 = \sigma_2^2 = \cdots = \sigma_k^2 = \sigma^2$ and σ^2 is known, and $Y_j(\mathbf{x}_i)$ is independent of $Y_n(\mathbf{x}_m)$ whenever $i \neq m$ (different solutions) or $j \neq n$ (different observations) or both.

Bechhofer's procedure determines the sample sizes required for all k solutions based on the variance of the solutions and by assuming $g(\mathbf{x}_1) + \delta = g(\mathbf{x}_2) = \cdots = g(\mathbf{x}_k)$—the most difficult case—which frees it from needing to know anything about the true means. This procedure can be thought of as a hypothesis test with controlled power for detecting that one solution is $\geq \delta$ better than the others; it provides an experiment design that leads to the indifference-zone definition of optimality being satisfied with prespecified probability.

Bechhofer's Procedure

Step 1. Determine the constant h that satisfies $\mathsf{P}(Z_i \leq h, i = 1,2,\ldots,k-1) = 1 - \alpha$ where $(Z_1, Z_2, \ldots, Z_{k-1})$ has a multivariate normal distribution with means 0, variances 1, and common pairwise correlations $1/2$. Let

$$n = \left\lceil \frac{2h^2\sigma^2}{\delta^2} \right\rceil.$$

Step 2. Take n observations from each solution and calculate $\overline{Y}(\mathbf{x}_i;n)$ for all $i = 1,2,\ldots,k$, where $\overline{Y}(\mathbf{x}_i;n)$ denotes the sample mean of $Y_j(\mathbf{x}_i), j = 1,2,\ldots,n$.

Step 3. Select the solution with the smallest sample mean $\overline{Y}(\mathbf{x}_i;n)$ as the best.

A feature of Bechhofer's procedure and its descendants is that they do not try to exploit information provided by the sample mean of each alternative until the very end. When the samples are collected one at a time, as they are in simulation experiments, it is possible to evaluate the selection decision at intermediate stages. Fully-sequential procedures evaluate after every sample is taken. The simplest fully-sequential procedure is Paulson's procedure [54], which makes the same assumptions as Bechhofer's procedure.

Paulson's Procedure

Step 1. Let $0 < \lambda < \delta$ and

$$a = \ln\left(\frac{k-1}{\alpha}\right)\frac{\sigma^2}{\delta - \lambda}.$$

Let $I = \{\mathbf{x}_1, \mathbf{x}_2, \ldots, \mathbf{x}_k\}$ and $r = 0$.

Step 2. Let $r = r+1$. Take one observation from each solution that is in I and compute $\overline{Y}(\mathbf{x}_i;r)$ for all $\mathbf{x}_i \in I$.

Step 3. Let $I^{\text{old}} = I$ and

$$I = \left\{\mathbf{x}_i \in I^{\text{old}} : \overline{Y}(\mathbf{x}_i;r) \leq \min_{\ell \in I^{\text{old}}} \overline{Y}(\mathbf{x}_\ell;r) + \left(\frac{a}{r} - \lambda\right)^+\right\}, \text{ where } (x)^+ \equiv \max\{0,x\}.$$

If $|I| > 1$, then go to Step 2; otherwise, select the solution in I as the best.

Paulson's procedure uses a large deviations bound to account for taking multiple looks at the data. Descendants of Paulson's procedure use bounds based on Brownian motion crossing boundaries. The efficiency that comes from eliminating noncompetitive solutions early can be substantial.

Relatively large numbers of alternatives typically means that (a) they were created by taking all feasible combinations of some more basic decision variables, and (b) many of them are not really competitive. Both Bechhofer-like and Paulson-like procedures can benefit from a prescreening step that eliminates many of the uncompetitive solutions while retaining the best with a guaranteed probability. Subset selection procedures trace back to a procedure due to Gupta [22, 23]:

Gupta's Procedure

Step 1. Determine the constant h that satisfies $\mathsf{P}(Z_i \le h, i = 1,2,\ldots,k-1) = 1-\alpha$ where $(Z_1, Z_2, \ldots, Z_{k-1})$ has a multivariate normal distribution with means 0, variances 1, and common pairwise correlations $1/2$. Select $n \ge 1$.

Step 2. Take n observations from each solution and calculate $\overline{Y}(\mathbf{x}_i;n)$ for all $i = 1,2,\ldots,n$.

Step 3. Let

$$I = \left\{ \mathbf{x}_i : \overline{Y}(\mathbf{x}_i;n) \le \min_{\ell \ne i} \overline{Y}(\mathbf{x}_\ell, n) + h\sigma\sqrt{\frac{2}{n}} \right\}.$$

Step 4. Return I.

Under the same assumptions as Bechhofer's and Paulson's procedures, Gupta's procedure guarantees that $\mathsf{P}(\mathbf{x}_1 \in I) \ge 1-\alpha$. No indifference-zone parameter is specified, and it is possible that $|I| = k$, i.e., no solution is eliminated. In practice, when k is large many solutions are screened out so that a selection procedure like Bechhofer's can be applied to a much smaller set of solutions. By appropriately spending the allowable error α between screening and selection, the desired indifference-zone optimality condition can be attained.

We now present two procedures that are based on the principles of Bechhofer, Paulson and Gupta, but have been extended to be relevant for simulation. NSGS (Nelson et al. [49]) combines subset selection like Gupta with ranking like Bechhofer. It allows unknown and unequal variances.

NSGS Procedure

Step 1. Specify a common first-stage number of replications from each solution $n_0 \ge 2$; further, set

$$t = t_{n_0-1,(1-\alpha/2)^{\frac{1}{k-1}}}$$

the $(1-\alpha/2)^{\frac{1}{k-1}}$ quantile of the t distribution with $n_0 - 1$ degrees of freedom, and obtain Rinott's constant $h = h(n_0, k, 1-\alpha/2)$ from the tables in Wilcox [72], Bechhofer et al. [5] or Goldsman and Nelson [21].

Step 2. Take n_0 replications from each feasible solution. Calculate the first-stage sample means $\overline{Y}(\mathbf{x}_i;n_0)$ and marginal sample variances

$$S(\mathbf{x}_i)^2 = \frac{1}{n_0-1}\sum_{j=1}^{n_0}\left(Y_j(\mathbf{x}_i) - \overline{Y}(\mathbf{x}_i;n_0)\right)^2,$$

for $i = 1,2,\ldots,k$.

Step 3. Calculate the quantity

$$W_{i\ell} = t\left(\frac{S(\mathbf{x}_i)^2 + S(\mathbf{x}_\ell)^2}{n_0}\right)^{1/2}$$

for all $i \neq \ell$. Form the screening subset

$$I = \{\mathbf{x}_i : \overline{Y}(\mathbf{x}_i;n_0) \leq \overline{Y}(\mathbf{x}_\ell;n_0) + W_{i\ell} \quad \text{for all } \ell \neq i\}.$$

Step 4. If $|I| = 1$, then stop and return the solution in I as the best. Otherwise, for all $\mathbf{x}_i \in I$, compute the second-stage sample sizes

$$N_i = \max\left\{n_0, \left\lceil \left(\frac{hS(\mathbf{x}_i)}{\delta}\right)^2 \right\rceil\right\}.$$

Step 5. Take $N_i - n_0$ additional replications from all solutions $\mathbf{x}_i \in I$.

Step 6. Compute the overall sample means $\overline{Y}(\mathbf{x}_i;N_i)$ for all $\mathbf{x}_i \in I$. Select the solution $\mathbf{x}_B = \text{argmin}_{\mathbf{x}_i}\overline{Y}(\mathbf{x}_i;N_i)$ as best.

NSGS guarantees that $\mathbf{x}_B = \mathbf{x}_1$, or $g(\mathbf{x}_B)$ is within δ of $g(\mathbf{x}_1)$, and also that $g(\mathbf{x}_B) \in \overline{Y}(\mathbf{x}_B;N_B) \pm \delta$, all w.p. $\geq 1-\alpha$. NSGS has been applied to problems with more than 1,000 feasible solutions, and tends to be very efficient when there are a few competitive solutions and many non-competitive ones. The procedure works even if this is not the case, but may be computationally expensive if the subset-selection Step 3 cannot screen out a significant number of feasible solutions.

Procedure KN (Kim and Nelson [39]) below is a descendant of Paulson that allows unknown and unequal variances, and the use of common random numbers.

KN Procedure

Step 1. Specify common first-stage number of replications $n_0 \geq 2$. Set

$$\eta = \frac{1}{2}\left[\left(\frac{2\alpha}{k-1}\right)^{-2/(n_0-1)} - 1\right].$$

Step 2. Let $I = \{\mathbf{x}_1, \mathbf{x}_2, \ldots, \mathbf{x}_k\}$ be the set of solutions still in contention, and let $h^2 = 2\eta(n_0-1)$. Obtain n_0 observations $Y_j(\mathbf{x}_i), j = 1,2,\ldots,n_0$ from each solution $\mathbf{x}_i \in I$ and compute $\overline{Y}(\mathbf{x}_i;n_0)$. For all $i \neq \ell$ calculate

$$S_{i\ell}^2 = \frac{1}{n_0 - 1} \sum_{j=1}^{n_0} \left(Y_j(\mathbf{x}_i) - Y_j(\mathbf{x}_\ell) - \left[\overline{Y}(\mathbf{x}_i; n_0) - \overline{Y}(\mathbf{x}_\ell; n_0) \right] \right)^2,$$

the sample variance of the difference between solutions i and ℓ. Set $r = n_0$.

Step 3. Set $I^{\text{old}} = I$. Let

$$I = \left\{ \mathbf{x}_i : \mathbf{x}_i \in I^{\text{old}} \text{ and } \overline{Y}(\mathbf{x}_i; r) \le \overline{Y}(\mathbf{x}_\ell; r) + W_{i\ell}(r), \forall \ell \in I^{\text{old}}, \ell \ne i \right\},$$

where

$$W_{i\ell}(r) = \frac{\delta}{2r} \left(\frac{h^2 S_{i\ell}^2}{\delta^2} - r \right)^+.$$

Step 4. If $|I| = 1$, then stop and select the solution whose index is in I as the best. Otherwise, take one additional observation $Y_{r+1}(\mathbf{x}_i)$ from each solution $\mathbf{x}_i \in I$, set $r = r + 1$ and go to Step 3.

Many extensions and variations of KN have appeared in the literature. Kim and Nelson [41] proposed KN++, which is asymptotically valid even when observations are non-normal and dependent. A drawback of a fully sequential procedure, such as KN, relative to a two-stage procedure, like NSGS, is that fully sequential procedures frequently switch among the simulations of different solutions. Switching can be computationally much more costly than running simulation experiments, depending on the computing environment, offsetting the efficiency gain of the fully sequential procedures compared to two-stage procedures that require a minimum number of switches. Hong and Nelson [27] and Osogami [53] designed sequential procedures that reduce the number of switches dramatically while still maintaining the benefit of being sequential.

There are two basic paradigms for solving the selection-of-the-best problem: frequentist (described above) and Bayesian. A comprehensive reference that covers the basic theory upon which frequentist R&S procedures are based is Kim and Nelson [40]. We give a brief overview of the Bayesian approach below, based largely on Frazier and Powell [17]. For more comprehensive treatments, see [11, 16].

A Bayesian procedure consists of a sequence of decisions; the decisions include which solution $\mathbf{x}$ to simulate next, and possibly whether or not to stop the procedure and select a solution. Let $\mathbf{x}^{(j)}, j = 0, 1, 2, \ldots$ be the jth decision of which solution to simulate, which leads to obtaining the $(j+1)$st simulation observation $Y_{j+1}(\mathbf{x}^{(j)})$. The linking of the jth decision to the $(j+1)$st observation emphasizes that in a Bayesian framework we may have informative prior distributions on the values of $g(\mathbf{x}_1), g(\mathbf{x}_2), \ldots, g(\mathbf{x}_k)$ that could be used to make an intelligent decision $\mathbf{x}^{(0)}$ even when no data have yet been obtained.

Let $\mathscr{H}_j = \{\mathbf{x}^{(0)}, Y_1(\mathbf{x}^{(0)}), \mathbf{x}^{(1)}, Y_2(\mathbf{x}^{(1)}) \ldots \mathbf{x}^{(j-1)}, Y_j(\mathbf{x}^{(j-1)})\}$ denote the history up through decision $j-1$ (observation j), with $\mathscr{H}_0 = \emptyset$. The procedure is controlled by a policy π and a stopping time τ where $\tau(\mathscr{H}_j)$ yields a binary decision to stop the procedure or to apply

$$\mathbf{x}^{(j)} = \pi(\mathscr{H}_j)$$

to decide which solution $\mathbf{x}^{(j)}$ to simulate next.

The key to seeking optimal policies in the Bayesian formulation is that uncertainty about $g(\mathbf{x})$ is represented as a prior probability distribution on its value, which is updated to a posterior distribution using Bayes rule as observations are obtained. A typical choice of prior distribution is a non-informative normal-gamma prior ($g(\mathbf{x})$ is normally distributed given its posterior variance $\sigma^2(\mathbf{x})$, and $1/\sigma^2(\mathbf{x})$ has a gamma distribution); see [17] or [15]. This choice of prior leads to a generic Bayes procedure of the following form:

Procedure Generic Bayes

Step 1. Set $n(\mathbf{x}) = 0$, $\mathscr{H}_0 = \emptyset$ and $\overline{Y}(\mathbf{x}) =$ null for all $\mathbf{x} \in \Theta$, and $j = 0$.
Step 2. Let $\mathbf{x}^{(j)} = \pi(\mathscr{H}_j)$.
Step 3. Obtain observation $Y_{j+1}(\mathbf{x}^{(j)})$ and update

$$\begin{aligned} n(\mathbf{x}^{(j)}) &= n(\mathbf{x}^{(j)}) + 1 \\ \overline{Y}(\mathbf{x}^{(j)}) &= \frac{1}{n(\mathbf{x}^{(j)})} \sum_{i:\mathbf{x}^{(i)}=\mathbf{x}^{(j)}} Y_{i+1}(\mathbf{x}^{(i)}). \end{aligned}$$

Step 3. If $\tau(\mathscr{H}_{j+1})$ indicates time to stop, then return $\mathbf{x}_B = \operatorname{argmin}_{\mathbf{x}\in\Theta} \overline{Y}(\mathbf{x})$.
Else $j = j+1$ and go to Step 2.

Step 2 as stated above masks what is really happening: The decision $\mathbf{x}^{(j)}$ is actually a function of the posterior distributions as calculated from $\mathscr{H}_j$, and these posterior updates may be easy (in the case of a conjugate prior) or numerically challenging to obtain [11, 16].

The best policy π and stopping time τ depend on the objective. For instance, a natural objective is

$$\inf_{\pi} \mathsf{E}^{\pi}\left[g(\mathbf{x}_N^{\star})\right] = \inf_{\pi} \mathsf{E}^{\pi}\left[\min_{\mathbf{x}\in\Theta} \overline{Y}_N(\mathbf{x})\right] \tag{2.5}$$

where N is a fixed simulation budget. This is equivalent to minimizing the expected opportunity cost for the selected solution within a given simulation budget. Another objective is

$$\inf_{\pi,\tau} \mathsf{E}^{\pi}\left[g(\mathbf{x}_\tau^{\star}) - c(\tau)\right] = \inf_{\pi,\tau} \mathsf{E}^{\pi}\left[\min_{\mathbf{x}\in\Theta} \overline{Y}_\tau(\mathbf{x}) + c(\tau)\right] \tag{2.6}$$

where $c(j)$ is the cost of running j simulations; this policy and stopping rule balance selection loss with the cost of additional simulation. The expectations in (2.5)–(2.6) are with respect to the posterior distributions of $g(\mathbf{x})$.

The optimal policies for (2.5) and (2.6) are the solutions to dynamic programming problems. Unfortunately, it is computationally difficult or impossible to actually achieve the optimal policy; therefore, research has focused on heuristics that are implementable and effective. These include the optimal computing budget allocation (OCBA) procedures [9], the expected value of information (EVI) procedures [12], and the knowledge gradient (KG) methods [17, 18], which are the topics of Chap. 3.

An advantage of the Bayesian formulation is its flexibility; many kinds of information or knowledge about the problem can be incorporated into the prior beliefs, leading to substantial gains in efficiency. For instance, the knowledge that two similar solutions ($\mathbf{x}_1 \approx \mathbf{x}_2$) will probably have similar values ($g(\mathbf{x}_1) \approx g(\mathbf{x}_2)$) can be exploited (e.g., [19]).

Branke et al. [7] conducted a comprehensive set of experiments to compare the performance of different R&S procedures on thousands of combinations of problem structures. They found that no R&S procedure can dominate in all situations. They also found that the Bayesian procedures are often more efficient in terms of the total number of samples required to make a decision. However, they do not provide the type of correct-selection optimality guarantee that the frequentist procedures provide.

2.4 Ordinal Optimization

Ordinal optimization (OO), introduced by Ho et al. [25] and treated in detail in the book by Ho et al. [26], proposes "soft optimization" for OvS problems when the number of feasible solutions $k = |\Theta|$ is too large for R&S methods. Ordinal optimization selects a subset S from Θ and limits further analysis to S. If we define a set T of good enough solutions in Θ, which are often the top t solutions in Θ, we are interested in the probability that at least l solutions in T are in S, i.e., $\mathrm{P}(|T \cap S| \geq l)$. This probability is referred to as the *alignment probability* (AP) and l is called the alignment level.

There are two basic ideas behind ordinal optimization:

1. Estimating the order among solutions is much easier than estimating the absolute objective values of each solution.
2. Softening the optimization goal and accepting good enough solutions leads to an exponential reduction in computational burden.

To understand the first idea, recall that estimating $g(\mathbf{x})$ with $\overline{Y}(\mathbf{x})$ only has a convergence rate of $1/\sqrt{n}$ according to the Central Limit Theorem. By comparison, if one is only interested in identifying the set of optimal solutions Θ^*, one can achieve exponential convergence rate with respect to order using results from large deviation theory. Specifically, if $Y(\mathbf{x})$ has a finite moment generating function $M(\lambda) = \mathsf{E}[e^{\lambda Y(\mathbf{x})}]$, then for any positive constant $\delta > 0$, there exists a positive constant β such that

$$\mathsf{P}(|\overline{Y}(\mathbf{x}) - g(\mathbf{x})| > \delta) \le e^{-n\beta}. \tag{2.7}$$

Based on this result, for any alignment level l, the misalignment probability decays exponentially, as shown in [13, 14, 26, 68]. Without loss of generality, we assume that $g(\mathbf{x}_1) < g(\mathbf{x}_2) < \ldots < g(\mathbf{x}_k)$. For simplicity, we assume that all solutions receive the same number of simulation replications n. Let $\Delta = \min_{i=1,2,\ldots,k-1}(g(\mathbf{x}_{i+1}) - g(\mathbf{x}_i))$, and $\delta = \Delta/2$. Clearly, if no sample mean $\overline{Y}(\mathbf{x}_i)$ deviates from its true mean $g(\mathbf{x}_i)$ by more than δ, then all sample means are in the same order as the true means, and thus all solutions are aligned. Therefore, for misalignment to happen, there must exist some $\mathbf{x}_i$ such that $|\overline{Y}(\mathbf{x}_i) - g(\mathbf{x}_i)| \ge \delta$. We thus have the following inequality to bound the misalignment probability

$$\begin{aligned}
\mathsf{P}(|T \cap S| < l) &\le \mathsf{P}(\exists \mathbf{x}_i \text{ s.t. } |\overline{Y}(\mathbf{x}_i) - g(\mathbf{x}_i)| \ge \delta) \\
&= \mathsf{P}\left(\bigcup_{i=1,\ldots,k} [|\overline{Y}(\mathbf{x}_i) - g(\mathbf{x}_i)| \ge \delta] \right) \\
&\le \sum_{i=1}^{k} \mathsf{P}(|\overline{Y}(\mathbf{x}_i) - g(\mathbf{x}_i)| \ge \delta) \\
&\le k e^{-n\beta}.
\end{aligned}$$

The last inequality follows from (2.7). Therefore, the misalignment probability decays exponentially fast as simulation replication n increases, confirming the first basic idea that estimating order is easier than estimating the absolute objective value.

It is also worthwhile noticing that the analysis above does not impose the normality assumption as in the R&S algorithms in Sect. 2.3. A finite moment generating function is both sufficient and necessary to achieve the asymptotic exponential convergence rate with respect to order [20]. However, common distributions such as lognormal and certain Gamma distributions do not have finite moment generating functions. In such cases, proper truncations can be used to recover the exponential convergence rate [20].

To understand the benefit of goal softening, we assume that we randomly (and thus blindly) pick the set S from all k solutions. For simplicity, we only consider $l = 1$. Let $s = |S|$ and $t = |T|$. The misalignment probability is $\mathsf{P}(|T \cap S| = 0) = \binom{k-t}{s}/\binom{k}{s}$. So the AP is given by

$$\mathsf{P}(|T \cap S| \ge 1) = 1 - \mathsf{P}(|T \cap S| = 0)$$

$$= 1 - \binom{k-t}{s} \Big/ \binom{k}{s}$$

$$= 1 - \frac{(k-t)(k-t-1)\cdots(k-t-s+1)}{k(k-1)\cdots(k-s+1)}. \tag{2.8}$$

We can bound (2.8) by using the fact that

$$\frac{k-t-i}{k-i} = 1 - \frac{t}{k-i} \leq 1 - \frac{t}{k},$$

for all $i = 0, 1, \ldots, s-1$. So we have

$$\mathsf{P}(|T \cap S| \geq 1) \geq 1 - \left(1 - \frac{t}{k}\right)^s. \tag{2.9}$$

Since $1 - t/k \leq e^{-t/k}$, we can further bound (2.9) with the following inequality

$$\mathsf{P}(|T \cap S| \geq 1) \geq 1 - e^{-\frac{ts}{k}}. \tag{2.10}$$

The righthand side of (2.10) establishes the fact that as we soften our goal, i.e., increase t to make the "good enough" set T larger, or increase the size of the selected set s, the alignment probability converges exponentially to 1.

Instead of blindly picking the set S, one may run a few simulation replications on all $\mathbf{x} \in \Theta$ and choose S according to the sample mean $\overline{Y}(\mathbf{x})$. Under the assumption of a common additive Gaussian simulation error term $\mathcal{N}(0, \sigma^2)$ for all $\mathbf{x}$, when an equal number of simulation replications n is allocated to every $\mathbf{x} \in \Theta$, selecting the top s solutions ranked by $\overline{Y}(\mathbf{x})$ has the same lower bound on AP given in (2.10) as in the blind picking case under a Least Favorable Configuration (LFC) [42]. In an LFC, without loss of generality, we have $g(\mathbf{x}) = 0$ for all $\mathbf{x} \in S$ and $g(\mathbf{x}) = \Delta$ for all $\mathbf{x} \in \Theta \setminus S$. Notice that when $\Delta = 0$, it is equivalent to the blind picking case. Furthermore, if $\Delta > \overline{\Delta}$, where

$$\overline{\Delta} = \frac{\sigma}{\sqrt{n}} \left[\frac{1}{2} + \log\left(\frac{k-1}{\sqrt{2\pi}} \right) \right],$$

then a tighter lower bound on AP is [42]

$$\mathsf{P}(|T \cap S| \geq 1) \geq 1 - \frac{\max(t,s)}{|t-s|} e^{\left[-\min(t,s)(\Delta - \overline{\Delta})\frac{\sqrt{n}}{\sigma}\right]}. \tag{2.11}$$

The new lower bound (2.11) shows that AP converges exponentially to 1 when

- the optimization goal is softened by increasing $\min(t,s)$;
- the difference in solution quality Δ between good and bad solutions is larger;
- simulation replications n increases.

The LFC represents the worst case scenario of AP and thus (2.10) is a universal lower bound on all problems. Given that the least favorable configuration is hardly the case in real problems, using $\overline{Y}(\mathbf{x})$ to choose the set S can achieve much higher alignment probability than the probability given in (2.8) when S is chosen randomly. However, unless there is structural information about the problem, such as an LFC configuration with Δ difference, there is no readily available tighter lower bound on AP other than (2.10). Nevertheless, for many practical engineering problems, one may conduct pilot experiments to gain knowledge about the problem and use the tables in [26] to identify the s that approximately achieves the required AP.

Therefore, to implement OO, one first determines a target AP and picks the set S. Once the set S is determined, one can then apply a R&S procedure to select the best solution in S. Because the set S is often much smaller than the feasible solution space Θ, R&S procedures are both effective and efficient. Furthermore, because the set S contains at least one good enough solution with a given AP and a R&S procedure can select the best solution from the set S with a given PCS, one can thus select the appropriate AP and PCS to ensure that the solution finally selected is a good enough solution of the original DOvS problem with a target probability.

2.5 Globally Convergent Random Search Algorithms

In this section we consider globally convergent algorithms designed for large, but still finite, feasible regions Θ. We focus on algorithms that exploit no structural information other than $|g(\mathbf{x})| < \infty$ and that we have a consistent estimator $\hat{g}(\mathbf{x})$ of $g(\mathbf{x})$ for all $\mathbf{x} \in \Theta$.

The methods described here can be broadly characterized as globally convergent adaptive random search (GCARS). We start by defining a high-level GCARS algorithm that contains the key features found in the research literature; we then discuss some of the possible choices for these features and their consequences. Finally, we present several specific algorithms that illustrate these choices.

Let $\Theta^\star \subset \Theta$ be the set of globally optimal solutions, which is guaranteed to be non-empty since $|\Theta| < \infty$. Further, let $m = 0, 1, 2, \ldots$ be the index of the number of iterations of the algorithm, and let $\mathbf{x}_m^\star$ be the solution that the algorithm would report as optimal if stopped at the end of iteration m. We are interested in algorithms that provide one of the following convergence guarantees:[1]

$$\lim_{m \to \infty} \mathsf{P}(\mathbf{x}_m^\star \in \Theta^\star) = 1 \text{ or}$$

$$\mathsf{P}\left(\lim_{m \to \infty} \mathbf{1}\{\mathbf{x}_m^\star \in \Theta^\star\} = 1\right) = 1,$$

i.e., convergence in probability or w.p.1, respectively.

[1]One can also define convergence of $\hat{g}(\mathbf{x}_m^*)$, the estimated optimal value, but we do not do so here. See for instance Andradóttir [1].

On iteration m of our generic algorithm, there is an estimation set $\mathscr{E}_m \subset \Theta$ containing the solutions that will be simulated to estimate, or refine the estimate of, $g(\mathbf{x})$ for all $\mathbf{x} \in \mathscr{E}_m$. From iteration to iteration the algorithm retains some memory of what is observed through estimation; let $\mathscr{M}_m \subset \Theta$ denote the set of solutions on which information is retained through iteration m. For the moment we are intentionally vague about what the information is, but it could be as little as the identity of $\mathbf{x}_m^\star$, or as much as a record of all solutions that have ever been estimated, the order in which they were encountered, and all of the observations collected on each of them.

Solutions in the estimation set are simulated; how much simulation effort is expended depends on a simulation allocation rule, denoted $\mathrm{SAR}_m(\mathscr{E}_m|\mathscr{M}_m)$, which may depend on the estimation set, the solutions on which we retain information, and the iteration number. The result of estimation is that each solution $\mathbf{x} \in \mathscr{M}_m$ has a value, denoted $V(\mathbf{x})$, which may be an estimate of $g(\mathbf{x})$ or an indicator that $\mathbf{x}$ is, or is not, the current estimated optimal solution $\mathbf{x}_m^\star$.

To represent the "random" aspect of adaptive random search, let $F_m(\cdot|\mathscr{M}_m)$ be a probability distribution on $\mathbf{x} \in \Theta$ that may depend on the iteration m and information on the solutions in $\mathscr{M}_m$. Given these components, the generic GCARS algorithm is as follows:

Generic GCARS Algorithm

Initialization: Set $\mathscr{M}_0 = \emptyset$ and choose feasible solution $\mathbf{x}_0^\star$. Set the iteration index $m = 0$.

Sampling: Choose the estimation set $\mathscr{E}_m$ where some or all of the solutions are sampled from Θ according to $F_m(\cdot|\mathscr{M}_m)$.

Estimation: Apply the $\mathrm{SAR}_m(\mathscr{E}_m|\mathscr{M}_m)$ to solutions $\mathbf{x} \in \mathscr{E}_m$ in the estimation set.

Iteration: Update $V(\mathbf{x})$ for all $\mathbf{x} \in \mathscr{E}_m$ and choose $\mathbf{x}_{m+1}^\star$ as the solution with the best $V(\mathbf{x})$ value. Update the set $\mathscr{M}_{m+1}$, let $m = m+1$ and go to Sampling.

Notice that the generic GCARS algorithm contains no stopping rule, which is appropriate for proving asymptotic convergence. In practice stopping may occur, for instance, when a computation budget is exhausted or when progress appears too slow. Unless all solutions in Θ are actually simulated, or we have structural information on g, it is not possible to stop a GCARS algorithm with any statistical guarantee that $\mathbf{x}_m^\star$ is an optimal solution.

We now describe several different ways that the steps of GCARS might be accomplished. Let $\mathscr{V}_m$ be the set of solutions that have been visited by the algorithm through iteration m. By "visited" we mean that the solutions in $\mathscr{V}_m$ have been simulated during one or more iterations; therefore, $\mathscr{V}_m = \cup_{j=0}^m \mathscr{E}_m$. A key requirement for GCARS that exploit no structural information about g is that

$$\mathscr{V}_m \stackrel{m\to\infty}{\longrightarrow} \Theta \text{ w.p.1}, \tag{2.12}$$

i.e., all solutions will eventually be visited. Clearly this is a strong condition, and one that will not be realized in practice when Θ is very large. Therefore, the focus of algorithm design is often on being as aggressive as possible in exploring promising areas of Θ while still maintaining Condition (2.12) as well as others.

The burden of insuring (2.12) falls primarily on F_m. Three types of distributions are common.

- A distribution $F_m(\cdot|\mathscr{M}_m)$ that puts positive probability on a small number of feasible solutions in a neighborhood of $\mathbf{x}_m^\star$. When this is the case, then F_m and the neighborhood structure on which it is defined must connect Θ so that any solution is reachable from any other solution after a sufficient number of iterations.
- A distribution $F_m(\cdot|\mathscr{M}_m)$ that puts positive probability on a "promising" subset $\Theta^m \subset \Theta$ that may be large or small, but is not necessarily a neighborhood of $\mathbf{x}_m^\star$. Typically these distributions attempt to use the memory $\mathscr{M}_m$ in an intelligent way to concentrate the search.
- A distribution $F_m(\cdot|\mathscr{M}_m)$ that puts positive probability on all of Θ, but may change as a function of m and $\mathscr{M}_m$. In this case the search is always global, although it may become probabilistically focused on promising regions.

For instance, the Stochastic Ruler algorithm [73] (described below) takes $\mathscr{M}_m = \{\mathbf{x}_m^\star\}$, and only the identity of the current estimated optimal solution is retained. The estimation set is $\mathscr{E}_m = \{\mathbf{x}_m^\star, \mathbf{x}'\}$, where $\mathbf{x}' \sim F_m(\cdot|\mathbf{x}_m^\star)$, and $F_m(\cdot|\mathbf{x}_m^\star)$ puts positive probability only on a neighborhood of $\mathbf{x}_m^\star$. When the support of F_m is a small local neighborhood of $\mathbf{x}_m^\star$, then solution sampling is typically easy. And because the estimation set is a single neighbor, $\mathbf{x}_{m+1}^\star$ is one of $\mathbf{x}_m^\star$ and $\mathbf{x}'$. Algorithms built in this way require very low memory, but need increasing effort per iteration to converge to an optimal solution. In other words, $\mathrm{SAR}_m(\mathscr{E}_m|\mathbf{x}_m^\star)$ must prescribe longer and longer simulation runs as m increases.

The Nested Partitions algorithm [63, 64] (also described below) takes $\mathscr{M}_m = \mathscr{V}_m$, and also retains some measure of the value of each solution visited. Typical choices for the value are $V(\mathbf{x}) = C(\mathbf{x})$, a count of the number of times that $\mathbf{x}$ has been visited by the algorithm through iteration m; or $V(\mathbf{x}) = \overline{Y}(\mathbf{x}; n(\mathbf{x}))$, the cumulative sample mean of the $n(\mathbf{x})$ observations of solution $\mathbf{x}$ for all $\mathbf{x} \in \mathscr{V}_m$. In these algorithms,

$$\mathbf{x}_{m+1}^\star = \mathrm{argmax}_{\mathbf{x}\in\mathscr{V}_m} C(\mathbf{x}) \text{ or} \tag{2.13}$$

$$\mathbf{x}_{m+1}^\star = \mathrm{argmin}_{\mathbf{x}\in\mathscr{V}_m} \overline{Y}(\mathbf{x}; n(\mathbf{x})). \tag{2.14}$$

In the Nested Partitions algorithm, $F_m(\cdot|\mathscr{M}_m)$ assigns substantial probability to a subset $\Theta^m \subset \Theta$ that shows promise of containing good solutions based on previous iterations, but also some to $\Theta \backslash \Theta^m$ to insure convergence. Depending on how these subregions are formed, sampling $\mathbf{x}$ from Θ^m may be easy or computationally challenging. When the estimated optimal is defined by (2.14) or (2.13) it is not essential that SAR_m increase the simulation effort in m. For instance, when the cumulative average (2.14) is employed, and $\overline{Y}(\mathbf{x}; n(\mathbf{x}))$ satisfies a strong law of large numbers, then SAR_m need only assure that if $\mathbf{x}$ is in the estimation set infinitely often, then it will receive an infinite amount of simulation effort.

We now look at four GCARS algorithms in more detail to gain insight into the strengths and weaknesses of various approaches. To simplify the presentation and statement of results, we assume that there is a unique globally optimal solution $\mathbf{x}^\star$.

For the Stochastic Ruler algorithm, we need an unbiased estimator $\hat{g}(\mathbf{x})$ of $g(\mathbf{x})$, whose distribution need not change as a function of the iteration m; e.g., it could be $\overline{Y}(\mathbf{x};n)$ with n fixed. We also need constants a and b such that $\mathsf{P}(a \leq \hat{g}(\mathbf{x}) \leq b) = 1$ for all $\mathbf{x} \in \Theta$.

Stochastic Ruler Algorithm

Step 0. Choose a and b such that $\mathsf{P}(a \leq \hat{g}(\mathbf{x}) \leq b) = 1$ for all $\mathbf{x} \in \Theta$, an irreducible Markov chain transition matrix $\mathbf{R}$ on Θ such that $\mathbf{R}(\mathbf{x},\mathbf{x}') = \mathbf{R}(\mathbf{x}',\mathbf{x})$ for all solutions $\mathbf{x},\mathbf{x}' \in \Theta$, and a sequence of positive integers t_m such that $t_m \to \infty$ as $m \to \infty$. Select an initial solution $\mathbf{x}_0^\star$ and set $m = 0$.

Step 1. Generate a candidate solution $\mathbf{x}'$ from $\mathbf{R}(\mathbf{x}_m^\star, \cdot)$; in other words, randomly select a solution using the $\mathbf{x}_k^\star$ row of $\mathbf{R}$ as the probability distribution on solutions.

Step 2. For $i = 1$ to t_m do:

  Generate an independent estimate $\hat{g}(\mathbf{x}')$ of $g(\mathbf{x}')$

  Generate $U \sim U(a,b)$

  If $\hat{g}(\mathbf{x}') > U$, then

    $\mathbf{x}_{m+1}^\star = \mathbf{x}_m^\star$; go to Step 3

  endif

 Next i

 $\mathbf{x}_{m+1}^\star = \mathbf{x}'$

Step 3. $m = m + 1$; go to Step 1

The Stochastic Ruler algorithm works because it insures that the search is attracted to $\mathbf{x}^\star$ from which it is difficult to leave. Specifically, the probability of rejecting the candidate solution $\mathbf{x}'$ at Step 2 is minimized at $\mathbf{x}^\star$; furthermore, the transition probability into $\mathbf{x}^\star$ is greater than out of $\mathbf{x}^\star$. And even though candidate solutions are generated from a stationary discrete-time Markov chain, the implied transition matrix that describes the movement from solution to solution is irreducible, aperiodic and finite, and its steady-state probabilities degenerate to a distribution putting probability 1 on $\mathbf{x}^\star$ as $m \to \infty$. Thus the convergence is in probability.

The algorithm is elegant and compact (since it retains no past data), but it is not adaptive and requires increasing effort from iteration to iteration in Step 2. As a result its performance in practice is often poor.

When memory of visited solutions is not a limitation, Andradóttir [1] showed that there are significant advantages to using the cumulative sample mean to estimate the value of the optimal solution, even if it is not used for guiding the search. This approach makes almost sure convergence of the algorithms easy to prove via the strong law of large numbers, provides a better estimate of the true value of the selected solution whenever the algorithm terminates since information on it is accumulated, and tends to yield better empirical performance. Since it is now computationally possible to store sample mean information on a very large number of solutions, this insight has had a profound impact on algorithm design. The next two algorithms we describe provide better performance than the Stochastic Ruler algorithm by being more adaptive and retaining the cumulative sample means.

The principles of branch and bound have had an important influence on DOvS, even though classical branch and bound techniques such as relaxing integrality constraints to bound the potential of a partition of the feasible region are not possible. We first describe the stochastic branch-and-bound method (SB&B) of Norkin et al. [51,52], and then show how it leads to the widely used Nested Partitions (NP) algorithms of Shi and Ólafsson [63, 64] and Pichitlamken and Nelson [58]. This development is based on Nelson [48].

To describe a simplified version of SB&B, let $\{\Theta^p\}$ be subsets of Θ creating a partition $\mathscr{P}$. Define the value of the optimal solution restricted to Θ^p by

$$g^\star(\Theta^p) = \min_{\mathbf{x}\in\Theta^p} g(\mathbf{x}).$$

Clearly $g(\mathbf{x}^\star) = \min_{\Theta^p\in\mathscr{P}} g^\star(\Theta^p)$. Suppose that there exist two bounding functions ℓ and u defined on subsets of Θ such that

- $\ell(\Theta^p) \le g^\star(\Theta^p) \le u(\Theta^p)$
- $u(\Theta^p) = g(\mathbf{x}')$ for some $\mathbf{x}' \in \Theta^p$
- If $|\Theta^p| = 1$ then $\ell(\Theta^p) = g^\star(\Theta^p) = u(\Theta^p)$.

If we knew ℓ and u then we could directly apply branch and bound. Instead, suppose that there are estimators L_k and U_k defined on subsets Θ^p such that w.p.1,

$$\begin{aligned}\lim_{m\to\infty} L_m(\Theta^p) &= \ell(\Theta^p),\\ \lim_{m\to\infty} U_m(\Theta^p) &= u(\Theta^p).\end{aligned}$$

Under these assumptions, a SB&B algorithm is the following:

Stochastic Branch and Bound Algorithm

Step 1. Set $m = 0$, $\mathscr{P}_0 = \Theta$ and generate $L_m(\Theta)$ and $U_m(\Theta)$.
Step 2. Set

$$\begin{aligned}\Theta_m &= \operatorname{argmin}\{L_m(\Theta^p) : \Theta^p \in \mathscr{P}_m\}\\ \mathbf{x}_m^\star &\in \operatorname{argmin}\{U_m(\Theta^p) : \Theta^p \in \mathscr{P}_m\}.\end{aligned}$$

Step 3. If $|\Theta_m| = 1$ then $\mathscr{P}_{m+1} = \mathscr{P}_m$ and go to Step 4.
Else let $\mathscr{P}'_m$ be a partition of Θ_m and let $\mathscr{P}_{m+1} = (\mathscr{P}_m\backslash\Theta_m)\cup\mathscr{P}'_m$.
Step 4. For all $\Theta^p \in \mathscr{P}_{m+1}$ generate $L_{m+1}(\Theta^p)$ and $U_{m+1}(\Theta^p)$, set $m = m+1$ and go to Step 2.

As the algorithm progresses, better estimates are obtained of the bounding functions, and the partition with the best lower bound is partitioned finer and finer. Under mild conditions $\mathbf{x}_m^\star$ converges w.p.1 to $\mathbf{x}^\star$.

There are two practical barriers to the application of SB&B. First, there needs to be bounding functions ℓ and u and convergent estimators of them; see [24, 51, 52] for some specific stochastic optimization problems where this is the case. A second barrier is the computing overhead needed to retain and refine a larger and larger partition structure as the algorithm progresses, since no partition is ever eliminated from consideration as in deterministic branch and bound.

Notice, however, that for any subset Θ^p, it is trivially true that

$$\ell(\Theta^p) = \min_{\mathbf{x}\in\Theta^p} g(\mathbf{x})$$

$$L_m(\Theta^p) = \min_{\mathbf{x}\in\Theta^p} \overline{Y}(\mathbf{x};n(\mathbf{x}))$$

satisfy the required conditions, provided $n(k)$, the cumulative number of replications of solution $\mathbf{x}$ through iteration k, increases. In other words, the smallest objective function value in a partition is a (tight) lower bound, and a consistent estimator of it is the smallest sample mean provided all solutions in the partition are simulated infinitely many times. But it is also true that the estimator

$$\hat{L}_m(\Theta^p) = \min_{\mathbf{x}\in\mathscr{X}^p(m)} \overline{Y}(\mathbf{x};n(\mathbf{x}))$$

works provided $\mathscr{X}^p(m)$ is a randomly sampled subset of solutions from Θ^p that converges to Θ^p as $m \to \infty$. Therefore, $\hat{L}_m(\Theta^p)$ is a sampling-based lower bound that is available for any problem, which addresses the first drawback of SB&B.

To avoid the need to carry along information on an increasing number of partitions, we can modify the definition of the new partition, $\mathscr{P}_{m+1}$, to be

$$\mathscr{P}_{m+1} = (\Theta \backslash \Theta_m) \cup \mathscr{P}'_m.$$

In words, we maintain only the most recently refined partition, and aggregate all other feasible solutions into a single "surrounding region." With these two refinements SB&B becomes a version of the NP method that is similar to Pichitlamken and Nelson [58].

NP uses a very straightforward adaptation: sample solutions more intensely in the partition that has most recently provided an apparently good solution, but continue to sample solutions from the surrounding region in case the global optimal is in it. The effectiveness of both SB&B and NP can be enhanced by making good decisions about which region to partition further, and how many solutions to sample from each partition. Shi and Ólaffson [64] describe embedding ranking and selection (Sect. 2.3) or ordinal optimization (Sect. 2.4) into NP to increase the likelihood that it partitions a region with good solutions. Xu and Nelson [71] suggest using a more sophisticated sampling-based bound than $\hat{L}_m$, one that is based on an empirical Chebyshev inequality, to guide the solution sampling effort allocated to each partition in SB&B.

A typical strategy for GCARS is to exploit (search intensively) regions of Θ that appear to have good solutions, while still maintaining enough global exploration to be sure to capture $\mathbf{x}^\star$ in the limit. And since $g(\mathbf{x})$ can only be estimated, solutions must not only be visited, they must be estimated with less and less error in the limit to allow convergence. Prudius and Andradóttir [60] proposed a framework called balanced explorative and exploitative search with estimation (BEESE), which keeps global exploration, local exploitation and solution estimation in play by switching back and forth among them. Specifically, BEESE uses a `Global` probability distribution that places positive probability on all elements of Θ for exploration; a family of `Local` probability distributions that assigns probability only to solutions that are close (in some sense) to the current sample best solution in Θ for exploitation; and an estimation scheme that allocates replications to a solution $\mathbf{x}$ to estimate $g(\mathbf{x})$. The probability 1 global convergence of an algorithm that falls into the BEESE framework can be proved provided the `Global` search distribution satisfies certain conditions. The simplest version of BEESE, known as R-BEESE, has the following high-level structure:

R-BEESE

Step 1. Sample a solution $\mathbf{x}' \sim \texttt{Global}(\Theta)$ and estimate $g(\mathbf{x}')$.

Step 2. With probability q, take additional replications of the current sample best solution $\mathbf{x}_m^\star$ to refine the estimate of $g(\mathbf{x}_m^\star)$.
Else w.p. p sample a solution $\mathbf{x}' \sim \texttt{Global}(\Theta)$ and estimate $g(\mathbf{x}')$ or refine the estimate of $g(\mathbf{x}')$ if $\mathbf{x}'$ has been visited before.
Otherwise sample a solution $\mathbf{x}' \sim \texttt{Local}(\mathbf{x}_m^\star)$ and estimate or refine the estimate of $g(\mathbf{x}')$.

Step 3. Update current sample best solution and go to Step 2.

The `Global` and `Local` distributions on Θ, and the switching probabilities p and q, have an impact on performance. Since p and q are hard to choose (and should probably evolve), Prudius and Andradóttir [60] also describe A-BEESE which makes the switching decisions dynamic and adaptive to the progress of the search. The BEESE framework provides a structure and conditions that guarantee global convergence, but within which smart heuristics can be employed.

The Stochastic Ruler algorithm generates new candidate solutions directly from a neighborhood of the previous candidate; NP samples solutions intensely from a promising region defined by constraints; while R-BEESE switches between sampling solutions from a static global distribution on Θ and a local distribution that concentrates around the current sample best. Another approach is to always generate candidate solutions from a global probability distribution over Θ, but one that adapts based on the performance of previous candidates. Therefore, the search is always global, but concentrates on promising areas by changing the global distribution. The final GCARS algorithm we describe is the Model Reference Adaptive Search (MRAS) algorithm of Hu et al. [34, 35], and in particular its stochastic simulation counterpart SMRAS [36]. MRAS and SMRAS are closely related to the cross-entropy method [62] and the estimation of distribution algorithm [46], and thus demonstrates the principles of those algorithms as well.

To make the algorithm easier to state, consider a problem where the goal is maximization, $g(\mathbf{x}) > 0$ and (for the moment) g can be evaluated exactly, i.e., we have a zero-variance estimator of $g(\mathbf{x})$, so the optimization problem is deterministic.[2] Let $r(\cdot)$ be a probability mass function over $\mathbf{x} \in \Theta$, which defines a random variable $\mathbf{X} \sim r$; in other words, $\mathbf{X}$ is a randomly sampled solution from Θ, sampled according to distribution r. Therefore r induces a distribution on the random variable $g(\mathbf{X})$, the value of the objective function at $\mathbf{X}$, making quantities such as $\mathsf{E}_r[g(\mathbf{X})]$ well defined.

Under appropriate conditions, there exists a recursive sequence of reference distributions $\{r_m; m = 0, 1, 2, \ldots)$ on Θ with the property that [35]

$$\lim_{m \to \infty} \mathsf{E}_{r_m}[g(\mathbf{X})] = g(\mathbf{x}^\star).$$

Since $\mathbf{x}^\star$ is unique, this sequence of distributions converges to a distribution that concentrates all probability on $\mathbf{x}^\star$. Specifically, starting from some initial distribution $r_0(\mathbf{x})$ that assigns positive probability to all $\mathbf{x} \in \Theta$,

$$r_{m+1}(\mathbf{x}) = \frac{g(\mathbf{x}) r_m(\mathbf{x})}{\sum_{\mathbf{x}' \in \Theta} g(\mathbf{x}') r_m(\mathbf{x}')}.$$

If we could generate samples from r_m, then we could empirically estimate r_{m+1}, and continue to do this until the reference distribution essentially degenerates onto $\mathbf{x}^\star$. Unfortunately, r_m may have no special structure, making sampling from it computationally difficult. Therefore, MRAS samples solutions from Θ using a convenient parametric distribution $f(\cdot; \boldsymbol{\beta}_m)$, where at each iteration, $\boldsymbol{\beta}_m$ minimizes the Kullback–Leibler divergence between the parametric distribution and the reference distribution. SMRAS adapts MRAS to DOvS problems by substituting simulation estimators for $g(\mathbf{x})$, increasing the precision of these estimators as the algorithm closes in on $\mathbf{x}^\star$. A high-level description of SMRAS is given below.

SMRAS Algorithm

Step 1. Choose initial distribution $f(\cdot; \boldsymbol{\beta}_0)$ that assigns positive probability to all of Θ; a mixing coefficient $\lambda \in (0, 1)$; an initial number of solutions to sample t_0; an initial simulation sample size $n_0 > 1$; a simulation allocation rule n_m; and set $m = 0$.

---

[2]Clearly any minimization problem on $g(\mathbf{x})$ can be formulated as a maximization problem. If an estimator $\hat{g}(\mathbf{x})$ of $g(\mathbf{x})$ could be negative, then MRAS/SMRAS maximizes $s(g(\mathbf{x}))$ instead, where s is a non-negative, strictly increasing function.

Step 2. Generate t_m candidate solutions from $\bar{f}(\cdot;\boldsymbol{\beta}_m) = (1-\lambda)f(\cdot;\boldsymbol{\beta}_m) + \lambda f(\cdot;\boldsymbol{\beta}_0)$ to fill the estimation set $\mathscr{E}_m = \left\{\mathbf{x}^{(1)}, \mathbf{x}^{(2)}, \ldots, \mathbf{x}^{(t_m)}\right\}$.
Step 3. Simulate n_m i.i.d. observations $Y_1(\mathbf{x}), Y_2(\mathbf{x}), \ldots, Y_{n_m}(\mathbf{x})$ for each solution $\mathbf{x} \in \mathscr{E}_m$ and compute its sample mean $\overline{Y}(\mathbf{x};n_m)$.
Step 4. Calculate a threshold γ for elite solutions.
Step 5. Determine new distribution parameter $\boldsymbol{\beta}_{m+1}$ by solving

$$\boldsymbol{\beta}_{m+1} = \arg\max_{\beta} \left\{ \frac{1}{t_m} \sum_{\mathbf{x}\in\mathscr{E}_m} \frac{[\overline{Y}(\mathbf{x};n_m)]^m}{f(\mathbf{x};\boldsymbol{\beta}_m)} w\left(\overline{Y}(\mathbf{x};n_m)\right) \ln f(\mathbf{x};\boldsymbol{\beta}) \right\}$$

where

$$w(y) = \begin{cases} 0, & \text{if } y \le \gamma - \varepsilon \\ \dfrac{y-\gamma+\varepsilon}{\varepsilon}, & \text{if } \gamma - \varepsilon < y < \gamma \\ 1, & \text{if } y \ge \gamma. \end{cases}$$

Step 6. Set $m = m+1$, choose new solution sample size t_m and go to Step 2.

Step 2.5 minimizes the empirical Kullback–Leibler divergence between the parametric distribution and the reference distribution. There are conditions on the growth of n_m and t_m necessary for convergence; and while the algorithm need not maintain memory of previously visited solutions (since β_{k+1} completely specifies the sampling distribution for the next iteration) simulation effort can be saved by retaining $\overline{Y}(\mathbf{x};n(\mathbf{x}))$ and $n(\mathbf{x})$ for each visited solution, so that if a solution is revisited then only $n_m - n(\mathbf{x})$ additional replications need to be obtained. The evolving threshold in Step 2.5 is also important for algorithm performance; see [36].

Notice that SMRAS employs the mixture distribution $\bar{f}(\cdot;\boldsymbol{\beta}_m) = (1-\lambda)f(\cdot;\boldsymbol{\beta}_m) + \lambda f(\cdot;\boldsymbol{\beta}_0)$, where $f(\cdot;\boldsymbol{\beta}_0)$ forces the algorithm to keep a global perspective. As a result, the distribution can never degenerate to the optimal solution. What can be shown is that for certain choices of parametric distributions

$$\lim_{m\to\infty} \mathsf{E}_{\beta_m}[\mathbf{X}] = \mathbf{x}^*.$$

2.6 Locally Convergent Random Search Algorithms

In this section, we consider locally convergent DOvS algorithms designed for a finite but potentially large feasible region Θ. As in Sect. 2.5, we focus on general-purpose algorithms that only assume $\mathrm{Var}[Y(\mathbf{x})] < \infty$ and that we have a consistent estimator $\hat{g}(\mathbf{x})$ of $g(\mathbf{x})$ for all $\mathbf{x} \in \Theta$.

Recall from Sect. 2.2 that we use $N(\mathbf{x}) \subset \Theta$ to denote the local neighborhood of a solution $\mathbf{x} \in \Theta$. We define $\mathbf{x}$ as a locally optimal solution if $g(\mathbf{x}) \leq g(\mathbf{y})$ for all $\mathbf{y} \in N(\mathbf{x})$. Since this definition of local optimality depends on the definition of $N(\mathbf{x})$, we may have different sets of locally optimal solutions when different neighborhood structures are used. For notational simplicity, we do not explicitly mention this dependence on the definition of $N(\mathbf{x})$ in this section.

Let $\mathscr{L}$ denote the set of locally optimal solutions for the DOvS problem. We again use $\mathbf{x}_m^*$ to denote the solution that the DOvS algorithm would report as optimal if terminated at the end of iteration m. We are interested in algorithms that provide local convergence in probability or local convergence w.p.1:

$$\lim_{m\to\infty} \mathsf{P}(\mathbf{x}_m^\star \in \mathscr{L}) = 1,$$

$$\mathsf{P}\left(\lim_{m\to\infty} \mathbf{1}\{\mathbf{x}_m^\star \in \mathscr{L}\} = 1\right) = 1,$$

Similar to GCARS, an LCARS algorithm has the following key components: an estimation set $\mathscr{E}_m \subset \Theta$ containing the solutions that will be simulated to estimate, or refine the estimate of, $g(\mathbf{x})$ for all $\mathbf{x} \in \mathscr{E}_m$; a memory set $\mathscr{M}_m \subset \Theta$ containing information on a set of solutions simulated up to iteration m; a simulation allocation rule, denoted $\mathrm{SAR}_m(\mathscr{E}_m|\mathscr{M}_m)$, determining how much simulation effort is expended on each solution $\mathbf{x} \in \mathscr{E}_m$, which may depend on the memory set $\mathscr{M}_m$ and the iteration number m; and a sampling probability distribution $F_m(\cdot|\mathscr{M}_m)$ to control the adaptive random search process.

Unlike $F_m(\cdot|\mathscr{M}_m)$ in GCARS algorithms, which may put positive probability on all of Θ, $F_m(\cdot|\mathscr{M}_m)$ in LCARS algorithms typically only puts positive probability on a "promising" subset $\Theta^m \subset \Theta$, which may be a small neighborhood of $\mathbf{x}_m^*$ or a larger area that is believed to contain good solutions according to information in $\mathscr{M}_m$. When global convergence is not required, focusing sampling on "promising" areas can speed up the progress of random search considerably.

Similar to GCRS, LCARS algorithms converge in an asymptotic sense as simulation effort goes to infinity. But unlike its globally convergent counterpart, an LCARS algorithm can be, and should be, combined with a statistical procedure to test the local optimality of $\mathbf{x}_m^*$. Formally, the statistical local optimality test of a solution $\mathbf{x}_m^*$ is

$$\mathrm{H}_0: \ g(\mathbf{x}_m^*) \leq \min_{\mathbf{x}\in N(\mathbf{x}_m^*)} g(\mathbf{x}) \quad \text{vs} \quad \mathrm{H}_1: \ g(\mathbf{x}_m^*) > \min_{\mathbf{x}\in N(\mathbf{x}_m^*)} g(\mathbf{x}).$$

A local optimality test procedure controls the type I and type II errors and provides an implementable stopping rule for LCRS algorithms to terminate the search when a locally optimal solution is found. Such a capability is a very desirable improvement over GCRS, for which it is not possible to stop with any statistical guarantee that $\mathbf{x}_m^*$ is an optimal solution. This test can be rewritten in the following form:

$$\mathsf{P}(\text{declare}\mathbf{x}_m^*\text{locally optimal}) \geq 1-\alpha_L \qquad \text{if } g(\mathbf{x}_m^*) \leq \min_{\mathbf{x}\in N(\mathbf{x}_m^*)} g(\mathbf{x}),$$

$$\mathsf{P}(\text{declare}\mathbf{x}_m^*\text{not locally optimal}) \geq 1-\alpha_L \qquad \text{if } g(\mathbf{x}_m^*) \geq \min_{\mathbf{x}\in N(\mathbf{x}_m^*)} g(\mathbf{x})+\delta_L,$$

where α_L is the type I error and δ_L is the indifference zone parameter. This test is thus a special case of comparison with a standard, which is $\mathbf{x}_m^*$. Therefore, efficient sequential procedures such as that of Kim [38] can be used to perform this test.

The generic LCARS algorithm is identical to the generic GCARS algorithm presented in Sect. 2.5. We now look at two specific LCARS algorithms in more detail to learn how different components of the generic LCARS algorithm can be designed to improve the practical performance of an LCARS algorithm while preserving its local convergence property. We will pay special attention to the design of the "promising" subset $\Theta^m \subset \Theta$ and the sampling probability distribution $F_m(\cdot|\mathscr{M}_m)$.

Convergent Optimization via Most-Promising-Area Stochastic Search (COMPASS) [28] proposes a unique structure of the most promising area Θ^m: it includes all solutions that are closer to $\mathbf{x}_m^*$ than any other simulated solution. Let $\mathscr{V}_m$ denote the set of all solutions simulated through iteration m, and $N_m(\mathbf{x})$ be the total number of i.i.d. simulation replications a solution $\mathbf{x}$ has received up to iteration m. COMPASS starts with an initial feasible solution $\mathbf{x}_0 \in \Theta$ provided by the user and randomly samples t_m solutions (duplicates allowed) from Θ_m at each iteration.

COMPASS

Step 1. Set $m=0, \mathscr{V}_0=\{\mathbf{x}_0\}, \mathbf{x}_0^*=\mathbf{x}_0$. Simulate $n_0(\mathbf{x}_0)$ i.i.d. observations for $\mathbf{x}_0$, set $N_0(\mathbf{x}_0)=n_0(\mathbf{x}_0)$, and calculate its sample mean $\overline{Y}_0(\mathbf{x}_0)$. Let $\Theta_0=\Theta$.

Step 2. Let $m=m+1$. Sample t_m candidate solutions $\mathbf{x}_m^{(1)},\mathbf{x}_m^{(2)},\ldots,\mathbf{x}_m^{(t_m)}$ uniformly and independently from Θ_{m-1}. Let $\mathscr{V}_m=\mathscr{V}_{m-1}\bigcup\{\mathbf{x}_m^{(1)},\mathbf{x}_m^{(2)},\ldots,\mathbf{x}_m^{(t_m)}\}$. Determine $n_m(\mathbf{x})$ according to the SAR for every $\mathbf{x}\in\mathscr{V}_m$ and simulate $n_m(\mathbf{x})$ i.i.d. replications for every $\mathbf{x}\in\mathscr{V}_m$. Update $N_m(\mathbf{x})$ and $\overline{Y}(\mathbf{x})$ for every $\mathbf{x}\in\mathscr{V}_m$.

Step 3. Let $\mathbf{x}_m^*=\arg\min_{\mathbf{x}\in\mathscr{V}_m}\overline{Y}(\mathbf{x})$. Let $\Theta_m=\{\mathbf{x}:\mathbf{x}\in\Theta, ||\mathbf{x}-\mathbf{x}_m^*||\leq||\mathbf{x}-\mathbf{y}||\ \forall\mathbf{y}\in\mathscr{V}_m, \mathbf{y}\neq\mathbf{x}_m^*\}$. Go to Step 2.

Since COMPASS does not aim for global convergence, it can focus search effort in the area Θ_m, which is changed adaptively at each iteration based on information collected on all simulated solutions $\mathbf{x}\in\mathscr{V}_m$. As the algorithm iterates, Θ_m shrinks quickly and guides the search towards a potential locally optimal solution. When more simulations reveal that $\mathbf{x}_{m-1}^*$ is no longer the optimal solution, COMPASS allows the construction of a new Θ_m and moves the search towards new areas.

COMPASS maintains the cumulative sample mean of each simulated solution $\mathbf{x}\in\mathscr{V}_m$, and thus the memory set $\mathscr{M}_m=\mathscr{V}_m$. The sampling distribution $F_m(\cdot|\mathscr{M}_m)$ is uniform on Θ_m and puts zero density on $\Theta\setminus\Theta_m$. The estimation set $\mathscr{E}_m$ is set to be $\mathscr{V}_m$, and the SAR$(\mathscr{E}_m|\mathscr{M}_m)$ requires that $n_m(\mathbf{x})\to\infty$ as $m\to\infty$. From a practical point of view, increasing $n_m(\mathbf{x})$ fast can slow down the progress significantly as

$\mathscr{E}_m$ includes every visited solution $\mathbf{x} \in \mathscr{V}_m$. The original COMPASS algorithm thus increases $n_m(\mathbf{x})$ logarithmically.

As COMPASS iterates, the most promising area Θ_m eventually will only contain $\mathbf{x}_m^*$. When this happens, we may perform a statistical local optimality test on $\mathbf{x}_m^*$ and its neighbors. Although COMPASS converges to a locally optimal solution w.p.1 as $m \to \infty$, the statistical local optimality test provides the capability to stop the optimization with a rigorous statistical guarantee of local optimality. This is another significant advantage of LCRS algorithms compared to GCRS algorithms that are typically stopped either when computation budget is exhausted or progress has been too slow, and thus no guarantee can be provided on the performance of $\mathbf{x}_m^*$ when the algorithm is stopped.

COMPASS sets the estimation set $\mathscr{E}_m$ to the entire set of visited solutions. As COMPASS iterates, the size of $\mathscr{E}_m$ keeps increasing and simulating all solutions in $\mathscr{E}_m$ becomes a big computational burden. It has been shown by Hong and Nelson [29] that for COMPASS to converge to a locally optimal solution w.p.1, it only requires the simulation effort for solutions that define the most promising area Θ_m to go to infinity as $m \to \infty$. The constraint defining Θ_m can be written in the following form:

$$(\mathbf{x}_m^* - \mathbf{x}_i)^\top \left(\mathbf{x} - \frac{\mathbf{x}_m^* + \mathbf{x}_i}{2}\right) \geq 0, \quad \mathbf{x}_i \in \mathscr{V}_m.$$

Some of these constraints are inactive in the sense that removing them will not change Θ_m. To determine if a solution $\mathbf{x}_i$ defines an active constraint, Xu et al. [69] propose to solve the following linear program (LP)

$$\begin{aligned} \min_{\mathbf{x}} \ & (\mathbf{x}_m^* - \mathbf{x}_i)^\top \left(\mathbf{x} - \frac{\mathbf{x}_m^* + \mathbf{x}_i}{2}\right) \\ \text{s.t.} \ & (\mathbf{x}_m^* - \mathbf{x}_j)^\top \left(\mathbf{x} - \frac{\mathbf{x}_m^* + \mathbf{x}_j}{2}\right) \geq 0 \qquad \forall \mathbf{x}_j \in \mathscr{V}_m \setminus \{\mathbf{x}_m^*\}, j \neq i. \end{aligned}$$

The solution $\mathbf{x}_i$ defines an active constraint if and only if the objective function value is negative. The LP needs to be solved for every $\mathbf{x}_i \in \mathscr{V}_m$ to find all active solutions. This step is referred to as constraint pruning [69]. When simulation is very time-consuming, there is still substantial saving in computational cost via constraint pruning. Numerical experiments conducted in [69] show that performing constraint pruning every 50 iterations seems to give good practical performance.

Hong and Nelson [29] introduce a generic LCRS algorithm framework with rather mild conditions on the sampling and estimation steps to ensure local convergence. Xu et al. [70] propose another set of conditions that facilitate implementation of fast LCRS algorithms. Their results show that to achieve local convergence, it is sufficient for an LCRS algorithm to satisfy the following conditions on

$F_m(\cdot|\mathscr{M}_m)$, $\mathscr{E}_m$, and the sample allocation rule $\text{SAR}_m(\mathscr{E}_m|\mathscr{M}_m)$ on every $\mathbf{x} \in \mathscr{E}_m$. In the following, let $\mathscr{S}_m$ be the set of solutions sampled from Θ_m at iteration m using the sampling distribution $F_m(\cdot|\mathscr{M}_m)$. Their conditions are:

1. The sampling distribution $F_m(\cdot|\mathscr{M}_m)$ guarantees that $\Pr\{\mathbf{x} \in \mathscr{S}_m\} \geq \varepsilon$ for all $\mathbf{x} \in \mathscr{N}(\mathbf{x}^*_{m-1})$ for some $\varepsilon > 0$ that is independent of m.
2. The estimation scheme satisfies the following requirements:
   (a) $\mathscr{E}_m$ is a subset of $\mathscr{V}_m$;
   (b) $\mathscr{E}_m$ contains $\mathbf{x}^*_{m-1}$ and $\mathscr{S}_m$;
   (c) $n_m(\mathbf{x})$ is allocated such that $\min_{\mathbf{x}\in\mathscr{E}_m} N_m(\mathbf{x}) \geq 1$ for all $m = 1,2,\ldots$ and $\min_{\mathbf{x}\in\mathscr{E}_m} N_m(\mathbf{x}) \to \infty$ w.p.1 as $m \to \infty$.

These flexible conditions allow the construction of alternative most promising areas Θ_m, which has critical influence on the practical performance of an LCRS algorithm. Xu et al. [70] propose a hyperbox-shaped Θ_m and call the algorithm the Adaptive Hyperbox Algorithm (AHA).

Let $x^{(k)}$ be the kth coordinate, $1 \leq k \leq d$, of a visited solution $\mathbf{x} \in \mathscr{V}_m$. Set $l_m^{(k)} = \max_{\mathbf{x}\in\mathscr{V}_m,\mathbf{x}\neq\mathbf{x}^*_m}\{x^{(k)} : x^{(k)} < x_m^{*(k)}\}$ if it exists; otherwise, let $l_m^{(k)} = -\infty$. Also, let $u_m^{(k)} = \min_{\mathbf{x}\in\mathscr{V}_m,\mathbf{x}\neq\mathbf{x}^*_m}\{x^{(k)} : x^{(k)} > x_m^{*(k)}\}$ if it exists; otherwise, let $u_m^{(k)} = \infty$. The hyperbox containing $\mathbf{x}^*_m$ is $\mathscr{H}_m = \{\mathbf{x} : l_m^{(k)} \leq x^{(k)} \leq u_m^{(k)}, 1 \leq k \leq d\}$.

In words, $u_m^{(k)}$ and $l_m^{(k)}$ give the boundaries, along the kth coordinate direction, of the largest hyperbox that encloses $\mathbf{x}^*_m$ but has all other visited solution $\mathbf{x} \in \mathscr{V}_m$ either on the boundary or outside. Note that $u_m^{(k)}$ and $l_m^{(k)}$ is $\pm\infty$ when there is no other solution to provide the hyperbox boundary along the kth coordinate direction. This may arise when $\mathbf{x}^*_m$ is on the boundary, or when AHA has not visited enough solutions yet. Let $\mathscr{L}_m = (l_m^{(1)}, \ldots, l_m^{(d)})$ and $\mathscr{U}_m = (u_m^{(1)}, \ldots, u_m^{(d)})$.

AHA constructs its most promising area Θ_m by finding $\mathscr{H}_m$ and setting $\Theta_m = \mathscr{H}_m \bigcap \Theta$. This construction of Θ_m allows AHA to shrink the volume of Θ_m exponentially fast and thus scales up to higher-dimensional DOvS problems. Another advantage is that it is much less computationally expensive to identify $\mathscr{H}_m$ than to identify the set of active solutions for the COMPASS algorithm. Again, we denote the starting solution as $\mathbf{x}_0$.

AHA

Step 1. Set $m = 0, \mathscr{V}_0 = \{\mathbf{x}_0\}, \mathscr{E}_0 = \{\mathbf{x}_0\}, \mathbf{x}^*_0 = \mathbf{x}_0$. Simulate $n_0(\mathbf{x}_0)$ i.i.d. observations for $\mathbf{x}_0$, set $N_0(\mathbf{x}_0) = n_0(\mathbf{x}_0)$, and calculate its sample mean $\overline{Y}_0(\mathbf{x}_0)$. Let $\mathscr{U}_0 = \mathscr{L}_0 = \mathscr{H}_0 = \emptyset$, and $\Theta_0 = \Theta$.

Step 2. Let $m = m+1$. Sample t_m candidate solutions $\mathbf{x}_m^{(1)}, \mathbf{x}_m^{(2)}, \ldots, \mathbf{x}_m^{(t_m)}$ uniformly and independently from Θ_{m-1}. Remove any duplicates from $\mathbf{x}_m^{(1)}, \mathbf{x}_m^{(2)}, \ldots, \mathbf{x}_m^{(t_m)}$, and let $\mathscr{S}_m$ be the remaining set. Let $\mathscr{E}_m = \mathscr{S}_m \bigcup \{\mathbf{x}^*_{m-1}\}$. Determine $n_m(\mathbf{x})$ according to the SAR for every $\mathbf{x} \in \mathscr{E}_m$ and simulate $n_m(\mathbf{x})$ i.i.d. replications for every $\mathbf{x} \in \mathscr{E}_m$. Update $N_m(\mathbf{x})$ and $\overline{Y}(\mathbf{x})$ for every $\mathbf{x} \in \mathscr{E}_m$.

Step 3. Let $\mathbf{x}_m^* = \arg\min_{\mathbf{x}\in\mathscr{E}_m} \overline{Y}(\mathbf{x})$. Identify $\mathscr{U}_m$ and $\mathscr{L}_m$ and thus $\mathscr{H}_m$. Let $\Theta_m = \mathscr{H}_m \bigcap \Theta$. Go to Step 2.

It is straightforward to verify that AHA satisfies the convergence conditions on the sampling distribution $F_m(\cdot|\mathscr{M}_m)$ and the estimation scheme. Therefore, AHA converges to the set of locally optimal solutions w.p.1.

Both COMPASS and AHA use a uniform sampling distribution with support on the most promising area Θ_m. This choice is reasonable when there is no structural information about the problem inside the most promising area Θ_m. As an alternative, Hong et al. [32] propose uniformly sampling along coordinate directions inside Θ_m and illustrate with a special case how coordinate sampling may help increase the chance of finding a locally optimal solution inside Θ_m. Yet another approach by Sun et al. [65] proposes to use a Gaussian process model as the sampling distribution to balance exploration and exploitation. It can be effectively combined with an LCRS algorithm like COMPASS or AHA to improve its practical performance.

Another category of locally convergent DOvS algorithms extend the problem to the continuous domain via linear interpolation. The advantage is by doing so, one can apply efficient gradient-based line search methods. The R-SPLINE algorithm of Wang, Pasupathy, and Schmeiser [66, 67] works within a retrospective framework [10,33,55–57]. At each iteration m, the retrospective framework converts a stochastic problem into a deterministic problem by averaging across k_m sample paths (generated using common random numbers). Given the deterministic sample-path problem, R-SPLINE uses piecewise linear interpolation to extend the problem into the continuous domain, and gradient estimates can then be computed to perform a line search. R-SPLINE also conducts a neighborhood enumeration search after every line search step. As $k_m \to \infty$ with at least a logarithmic pace, R-SPLINE converges to a locally optimal solution w.p.1.

Stochastic approximation [37, 61] uses stochastic estimates of the gradient directly to guide a line search. Lim [43] also extends the discrete problem $g(\mathbf{x})$ into a continuous problem $\tilde{g}(\mathbf{x})$ via piecewise linear interpolation. The basic idea is to find a $\tilde{g}(\mathbf{x})$ that has the following properties:

1. Both $g(\mathbf{x})$ and $\tilde{g}(\mathbf{x})$ have the same set of locally optimal solutions.
2. It is relatively easy to compute unbiased estimates of $\tilde{g}(\mathbf{x})$.
3. Stochastic approximation converges to a locally optimal solution of $\tilde{g}(\mathbf{x})$.

When these conditions are satisfied, stochastic approximation can be used to solve the original DOvS problem. The algorithms in [43] assume simulation noise has zero mean and finite variance. For a one-dimensional problem, the algorithm requires that $g(\mathbf{x})$ has a unique local minimizer. In the multidimensional case, the algorithms require that $g(\mathbf{x})$ is $L^\natural$-convex or multimodular [47]. Multimodular or $L^\natural$-convex functions arise naturally in many important problems in inventory systems and queueing networks.

2.7 Algorithm Enhancements

R&S procedures have also been used in conjunction with other DOvS algorithms to improve their efficiency or to make a correct decision at the end of the optimization process. Boesel et al. [6] proposed a "clean up" R&S selection procedure that selects the best solution among all solutions evaluated by a DOvS algorithm and provides a fixed-width confidence interval for the objective function value of the best solution. Many search-based OvS algorithms select the best solution from a neighborhood. Pichitlamken et al. [59] designed a sequential procedure for that purpose. Since a DOvS algorithm often runs for many iterations and the algorithm may stop at any iteration, it is desirable to have a R&S procedure that guarantees the solution of the current iteration is the best among all visited solution. Hong and Nelson [30] designed such a procedure. In Xu et al. [69], they also used the comparison-with-a-standard procedure of Kim [38] to test the local optimality of a solution when solving DOvS problems.

2.8 Using Commercial Solvers

This section is based on material in Hong and Nelson [31], Chap. 12 of Banks et al. [3], and Nelson [48].

Most commercial simulation modeling software also includes an OvS tool; however, to the best of our knowledge none of these tools are based on the DOvS algorithms presented in this chapter, with the exception of the ranking and selection procedures that are found in a number of simulation packages. In addition, a free version of COMPASS called "Industrial Strength COMPASS" can be obtained from `www.iscompass.net`; with some effort it could be used in conjunction with commercial simulation modeling software, although it is most suitable for use with a lower-level programming language such as `C++`.

Instead of provably convergent DOvS algorithms, robust metaheuristics are the most common foundation for integrated OvS tools. A "robust metaheuristic" is an OvS procedure that does not depend on strong problem structure to be effective, and is somewhat tolerant of some sampling variability. Examples include genetic algorithms and tabu search. These integrated tools can be applied to problems with continuous, integer and categorical decision variables. Robust metaheuristics have been observed to be effective on difficult *deterministic* optimization problems, but they usually provide no performance guarantees for deterministic problems, and certainly not for OvS problems. The following three simple ideas can make them more effective in practice and help avoid the three types of errors described in Sect. 2.1.4.

2.8.1 Preliminary Experiment to Control Sampling Variability

It will often be up to the simulation user to determine how many replications are needed at each feasible solution examined by the heuristic. We know that convergence requires that the number of replications should increase as the heuristic discovers better and better solutions because it is statistically much more difficult to distinguish solutions that are close in performance than ones that differ substantially. Therefore at the beginning of the search very little error control may be needed for the solver to identify good solutions and search directions, but later in the search this might not be the case. Unfortunately, some solvers use the same number of replications at all solutions visited, and do not revisit solutions to add more replications.

However, some OvS software does have an "adaptive" setting, meaning it adjusts the number of replications based on the variance of the simulation estimates. If this feature is available, then use it. When the user must specify a fixed number of replications per solution, then a preliminary experiment should be conducted: Simulate several solutions, some at the extremes of the feasible region and some in the interior. Compare the apparent best and apparent worst of these solutions. Find the minimum number of replications required to declare them to be statistically significantly different. This is the minimum number of replications that should be used, which may be substantially more than the default minimum number of replications specified by the OvS tool because the software designers want their tool to deliver results quickly. But remember, when the decision that will be based on the DOvS results really matters, then waiting hours or even days for the *best* decision may well be worth it.

2.8.2 Restarting the Optimization

Even with infinite effort, robust metaheuristics may provide no guarantee that they converge to the optimal solution. Therefore, the chances of finding a very good, or even the best, solution is increased if the solver is run multiple times. Each optimization run should use different random number seeds or streams, and should start from different initial solutions if possible. Be sure to select starting solutions on the extremes of the solution space, in the center of the space, or even randomly generated. If you suspect that certain solutions will be good, include them as starting solutions also. *These runs can be made in parallel on different computers.* The inconvenience of initializing several optimization runs is worth it if it leads to a much better solution, and if all runs lead to the same solution then you can have higher confidence that you have found the best.

2.8.3 *Statistical Clean Up After Search*

After the optimization run or runs have completed, it is critical to perform a second set of statistically designed experiments on the apparent best solutions identified by the heuristic; we call these "clean-up experiments."

In an OvS problem you can never be sure you have found the optimal solution; this is an error that cannot be avoided unless you exhaust Θ, although restarting helps. The two other types of errors are avoidable: failing to recognize the best solution that actually *was* visited, and poorly estimating the performance of the solution that was selected in the end. These errors occur because no optimization algorithm can hope to make any progress while at the same time maintaining statistical error control every step of the way, and because there is a natural bias toward solutions that, by chance, received favorable simulation estimates. Therefore, it is prudent to perform a rigorous statistical analysis, using a ranking-and-selection procedure such as those described in Sect. 2.3, to decide which are the best or near-best of the solutions visited during the search. Include at least the top 5 % of the solutions encountered during the search in this controlled experiment. The ranking-and-selection procedures built into the packages are ideal for this purpose.

The "clean up" concept was introduced in [6], which extended NSGS to be able to start with solutions having unequal numbers of observations, as one would expect at the end of a DOvS run.

In summary, the outcomes from using commercial OvS software can be improved by (a) doing some preliminary experiments to assess output variability; (b) making multiple optimization runs to improve the chances of identifying good solutions; and (c) performing a sound experiment on the top solutions to provide a statistical guarantee of selecting the best among them and estimating its performance precisely.

Acknowledgements This work was supported in part by the National Science Foundation 1099 under Grant CMMI-1233376, and by the Hong Kong Research Grants Council under Project 613011, 613012 and N_HKUST626/10.

References

1. S. Andradóttir. Accelerating the convergence of random search methods for discrete stochastic optimization. *ACM Transactions on Modeling and Computer Simulation*, 9:349–380, 1999.
2. S. Andradóttir. Simulation optimization: integrating research and practice. *INFORMS Journal on Computing*, 14:216–219, 2002.
3. J. Banks, J. S. Carson, B. L. Nelson, and D. M. Nicol. *Discrete-Event System Simulation*. Prentice Hall, Inc., Upper Saddle River, NJ, 5th edition, 2010.
4. R. E. Bechhofer. A single-sample multiple decision procedure for ranking means of normal populations with known variances. *Annals of Mathematical Statistics*, 25:16–39, 1954.
5. R. E. Bechhofer, T. J. Santner, and D. Goldsman. *Design and Analysis of Experiments for Statistical Selection, Screening and Multiple Comparisons*. John Wiley, New York, 1995.

6. J. Boesel, B. L. Nelson, and S.-H. Kim. Using ranking and selection to 'clean up' after simulation optimization. *Operations Research*, 51:814–825, 2003.
7. J. Branke, S. E. Chick, and C. Schmidt. Selecting a selection procedure. *Management Science*, 53:1916–1932, 2007.
8. J. A. Buzacott and J. G. Shantikumar. *Stochastic Models of Manufacturing Systems*. Prentice-Hall, Englewood Cliffs, NJ, 1993.
9. C.-H. Chen, J. Lin, E. Yücesan, and S. E. Chick. Simulation budget allocation for further enhancing the efficiency of ordinal optimization. *Discrete Event Dynamic Systems: Theory and Applications*, 10:251–270, 2000.
10. H. Chen and B. W. Schmeiser. Stochastic rooting-finding via retrospective approximation. *IIE Transactions*, 33: 259–275, 2001.
11. S. E. Chick. Subjective probability and Bayesian methodology. In S. G. Henderson and B. L. Nelson, editors, *Handbooks in Operations Research and Management Science: Simulation*. Elsevier, New York, 2006.
12. S. E. Chick and K. Inoue. New two-stage and sequential procedures for selecting the best simulated system. *Operations Research*, 49:732–743, 2001.
13. L. Dai. Convergence properties of ordinal comparison in simulation of discrete event dynamic systems. *Journal of Optimization Theory and Applications*, 91:363–388, 1996.
14. L. Dai and C.-H. Chen. Rates of convergence of ordinal comparison for dependent discrete event dynamic systems. *Journal of Optimization Theory and Applications*, 94:29–54, 1997.
15. M. H. DeGroot. *Optimal Statistical Decisions*. Wiley, New York, 1970.
16. P. I. Frazier. Decision-theoretic foundations of simulation optimization. *Wiley Encyclopedia of Operations Research and Management Sciences*. Wiley, New York, 2010.
17. P. I. Frazier and W. B. Powell. The knowledge-gradient stopping rule for ranking and selection. In S. J. Mason, R. R. Hill, L. Möench, O. Rose, T. Jefferson, and J. W. Fowler, editors, *Proceedings of the 2008 Winter Simulation Conference*, pages 305–312. IEEE, Piscataway, NJ, 2008.
18. P. I. Frazier, W. B. Powell, and S. Daynik. A knowledge-gradient policy for sequential information collection. *SIAM Journal on Control and Optimization*, 47:2410–2439, 2008.
19. P. I. Frazier, W. B. Powell, and S. Daynik. The knowledge gradient policy for correlated normal beliefs. *INFORMS Journal on Computing*, 21:599–613, 2009.
20. M. C. Fu and X. Jin. On the convergence rate of ordinal comparison of random variables. *IEEE Transactions on Automatic Control*, 46: 1950–1954, 2001.
21. D. Goldsman and B. L. Nelson. Comparing systems via simulation. In J. Banks, editor, *Handbook of Simulation*. John Wiley, New York, 1998.
22. S. S. Gupta. *On a Decision Rule for a Problem in Ranking Means*. PhD thesis, University of North Carolina, Chapel Hill, NC, 1956.
23. S. S. Gupta. On some multiple decision (ranking and selection) rules. *Technometrics*, 7:225–245, 1965.
24. W. J. Gutjahr, A. Hellmayr, and G. Ch. Pflug. Optimal stochastic single-machine-tardiness scheduling by stochastic branch-and-bound. *European Journal of Operational Research*, 117:396–413, 1999.
25. Y.-C. Ho, R. Sreenivas and P. Vakili. Ordinal optimization of discrete event dynamic systems. *Journal of Discrete Event Dynamic Systems*, 2:61–88, 1992.
26. Y.-C. Ho, Q.-C. Zhao and Q.-S. Jia. *Ordinal Optimization: Soft Optimization for Hard Problems*. Springer, New York, 2007.
27. L. J. Hong and B. L. Nelson. The tradeoff between sampling and switching: New sequential procedures for indifference-zone selection. *IIE Transactions*, 37:623–634, 2005.
28. L. J. Hong and B. L. Nelson. Discrete optimization via simulation using COMPASS. *Operations Research*, 54:115–129, 2006.
29. L. J. Hong and B. L. Nelson. A framework for locally convergent random search algorithms for discrete optimization via simulation. *ACM Transactions on Modeling and Computer Simulation*, 17:19/1–19/22, 2007.

30. L. J. Hong and B. L. Nelson. Selecting the best system when systems are revealed sequentially. *IIE Transactions*, 39:723–734, 2007.
31. L. J. Hong and B. L. Nelson. A brief introduction to optimization via simulation. In M. D. Rossetti, R. R. Hill, B. Johansson, A. Dunkin and R. G. Ingalls, editors, *Proceedings of the 2009 Winter Simulation Conference*, pages 75–85. IEEE, Piscataway, NJ, 2009.
32. L. J. Hong, B. L. Nelson and J. Xu. Speeding up COMPASS for high-dimensional discrete optimization via simulation. *Operations Research Letters*, 38:550–555, 2010.
33. J. Jin. *Retrospective Optimization of Stochastic Systems*. PhD thesis, Purdue University, West Lafayette, IN, 1998.
34. J. Hu, M. C. Fu, and S. I. Marcus. Stochastic optimization using model reference adaptive search. In M. E. Kuhl, N. M. Steiger, F. B. Armstrong, and J. A. Joines, editors, *Proceedings of the 2005 Winter Simulation Conference*, pages 811–818. IEEE, Piscataway, NJ, 2005.
35. J. Hu, M. C. Fu, and S. I. Marcus. A model reference adaptive search method for global optimization. *Operations Research*, 55:549–568, 2007.
36. J. Hu, M. C. Fu, and S. I. Marcus. A model reference adapative search method for stochastic global optimization. *Communications in Information and Systems*, 8:245–276, 2008.
37. J. Kiefer and J. Wolfowitz. Stochastic estimation of the maximum of a regression function. *Annals of Mathematical Statistics*, 23:462–466, 1952.
38. S.-H. Kim. Comparison with a standard via fully sequential procedure. *ACM Transactions on Modeling and Computer Simulation*, 15:1–20, 2005.
39. S.-H. Kim and B. L. Nelson. A fully sequential procedure for indifference-zone selection in simulation. *ACM Transactions on Modeling and Computer Simulation*, 11:251–273, 2001.
40. S.-H. Kim and B. L. Nelson. Selecting the best system. In S. G. Henderson and B. L. Nelson, editors, *Handbooks in Operations Research and Management Science: Simulation*. Elsevier, New York, 2006.
41. S.-H. Kim and B. L. Nelson. On the asymptotic validity of fully sequential selection procedures for steady-state simulation. *Operations Research*, 54:475–488, 2006.
42. L.-H. Lee, T.-W. E. Lau, and Y.-C. Ho. Explanation of goal softening in ordinal optimization. *IEEE Transactions on Automatic Control*, 44:94–99, 1999.
43. E. Lim. Stochastic approximation over multidimensional discrete sets with applications to inventory systems and admission control of queueing networks. *ACM Transactions on Modeling and Computer Simulation*, 22:19:1–19:23, 2012.
44. J. Luo, L. J. Hong, B. L. Nelson, and Y. Wu. Fully sequential procedures for large-scale ranking-and-selection problems in parallel computing environments. Working paper, Department of Industrial Engineering and Logistics Management, Hong Kong University of Science and Technology, 2013.
45. S. Mahajan and G. Van Ryzin. Stocking retail assortments under dynamic consumer substitution. *Operations Research*, 49:334–351, 2001.
46. H. Mühlenbein and G. Paaß. From recombination of genes to the estimation of distributions I. Binary parameters. In H.-M. Voigt, W. Ebeling, I. Rechenberg, and H.-P. Schwefel, editors, *Parallel Problem Solving from Nature – PPSN IV*, pages 178–197. Springer Verlag, Berlin, Germany, 1996.
47. K. Murota. Note on multimodularity and *L*-convexity. *Mathematics of Operations Research*, 30:658–661, 2005.
48. B. L. Nelson. Optimization via simulation over discrete decision variables. In J. J. Hasenbein, editor, *TutORials in Operations Research*, 7:193–207. INFORMS, Hanover, MD, 2010.
49. B. L. Nelson, J. Swann, D. Goldsman, and W.-M. Song. Simple procedures for selecting the best simulated system when the number of alternatives is large. *Operations Research*, 49:950–963, 2001.
50. J. Nocedal and S. Wright. *Numerical Optimization*. Springer, New York, 2nd edition, 2006.
51. V. I. Norkin, Y. M. Ermoliev, and A. Ruszczyński. On optimal allocation of indivisibles under uncertainty. *Operations Research*, 46:381–395, 1998.
52. V. I. Norkin, G. Ch. Pflug and A. Ruszczyński. A branch and bound method for stochastic global optimization. *Mathematical Programming*, 83:425–450, 1998.

53. T. Osogami. Finding probably best systems quickly via simulations. *ACM Transactions onModeling and Computer Simulation*, 19:12:1–12:18, 2009.
54. E. Paulson. A sequential procedure for selecting the population with the largest mean from k normal populations. *Annals of Mathematical Statistics*, 35:174–180, 1964.
55. R. Pasupathy. *Retrospective-Approximation Algorithms for the Multidimensional Stochastic Rooting-Finding Problem*. PhD thesis, Purdue University, West Lafayette, IN, 2005.
56. R. Pasupathy and B. W. Schmeiser. Retrospective-approximation algorithms for the multidimensional stochastic rooting-finding problem. *ACM Transactions on Modeling and Computer Simulation*, 19:2:1–2:36, 2009.
57. R. Pasupathy. On choosing parameters in retrospective-approximation algorithms for stochastic rooting-finding and simulation optimization. *Operations Research*, 58:889–901, 2010.
58. J. Pichitlamken and B. L. Nelson. A combined procedure for optimization via simulation. *ACM Transactions on Modeling and Computer Simulation*, 13:155–179, 2003.
59. J. Pichitlamken, B. L. Nelson, and L. J. Hong. A sequential procedure for neighborhood selection-of-the-best in optimization via simulation. *European Journal of Operational Research*, 173:283–298, 2006.
60. A. A. Prudius and S. Andradóttir. Balanced explorative and exploitative search with estimation for simulation optimization. *INFORMS Journal on Computing*, 21:193–208, 2009.
61. H. Robbins and S. Monro. A stochastic approximation method. *Annals of Mathematical Statistics*, 22:400–407, 1951.
62. R. Y. Rubinstein and D. P. Kroese. *The Cross-Entropy Method: A Unified Approach to Combinatorial Optimization, Monte Carlo Simulation, and Machine Learning*. Springer, New York, 2004.
63. L. Shi and S. Ólafsson. Nested partitions method for stochastic optimization. *Methodology and Computing in Applied Probability*, 2:271–291, 2000.
64. L. Shi and S. Ólafsson. *Nested Partitions Method, Theory and Applications*. Springer, New York, 2009.
65. L. Sun, L. J. Hong, and Z. Hu. Optimization via simulation using Gaussian process-based search. In S. Jain, R. R. Creasey, J. Himmelspach, K. P. White, and M. Fu, editors, *Proceedings of the 2011 Winter Simulation Conference*, pages 4139–4150. IEEE, Piscataway, NJ, 2011.
66. H. Wang. *Retrospective Optimization of Discrete Stochastic Systems Using Simplicial Linear Interpolation*. PhD thesis, Purdue University, West Lafayette, IN, 2009.
67. H. Wang, R. Pasupathy, and B. W. Schmeiser. Integer-ordered simulation optimization using R-SPLINE: Retrospective search with piecewise-linear interpolation and neighborhood enumeration. *ACM Transactions on Modeling and Computer Simulation*, 23:17:1–17:24, 2013.
68. X. Xie. Dynamics and convergence rate of ordinal comparison of stochastic discrete-event systems. *IEEE Transactions on Automatic Control*, 42: 586–590, 1998.
69. J. Xu, B. L. Nelson, and L. J. Hong. Industrial Strength COMPASS: A comprehensive algorithm and software for optimization via simulation. *ACM Transactions on Modeling and Computer Simulation*, 20:1–29, 2010.
70. J. Xu, B. L. Nelson, and L. J. Hong. An adaptive hyperbox algorithm for high-dimensional discrete optimization via simulation problems. *INFORMS Journal on Computing*, 25:133–146, 2013.
71. W. L. Xu and B. L. Nelson. Empirical stochastic branch-and-bound for optimization via simulation. *IIE Transactions*, 45:685–698, 2013.
72. R. R. Wilcox. A table for Rinott's selection procedure. *Journal of Quality Technology*, 16:97–100, 1984.
73. D. Yan and H. Mukai. Stochastic discrete optimization. *SIAM Journal of Control and Optimization*, 30:594–612, 1992.

Chapter 3
Ranking and Selection: Efficient Simulation Budget Allocation

Chun-Hung Chen, Stephen E. Chick, Loo Hay Lee, and Nugroho A. Pujowidianto

Abstract This chapter reviews the problem of selecting the best of a finite set of alternatives, where best is defined with respect to the highest mean performance, and where the performance is uncertain but may be estimated with simulation. This problem has been explored from several perspectives, including statistical ranking and selection, multiple comparisons, and stochastic optimization. Approaches taken in the literature include frequentist statistics, Bayesian statistics, related heuristics, and asymptotic convergence in probability. This chapter presents algorithms that are derived from Bayesian and related conceptual frameworks to provide empirically effective performance for the ranking and selection problem. In particular, we motivate the optimal computing budget allocation (*OCBA*) algorithm and expected value of information (*EVI*) approaches, give example algorithms, and provide pointers to the literature for detailed derivations and extensions of these approaches.

3.1 Introduction

This chapter deals with the problem of selecting the best of a finite set of alternatives. In the context of simulation optimization, each alternative represents a different potential configuration of (or settings for) relevant decision variables. The performance of each alternative is not known with certainty, but its value can be estimated using stochastic simulation. We consider the setting where the number

---

C.-H. Chen (✉)
George Mason University, Fairfax, VA, USA
e-mail: cchen9@gmu.edu

S.E. Chick
INSEAD, Fontainebleau, France
e-mail: stephen.chick@insead.edu

L.H. Lee
National University of Singapore, Singapore, Singapore
e-mail: iseleelh@nus.edu.sg

N.A. Pujowidianto
Hewlett-Packard Singapore, Singapore, Singapore
e-mail: nugroho@hp.com

M.C. Fu (ed.), *Handbook of Simulation Optimization*, International Series in Operations Research & Management Science 216, DOI 10.1007/978-1-4939-1384-8_3

of feasible solutions (called "alternatives") is small enough so that each alternative can be simulated at least a few times in order to estimate its performance. In the conclusion, we will point to recent work addressing settings where this is not the case.

In this chapter, we will refer to a replication of a stochastic simulation as a sample. When using simulation to compare the performance of multiple alternatives, one might gradually increase the number of samples for each alternative until the variance of the estimator is sufficiently small (i.e., the confidence intervals for estimation are satisfactorily narrow) so that a satisfactory amount of evidence exists to justify a selection of the "best" alternative. One very simple approach is to use an identical number of samples for each alternative. This approach can be inefficient: if one alternative has very low variance, then it may only require very few samples to accurately estimate its performance.

To improve the efficiency of selecting the best alternative, several approaches have been explored. Intuitively, to ensure a high probability of correctly selecting an optimal alternative, a larger portion of the sampling budget should be allocated to those alternatives that are more critical in the process of identifying good alternatives quickly. Those could be alternatives with high estimated mean performance in combination with a certain degree of uncertainty about the actual mean performance. On the other hand, one might wish to sample less often the alternatives whose estimated means are either poor or have a low degree of uncertainty about their values. Two questions remain. How should one allocate resources to sample from the different alternatives, as a function of their estimated mean performance, and the uncertainty about their mean performance? For how long should one sample until stopping to select an alternative as best?

This chapter focuses on two different approaches to answering these questions, based on the optimal computing budget allocation (*OCBA*) approach and the Bayesian expected value of information (*EVI*) approach. The approaches are motivated and basic algorithms are presented. Derivations can be found in the provided references. Empirical results from these two approaches show that they perform favorably relative to some other approaches that have been proposed, in the sense of providing a relatively high average-case performance over problem instances, for a given number of samples that are observed to estimate the mean performances of the alternatives.

3.1.1 Intuitive Explanations of Simulation Budget Allocation

Consider an inventory control problem where the goal is to maximize the expected profit during a certain horizon by determining the best among five inventory ordering policies. Each alternative inventory policy is specified by two numbers s and S, where $0 \leq s < S$. If the inventory level falls below s, an order is placed to increase the inventory level to S. Otherwise, no order is placed. The profit is estimated via simulation due to randomness in demands and the amount delivered. The goal is to find the alternative with the highest mean profit.

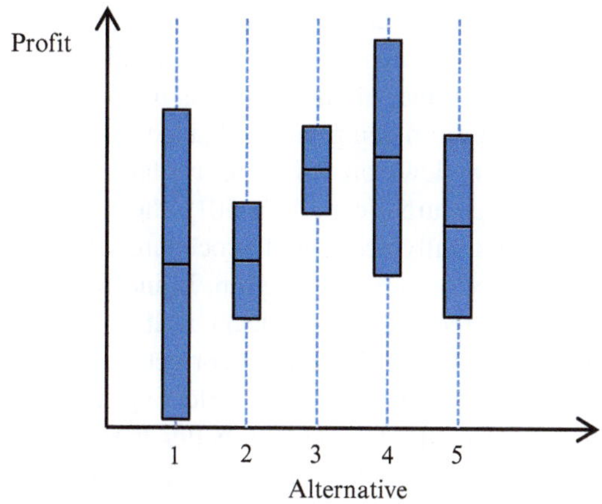

Fig. 3.1 Ninety-nine per cent confidence intervals for five alternatives in a given stage of sequential sampling

One advantage of simulation experiments is that a decision maker can collect samples in a sequential manner. After a given stage of sampling, estimates of the means of each alternative are available, along with an assessment of the uncertainty in the estimates. A selection procedure considers whether additional sampling is required before selecting an alternative, and if so, how to allocate a sampling budget to the different simulated alternatives in the next stage of sampling.

Figure 3.1 gives a representative scenario of possible results from the samples that have been collected through a given stage of sampling. It shows the 99 % confidence intervals along with the accompanying mean estimator (represented as the line in the middle of the confidence interval) for each alternative. Some alternatives seem better, but none are clearly better than all the others: all of the confidence intervals overlap. In situations such as this, it is not straightforward to determine which alternatives can be eliminated and which alternatives should receive more simulation budget.

Intuitively, the decision maker may want to allocate more samples to the alternatives with bigger half width such as alternatives 1 and 4 to reduce the variance of their estimators. On the other hand, it is sensible to allocate more samples to alternatives with larger means such as alternatives 3 and 4 as the objective is to maximize the profit. The question is how much these factors influence what gets sampled in the next stage of sampling. The next section gives an overview of the works attempting to select the best alternative.

3.1.2 Overview of Ranking and Selection (R&S)

R&S procedures aim to identify the best alternative. One R&S procedure might be considered to be better in some sense than another if it requires fewer samples, in expectation, to achieve the same level of evidence for correct selection than that other procedure does. In this chapter, we will focus on a common context: all alternatives are simulated. This is suitable for simulation optimization problems when the number of alternatives is finite and not so large that it would prevent each alternative from being sampled at least a few times for statistical inference [26].

Many reviews on R&S are available [1, 29, 39]. There are two-stage or few-stage procedures (e.g., [21, 50]), the two-stage procedures with screening (e.g., [48]), and fully-sequential procedures, (e.g., [38]) that can guarantee a desired probability of correct selection. In indifference zone (IZ) procedures, a difference is considered to be significant if it is larger than a specified indifference-zone parameter. The probability of correct selection guarantee in the IZ approach is with respect to the probability of selecting the true best, subject to the condition that the mean of the true best is better than the mean of all of the other alternatives by at least the indifference-zone parameter. Thus, this is based on a worst-case performance metric. This worst case approach can provide frequentist guarantees for correct selection, but might require more samples to be collected to obtain that guarantee than may be practically implementable. As such, they can be statistically conservative.

In this chapter, we present R&S procedures based on average case performance metrics that sample in a highly sequential manner. The goal is either to maximize evidence for correct selection subject to a constraint on the sampling budget or to reach a level of evidence for correct selection with the fewest expected number of samples. Using an average-case analysis rather than a worst-case bound of the IZ approach, we present two distinct approaches. The *OCBA* approach uses a thought experiment that attempts to sequentially maximize the probability that the best alternative can be correctly identified after the next stage of sampling. The *EVI* approach uses a Bayesian description of the uncertainty about the mean of each alternative, a loss function to describe the penalty for not correctly selecting the best alternative, and expected value of information ideas to minimize the expected loss from selecting an alternative after simulation. This expected loss can depend upon which alternative is selected for simulation at each stage. These procedures tend to require much less sampling to achieve the same or better empirical performance for correct selection than procedures which are statistically more conservative [4, 58].

The *EVI* approach differs from the *OCBA* in one key respect. In the *OCBA* approach, the effect of additional sampling is modeled using a Bayesian asymptotic normality result: the distribution that is used to describe uncertainty after the samples are observed is assumed to be normally distributed with the same mean and with a variance that shrinks as samples accumulate. In the *EVI* approach, a decision theoretic framework is used. It explicitly models the fact that the posterior mean will change after the samples are observed. It models the distribution of what the posterior mean will be and explicitly models how changes in the posterior mean will potentially change decisions as to which alternative is best. Changes in a decision from sampling imply a value of information from those samples. The *EVI* approach allocates samples in a way that maximizes, in some sense, a measure of the expected value of information of those samples.

3.1.3 *Organization*

Section 3.2 formulates the selection problem and provides the basic notation. In addition, it provides a generic algorithm for selection procedures together with explanations on the components of the algorithm. The main insights of the efficient R&S procedures are presented at the end of Sect. 3.2. Section 3.3 describes one class of efficient R&S procedures, called *OCBA*. It encompasses the objective of the allocation, the allocation rule, the algorithm, and how it has been extended in other settings. Similarly, Sect. 3.4 provides the details of the *EVI* procedures. In particular, three different procedures are presented together with their algorithms. Section 3.5 concludes this chapter.

3.2 Problem Formulation and Selection Procedures

The problem of selecting the best is formulated in Sect. 3.2.1, and a generic selection procedure algorithm is provided in Sect. 3.2.2. Section 3.2.3 provides an overview of *OCBA* and *EVI* procedures, which are presented in more detail in Sects. 3.3 and 3.4, respectively.

3.2.1 *Problem Formulation of Selecting the Best*

We consider the problem of selecting the best of several alternatives based on their means which have to be estimated via stochastic simulation. Without loss of generality, we define the alternative with the largest mean as the best. For readability, we use upper case for random variable, lower case for fixed value or realization, and bold face for vectors.

Let X_{ij} be a random variable whose realization x_{ij} is the output of the jth sample from alternative $i, j=1,2,\ldots$. There are k alternatives so that $i=1,2,\ldots,k$. Let w_i and σ_i^2 be the unknown mean and variance of alternative i. It will be easier at times to refer to the precision $\lambda_i = 1/\sigma_i^2$ instead of the variance. Let $w_{[1]} \leq w_{[2]} \leq \ldots \leq w_{[k]}$ be the ordered means. In practice, the ordering $[\cdot]$ is unknown, and the best alternative is to be identified by sampling.

A problem configuration is denoted by $\chi = (\boldsymbol{w}, \boldsymbol{\sigma}^2)$ where $\boldsymbol{w} = (w_1, w_2, \ldots, w_k)$ and $\boldsymbol{\sigma}^2 = (\sigma_1^2, \sigma_2^2, \ldots, \sigma_k^2)$. Let n_i be the number of samples from alternative i so far. Let $\bar{x}_i = \sum_{j=1}^{n_i} x_{ij}/n_i$ be the sample mean and $\widehat{\sigma}_i^2 = \sum_{j=1}^{n_i} (x_{ij} - \bar{x}_i)^2 / (n_i - 1)$ be the sample variance. The ordered sample means are $\bar{x}_{(1)} \leq \bar{x}_2 \leq \cdots \leq \bar{x}_{(k)}$. The quantity n_i depends on the decision maker while the quantities $\bar{x}_i$, $\widehat{\sigma}_i^2$ and (i) are updated as more samples are observed.

Let $\mathfrak{D}$ be the alternative that is selected as best by the selection procedure when sampling is completed. Each selection procedure generates estimates $\widehat{w}_i$ of w_i, for

$i = 1, 2, \ldots, k$. This chapter focuses on selection procedures where the estimates are based on the sample mean, i.e., $\widehat{w}_i = \overline{x}_i$. Thus, at the time sampling stops, $\mathfrak{D} = (k)$ and a correct selection occurs when the alternative with the best sample mean is the true best (i.e., $(k) = [k]$).

In Bayesian approaches, unknown quantities are represented as random variables. Let W_i be the random variable that represents the unknown mean of alternative i. The Bayesian framework uses the notion that we can update our knowledge using the conditional distribution of parameters, given the data. Once data are available, the posterior distribution describes the uncertainty of the unknown mean. If we assume that samples are independent and normally distributed with unknown mean and variance, and that a non-informative prior distribution is used for the unknown mean and variance of each alternative, then the posterior marginal distribution for the unknown mean W_i follows a Student's t-distribution $St\left(\overline{x}_i,\ n_i/\sigma_i^2,\ v_i\right)$ where $v_i = n_i - 1$ is the degrees of freedom [20]. The mean is $\overline{x}_i$ for $v_i > 1$ and the variance is $(\sigma_i^2/n_i)v_i/(v_i - 2)$ for $v_i > 2$.

In the above framework, we have assumed that the unknown means of each alternative are independently distributed. Some extensions that are mentioned below describe how to relax some of the assumptions made above.

It will be useful to define two figures of merit that are used by the *OCBA* and *EVI* procedures below. Given the data $\mathscr{E} = \{(x_{i1},\ x_{i2} \ldots,\ x_{in_i})$ for $i = 1,\ 2, \ldots,\ k\}$, the posterior probability of correct selection (*PCS*), or posterior probability that the best alternative is correctly selected, is

$$PCS = \mathrm{P}\left(W_{\mathfrak{D}} \geq W_{[k]} \middle| \mathscr{E}\right). \tag{3.1}$$

We could further write $PCS = \mathrm{P}(\mathfrak{D} = [k])$ in contexts where ties occur with probability 0, as is the case in this setting.

If the vector of means $\boldsymbol{w} = (w_1, w_2,\ \ldots,\ w_k)$, the opportunity cost of selecting alternative i is

$$\mathscr{L}_{LL}(i,\ w) = w_{[k]} - w_i. \tag{3.2}$$

Thus, the expected opportunity cost (*EOC*), given the data ε, when $\mathfrak{D} = (k)$ is selected as best is

$$EOC = \mathrm{E}\left[\mathscr{L}_{LL}(\mathfrak{D},\ W)\middle|\mathscr{E}\right] = \mathrm{E}\left[W_{[k]} - W_{\mathfrak{D}}\middle|\mathscr{E}\right]. \tag{3.3}$$

For both the *PCS* and *EOC*, the probability expectations are with respect to the posterior distribution of the unknown means, given all observed data.

3.2.2 A Generic Algorithm for Selection Procedures

We present a generic sequential selection procedure. Variations on this generic selection procedure result by making different assumptions about the nature of the evidence for correct selection and the approximations made to measure that evidence. In summary, an initialization step is used to provide initial estimates of the mean and variance of each alternative. Then, samples are collected sequentially until a stopping rule is activated. At each stage of sampling, an allocation rule specifies how the samples to be collected during that stage should be allocated among the alternatives.

The following are the steps for a generic algorithm for selection procedures.

1. Specify a first-stage sample size $n_0 > 2$, and a total number of samples $\tau > 0$ to allocate per subsequent stage. Specify stopping rule parameters.
2. Sample $X_{i1}, X_{i2}, \ldots, X_{in_0}$ independently and initialize the number of samples $n_i \leftarrow n_0$ so far for each alternative, $i = 1, 2, \ldots, k$.
3. Determine the sample statistics $\bar{x}_i$ and $\widehat{\sigma}_i^2$ and the order statistics so that $\bar{x}_{(1)} \leq \cdots \leq \bar{x}_{(k)}$.
4. WHILE stopping rule is not satisfied, DO another stage
   a. Use the allocation rule to identify which alternative to sample and determine τ_i, the number of samples to allocate to alternative i.
   b. Observe the additional samples, update the sample statistics and the order statistics.
   c. Update the number of samples collected so far for each alternative, $n_i \leftarrow n_i + \tau_i$.
5. Select the best alternative based on the selection rule.

As described in the algorithm, the decision maker needs to decide the allocation rule, the stopping rule, and the selection rule. The following three paragraphs describe each of the components of the algorithm in general. Sections 3.3 and 3.4 give specific examples of allocation rules and stopping rules.

An *allocation rule* is a mapping from the sampling statistics of the k alternatives to a vector of integers that represents the number of samples to observe from each alternative in the next stage of sampling. Let τ_i be the number of samples allocated to alternative i in the next stage of sampling, for $i = 1, 2, \ldots, k$. We require that there be a total of τ samples, so that $\sum_{i=1}^{k} \tau_i = \tau$. The τ_i's are recalculated at each stage of sampling. For example, in an equal allocation, the allocation rule is to sample evenly from each alternative: it assigns samples so that, at the end of the stage of sampling, the difference between the number of samples between the most-sampled and the least-sampled alternatives, is minimized. If $\tau = k$, then the equal allocation rule sets $\tau_i = 1$ for each i at each stage. If $\tau = 1$, then one equal allocation rule could set $\tau_i = 1$ for the alternative with the smallest index among those alternatives that have been sampled the least so far, and $\tau_i = 0$ for the other alternatives.

The *stopping rule* specifies the condition under which sampling is terminated so that an alternative can be selected as best. It can be based on a total sampling budget or the desired level of evidence for a correct selection. If a total sampling budget is chosen, this reflects a choice to take a deterministic number of samples before selecting an alternative as best independent of the samples seen: sampling continues if and only if $\sum_{i=1}^{k} n_i < \beta$, where β is a user-specified total sampling budget. A selection procedure with such a deterministic sampling budget is "better" if it provides a higher expected level of evidence for correct selection after sampling. Such evidence might be the posterior *PCS*.

If a desired level of evidence for correct selection is chosen as a criterion for stopping, such as a choice to stop sampling when the posterior *PCS* is above a pre-specified threshold, then a "better" procedure is one that requires a fewer number of samples, in expectation, to reach that threshold. Such a stopping rule is called an adaptive stopping rule, because the total number of samples may depend on the values of the samples.

The *selection rule* specifies which alternative to select as best when sampling is completed. A very common sampling rule is to pick the alternative with the largest sample mean, i.e., $\mathfrak{D} = (k)$, where (k) is the alternative with the largest sample mean when the sampling stops, and $\mathfrak{D}$ is the (random) decision variable that represents the selected alternative. This decision rule is known to be optimal in some situations (e.g., if the loss function is the expected opportunity cost [16, 30]). An alternative selection rule is to select the alternative with the largest posterior probability of being the best [2].

Throughout this chapter, the "default" approach is to use a non-informative prior distribution for the unknown parameters [20]. This implies that the decision maker does not favor any specific value for the unknown means. As a result, initial number of samples need to be collected. If there is additional information, the decision maker can use an informative prior distribution to describe that information. Branke et al. [4] and Chick and Frazier [14] show how to handle the analysis with informative prior distributions.

3.2.3 General Concepts for OCBA and EVI

Before going into the details in Sects. 3.3 and 3.4, the basic ideas of four allocation rules are presented. The first allocation rule is *OCBA* for unconstrained optimization, which will be further described in Sect. 3.3. The other three allocation rules attempt to maximize *EVI* and will be discussed in detail in Sect. 3.4.

Basic Idea of *OCBA*

The basic goal of *OCBA* is to maximize the probability of correct selection for a given stage of sampling. It aims to derive closed-form expressions that are easy to implement to allocate multiple samples to multiple alternatives.

This is done using an asymptotic framework to analyze the structure of the optimal allocation when the number of samples tends to infinity. For example, it uses the Bayesian asymptotic normality result where the posterior distribution of W_i follows a normal distribution, $W_i \sim \mathcal{N}\left(\overline{x}_i,\, \sigma_i^2/n_i\right)$. Based on asymptotic analysis, the simulation budget allocation problem can be formulated as a non-linear deterministic optimization problem that can be solved using classical techniques such as the Karush–Kuhn–Tucker (KKT) conditions. The resulting closed-form allocation rule can then be implemented using a sequential heuristic algorithm.

Basic Idea of *EVI*

The value of information is defined as the expectation of the reward obtained with additional information less the reward obtained without that information (that is, the expected opportunity cost). Ideas for the *EVI* approach have been developed independently for several different distributional assumptions and approaches for how to value the information [16, 22, 31, and others]. The three approaches to *EVI* discussed here include variations on whether the information to be valued is obtained by sampling from one or from multiple alternatives in a given stage of sampling, or whether the information from only a single stage (resulting in a so-called one-step lookahead policy) or from potentially multiple stages of sampling is modeled. The former is typically easier to do than the latter.

The Linear Loss procedure (*LL*) is an *EVI* procedure that can allocate multiple samples to multiple alternatives in each stage of sampling. It aims to minimize the expected opportunity cost (*EOC*), which is the difference in means between the selected alternative and the best alternative. Linear loss is another name for the *EOC* [30]. In the *LL* procedure, the additional information is one extra stage of sampling, and the reward is the posterior mean reward from the alternative that would be chosen as best.

The LL_1 *allocation* is like the *LL* in that it allocates samples to alternatives at each stage of sampling in order to minimize *EOC* of a potentially incorrect selection, and does so in a myopic one-step lookahead manner. The LL_1 differs from the *LL* allocation in the sense that it requires all samples within a given stage to be taken from a single alternative and so it is called LL_1. This restriction enables a closed form solution for the one-stage value of information when samples are normally distributed (with known or unknown means and variances). The LL_1 allocation was independently developed by Frazier et al. [22]. The resulting approach is called the knowledge gradient of which idea is extensively presented and discussed by Powell and Ryzhov [49].

Economics of Selection Procedure (*ESP*) looks at the case where the decision maker has the option to continue sampling or to stop and select the best. It specifies one alternative to be sampled like LL_1. However, it looks at the future streams of rewards (potentially multiple stages of learning and sampling costs, the benefit from implementing a selected alternative) instead of the one-step lookahead analysis as in the case of *OCBA*, *LL*, and LL_1 which only maximizes the evidence for correct selection by the end of one stage of sampling. As a result, it accounts for the value of the ability of further sampling after that stage and therefore takes a non-myopic view of the value of sampling.

3.3 Optimal Computing Budget Allocation (*OCBA*)

This section briefly introduces the *OCBA* approach. As n_i increases, $\widehat{w}_i = \overline{x}_i$ becomes a better approximation to w_i in the sense that its corresponding variance becomes smaller. At a given stage, we need to allocate the additional τ samples to each alternative. As motivated in Sects. 3.1 and 3.2, instead of equally simulating all alternatives, we want to choose $\tau_1, \tau_2, \ldots, \tau_k$ more intelligently so that the simulation efficiency can be enhanced.

3.3.1 Maximization of PCS

The simulation budget allocation problem that *OCBA* in Chen et al. [12] aims to maximize the probability of correct selection (*PCS*) subject to the sampling budget of a given stage of sampling τ,

$$\max_{\tau_1,\tau_2,\ldots,\ \tau_k} PCS\ s.t. \sum_{i=1}^{k} \tau_i = \tau, \tau_i \geq 0. \tag{3.4}$$

Here $\sum_{i=1}^{k} \tau_i = \tau$ denotes the total computational cost assuming the simulation execution times for different alternatives are roughly the same. Formally, τ_i is a non-negative integer. However, the allocation rule is derived assuming that τ_i is a continuous variable.

3.3.2 Asymptotic Allocation Rule

We use the Bayesian asymptotic normality result, $W_i \sim \mathcal{N}\left(\overline{x}_i,\ \sigma_i^2/n_i\right)$ [20]. After the simulation is performed, $\overline{x}_i$ can be calculated, σ_i^2 can be approximated by the sample variance; *PCS* can then be estimated using a Monte Carlo simulation. However, estimating *PCS* via Monte Carlo simulation is time-consuming. Since the purpose of budget allocation is to improve simulation efficiency, we need a relatively fast and inexpensive way of estimating *PCS* within the budget allocation procedure. Efficiency is more crucial than estimation accuracy in this setting. We adopt a common approximation procedure used in simulation and statistics literature. This approximation is based on the Bonferroni inequality. For brevity, we drop the notation ε.

$$\begin{aligned} PCS &= \mathrm{P}\left(\bigcap_{i:(i)\neq(k)} \left(W_{(k)} - W_{(i)} \geq 0\right)\right) \\ &\geq 1 - \sum_{i:(i)\neq(k)} \left[1 - \mathrm{P}\left(W_{(k)} - W_{(i)} \geq 0\right)\right] \\ &= 1 - \sum_{i:(i)\neq(k)} \mathrm{P}\left(W_{(k)} < W_{(i)}\right) \\ &= 1 - \sum_{i:(i)\neq(k)} \Phi_N\left(-d_{(j)(k)}\right) = APCS. \end{aligned} \tag{3.5}$$

where Φ_N is the cumulative distribution function of standard normal distribution and $d_{(j)(k)} = \bar{x}_{(k)} - \bar{x}_{(j)}$.

Consider an asymptotic case ($\tau \to \infty$ so that the total sampling budget $\beta \to \infty$ and $\tau_i \to n_i$), Chen et al. [12] show that the approximation of PCS given in (3.4) can be maximized when

$$\frac{n_{(i)}}{n_{(j)}} = \left(\frac{\sigma_{(i)}/d_{(i)(k)}}{\sigma_{(j)}/d_{(j)(k)}}\right)^2, \quad (i),(j) \in \{1, 2, \ldots, k\} \text{ and } (i) \neq (j) \neq (k), \tag{3.6}$$

$$n_{(k)} = \sigma_{(k)} \sqrt{\sum_{i:(i)\neq(k)} \frac{n_{(i)}^2}{\sigma_{(i)}^2}}. \tag{3.7}$$

It is interesting to see that (3.6) implies that the number of replications for alternative i is proportional to the square of a noise-to-signal ratio, where the noise refers to the sample standard deviation and the signal refers to the difference between alternative i's sample mean and the best sample mean.

3.3.3 *Sequential Heuristic Algorithm for Allocation*

With the asymptotic solution in (3.6) and (3.7), we now present a cost-effective sequential approach based on *OCBA* to select the best alternative from k alternatives with a user-specified total sampling budget β. Each alternative is initially simulated with n_0 samples in the first stage, and additional samples are allocated incrementally with τ samples to be allocated in each iteration. Let t be the stage number.

OCBA Algorithm

INPUT k, β, τ, n_0 ($\beta - kn_0$ is a multiple of τ and $n_0 \geq 5$);

INITIALIZE $t \leftarrow 0$;
Perform n_0 samples for all alternatives; $n_{1,t} = n_{2,t} = \ldots = n_{k,t} = n_0$.
LOOP WHILE $\sum_{i=1}^{k} n_{i,t} < \beta$ **DO**
UPDATE Calculate sample means $\bar{x}_i \sum_{j=1}^{n_{i,t}} x_{ij}/n_{i,t}$, and sample standard deviation $\hat{\sigma}_i = \sqrt{\sum_{j=1}^{n_{i,t}} (x_{ij} - \bar{x}_i)^2 / (n_{i,t} - 1)}$, $i = 1, 2, \ldots, k$, using the new simulation output; find $(k) = \arg\max_i \bar{x}_i$.
ALLOCATE Increase the sampling budget by τ and calculate the new budget allocation, $n_{1,t+1}, n_{2,t+1}, \ldots, n_{k,t+1}$, according to
i) $\frac{n_{(i),t+1}}{n_{(j),t+1}} = \left(\frac{\hat{\sigma}_{(i)}/d_{(i)(k)}}{\hat{\sigma}_{(j)}/d_{(j)(k)}}\right)^2$, for all $(i) \neq (j) \neq (k)$, and
ii) $n_{(k),t+1} = \hat{\sigma}_{(k)} \sqrt{\sum_{i:(i) \neq (k)} \frac{n_{(i),t+1}^2}{\hat{\sigma}_{(i)}^2}}$,
SIMULATE Perform $\tau_i = (n_{(i),t+1} - n_{(i),t})^+$ additional simulations for alternative $i = 1, 2, \ldots, k$, where $(x)^+ = \max(0, x)$;
$t \leftarrow t + 1$.
END OF LOOP

The resulting $n_{(i)}$ in the ALLOCATE step is a continuous number that must be rounded to an integer. In the numerical experiments in the next section, $n_{(i)}$ is rounded to the nearest integer such that the summation of additional simulation replications for all solutions equals τ. Note that there may not always exist a solution that satisfies all the three constraints. It actually occurs when at least one solution has been over simulated, i.e., $n_{(i),t+1} < n_{(i),t}$. In this case, we have to relax the constraint. For ease of control of the simulation experiment, we can choose to maintain the constraint $\sum_{i=1}^{k} n_{(i),t+1} = \tau + \sum_{i=1}^{k} n_{(i),t+1}$ and apply some heuristics to round $n_{(i),t+1}$ for all i to nearest integers. Chen and Lee [8] have found numerically that the performance is not sensitive to how we round $n_{(i)}$, probably due to the robustness of a sequential procedure.

Alternative Simpler *OCBA* Procedure

When the computational cost of samples (simulation replications) is relatively large as compared to the computational cost of the ALLOCATE step, we recommend that τ should be small, or even set to 1. When $\tau = 1$ the ALLOCATE and SIMULATE steps can be simplified as follows.

ALLOCATE Increase the sampling budget by $\tau = 1$ and calculate a tentative allocation $n_{1,t+1}, n_{2,t+1}, \ldots, n_{k,t+1}$, according to
i) $\frac{n_{(i),t+1}}{n_{(j),t+1}} = \left(\frac{\hat{\sigma}_{(i)}/d_{(i)(k)}}{\hat{\sigma}_{(j)}/d_{(j)(k)}}\right)^2$, for all $(i) \neq (j) \neq (k)$, and
ii) $n_{(k),t+1} = \hat{\sigma}_{(k)} \sqrt{\sum_{i:(i) \neq (k)} \frac{n_{(i),t+1}^2}{\hat{\sigma}_i^2}}$,
leave $n_{(i),t+1}$ as a decimal number and find $(i^*) = \arg\max_i \left(n_{(i),t+1} - n_{(i),t}\right)$.

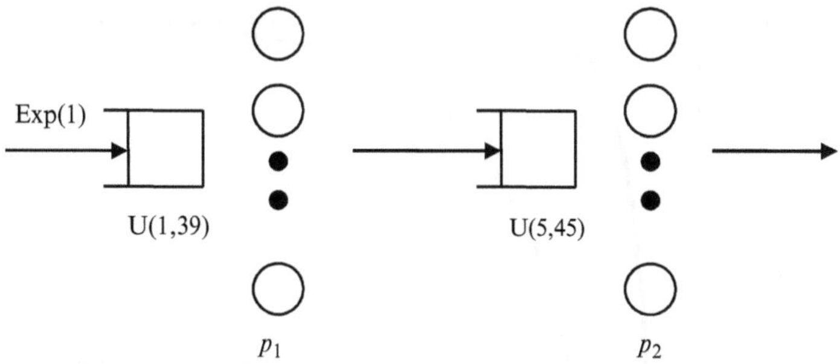

Fig. 3.2 A two-stage queuing system where both p_1 and p_2 must be greater than 10

SIMULATE Perform additional one sample for alternative (i^*);
$n_{(i^*),t+1} = n_{(i^*),t} + 1$; $n_{(i),t+1} = n_{(i),t}$ for $(i) \neq (i^*)$;
$t \leftarrow t+1$.

Intuitively, we determine which alternative is the most starving one in terms of the need of additional simulation, and then simulate that alternative for one additional replication. This procedure is iteratively continued until the total budget β is exhausted or the estimated *APCS* is sufficiently high. As shown in Chen and Lee [8], this simpler procedure performs equally well in our numerical testing.

3.3.4 Numerical Results

Chen and Lee [8] provide extensive numerical results for *OCBA*. Among them, Fig. 3.2 gives an example of a two-stage queuing system, where we want to allocate 31 parallel servers within a two-stage queue where each stage of the queue can contain no less than 11 servers.

Denote p_1 and p_2 as the numbers of workers allocated to nodes 1 and 2 respectively. Thus, $p_1 + p_2 = 31$, $p_1 \geq 11$, and $p_2 \geq 11$. There are ten alternative combinations of (p_1, p_2). We want to choose the best alternative of (p_1, p_2) so that the average system time for the first 100 customers is minimized. Since there is no closed-form analytical solution for the estimation of the system time, stochastic (discrete-event) simulation can be performed to find the best alternative.

To characterize the performance of different procedures as a function of β, we vary β between 200 and 8,000 for all of the sequential procedures and the estimated achieved *PCS* as a function of β is shown in Fig. 3.3. We estimate the *PCS* by estimating the fraction of the event of correct selection out of the independent experiments that are conducted.

We see that all procedures obtain a higher *PCS* as the available sampling budget increases. We can then record the number of samples that correspond to where the curve crosses at a certain level of *PCS* that we are interested in. It can be seen that

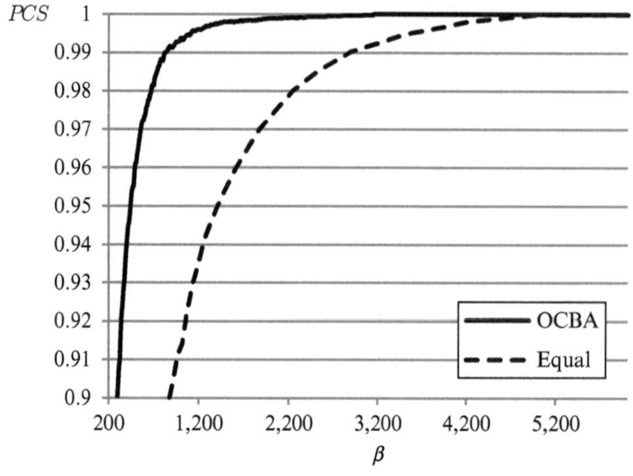

Fig. 3.3 *PCS* vs. β using three sequential allocation procedures

Table 3.1 The sampling budget to attain $PCS = 0.95$ or 0.99

PCS	*OCBA*	Equal allocation
0.95	470	1,450
0.99	850	2,890

Table 3.2 Number of maximum allowable workers to simulate varying numbers of alternatives

Maximum number of workers	Number of alternatives
31	10
41	20
51	30
61	40
71	50
81	60
91	70
101	80
111	90
121	100

OCBA achieves a same *PCS* using the lowest amount of sampling budget. Table 3.1 shows the sampling budget to attain $PCS = 0.95$ and 0.99 for *OCBA* and Equal Allocation.

It is not surprising that the actual sampling cost using *OCBA* depends on the specific problem and the corresponding *PCS* requirement. However, the speedup factor of using *OCBA* versus equal allocation is not very sensitive to problem specifics, except for the number of alternatives.

Instead of providing only 31 servers, we increase the number of servers up to 121, where there must be at least 11 servers at each station. As a result, the number of possible alternatives varies from 10 to 100. Table 3.2 lists some of the possibilities.

Table 3.3 Speedup factor of using *OCBA* compared with the use of equal allocation

Number of alternatives (k)	4	10	20	50	75	100
Speedup factor	1.75	3.42	6.45	12.8	16.3	19.8

In this test, we compare *OCBA* and equal allocation, and focus on the "speedup factors" under *OCBA*. For both procedures, we record the minimum sampling budget where the curve plotting the estimated *PCS* vs. β crosses $PCS = 0.99$: β_{OCBA} and β_{EA}. The "speedup factor" using *OCBA* is given by the ratio β_{EA}/β_{OCBA}. Table 3.3 shows the numerical results for different number of alternatives. We see that *OCBA* is even more efficient as the number of alternatives increases. The higher efficiency is obtained, because a larger alternative space gives the *OCBA* algorithm more flexibility in allocating the sampling budget.

3.3.5 Minimization of EOC

Instead of maximizing *PCS*, we turn our attention to the expected opportunity cost (*EOC*). From the simulation efficiency perspective, one has the same question to ask: how should we allocate the simulation samples so that we can select an alternative within the given sampling budget while *EOC* is minimized, instead of maximizing *PCS* as in previous sections?

Deriving an asymptotic solution for minimizing *EOC* is much more complicated than its counterpart for *PCS*. Following the same notion of the greedy approach given in Chen et al. [11] and Hsieh et al. [36], He et al. [33] present a greedy selection procedure, called $OCBA_{LL}$ (or $OCBA - EOC$), to reduce the *EOC* of a potentially incorrect selection by taking a similar *OCBA* approach to selection.

A critical component in the proposed procedure is to estimate how *EOC* changes as n_i changes. Let τ_i be a nonnegative integer denoting the number of additional simulation samples allocated to alternative i in the next stage of sampling. We are interested in assessing how *EOC* would be affected if alternative i was simulated for τ additional replications. He et al. [33] present an *Estimated Expected Opportunity Cost* (*EEOC*), which is an upper bound of *EOC*, as follows

$$EEOC = \sum_{i:(i)\neq(k)} \mathrm{P}\left(W_{(i)} \geq W_{(k)}\right) \mathrm{E}\left[W_{(i)} - W_{(k)} \middle| W_{(i)} \geq W_{(k)}\right]. \tag{3.8}$$

The *OCBA* approach in this case aims to sequentially minimize *EEOC*. A critical component is to estimate how *EEOC* changes if alternative i is allocated with τ_i additional replications in a given stage. A heuristic approach to the approximation of the predictive posterior distribution yields $W_i \sim \mathcal{N}\left(\frac{1}{n_i}\sum_{j=1}^{n_i} x_{ij},\ \frac{\widehat{\sigma}_i^2}{n_i+\tau_i}\right)$.

The *EEOC* can then be determined by using the original distributions for the unknown means of alternatives other than i as follows

$$EEOC(i) = \sum_{j:(j)\neq(k)} \int_{0}^{+\infty} x f^{*}_{(k),\,(j),(i)}(x)dx, \tag{3.9}$$

where $f^{*}_{(k),\,(j),(i)}(x)$ is the probability density function (p.d.f.) of the difference between alternative (k) and (j), given that $\tau_{(i)}$ additional replications are allocated to alternative (i), and none is allocated to others. If $(i) = (k)$, then $f^{*}_{(k),\,(j),(i)}(x)$ is the p.d.f. of $\mathcal{N}\left(\overline{x}_{(j)} - \overline{x}_{(k)}, \frac{\widehat{\sigma}^2_{(k)}}{n_{(k)}+\tau_{(k)}} + \frac{\widehat{\sigma}^2_{(i)}}{n_{(i)}}\right)$. If $(i) = (j)$, then $f^{*}_{(k),\,(j),(i)}(x)$ is the p.d.f. of $\mathcal{N}\left(\overline{x}_{(i)} - \overline{x}_{(k)}, \frac{\widehat{\sigma}^2_{(k)}}{n_{(k)}} + \frac{\widehat{\sigma}^2_{(i)}}{n_{(i)}+\tau_{(i)}}\right)$. Otherwise, no new information is available to distinguish alternatives (k) and (j) and $f^{*}_{(k),\,(j),(i)}(x)$ is the p.d.f. of $\mathcal{N}\left(\overline{x}_{(j)} - \overline{x}_{(k)}, \frac{\widehat{\sigma}^2_{(k)}}{n_{(k)}} + \frac{\widehat{\sigma}^2_{(i)}}{n_{(i)}}\right)$.

Since we want to minimize *EOC*, $OCBA_{LL}$ sequentially allocates additional samples to the alternatives that lead to the lowest *EEOC* at each stage. Let r be the number of alternatives to simulate in each stage. The ALLOCATE steps are revised as follows

ALLOCATE Find the set $S(r) \equiv \{j: EEOC(j)$ is among the r lowest values$\}$. Increase the sampling budget by $\tau_{(i)} = \tau/|S(r)|$ for alternative $(i) \in S(r)$, i.e., $n_{(i),t+1} = n_{(i),t} + \tau_{(i)}$ if $(i) \in S(r)$, $n_{(i),t+1} = n_{(i),t}$ otherwise. If $\tau = 1$, only a single replication is allocated to alternative j which minimizes $EEOC(j)$.

3.3.6 Other Variants

This section discusses some extensions of the *OCBA* approach. We begin by considering an efficient budget allocation procedure for selecting an optimal subset of top-r alternatives rather than the single best alternative [10]. Then we consider problems with multiple performance measures, which can be formulated either as constrained optimization or multi-objective optimization. Other recent developments are then discussed and finally concluded with examples of generalizations of *OCBA* notions.

Subset Selection Problem

Instead of selecting the best alternative as in previous section, we consider a class of subset selection problems in simulation optimization or ranking and selection. In some cases, it is more useful to provide a set of good alternatives than a single best alternative for decision maker to choose, because he/she may have other concerns which are not modeled in the simulation. Such efficient subset selection procedures are also beneficial to some recent developments in simulation optimization that require the selection of an "elite" subset of good candidate solutions in each iteration of the algorithm, such as evolutionary population-based algorithm. A subset with good performing solutions will result in an update that leads the search in a promising direction.

Specifically, our objective is to find *all* top-r alternatives, where $r > 1$ is the number of alternatives to select. Koenig and Law [40] develop a two-stage procedure for selecting all the r best alternatives along the lines of the procedure in Dudewicz and Dalal [21] (see also Sect. 10.4 of Law [41] for an extensive presentation of the problem and procedure). However, the number of additional samples for the second stage is computed based on a least favorable configuration, resulting in very conservative allocations, so that the required computational cost is much higher than actually needed.

This problem can be easily handled by changing the correct selection definition in *OCBA*. Specifically, $PCS = \mathrm{P}(W_{(i)} \geq W_{(j)}, \forall\, (i) \in S_r, (j) \notin S_r)$ where S_r is the optimal subset. Chen et al. [10] show that the allocation rule is similar to (3.6), that is $\frac{n_{(i)}}{n_{(j)}} = \left(\frac{\sigma_{(i)}/(\overline{x}_{(i)}-q)}{\sigma_{(j)}/(\overline{x}_{(j)}-q)}\right)^2$, where q is a value between $\overline{x}_{(r)}$ and $\overline{x}_{(r+1)}$. A suggested value of q, which asymptotically maximizes both $\mathrm{P}(W_{(r)} \leq q)$ and $\mathrm{P}(W_{(r+1)} \geq q)$, is given by $q = \frac{\widehat{\sigma}_{(r+1)}\overline{x}_{(r)} + \widehat{\sigma}_{(r)}\overline{x}_{(r+1)}}{\widehat{\sigma}_{(r)} + \widehat{\sigma}_{(r+1)}}$.

Handling Optimization with Multiple Performance Measures

OCBA has also been extended to tackle other simulation-based optimization problems. The needs of optimization problems with multiple performance measures become more evident. We can categorize these problems according to whether there are any constraints and whether the constraints are stochastic or deterministic.

In the case of multi-objective optimization where no performance measures are constrained, Lee et al. [43, 44] consider the problem of finding the non-dominated Pareto set where the evidences for correct selection used are type I and type II errors.

There are cases where the secondary stochastic performance measures act as constraints. In this case, the simulation budget allocation can be allocated based on the optimality only, feasibility only, or both. Lee et al. [45] propose an *OCBA* approach that maximizes a lower bound of *PCS*. The procedure is applicable for both the independent case and the case with correlated performance measures. In some situations, the decision makers are only interested in differentiating the feasible alternatives from the infeasible ones, which is called feasibility determination problem as addressed by Szechtman and Yücesan [55]. In this case, there is no need to select the best alternative.

It is also possible to consider the descriptive complexity preference. In this case, an alternative that is simpler, i.e., having smaller descriptive complexity is preferred compared to a complex one if they have similar performance. The decision makers therefore want to select the top-r simplest alternatives of which performance measures are above certain desired level [37, 60].

Other Recent Developments

There are several other related works along the lines of the *OCBA* research. This section gives some examples, which is by no means an exhaustive list. The *OCBA* rule presented in Sect. 3.3.2 assume that the variances are known. In the case of unknown variances, Chen et al. [13] proposed the *OCBA* algorithm based on t distribution. It is found that the differences between the results obtained based on the t distribution model and the normal distribution model is not significant. Instead of finding the alternative with the best mean, Trailovic and Pao [57] develop an *OCBA* approach for finding an alternative with minimum variance. Unlike the independence assumption of simulation samples required in this book, Fu et al. [27] extend the *OCBA* to problems in which the simulation outputs between alternatives are correlated. Chen et al. [9] study of the benefit of dynamic allocation. Glynn and Juneja [28] extend the *OCBA* to problems in which the simulation output is no longer normally distributed by utilizing large deviation theory. Blanchet et al. [3] further extend the work to heavy-tailed distributions, also utilizing large deviation theory.

Brantley et al. [5, 6] enhance *OCBA* efficiency by incorporating information from across the domain into a regression equation. Morrice et al. [46, 47] further extend the concepts to a method for selecting the best alternative based on a transient mean performance measure.

Generalized *OCBA* Notions

The different extensions presented earlier indicate that there exists a consistent notion where an optimization model is used to determine the best allocation scheme to maximize a certain desired quality of the outcome given a fixed budget. The *OCBA* notion has been generalized for different purposes well beyond selecting the best alternative as presented in previous subsections. The main idea is that it is possible to replace the objective function *PCS* with other objectives. In addition, the budget to be allocated is not necessarily in terms of computer time or simulation replications. We give three examples in this section.

The Cross-Entropy (CE) method introduced by Rubinstein [51] belongs to a class of global optimization algorithms called estimation of distribution algorithms, which work with a probability distribution over the solution space. In every iteration of CE, we will first generate a population of solutions from a probability density function (p.d.f.) with a certain parameter. After these generated solutions are simulated, the parameters of the distribution are updated by minimizing the Kullback–Leibler (KL) divergence (or the cross entropy) between the parameterized p.d.f. and the target optimal p.d.f.. The CE method has shown to be promising in solving difficult global optimization problems, but its main focus has been on deterministic optimization problems. For the stochastic setting, He et al. [34] develop the *OCBA–CE* procedure that integrates the objectives of minimizing the KL divergence from a parameterized distribution that generates the candidate solutions in the CE method with that of minimizing the total computing budget per iteration. Numerical testing indicates that the *OCBA–CE* is promising, resulting in substantial computational efficiency gains over the CE method with equal allocation.

The second example is the work by Shortle et al. [54], which uses the notion of *OCBA* to the problem of estimating a rare-event probability using splitting simulation. Multi-level splitting is an effective variance reduction technique. The basic idea is to create separate copies (splits) of the simulation whenever it gets close to the rare event. Each level is smaller and much easier to simulate than the original. Note that this problem is not an optimization problem as the decision maker does not need to select the best alternative. The problem of determining the number of splits is formulated as an optimal computing budget allocation problem. In this context, the objective is to minimize the variance of the rare-event probability estimator. The budget is the total computation time, which needs to be allocated to different levels.

OCBA can also be extended to problems without simulation or optimization. For example, Wong et al. [59] propose an *OCBA* approach for Data Envelopment Analysis (DEA), which is a mathematical programming approach by Charnes et al. [7] for measuring efficiency for decision-making units with multiple inputs and multiple outputs. The idea is to compare different decision-making units in terms of how many inputs they have used in achieving the outputs. In this case, the objective is to minimize the expected mean square error for the prediction of the efficiency score in DEA. The budget refers to the total budget for data allocation. Therefore, the budget allocation problem is to determine how the data should be collected, i.e., finding the optimal number of data points allocated for different unknown variables to maximize the predicted efficiency score.

3.4 Expected Value of Information (*EVI*)

The *EVI* approach is based on a Bayesian decision theoretic approach rather than a frequentist statistical approach. It is Bayesian in the sense that it presumes that uncertainty about all unknown parameters, such as the unknown means of each alternative, be described with probability distributions. It is decision theoretic in the sense that sampling allocation decisions, as well as decisions to select a given alternative as best, are based on maximizing an expected reward (or equivalently, minimizing an expected loss).

There are several potential loss functions of interest in this framework. One loss function is the $0-1$ loss function: a loss of 1 is incurred if the true best alternative is not selected as best, and a loss of 0 is incurred if the best alternative is best. With the $0-1$ loss function, the expected loss is the *PCS* in (3.1). Another loss function of interest is the opportunity cost. The opportunity cost is 0 if the best alternative is selected as best. Otherwise, the opportunity cost is the difference between the performance of the best alternative and the performance of the alternative selected as best. The expected opportunity cost (*EOC*), averaged over sampling decisions and realizations of the unknown true performance, has particular significance when the outputs of simulated alternatives are linear measures of financial (profit) outcomes for the alternatives they simulate. Minimizing the *EOC* in that case is equivalent to maximizing the expected profit. The *EOC* is therefore particularly interesting when using a selection procedure to select an alternative with the greatest financial benefit.

Independent of whether the loss function is the $0-1$ loss function, the opportunity cost, or some other loss function, the *EVI* approach allocates samples in each stage of sampling with the goal of reducing the expected loss obtained after the samples are observed. This problem can be solved analytically in a few special cases (e.g., normal distributed output with known sampling variances, $k=2$ alternatives); otherwise analytically motivated approximations that have attractive theoretical properties can be assembled to provide good heuristics. Some of these are described below.

We highlight three *EVI* procedures that focus on information with respect to the *EOC* loss function. Section 3.4.1 presents the *LL* procedure, which allocates samples to multiple alternatives in each stage. Section 3.4.2 focuses on a simplification where only one additional sample to a single alternative can be collected in each stage. Section 3.4.5 presents the multi-step valuation of information where the proposed procedure that looks at the value of multiple stages of sampling in order to further improve effectiveness of an allocation in a given stage.

3.4.1 Linear Loss (LL)

Chick and Inoue [16] proposed a procedure to determine the number of samples to minimize the expected opportunity cost (*EOC*).

To find the expected value of sampling an alternative in the next stage of sampling, we introduce some additional notations to account for the random output that will be observed. Let Y_{ij} be a random variable where y_{ij} is the output of the *j*th sample from alternative i, $j=1, 2, \ldots,$ that will be observed in the next stage of sampling. Let $\overline{y}_i$ be the sample average of the output in the additional stage based on the additional τ_i samples, $\overline{y}_i = \sum_{j=1}^{\tau_i} y_{ij}/\tau_i$. As defined in Sect. 3.2.1, $\overline{x}_i$ is the sample mean based on n_i, the number of samples from alternative i so far. The overall sample mean for alternative i, denoted as z_i is therefore

$$z_i = \frac{n_i\overline{x}_i + \tau_i\overline{y}_i}{n_i + \tau_i}. \tag{3.10}$$

Before the sampling is done, $Z_i = \frac{n_i\bar{x}_i+\tau_i\bar{Y}_i}{n_i+\tau_i}$ is random. Thus, Z_i is the posterior mean of the unknown mean for alternative i given that ε is the information seen so far and the fact that τ_i samples from alternative i will be observed but have not yet been observed. The distribution of Z_i depends on the value of τ_i.

Minimization of *EOC*

Conceptually, at each stage of sampling, the *LL* procedure seeks to allocate samples to several alternatives so that the expected *EOC* that will be obtained will be minimized after the next stage of sampling. That so-called predictive *EOC* depends on the samples taken in the next stage as they will determine the $\{Z_i\}$, which in turn determine which alternative will be selected.

Assessing that predictive *EOC* is analytically intractable when $k > 2$ or when variances are unknown, even when samples are normally distributed. The *LL* procedure attempts to minimize an upper bound on that predictive *EOC*. In particular, we minimize (for asymptotically large τ) the term $EOC_{bnd}(\tau_1, \tau_2, \ldots, \tau_k)$, where

$$EOC_{bnd} = \sum_{i:(i)\neq(k)} \mathrm{E}\left[\left(W_{(i)} - W_{(k)}\right)^+\right] - \mathrm{E}\left[\left(Z_{(i)} - Z_{(k)}\right)^+ \middle| \mathscr{E}\right]. \tag{3.11}$$

depends on the $\tau_{(i)}$ via the $Z_{(i)}$, is an upper bound on the predictive *EOC*, where (k) is the alternative with the highest mean given the data ε observed so far, and $(x)^+ = \max(0, x)$.

Conceptually, the difference in the summand in the equation for EOC_{bnd} is the expected opportunity cost of selecting with no additional information, $\mathrm{E}[(W_{(i)} - W_{(k)})^+|\varepsilon]$, minus the expected value of information of sampling, $\mathrm{E}[(Z_{(i)} - Z_{(k)})^+|\varepsilon]$, in a pairwise comparison between alternatives (i) and (k). When no additional samples are taken, the term $\mathrm{E}[(Z_{(i)} - Z_{(k)})^+|\varepsilon]$ is 0. When an infinite number of samples are taken for both (i) and (k), the difference in the summand is 0 (an infinite number of samples gives perfect information).

At each stage of sampling, ε is updated, and the goal of the *LL* procedure is to find an allocation that minimizes EOC_{bnd}.

$$\min_{\tau_1, \tau_2, \ldots, \tau_k} EOC_{bnd} \ s.t. \sum_{i=1}^{k} \tau_i = \tau, \ \tau_i \geq 0. \tag{3.12}$$

A solution to the problem in (3.12) for the case of known variances is not known other than by searching on a lattice, as the τ_i are non-negative integers. We therefore derive an asymptotically optimal solution to (3.12) which assumes that τ is very large and allows for real-valued τ_i. When sampling variances are also unknown, an additional approximation is needed to account for the fact that closed form solutions for the distribution of differences like $W_{(k)} - W_{(j)}$ are not available (the Behrens–Fisher problem). Thus, we use the so-called Welch approximation to describe those differences. To do so, we define

$$d^*_{jk} = \frac{d_{(j)(k)}}{\widehat{\lambda}_{jk}^{-1/2}}, \tag{3.13}$$

where $d_{(j)(k)} = \overline{x}_{(k)} - \overline{x}_{(j)}$ and $\widehat{\lambda}_{jk}^{-1/2} = \sqrt{\frac{\widehat{\sigma}^2_{(j)}}{n_{(j)}} + \frac{\widehat{\sigma}^2_{(k)}}{n_{(k)}}}$. We also describe the standardized statistics for the difference $Z_{(k)} - Z_{(j)}$ with

$$d^*_{\{jk\}} = \frac{d_{(j)(k)}}{\widehat{\lambda}_{\{jk\}}^{-1/2}}, \tag{3.14}$$

where $\widehat{\lambda}_{\{jk\}}^{-1/2} = \sqrt{\frac{\tau_{(j)}\widehat{\sigma}^2_{(j)}}{n_{(j)}\left(n_{(j)}+\tau_{(j)}\right)} + \frac{\tau_{(k)}\widehat{\sigma}^2_{(k)}}{n_{(k)}\left(n_{(k)}+\tau_{(k)}\right)}}$. The notation $d^*_{\{jk\}}$ differs slightly from the standardized statistics for the difference $W_{(k)} - W_{(j)}$ that is d^*_{jk}.

We denote the Student t distribution by $St(\mu, \kappa, \nu)$, the cumulative distribution function of the standard Student t distribution ($\mu = 0$, $\kappa = 1$) by $\Phi_\nu(\cdot)$, and the probability density function by $\phi_\nu(\cdot)$. The posterior marginal distribution for the unknown mean W_i has a Student t-distribution, $St\left(\overline{x}_i,\ n_i/\sigma_i^2,\ \nu_i\right)$. The standard *EOC* function $\Psi_\nu[m]$ gives the *EOC* when an alternative with known mean m is selected in preference to a single alternative whose unknown mean has a $St(0, 1, \nu)$ distribution,

$$\Psi_\nu\left[m\right] = \frac{\nu + m^2}{\nu - 1}\phi_\nu(m) - m\Phi_\nu\left(-m\right). \tag{3.15}$$

Welch's approximation for the degrees of freedom of $W_{[k]} - W_{[j]}$ is

$$\nu_{(j)(k)} = \frac{\left[\widehat{\sigma}^2_{(j)}/n_{(j)} + \widehat{\sigma}^2_{(k)}/n_{(k)}\right]^2}{\left[\widehat{\sigma}^2_{(j)}/n_{(j)}\right]^2/\left(n_{(j)}-1\right) + \left[\widehat{\sigma}^2_{(k)}/n_{(k)}\right]^2/\left(n_{(k)}-1\right)}. \tag{3.16}$$

With this notation, and with the Welch approximation for the differences, we can approximate EOC_{bnd} by

$$EOC_{bnd} \approx \sum_{j:(j)\neq(k)} \widehat{\lambda}_{jk}^{-1/2}\Psi_{\nu_{(j)(k)}}\left[d^*_{jk}\right] - \widehat{\lambda}_{\{jk\}}^{-1/2}\Psi_{\nu_{(j)(k)}}\left[d^*_{\{jk\}}\right]. \tag{3.17}$$

When all of the $\tau_i = 0$, then $\widehat{\lambda}_{\{jk\}}^{-1/2}\Psi_{\nu_{(j)(k)}}\left[d^*_{\{jk\}}\right]$ becomes 0. We define

$$EOC_{Bonf} \triangleq EOC_{bnd}\left(0,0,\ldots,0\right) = \sum_{j:(j)\neq(k)} \widehat{\lambda}_{jk}^{-1/2}\Psi_{\nu_{(j)(k)}}\left[d^*_{jk}\right]. \tag{3.18}$$

EOC_{Bonf} is an upper bound on the posterior EOC given ε (without considering further sampling). Closed form solutions for the EOC are only available in special cases, so the LL procedure is derived by myopically minimizing an upper bound on the EOC at each stage of sampling. The upper bound is essentially a Bonferroni-type bound that emerges from considering the EOC in a comparison between the current best alternative, alternative (k), relative to each of the $k-1$ alternatives. See Chick and Inoue [16] and Branke et al. [4] for further details.

Allocation Rules

Given that there are τ samples to be allocated to k alternatives, the allocation rule for LL is given by

$$\tau_{(i)} = \frac{\left(\tau + \sum_{j=1}^{k} n_j\right)\left(\widehat{\sigma}_{(i)}^2 \gamma_{(i)}\right)^{1/2}}{\sum_{j=1}^{k}\left(\widehat{\sigma}_j^2 \gamma_j\right)^{1/2}} - n_{(i)}, \tag{3.19}$$

where

$$\gamma_{(i)} = \begin{cases} \widehat{\lambda}_{ik}^{1/2} \frac{\nu_{(i)(k)} + \left(d_{ik}^*\right)^2}{\nu_{(i)(k)} - 1} \phi_{\nu_{(i)(k)}}\left(d_{ik}^*\right), & \text{for } (i) \neq (k) \\ \sum_{j:(j)\neq(k)} \gamma_{(j)}, & \text{for } (i) = (k). \end{cases} \tag{3.20}$$

This formula asymptotically minimizes (3.17) when τ is arbitrarily large (so that all of the $\tau_{(i)}$ are nonnegative). If τ is not sufficiently large, then (3.19) might prescribe a nonpositive number of samples for some $\tau_{(i)}$. If that is the case, then a better approximation to the distribution of the posterior mean $Z_{(i)}$ should be used. Fortunately, one is available: there will be no change in the posterior mean following samples if no new samples are observed. The formulas in (3.19) and (3.20) can therefore be adapted to the case of small τ by making use of that observation. The following steps which check nonnegativity of the $\tau_{(i)}$ implement the necessary computations.

a. Initialize the set of alternatives considered for additional samples, $\mathscr{S} \leftarrow \{1,\ldots,k\}$.
b. For each (i) in $\mathscr{S}\backslash\{(k)\}$: If $(k) \in \mathscr{S}$ then set $\widehat{\lambda}_{ik}^{-1} \leftarrow \widehat{\sigma}_{(i)}^2/n_{(i)} + \widehat{\sigma}_{(k)}^2/n_{(k)}$, and set $\upsilon_{(i)(k)}$ with Welch's approximation. If $(k) \notin \mathscr{S}$ then set $\widehat{\lambda}_{ik} \leftarrow n_{(i)}/\widehat{\sigma}_{(i)}^2$ and $\upsilon_{(i)(k)} \leftarrow n_{(i)} - 1$.
c. Tentatively allocate a total of τ samples to alternatives $(i) \in \mathscr{S}$ $\left(\text{set } \tau_{(j)} \leftarrow 0 \text{ for } (j) \notin \mathscr{S}\right)$:

$$\tau_{(i)} \leftarrow \frac{\left(\tau + \sum_{j\in\mathscr{S}} n_j\right)\left(\widehat{\sigma}^2_{(i)}\gamma_{(i)}\right)^{\frac{1}{2}}}{\sum_{j\in\mathscr{S}}\left(\widehat{\sigma}^2_{(j)}\gamma_j\right)^{\frac{1}{2}}} - n_{(i)},$$

$$\text{where } \gamma_{(i)} \leftarrow \begin{cases} \lambda_{ik}^{1/2}\frac{\upsilon_{(i)(k)}+\left(d^*_{ik}\right)^2}{\upsilon_{(i)(k)}-1}\phi_{\upsilon_{(i)(k)}}\left(d^*_{ik}\right) & \text{for } (i)\neq(k) \\ \sum_{(j)\in\mathscr{S}\backslash\{(k)\}}\gamma_{(j)} & \text{for } (i)=(k). \end{cases}$$

d. If any $\tau_{(i)} < 0$ then fix the nonnegativity constraint violation: remove (i) from $\mathscr{S}$ for each (i) such that $\tau_{(i)} \leq 0$, and go to Step 4b. Otherwise, round the τ_i so that $\sum_{i=1}^k \tau_i = \tau$ (the allocation is determined).

Chick and Inoue [16] derive this algorithm, and generalize it to minimize the CPU time (rather than the number of samples) if the CPU time per sample differs from one alternative to the next. The sequential algorithm is the same as that in Sect. 3.2.2 except that we use the allocation rule in (3.19).

3.4.2 *Small-Sample EVI Allocation Rule (LL_1)*

The somewhat cumbersome check for nonnegativity of the allocation in the *LL* allocation rule above can be avoided if all samples in a given stage are allocated to a single alternative (the alternative simulated in a given stage can still change from one stage to the next). The LL_1 allocation rule does allocate all samples to one alternative in a given stage, and therefore simplifies the computation of the optimal allocation. This has been done for the simulation context by Chick et al. [18] and independently by Frazier et al. [22]. The latter paper used the term knowledge gradient to describe the idea of one-stage lookahead for the value of sampling from one alternative at each stage of sampling. The LL_1 allocation rule has therefore also been referred to as KG or KG_1 in the literature.

Small-Sample *EVI*

As there is only one alternative to be sampled, the small-sample *EVI* procedures avoid the asymptotic approximation, the use of Bonferroni's inequality and the Welch approximation which were employed in the previous *EVI* procedure.

In the small-sample *EVI* that seeks to minimize the posterior *EOC*, the alternative to be sampled is the one with highest $EVI_{LL,(i)}$ where

$$EVI_{LL,(i)} = \begin{cases} \lambda_{\{ik\}}^{-1/2} \Psi_{n_{(i)}-1} \left[d_{\{ik\}}^* \right] & \text{if } (i) \neq (k) \\ \lambda_{\{k-1,k\}}^{-1/2} \Psi_{n_{(k)}-1} \left[d_{\{k-1,k\}}^* \right] & \text{if } (i) = (k) \end{cases}. \tag{3.21}$$

Note that $\bar{x}_{(k)} + EVI_{LL,\,(i)}$ is the expected reward when only taking $\tau_{(i)}$ samples for alternative (i), and then selecting the alternative with the best sample mean. Thus, $EVI_{LL,(i)}$ relates to the expected value of information when comparing the alternative selected for sampling and the best of the other $k-1$ alternatives.

Sequential Algorithm

The following is the procedure for LL_1. It is similar to that for *LL* except in step 4 where the allocation rule of LL_1 states that only one alternative with the highest *EVI* will be sampled:

a. Set $\tau_{(i)} \leftarrow \tau$ for the alternative that maximizes $EVI_{LL,(i)}$, and $\tau_\ell \leftarrow 0$ for the others.

3.4.3 Stopping Rules

We now introduce some notation for some stopping rules and more formally describe them. They were originally proposed in the context of *EVI* procedures, but are equally applicable to both *OCBA* and *EVI* contexts.

The deterministic sampling rule, which was used in the *OCBA* algorithm presented in Sect. 3.3.3, continues sampling until a predefined sampling budget has been exhausted. That is, sampling continues if and only if $\sum_{i=1}^{k} n_i < \beta$ for some user-specified β. This is denoted here as the *S* stopping rule.

There are several adaptive stopping rules that may be used in either the *OCBA* or *EVI* selection procedures—although they were introduced in the context of the *EVI* approach. The EOC_{Bonf} stopping rule continues sampling until the posterior *EOC* is sufficiently small. In particular, sampling continues until an upper bound, EOC_{Bonf} in (3.18), drops below a user-specified threshold $\varepsilon^* > 0$. The threshold ε^* can be chosen to be the highest acceptable expected opportunity cost associated with a possibly incorrect selection. (Someone comfortable with the indifference zone approach might select ε^* to be the indifference zone parameter times the maximal acceptable probability of incorrect selection [19]).

The PCS_{Slep} stopping rule continues sampling until a user-specified lower bound is exceeded by a lower bound on the posterior *PCS*, where the lower bound is due to Slepian. That is, sampling continues until $PCS_{Slep} = \prod_{(i)\neq(k)} \Phi_{\nu_{(i)(k)}} \left(d_{jk}^* \right)$ is at least as great as a user-specified threshold for the posterior *PCS*, $1-\alpha^*$, where $\alpha^* > 0$ is a user-specified acceptable level of probability of incorrect selection.

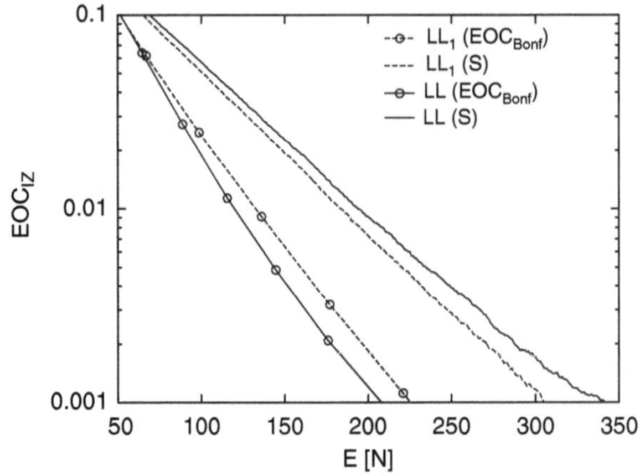

Fig. 3.4 EOC_{IZ} efficiency for LL and LL_1 in a slippage configuration with five alternatives

3.4.4 Numerical Results for LL and LL_1

Branke et al. [4] and Chick et al. [18] provide extensive numerical results for procedures derived from the *OCBA* and *EVI* approaches. This includes procedures based on the *OCBA*, *LL*, and LL_1 allocation rules. They also assess *EVI*-based allocation rules that focus on improving the posterior *PCS*, rather than the posterior *EOC*, and which are named the $0-1$ allocation rule and the $0-1_1$ allocation rule. These allocation rules seek to improve the value of information for the $0-1$ loss function, whose expected value is the posterior *PCS*, when allocating samples to multiple alternatives and to one alternative, respectively, per stage of sampling. The papers also assess how different stopping rules assess the efficiency of the sampling procedures. This section recalls some of the key results from those papers.

In making empirical assessments of selection procedures, allocation rules and stopping rules were combined into a procedure, and were repeatedly applied to different classes of selection problems. Curves like those in Fig. 3.4 were plotted for each selection procedure and for each selection problem. The x-axis of the curve gives the expected total number of samples when the procedure stops, $E[N]$. The y-axis plots the empirical evidence for correct selection. In Fig. 3.4, this evidence is the empirical expected opportunity cost EOC_{IZ}, the average, over repeated applications of the procedure to a selection problem, of the true opportunity cost, $w_{[k]} - w_{\mathfrak{D}}$, when alternative $\mathfrak{D}$ is selected as best. In Fig. 3.5, this evidence is the empirical fraction $PICS_{IZ} = 1 - PCS_{IZ}$, where PCS_{IZ} is the probability of correctly selecting the true best alternative, the fraction estimated over repeated applications of the procedure to a selection problem.

Figure 3.4 presents a representative graph that supports a claim that was systematically found to be true in numerical experiments: the adaptive stopping rules, EOC_{Bonf} and PCS_{Slep}, perform much more effectively than the deterministic stopping rule. This claim is supported in the graph because the curves with the EOC_{Bonf} stopping rule are lower (have a lower loss for any given expected number

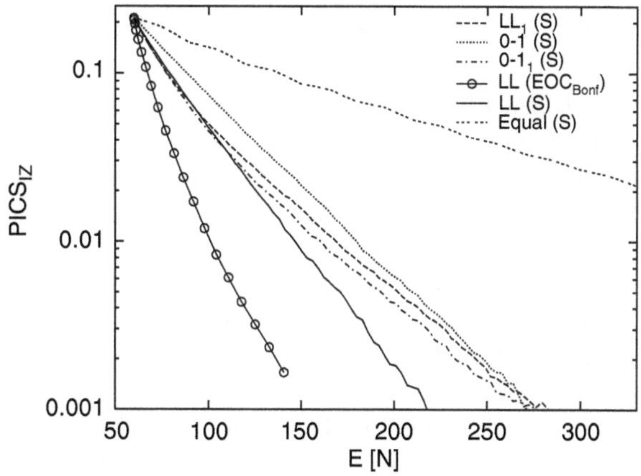

Fig. 3.5 $PICS_{IZ}$ efficiency for LL and LL_1 in a monotonically decreasing means configuration with ten alternatives (means of each alternative evenly spaced)

of samples) than for the (deterministic) S stopping rule. In Fig. 3.4, there are five alternatives in a so-called slippage configuration (the mean of the best alternative is exactly the same value better than the means of the other alternatives).

In Fig. 3.4, we also observe that for a given stopping rule, that the LL allocation rule was at least as good as the LL_1 allocation rule. This was not systematically true. Both of these allocations were equally good for $k=2$ alternatives. The LL_1 allocation was sometimes better than the LL allocation when there were a relatively small number of total samples (say, 50–120 depending on the problem structure) or when the true means of all competing systems were close to the true mean of the best alternative. As k increases, LL tends to improve in performance relative to LL_1.

In Fig. 3.5, we see that the LL allocation rule can perform even better than the $0-1$ allocation rule for PCS-based figures of merit, even though the latter is specifically designed to improve PCS. This is apparently due to the additional approximations that the $0-1$ loss would seem to impose to obtain analytical results for an easily computable allocation.

Overall, the empirical results showed that the small-sample procedures (LL_1) are competitive if either the number of additional samples allocated is very small, a fixed budget stopping rule is used (as opposed to an adaptive stopping rule such as EOC_{Bonf}), or the number of alternatives is small. This may be the case when alternatives are costly to sample, as when sampling time is very long. In most other settings, the LL performed better. As a general rule, the $OCBA$, the LL and the LL_1 allocations were found to be superior to the equal allocation, many times substantially so.

In terms of overall robustness, Branke et al. [4] and Chick et al. [18] found that the LL allocation or $OCBA_{LL}$ allocation rule (an allocation rule based on the EOC loss function instead of the PCS-based loss function used for the $OCBA$ allocation rule from Sect. 3.3 above; see [33]), with the EOC_{Bonf} stopping rule, proved to be an effective combination in a procedure that works best over a broad class of selection problem structures.

3.4.5 Economics of Selection Procedures (ESP)

The *OCBA*, *LL*, and LL_1 allocations above use criteria to allocate samples during a given stage of sampling that essentially are myopic and greedy: they try to maximize (up to approximations) the evidence for correct selection by the end of the stage. *ESP* procedures account for the value of the ability to continue to sample further after that stage. Thus, it better accounts for the information that a sequential selection procedure can provide.

The *ESP* procedures have been developed assuming a linear loss reward function, as the economic benefit of implementing the alternative selected as best is used to drive both the allocation of samples, and the decision of when to stop sampling. In making the decision for when to stop sampling, samples are assumed to come at a cost. There are at least two ways in which samples might come at a cost. One way is that additional simulation causes delays in implementation. In large scale business decisions, that may cause a discounting penalty due to delays in decisions. Another way that costs may be incurred is from the marginal cost of samples: computer time and resources cost money. Chick and Gans [15] explored the case of discounted sampling, either with or without additional marginal costs per sample. Chick and Frazier [14] handled the case of marginal costs of simulation without discounting.

This section describes the latter case: where simulation samples have costs but there is no additional penalty due to discounting. Thus, to implement them, one ideally needs to estimate the financial cost of sampling (e.g., the cost of run times on a bank of servers) and to have simulation output that expresses financial value. Alternatively, one might use the resulting selection procedure in the absence of those financial interpretations, and use a notional value of the cost to sample in order to determine how many simulation samples to run, and in which order. A reduction of the notional value of the cost to sample would increase the number of samples, if more samples were desired before selecting an alternative.

By its nature, both allocation rules and stopping rules are derived with this approach. The focus on economic criteria to determine sampling plans and when to stop sampling led to the choice of name Economics of Sampling Procedure (*ESP*).

Maximizing Expected Reward

Similar to the previous procedures, we want to maximize the expected reward. However, we go beyond a one-step lookahead for the value of sampling and seek to choose a sequence of alternatives to sample from so that the stream of costs and terminal reward together maximize the expected net reward from the start to the time of selecting an alternative as best.

To describe how to do so, it is useful to introduce some new concepts. A selection policy π is a dynamic method of choosing at each stage t whether to sample an alternative or to stop and select the alternative. The policy, at stage t, can use all information obtained up until stage t. Let $T \in \{t = 0, 1, 2, \ldots\}$ be the stage when the decision maker decides to stop and select the best alternative to implement. For $t < T$, let $i(t)$ be the index of the alternative to be simulated at stage t and $I(T)$ be the index of the selected alternative. Then the selection policy $\pi = (i(\cdot), T, I(T))$ is the choice of a sequence of alternatives to sample from, the stopping time, and the selected alternative.

Let X_i be the random variable of the unknown reward of alternative i and c_i is the cost per sample of alternative i. Note that c_i and the output X_i are in terms of monetary values. The sampling selection problem is the problem of maximizing expected value of the cost of sequential sampling plus the reward of implementing the alternative selected as best defined by

$$\sup_{\pi} V^{\pi} = \mathrm{E}_{\pi}\left[\sum_{t=0}^{T-1} - c_{i(t)} + X_{I(T),\, T+1} \middle| \mathscr{E}\right]. \tag{3.22}$$

It can be shown that, under certain technical conditions, a policy π maximizes V^{π} if it minimizes the sum of the expected total sampling cost ($\mathrm{E}_{\pi}[\sum_{t=0}^{T-1} - c_{i(t)} | \varepsilon]$) and the expected opportunity cost when an alternative is selected ($\mathrm{E}_{\pi}[\mathscr{L}_{LL}(I(T), W) | \varepsilon]$).

Finding a policy π that achieves the maximum in (3.22) is challenging to solve optimally in general, but Chick and Frazier [14] provided a solution, to an asymptotic approximation, for the special case of comparing one alternative with an unknown mean to a given standard alternative whose mean reward is known to be m. From that special case, they handle the case of $k > 1$ alternatives by using the following heuristic: At each stage, each alternative is assessed to see if it is worth performing more simulations if one were to compare that alternative (with unknown mean) to the mean of the best other alternative (presuming its mean to be known, with m set to the current estimate of the mean of the best other alternative).

Optimal Stopping Problem for the Special Case of $k = 1$ Alternative

Consider comparing one alternative with a known mean reward, m. The sampling selection problem in (3.22) then becomes an optimal stopping problem. Let $c, \bar{x}_t$ and n_t be the cost per sample, the sample mean, and the number of samples seen so far up to time t for the one alternative. The Bellman's recursion for this problem is

$$V^*(m,\, \bar{x}_t, n_t) = \max\left\{ m, -c + \mathrm{E}\left[V^*\left(m,\, \frac{n_t \bar{x}_t + X_{i,t+1}}{n_t + 1}, n_t + 1 \right) \middle| \bar{x}_t, n_t \right], \bar{x}_t \right\}. \tag{3.23}$$

The first maximand in Bellman's recursion indicates selecting the known alternative. The second maximand includes the cost of observing one sample from the alternative with the unknown mean and thenhaving the option to select an alternative or to continue sampling, given the information from that sample. The third maximand is associated with stopping to sample and to implement the alternative with the unknown mean.

Chick and Frazier [14] show how to convert the discrete time dynamic program in (3.23) into a continuous time problem. That continuous time problem is a free boundary problem for a heat equation, and whose solution gives an asymptotic approximation to the expected value of the option to continue sampling to learn more (a non-myopic, multistage version of the one-stage value of information calculations used in the *LL* and LL_1 procedures), and upper and lower optimal stopping boundaries. If the statistics of the alternative with the unknown mean are below the lower stopping boundary, it would be optimal to stop sampling and to select the known alternative for reward m. If the statistics of the alternative with an unknown mean are above the upper boundary, the alternative with the unknown mean should be selected as best with no further sampling. If the statistics of the alternative with the unknown mean are between the upper and lower stopping boundaries, then one additional sample should be taken for that alternative, and the process should be repeated.

Solving this problem for practical use in a procedure could require computing the solution of a partial differential equation with a free boundary. Fortunately, Chick and Frazier [14] provide a numerical approximation to the solution to the free boundary problem which has shown to be useful in problems and which does not require heavy mathematical machinery for implementations. In particular, they show that the upper and lower optimal stopping boundaries are given by

$$m \pm c_i^{1/3}\sigma_i^{2/3} b\left(\sigma_i^{2/3}/c_i^{2/3} n_i\right), \tag{3.24}$$

for some function $b(s) \geq 0$ for $s \geq 0$, and that a useful approximation to $b(s)$ is

$$\tilde{b}(s) = \begin{cases} 0.233s^2 & \text{if } 0 \leq s \leq 1, \\ 0.00537s^4 - 0.06906s^3 + 0.3167s^2 - 0.02326s & \text{if } 1 < s \leq 3, \\ 0.705s^{1/2}\ln(s) & \text{if } 3 < s \leq 40, \\ 0.642\left[s(2\ln(s))^{1.4} - \ln(32\pi)\right]^{1/2} & \text{if } 40 < s. \end{cases} \tag{3.25}$$

The region that is between the upper and lower stopping boundaries in (3.24) for $n_i > 0$ is called the continuation region (because it determines where sampling should continue).

ESP Allocation Rule and Stopping Rule

The approach for the optimal stopping problem can be extended in a heuristic way to provide a new allocation rule and a new stopping rule when there are $k > 1$ alternatives.

The stopping rule is motivated and presented first. From the previous subsection, a given alternative warrants continued simulation if it is between the upper and lower stopping boundaries in (3.24). Adapted to the case of $k > 1$ alternatives, each with unknown mean, one can substitute the value of m in (3.24) with the mean of the alternative with the best estimated mean (specifically, the highest posterior mean of the other alternatives). This motivates what we will call the *ESP_b stopping rule*: sampling continues as long as there is at least one alternative that would merit additional sampling if it were considered in comparison with the best of the other alternatives (at least up to asymptotic approximations). More formally, sampling continues as long as there is at least one alternative i such that

$$c_i^{1/3}\sigma_i^{2/3} b\left(\sigma_i^{2/3} / \left(c_i^{2/3} n_{i,\,t}\right)\right) > \widehat{\Delta}_{i,\,t}, \tag{3.26}$$

where $\widehat{\Delta}_{i,\,t} = \left|\overline{x}_{i,t} - \max_{j \neq i} \overline{x}_{j,t}\right|$ is the difference in expected value between the alternative i and the best of other alternatives (including the known mean standard m) conditional on information through stage t, and $n_{i,\,t}$ is the number of samples from alternative i through stage t. When the sampling variance σ_i^2 is unknown, it is appropriate in implementations to substitute the sample variance $\widehat{\sigma}_i^2$ for it in (3.26), as well as in (3.27) below [14].

The *ESP_b allocation rule* allocates one sample to the alternative that is 'furthest inside' the continuation region, in a standardized coordinate alternative that is natural to consider in this application [14]. In particular, one should sample from the alternative that is the solution to

$$\arg\max_i b\left(\sigma_i^{2/3} / \left(c_i^{2/3} n_{i,\,t}\right)\right) - \widehat{\Delta}_{i,\,t} / \left(c_i^{1/3}\sigma_i^{2/3}\right). \tag{3.27}$$

The term $c_i^{1/3}\sigma_i^{2/3}$ in (3.26) is the cube root of the product of the sampling costs and sampling variance, which is inversely proportional to the sampling efficiency [32].

Chick and Frazier [14] provide another allocation rule that is based more directly on the diffusion approximation for the expected reward of continuing to sample. But that other allocation rule requires access to the full solution of the free boundary problem, and is therefore less attractive from an implementation standpoint.

They also provide numerical results that illustrate the performance of a selection procedure with various allocation rules and stopping rules. The *LL* allocation rule was not tested with the *ESP_b* stopping rule, because *ESP* is based on approximations that presume one sample is taken at a time. Table 3.4, adapted from Chick and Frazier [14], shows the expected total penalty for not knowing the mean rewards $\mathrm{E}[cT + OC]$ for five different combinations between allocation rules and stopping rules calculated using Monte Carlo Simulation with 10^6 samples for $k = 2$, 5, 10, 20 and 10^5 samples for $k = 100$. Here, the sampling cost was assumed to be $c = 1$, T is the number of samples observed when an alternative is selected as best, and OC is the realized opportunity cost of selecting that alternative.

Table 3.4 Expected total penalty for not knowing the mean rewards $E[cT + OC]$ for several allocation and stopping rules

		k				
Allocation rule	Stopping rule	2	5	10	20	100
LL	EOC_{Bonf}	320	577	821	1,095	2,168
Equal allocation	EOC_{Bonf}	321	629	1,040	1,815	8,425
LL_1	EOC_{Bonf}	318	546	728	916	1,577
LL_1	ESP_b	231	506	694	875	1,516
ESP_b	ESP_b	233	505	700	856	1,308

In Table 3.4, the stopping rule appears to be more influential than the allocation rule. Specifically, the ESP_b stopping rule performs better than the EOC_{Bonf} stopping rule for each given allocation rule. Performance with the ESP_b stopping rule improves even upon the performance from the EOC_{Bonf} stopping rule with these allocations in all experiments run so far (the ESP_b stopping rule has been subject to less testing to date than has the EOC_{Bonf} stopping rule). This appears to be because the ESP_b stopping rule considers the benefit of multiple future stages of sampling when considering when to stop, whereas the EOC_{Bonf} stopping rule only looks one step ahead.

The ESP_b stopping rule works very well with the LL_1 and ESP_b allocation rules, and either of those options can be recommended for the selection problem.

3.4.6 *Other Variants of EVI*

Apart from ESP_b, Chick and Gans [15] propose another economics of selection procedure where there is a positive discount cost. That algorithm, which has somewhat different stopping boundaries due to the discounting, also has a stopping boundary which is approximated with an easy to compute function. That stopping boundary was shown to be related to the optimal stopping boundary for the Bayesian bandit problem when samples from each bandit are independent Gaussian whose means are unknown (and are inferred through sampling).

In addition, Chick and Gans [15] provide a discussion on whether the decision maker should develop a simulation platform. They proposed a view of simulation as an option to learn more about alternatives before making a selection, and a mechanism to quantify the value of that learning. That value could be compared with the time and financial costs of developing the simulation model in the first place: responding to the question "To simulate, or not to simulate?" when a Bayesian prior distribution can be used to quantify uncertainty about the potential financial benefit of various simulated alternatives, at least in some contexts. The approach is therefore different from most R&S work in simulation which assumes that a simulation platform has been built and is available.

There are also several other one-step lookahead policies to myopically maximize *EVI*. Frazier et al. [23] consider the case where the prior beliefs about the rewards are correlated and Ryzhov et al. [52] apply it to propose a new type of online learning policy. Frazier et al. [24] consider the case with both correlations in the prior belief and correlation in sampling which is achieved through common random numbers. The use of common random number is also considered in the work by Chick and Inoue [17], which minimizes the expected opportunity cost in two stages.

3.5 Conclusion

This chapter discussed some Bayesian approaches to the problem of selecting the best from a small- to medium-sized set of alternatives, where best is the largest mean and the means are to be estimated through sampling. The samples are presumed to be independent, as are the values of the unknown means. The approaches reviewed include *OCBA* and *EVI*, which have variations that are among the most effective known today. The variations are with different allocation rules and stopping rules. The *OCBA* algorithm is very effective in cases where *PCS* is of particular interest. To date, it appears that the most effective procedures in a very broad range of tests are based either on the *LL* or $OCBA_{LL}$ allocation rules in conjunction with the EOC_{Bonf} stopping rule if an adaptive stopping rule is allowed, or with a fixed budget stopping rule if that is required. It appears that the LL_1 and ESP_b allocation rule are even more effective when used in conjunction the ESP_b stopping rule, in cases where sampling costs can be estimated and the output of the simulation has a financial impact. An analysis of tweaking a notional cost of sampling when such conditions do not hold is yet to be done, but it would appear that such an approach could be more broadly applicable. More testing on that would be useful. The papers cited above also explain how a number of other practicalities might be addressed, such as the question of including common random numbers. The codes of the algorithms presented in this chapter can be obtained from the authors.

In cases where the number of alternatives is so large that it is not possible to simulate all alternatives at least a few times, some other techniques may be required. The simulation optimization literature addresses this problem. For example, see Chap. 2 and [25, 35, 56] for reviews. In such cases, one might integrate an R&S procedure with a search algorithm [42, 53]. He et al. [34] extend the *OCBA* notion to the cross-entropy method for combinatorial problems. Or one might use a response surface technique so that information observed for one alternative can be informative for the mean values of other alternatives. Initial work on this approach that is related to the *EVI* approach above includes the use of Kriging models for the unknown means of the alternatives when samples are either independent [22, 23] or correlated such as with common random numbers [24]. Brantley et al. [5, 6] develop a new *OCBA* procedure under the use of a regression equation and partitioning of the decision domain.

Acknowledgements This work has been supported in part by National Science Foundation under Award CMMI-1233376, Department of Energy under Award DE-SC0002223, NIH under Grant 1R21DK088368-01, National Science Council of Taiwan under Award NSC-100-2218-E-002-027-MY3, and the Program for Professor of Special Appointment (Eastern Scholar) at Shanghai Institutions of Higher Learning.

References

1. R. E. Bechhofer, T. J. Santner, and D. M. Goldsman. *Design and Analysis of Experiments for Statistical Selection, Screening, and Multiple Comparisons*. John Wiley & Sons, New York, 1995.
2. J. O. Berger and J. Deely. A Bayesian approach to ranking and selection of related means with alternatives to analysis-of-variance methodology. *Journal of the American Statistical Association*, 83(402):364–373, 1988.
3. J. Blanchet, J. Liu, and B. Zwart. Large deviations perspective on ordinal optimization of heavy-tailed systems. In S. J. Mason, R. R. Hill, L. Mönch, O. Rose, T. Jefferson, J. Fowler, editors, *Proceedings of the 2008 Winter Simulation Conference*, 489–494, 2008.
4. J. Branke, S. E. Chick, and C. Schmidt. Selecting a selection procedure. *Management Science*, 53:1916–1932, 2007.
5. M. W. Brantley, L. H. Lee, C. H. Chen, and A. Chen. Efficient simulation budget allocation with regression. *IIE Transactions*, 45(3):291–308, 2013.
6. M. W. Brantley, L. H. Lee, C.-H. Chen, and J. Xu. An efficient simulation budget allocation method incorporating regression for partitioned domains. *Automatica*, 50(5):1391–1400, 2014.
7. A. Charnes, W. W. Cooper, and E. Rhodes. Measuring the efficiency of decision making units. *European Journal of Operational Research*, 2:429–444, 1978.
8. C. H. Chen and L. H. Lee. *Stochastic Simulation Optimization: An Optimal Computing Budget Allocation*. World Scientific Publishing Co, 2011.
9. C. H. Chen, D. He, and M. C. Fu. Efficient dynamic simulation allocation in ordinal optimization. *IEEE Transactions on Automatic Control*, 51(12):2005–2009, 2006.
10. C. H. Chen, D. He, M. C. Fu, and L. H. Lee. Efficient simulation budget allocation for selecting an optimal subset. *INFORMS Journal of Computing*, 20:579–595, 2008.
11. C. H. Chen, V. Kumar, and Y.C. Luo. Motion planning of walking robots in environments with uncertainty. *Journal of Robotic Systems*, 16(10):527–545, 1999.
12. C. H. Chen, J. Lin, E. Yücesan, and S. E. Chick. Simulation budget allocation for further enhancing the efficiency of ordinal optimization. *Discrete Event Dynamic Systems: Theory and Applications*, 10:251–270, 2000.
13. C. H. Chen, E. Yücesan, L. Dai, and H. C. Chen. Efficient computation of optimal budget allocation for discrete event simulation experiment. *IIE Transactions*, 42(1):60–70, 2010.
14. S. E. Chick, and P. Frazier. Sequential sampling with economics of selection procedures. *Management Science*, 58(3):550–569, 2012.
15. S. E. Chick, and N. Gans. Economic analysis of simulation selection problems. *Management Science*, 55(3):421–437, 2009.
16. S. E. Chick and K. Inoue. New two-stage and sequential procedures for selecting the best simulated system. *Operations Research*, 49(5):732–743, 2001.
17. S. E. Chick and K. Inoue. New procedures to select the best simulated system using common random numbers. *Management Science*, 47(8):1133–1149, 2001.
18. S. E. Chick, J. Branke, and C. Schmidt. Sequential sampling to myopically maximize the expected value of information. *INFORMS Journal of Computing*, 22:71–80, 2010.
19. S. E. Chick and Y. Z. Wu. Selection procedures with frequentist expected opportunity cost bounds. *Operations Research*, 53(5):867–878, 2005.
20. M. H. de Groot. *Optimal Statistical Decisions*. McGraw-Hill, New York, 1970.

21. E. J. Dudewicz and S.R. Dalal. Allocation of observations in ranking and selection with unequal variances. *The Indian Journal of Statistics,* 37B, 1:28–78, 1975.
22. P. Frazier, W. B. Powell, and S. Dayanik. A knowledge-gradient policy for sequential information collection. *SIAM Journal on Control and Optimization*, 47(5):2410–2439, 2008.
23. P. Frazier, W. B. Powell, and S. Dayanik. The knowledge-gradient policy for correlated normal beliefs. *INFORMS Journal of Computing*, 21:599–613, 2009.
24. P. Frazier, J. Xie, and S. E. Chick. Bayesian optimization via simulation with correlated sampling and correlated prior beliefs. In S. Jain, R. R. Creasey, J. Himmelspach, K. P. White, M. Fu, editors, *Proceedings of the 2011 Winter Simulation Conference*, 3979–3911, 2011.
25. M. C. Fu, C. H. Chen, and L. Shi. Some topics for simulation optimization. In S. J. Mason, R. R. Hill, L. Mönch, O. Rose, T. Jefferson, J. Fowler, editors, *Proceedings the 2008 Winter Simulation Conference*, 27–38, 2008.
26. M. C. Fu, F. W. Glover, and J. April. Simulation optimization: A review, new developments, and applications. In M. E. Kuhl, N. M. Steiger, F. B. Armstrong, J. A. Joines, editors, *Proceedings the 2005 Winter Simulation Conference*, 83–95, 2005.
27. M. C. Fu, J. Q. Hu, C. H. Chen, and X. Xiong. Simulation allocation for determining the best design in the presence of correlated sampling. *INFORMS Journal of Computing*, 19:101–111, 2007.
28. P. Glynn and S. Juneja. A large deviations perspective on ordinal optimization. In R. G. Ingalls, M. D. Rosetti, J. S. Smith, B. A. Peters, editors, *Proceedings of the 2004 Winter Simulation Conference,* 577–585, 2004.
29. D. Goldsman and B. L. Nelson. Comparing systems via simulation. In J. Banks, editor, *Handbook of Simulation: Principles, Methodology, Advances, Applications, and Practice*. John Wiley & Sons, New York, 273–306, 1998.
30. S. S. Gupta and K. J. Miescke. Bayesian look ahead one stage sampling allocations for selecting the largest normal mean. *Statistical Papers*, 35:169–177, 1994.
31. S. S. Gupta and K. J. Miescke. Bayesian look ahead one-stage sampling allocations for selecting the best population. *Journal of Statistical Planning and Inference*, 54(2):229–244, 1996.
32. J. M. Hammersley and D.C. Hanscomb. *Monte Carlo Methods*. Methuen, London, 1964.
33. D. He, S. E. Chick, and C. H. Chen. The opportunity cost and OCBA selection procedures in ordinal optimization. *IEEE Transactions on Systems, Man, and Cybernetics—Part C (Applications and Reviews)*, 37(4):951–961, 2007.
34. D. He, L. H. Lee, C. H. Chen, and M. C. Fu, and S. Wasserkrug. Simulation optimization using the cross-entropy method with optimal computing budget allocation. *ACM Transactions on Modeling and Computer Simulation*, 20(1):Article 4, 2010.
35. L. J. Hong and B. L. Nelson. A brief introduction to optimization via simulation. In M. D. Rosetti, R. R. Hill, B. Johansson, A. Dunkin, R. Ingalls, editors, *Proceedings of the 2009 Winter Simulation Conference*, 75–85, 2009.
36. B. W. Hsieh, C. H. Chen, and S.C. Chang. Scheduling semiconductor wafer fabrication by using ordinal optimization-based simulation. *IEEE Transactions on Robotics and Automation*, 17(5):599–608, 2001.
37. Q. S. Jia, E. Zhou, and C. H. Chen. Efficient computing budget allocation for finding simplest good designs. *IIE Transactions*, 45(7):736–750, 2013.
38. S.-H. Kim and B. L. Nelson. A fully sequential procedure for indifference-zone selection in simulation. *ACM Transactions on Modeling and Computer Simulation*, 11:251–273, 2001.
39. S.-H. Kim. and B. L. Nelson. Selecting the best system. In S. G. Henderson, B. L. Nelson, editors, *Handbooks in Operations Research and Management Science: Simulation*. Elsevier Science, Oxford, 501–534, 2006.
40. L. W. Koenig and A. M. Law. A procedure for selecting a subset of size m containing the one best of k independent normal populations, with applications to simulation. *Communications in Statistics* B14, 3:719–734, 1985.
41. A. M. Law. *Simulation Modeling and Analysis*, 5th Edition. McGraw-Hill, New York, 2013.

42. L. H. Lee, E. P. Chew, and P. Manikam. A general framework on the simulation-based optimization under fixed computing budget. *European Journal of Operational Research*, 174:1828–1841, 2006.
43. L. H. Lee, E. P. Chew, S. Teng, and D. Goldsman. Optimal computing budget allocation for multi-objective simulation models. In R. G. Ingalls, M. D. Rosetti, J. S. Smith, B. A. Peters, editors, *Proceedings of the 2004 Winter Simulation Conference*, 586–594, 2004.
44. L. H. Lee, E. P. Chew, S. Teng, and D. Goldsman. Finding the non-dominated Pareto set for multi-objective simulation models. *IIE Transactions*, 42:656–674, 2010.
45. L. H. Lee, N. A. Pujowidianto, L. W. Li, C. H. Chen, and C. M. Yap. Approximate simulation budget allocation for selecting the best system in the presence of stochastic constraints. *IEEE Transactions on Automatic Control*, 57(11):2940–2945, 2012.
46. D. J. Morrice, M. W. Brantley, and C. H. Chen. An efficient ranking and selection procedure for a linear transient mean performance measure. In S. J. Mason, R. R. Hill, L. Mönch, O. Rose, T. Jefferson, J. Fowler, editors, *Proceedings of the 2008 Winter Simulation Conference*, 290–296, 2008.
47. D. J. Morrice, M. W. Brantley, and C. H. Chen. A transient means ranking and selection procedure with sequential sampling constraints. In M. D. Rosetti, R. R. Hill, B. Johansson, A. Dunkin, R. Ingalls, editors, *Proceedings of the 2009 Winter Simulation Conference*, 590–600, 2009.
48. B. L. Nelson, J. Swann, D. Goldsman, and W. M. Song. Simple procedures for selecting the best simulated system when the number of alternatives is large. *Operations Research*, 49(6):950–963, 2001.
49. W. B. Powell and I. O. Ryzhov. *Optimal Learning*. Wiley, 2012.
50. Y. Rinott. On two-stage selection procedures and related probability inequalities. *Communications in Statistics* A7:799–811, 1978.
51. R. Rubinstein. The cross-entropy method for combinatorial and continuous optimization. *Methodology and Computing in Applied Probability*, 1(2): 127–190, 1999.
52. I. O. Ryzhov, W. B. Powell, and P. Frazier. The knowledge-gradient algorithm for a general class of online learning problems. *Operations Research*, 60(1):180–195, 2012.
53. L. Shi and C. H. Chen. A new algorithm for stochastic discrete resource allocation optimization. *Discrete Event Dynamic Systems: Theory and Applications*, 10:271–294, 2000.
54. J. F. Shortle, C. H. Chen, B. Crain, A. Brodsky, and D. Brod. Optimal splitting for rare-event simulation. *IIE Transactions*, 44(5):352–367, 2012.
55. R. Szechtman and E. Yücesan. A new perspective on feasibility determination. In S. J. Mason, R. R. Hill, L. Mönch, O. Rose, T. Jefferson, J. Fowler, editors, *Proceedings of the 2008 Winter Simulation Conference*, 273–280, 2008.
56. E. Tekin and I. Sabuncuoglu. Simulation optimization: A comprehensive review on theory and applications. *IIE Transactions*, 36:1067–1081, 2004.
57. L. Trailovic and L. Y. Pao. Computing budget allocation for efficient ranking and selection of variances with application to target tracking algorithms. *IEEE Transactions on Automatic Control*, 49:58–67, 2004.
58. R. Waeber, P. Frazier, and S. G. Henderson. Performance measures for ranking and selection procedures. In B. Johansson, S. Jain, J. Montoya-Torres, J. Hugan, E. Yücesan, editors, *Proceedings of the 2010 Winter Simulation Conference*, 1235–1245, 2010.
59. W. P. Wong, W. Jaruphongsa, and L. H. Lee. Budget allocation for effective data collection in predicting an accurate DEA efficiency score. *IEEE Transactions on Automatic Control*, 56(6):1235–1245, 2011.
60. S. Yan, E. Zhou, and C. H. Chen. Efficient selection of a set of good enough designs with complexity preference. *IEEE Transactions on Automation Science and Engineering*, 9(3):596–606, 2012.

Chapter 4
Response Surface Methodology

Jack P.C. Kleijnen

Abstract This chapter first summarizes Response Surface Methodology (RSM), which started with Box and Wilson's 1951 article on RSM for real, non-simulated systems. RSM is a stepwise heuristic that uses first-order polynomials to approximate the response surface locally. An estimated polynomial metamodel gives an estimated local gradient, which RSM uses in steepest ascent (or descent) to decide on the next local experiment. When RSM approaches the optimum, the latest first-order polynomial is replaced by a second-order polynomial. The fitted second-order polynomial enables the estimation of the optimum. This chapter then focuses on simulated systems, which may violate the assumptions of constant variance and independence. A variant of RSM that provably converges to the true optimum under specific conditions is summarized, and an adapted steepest ascent that is scale-independent is presented. Next, the chapter generalizes RSM to multiple random responses, selecting one response as the goal variable and the other responses as the constrained variables. This generalized RSM is combined with mathematical programming to estimate a better search direction than the steepest ascent direction. To test whether the estimated solution is indeed optimal, bootstrapping may be used. Finally, the chapter discusses robust optimization of the decision variables, while accounting for uncertainties in the environmental variables.

4.1 Introduction

Response Surface Methodology (RSM) is one of the more popular methods in simulation optimization. Originally, RSM was developed for the optimization of real (physical) systems; see the classic 1951 article by Box and Wilson [12], the overview of RSM during the period 1966–1988 by Myers et al. [32], the recent third edition of the popular handbook by Myers et al. [33], and recent RSM software on the Web; see also [4]. RSM in simulation was discussed in the 1975 monograph by

J.P.C. Kleijnen (✉)
Tilburg University, Tilburg, The Netherlands
e-mail: kleijnen@tilburguniversity.edu

M.C. Fu (ed.), *Handbook of Simulation Optimization*, International Series in Operations Research & Management Science 216, DOI 10.1007/978-1-4939-1384-8_4

Kleijnen [25]. One of the first case studies on RSM in simulation is a 1977 report on a computer center with two servers and three priority classes—namely, small, medium, and large jobs—estimating the 90 % quantiles of the waiting times per class for different class limits, and applying RSM to find the optimal class limits; see [42]. RSM in simulation is also discussed in [5,6,31,39]. Google returned more than two million results for the term "Response Surface Methodology", on February 4, 2014; however, as of this writing, RSM (unlike search heuristics such as OptQuest) has not yet been implemented as an add-on to commercial simulation software.

RSM treats the simulation model as a black box, i.e., RSM observes the input/output (I/O) of the simulation model, but not the internal variables and specific functions implied by the simulation's computer modules. RSM is a *sequential* heuristic, i.e., it uses a sequence of local experiments that is meant to lead to the optimum input combination. Note that an input combination is also called a point or a scenario. RSM uses design of experiments (DOE) and the concomitant linear regression analysis. Although RSM is only a heuristic, it has gained a good track record, as we shall see in the next sections. Moreover, practitioners may not be interested in convergence proofs, because realistic experiments take so much time that large sample sizes are impossible; practitioners may be more interested in finding better solutions than the current one (a French expression claims that "the best is the enemy of the better").

More specifically, RSM uses a sequence of local metamodels (approximations) that are first-order polynomials in the inputs; once the optimum seems close, RSM augments the latest first-order polynomial to a second-order polynomial. Note that terminology may be misleading, because many authors use the term "response surface" instead of "metamodel" (e.g., [38]). Note that RSM may be combined with white-box methods that give estimated gradients, which may improve RSM, e.g., [21,37].

We focus on stochastic discrete-event simulation, though RSM has also been applied to deterministic simulation and the simulation of sets of random nonlinear difference equations. By definition, stochastic simulation (including discrete-event simulation) uses pseudorandom numbers (PRN). We do not discuss deterministic simulation, which may also have numeric noise.

Furthermore, we assume that the inputs are continuous variables. We focus on simulation that has a mean (expected value) as the response (output). Nevertheless, RSM has also been applied to outputs that are quantiles; see again [42]. The response may also be a probability (e.g., the overflow probability in a queueing system), which we may formulate as the expected value of a binary variable using the indicator function.

The optimum solution for the decision variables may turn out to be inferior when ignoring uncertainties in the non-controllable environmental variables, i.e., these uncertainties create a risk. Taguchi introduced robust optimization for the design of products; see [41] and the update in [33, pp. 483–555]. Ben-Tal et al. developed robust optimization in mathematical programming with uncertain coefficients; see [8] and the update in [43]. We shall discuss both approaches (due to Taguchi and Ben-Tal et al.).

We assume that RSM starts after the important factors (inputs) and their experimental area have been identified, i.e., before RSM starts we may need to use *factor screening* to identify the really important factors among the many conceivably important factors. A recent survey of various screening methods is [40]. However, [14] combines RSM with screening in a single method (called STRONG-S); we point out that RSM without a preceding screening phase may imply the simulation of extremely many combinations of simulation inputs, as we shall see in Sect. 4.2.

The remainder of this chapter is organized as follows. Section 4.2 summarizes classic RSM, developed for real-life experimentation. Section 4.3 adapts this RSM to the needs of simulation. Section 4.4 presents the adapted steepest descent (ASD) search direction, originally developed in [29]. Section 4.5 summarizes generalized RSM (GRSM) for simulation with multiple responses, developed in [2]. Section 4.6 summarizes a procedure for testing whether an estimated optimum is truly optimal—using the Karush–Kuhn–Tucker (KKT) conditions—developed in [10]. Section 4.7 discusses robust optimization. Section 4.8 presents conclusions. The chapter ends with 43 references, enabling further study of details. This chapter updates and extends the 2008 monograph [27].

4.2 RSM Basics

As previously mentioned, the basics of RSM were developed in real-life experiments for the simplest type of optimization problems, i.e., to minimize the expected value of a single output, with continuous inputs and without any constraints:

$$\min_{\mathbf{z}} \mathsf{E}[w_0|\mathbf{z}], \tag{4.1}$$

where $\mathsf{E}[w_0|\mathbf{z}]$ is the goal or objective output, which is to be minimized through the choice of the input combinations $\mathbf{z} = (z_1,\ldots,z_k)^\top$, where $z_j,\ j = 1,\ldots k$, denotes the jth "original" input, i.e., the inputs are not standardized such that they lie between -1 and 1.

Note: We use z for the original input and x for the standardized input; RSM is scale dependent as we shall see. For the output we use w, which may be a mnemonic referring to "waiting time," which is an output of many simulation models.

RSM has the following characteristics:

- RSM is an *optimization heuristic* that tries to estimate the input combination that minimizes a given goal function as in (4.1). Because RSM is a heuristic, it does not guarantee success.
- RSM is a *stepwise* (multi-stage) method, as summarized in the steps below.
- In each step, RSM fits a local first-order *polynomial* regression (meta)model—except for the last step, in which RSM fits a second-order polynomial.
- To fit (estimate, calibrate) these first-order polynomials, RSM uses I/O data obtained through so-called *resolution-III (R-III) designs*; for the second-order

polynomial, RSM uses a *central composite design (CCD)*; we shall define these R-III designs and CCDs later in this section.

- Each step—except the last one—selects the direction for changing the inputs through the *gradient* implied by the first-order polynomial fitted in that step. This gradient is used in the mathematical (not statistical) technique of *steepest descent*—or steepest ascent, in case the output is to be maximized.
- In the final step, RSM takes the *derivatives* of the locally fitted *second-order polynomial* to estimate the optimum input combination. RSM may also apply the mathematical technique of *canonical analysis* to this polynomial, to examine the shape of the optimal subregion: does that region have a unique minimum, a saddle point, or a ridge with stationary points?

We now provide more details for each of the steps described above. Figure 4.1 in Sect. 4.5 gives a specific example with a single goal function and two constrained random outputs; such constraints are not present in classic RSM.

RSM must be initialized, i.e., it needs a starting point. That point may be the input combination currently used in practice if the system already exists. Otherwise, we should use intuition and prior knowledge—as in many other heuristics.

For the neighborhood of this starting point, RSM explores the I/O behavior of the observed system. RSM approximates this behavior through a metamodel that is a local first-order polynomial—as Taylor's series expansion suggests—augmented with additive white noise e:

$$y = \beta_0 + \sum_{j=1}^{k} \beta_j z_j + e, \tag{4.2}$$

where white noise means $e \sim \mathcal{N}(0, \sigma^2)$ so e is independently and identically distributed (IID) in the local experimental area. We denote the regression parameters by $\boldsymbol{\beta} = (\beta_0, \beta_1, \ldots, \beta_k)^{\mathsf{T}}$ with the intercept β_0 and the k first-order or main effects β_j ($j = 1, \ldots, k$). When the next step moves to a new local area, the variance may change. Obviously, (4.2) is a linear regression model with white noise, so least squares (LS) gives the best linear unbiased estimator (BLUE) of β:

$$\hat{\beta} = (\mathbf{Z}^{\mathsf{T}}\mathbf{Z})^{-1}\mathbf{Z}^{\mathsf{T}}\mathbf{w}, \tag{4.3}$$

where the hat denotes the LS estimator, $\mathbf{Z}$ denotes the $N \times (k+1)$ matrix that is determined by the R-III design and the m_i replications of point i, and $\mathbf{w}$ denotes the vector with the N simulation outputs; obviously, $N = \sum_{i=1}^{n} m_i$, so $\mathbf{Z}$ has $m_i \geq 1$ identical rows, where the first element of each row is 1 and corresponds with the intercept β_0.

To select the n input combinations needed to estimate these $k+1$ effects, RSM uses a R-III design with $n \geq k+1$. In classic R-III designs this n is a multiple of four, e.g., if $k = 7$, then Table 4.1 gives a 2^{7-4} fractional factorial design in the standardized inputs x_j so '−' means −1 and '+' means +1; the last column has the so-called generator **7** = **1.2.3**, which means $x_{i;7} = x_{i;1}x_{i;2}x_{i;3}$; the generators in the columns 4, 5, and 6 are defined analogously. The design is "balanced," i.e., each

Table 4.1 A fractional factorial design for seven factors, labeled 1 through 7, with eight combinations

Combination	1	2	3	4 = 1.2	5 = 1.3	6 = 2.3	7 = 1.2.3
1	−	−	−	+	+	+	−
2	+	−	−	−	−	+	+
3	−	+	−	−	+	−	+
4	+	+	−	+	−	−	−
5	−	−	+	+	−	−	+
6	+	−	+	−	+	−	−
7	−	+	+	−	−	+	−
8	+	+	+	+	+	+	+

column has $n/2$ values equal to -1. Moreover, all pairs of columns are orthogonal. If n is not a power of 2, then the R-III design is called a Plackett–Burman design, and is tabulated in many publications and provided by DOE software, e.g., [28] gives the design for $k = 11$ inputs and $n = 12$ combinations.

Unfortunately, there are no general guidelines for determining the appropriate size of the local area in each step; again, intuition and prior knowledge are important. However, [13] decides on the size of the local area, using a so-called trust region; we shall give some details in Sect. 4.3.

To decide on the next subarea, RSM follows the steepest descent path, e.g., if the estimated first-order effects are such that $\hat{\beta}_1 \gg \hat{\beta}_2 > 0$, then RSM decreases input z_1 much more than input z_2. Unfortunately, the steepest-descent method is scale dependent, i.e., linear transformations of the inputs affect the search direction; see [33]. We shall present a scale-independent variant in Sect. 4.4.

The steepest descent technique does not quantify the step size along its path. We may therefore try some intuitively selected value for the step size. If that value yields an output that is significantly higher instead of lower, then we may reduce the step size. Otherwise, we take one more step in the current steepest descent direction. A more sophisticated mathematical procedure for selecting the step size will be presented in Sect. 4.5.

After a number of steps along the steepest descent path, the output will increase instead of decrease: the first-order polynomial is only a local approximation of the true I/O function. When such deterioration occurs, RSM observes the $n > k$ combinations specified by a R-III design centered around the best point found so far, i.e., RSM may use the same design as in step 2 (see Table 4.1), but translate the standardized inputs x_j into different values for the original inputs z_j. The best combination found so far, may be one of the corner points of the design; see again Fig. 4.1. Next RSM estimates the first-order effects in the new local polynomial approximation using LS via (4.3). And so the search continues.

It is intuitively clear that the plane implied by the most recent local first-order polynomial cannot adequately represent a hill top when searching to maximize a function or—equivalently—minimize (4.1). So in the neighborhood of the optimum, a first-order polynomial shows serious lack of fit. A popular and simple diagnostic measure is the coefficient of determination R^2. A related diagnostic uses an F-statistic to test whether all estimated first-order effects—and hence the gradient—are zero. These diagnostics are given by all modern regression software packages. Instead of these diagnostics, RSM might use "cross-validation." More details including references are given in [28].

If the most recently fitted first-order polynomial turns out to be inadequate, then RSM fits a second-order polynomial:

$$y = \beta_0 + \sum_{j=1}^{k} \beta_j z_j + \sum_{j=1}^{k} \sum_{j'=j}^{k} \beta_{j;j'} z_j z_{j'} + e. \tag{4.4}$$

To estimate the $q = 1 + 2k + k(k-1)/2$ coefficients of this polynomial, RSM uses a CCD with $n > q$ combinations More precisely, a CCD starts with a *resolution-V* (R-V) design, which by definition gives unbiased estimators of the intercept, the k first-order effects, and the $k(k-1)/2$ two-factor interactions in (4.4). For example, if k = 7, then the 2^{7-4} design in Table 4.1 is replaced by a 2^{7-1} design with the generator **7** = **1.2.3.4.5.6** so 64 combinations are used to estimate 29 effects. A CCD augments this R-V design such that the purely quadratic effects $\beta_{j;j}$ can also be estimated. More specifically, a CCD adds the central point and $2k$ axial points that form a star design, where—for the standardized factors—the central point is $(0, \ldots 0)^{\mathsf{T}}$ and the "positive" axial point for factor j is $x_j = c$ and all other $(k-1)$ factors are fixed at the center so $x_{j'} = 0$ with $j' = 1, \ldots, k$ and $j' \neq j$; the "negative" axial point for factor j is $x_j = -c$ and $x_{j'} = 0$. A detailed discussion of CCDs is given in [28], including more efficient so-called "saturated" designs; by definition, the latter designs have n = q. Using this fitted second-order polynomial, RSM estimates the optimal values of the inputs by straightforward differentiation or by more sophisticated *canonical analysis*; see the monograph [33].

If time permits, then RSM may try to escape from a possible local minimum and restart the search from a different initial local area—which brings RSM back to its initial step.

We recommend not eliminating inputs that have non-significant effects in a first-order polynomial fitted within a local experimental area, because these inputs may have significant effects in a next experimental area. A related issue is the determination of the number of replications. Indeed, determining whether the signal-to-noise ratio is big enough is a moot issue in metamodeling; see [28]. For the time being, we recommend estimating the true mean response for a given input combination such that a relative precision of (say) 10 % has a 90 % probability, using the method detailed in [31].

The Taylor series argument suggests that a higher-order polynomial is more accurate than a lower-order polynomial. A statistical counterargument, however, is that overfitting gives less accurate estimators of the polynomial coefficients. Consequently, the higher-order polynomial may give a predictor $\hat{y}$ with lower bias but higher variance such that its mean squared error (MSE) increases. Moreover, a higher-order polynomial requires the simulation of more input combinations.

4.3 RSM in Simulation

We consider the following two characteristics of simulation experiments:

1. The constant variance assumption does not hold.
2. The independence assumption does not hold if common random numbers (CRN) are applied.

Many simulation experiments treat queueing systems; examples are supply chains and telecommunication networks. The simplest queueing model is the $M/M/1$ queue, which has one server and i.i.d. exponential interarrival and service times, so it can be modeled as a Markov chain. It is well known that as the traffic rate of the $M/M/1$ queue increases, the mean steady-state waiting time increases and the variance increases even more; consequently, the constant-variance assumption does not hold.

CRN are often applied in simulation experiments, because it is the default option in many simulation software packages (e.g., Arena); moreover, CRN are a simple and intuitive variance reduction technique that gives more accurate estimators of the regression parameters—except for the intercept β_0. The outputs of all input combinations that use CRN are statistically dependent; actually, we expect these outputs to be positively correlated. CRN are related to "blocking" in real-life experiments. In simulation experiments, we may use blocking when combining CRN and antithetic random numbers through the so-called Schruben–Margolin strategy; this strategy is detailed in [15].

The preceding two characteristics imply that the ordinary LS (OLS) does not give the BLUE. Actually, weighted LS (WLS) does give the BLUE if the variances are not constant and no CRN are used, but assumes known response variances. Generalized LS (GLS) gives the BLUE if CRN are used, but assumes known response variances and covariances. We therefore recommend the following simple estimator, which is detailed in [28]: assuming a constant number of replications m (as is usual in CRN), we compute the OLS per replication replacing $\mathbf{w}$ in (4.3) by $\mathbf{w}_r$ to get $\hat{\boldsymbol{\beta}}_r$ $(r = 1, \ldots, m)$. Replication r then gives an estimate of the steepest descent direction if a first-order polynomial is used or the optimum input combination if a second-order polynomial is used. Together, the m replications give an estimate of the accuracy of this estimated direction or optimum; if we find this accuracy too low, then we may simulate additional replications. We have not yet any experience with this simple sequential approach.

Actually, if we have $m_i > 1$ $(i = 1, \ldots, n)$ replications, then we can further explore the statistical properties of the LS estimator of β through bootstrapping. Efron and Tibshirani wrote the classic textbook on bootstrapping in general; see [20]. Many more references are given in [28]. To explain distribution-free bootstrapping in

RSM, we first consider the case of no CRN. We resample—with replacement—the m_i simulation outputs $w_{i;r}$ ($r = 1, \ldots, m_i$) for input combination i ($i = 1, \ldots, n$), and obtain the bootstrapped simulation outputs $w^*_{i;r}$, where the superscript $*$ is the general symbol for bootstrapped observations. Resampling for each input combination i, we obtain the bootstrapped I/O data $(\mathbf{Z}, \mathbf{w}^*)$ with the N-dimensional vector $\mathbf{w}^* = (w^*_{1;1}, \ldots, w^*_{1;m_1}, \ldots, w^*_{n;1}, \ldots, w^*_{nm_n})^\top$ (so $N = \sum_{i=1}^n m_i$) and the original $N \times q$ matrix of input combinations $\mathbf{Z}$. Next we consider the case of CRN. The n elements of $\mathbf{w}_r = (w_{1;r}, \ldots, w_{n;r})^\top$, $r = 1, \ldots, m$, are then not IID, so we resample the m IID vectors $\mathbf{w}_r$. This case also gives $(\mathbf{Z}, \mathbf{w}^*)$. In both cases (CRN or no CRN), we use $(\mathbf{Z}, \mathbf{w}^*)$ to compute the bootstrapped regression parameter estimator $\hat{\boldsymbol{\beta}}^*$ replacing $\mathbf{w}$ in (4.3) by $\mathbf{w}^*$. This bootstrapping we repeat (say) B times; popular choices are $B = 100$ or $B = 1{,}000$. This bootstrap sample gives $\hat{\boldsymbol{\beta}}^*_b$ $(b = 1, \ldots, B)$, where we replace $\mathbf{w}$ in (4.3) by $\mathbf{w}^*_b$. After we sort these $\hat{\boldsymbol{\beta}}^*_b$ in ascending order, we obtain the estimated density function (EDF) of $\hat{\boldsymbol{\beta}}^*$. We can also use $\hat{\boldsymbol{\beta}}^*_b$ to derive the corresponding estimated steepest ascent direction and optimum.

Note: Instead of distribution-free bootstrapping, we can apply parametric bootstrapping, which assumes a specific type of distribution, e.g., a Gaussian distribution (also see Sect. 4.6). Parametric bootstrapping may be attractive if m_i is small and no CRN are used, e.g., the n means and n variances can be estimated if the weak condition $m_i > 1$ holds. If CRN are used, then the covariance matrix $\mathrm{Cov}(w_r, w_{r'})$ with r, $r' = 1, \ldots, m$ needs to be estimated; this estimation requires $m > n$, as proven in [19]. So parametric bootstrapping may require fewer replications, but the assumed distribution may not hold for the simulated outputs.

Recently, [13] developed STRONG, which is a completely automated variant of RSM combined with so-called *trust regions*. STRONG is proven to converge to the true optimum. Originally, trust regions were developed in [16] for deterministic nonlinear optimization. The trust region is a subregion in which the objective function is approximated such that if an adequate approximation is found within the trust region, then the region is expanded; else the region is contracted. In STRONG these trust regions replace the "local" regions of classic RSM. STRONG includes statistical tests to decide whether trust regions should be expanded or shrunken in the various steps, and how much these areas should change. If necessary, the trust region is small and a second-order polynomial is used. Next, [14] combines STRONG with screening, and calls the resulting procedure STRONG-S; this method is applied to several test functions with multiple local minima. Contrary to the Taylor series argument, STRONG may have a relatively large trust region that does not require a second-order polynomial metamodel but only a first-order polynomial metamodel.

4.4 Adapted Steepest Descent (ASD)

ASD accounts for the covariances between the elements of the estimated gradient $\hat{\boldsymbol{\beta}}_{-0} = (\hat{\beta}_1, \ldots, \hat{\beta}_k)^\top$, where the subscript -0 means that the intercept $\hat{\beta}_0$ of the estimated first-order polynomial vanishes in the estimated gradient, i.e., $\hat{\boldsymbol{\beta}} = (\hat{\beta}_0, \hat{\boldsymbol{\beta}}_{-0})^{\mathsf{T}}$ with $\hat{\boldsymbol{\beta}}$ defined in (4.3). Obviously, the white-noise assumption implies

$$\mathrm{Cov}(\hat{\boldsymbol{\beta}}) = \sigma_w^2(\mathbf{Z}^{\mathsf{T}}\mathbf{Z})^{-1} = \sigma_w^2 \begin{pmatrix} a & \mathbf{b}^{\mathsf{T}} \\ \mathbf{b} & \mathbf{C} \end{pmatrix}, \tag{4.5}$$

where σ_w^2 denotes the variance of the simulation output w; $\mathbf{Z}$ is the $N \times (1+k)$ matrix of explanatory regression variables including the column with N one's; $N = \sum_{i=1}^{n} m_i$ is the total number of simulation runs; n is the number of different simulated input combinations; m_i is the number of IID. replications for combination i; a is a scalar; $\mathbf{b}$ is a k-dimensional vector; and $\mathbf{C}$ is a $k \times k$ matrix such that $\mathrm{Cov}(\hat{\boldsymbol{\beta}}_{-0}) = \sigma_w^2\mathbf{C}$.

We estimate the variance σ_w^2 in (4.5) through the mean squared residuals (MSR):

$$\hat{\sigma}_w^2 = \frac{\sum_{i=1}^{n}\sum_{r=1}^{m_i}(w_{i;r} - \hat{y}_i)^2}{N - (k+1)}, \tag{4.6}$$

where $\hat{y}_i = \mathbf{z}_i^{\mathsf{T}}\hat{\boldsymbol{\beta}}$; also see [28].

We can easily prove that the predictor variance $\mathrm{Var}[\hat{y}|\mathbf{z}]$ increases as $\mathbf{z}$—the point to be predicted—moves further away from the local area where the gradient is estimated. The point with the minimum predictor variance is $-\mathbf{C}^{-1}\mathbf{b}$, where $\mathbf{C}$ and $\mathbf{b}$ were defined below (4.5). ASD means that the new point to be simulated is

$$\mathbf{d} = -\mathbf{C}^{-1}\mathbf{b} - \lambda\mathbf{C}^{-1}\hat{\boldsymbol{\beta}}_{-0},$$

where $-\mathbf{C}^{-1}\mathbf{b}$ is the point where the local search starts, i.e., the point with minimum local variance; λ is the step size; and $\mathbf{C}^{-1}\hat{\boldsymbol{\beta}}_{-0}$ is the classic steepest descent direction $\hat{\boldsymbol{\beta}}_{-0}$ adapted for $\mathrm{Cov}(\hat{\boldsymbol{\beta}}_{-0})$.

Accounting for $\mathrm{Cov}(\hat{\boldsymbol{\beta}}_{-0})$ gives a scale-independent search direction. Experimental results are presented in [29, 30]. These results show that ASD performs "better" than classic steepest descent, i.e., the angle between the search direction based on the true $\boldsymbol{\beta}_{-0}$ and the search direction estimated in ASD is smaller. In one example, this angle reduces from 89.87 for classic steepest descent to 1.83 for ASD.

4.5 Multiple Responses: Generalized RSM

In practice, simulation models have multiple responses (multivariate output), e.g., many realistic inventory models require the inventory system to minimize the sum of all inventory costs excluding the hard to quantify out-of-stock costs, and to result in at least a guaranteed lower threshold for the service rate or fill rate. Simulation software facilitates the collection of multiple outputs. There are several approaches to solve the resulting issues; see the survey in [39]. The RSM literature also offers several approaches for such situations; see the surveys in [1, 24, 35].

A novel heuristic is generalized RSM (GRSM) that combines the estimated gradients of RSM with a search procedure in mathematical programming. GRSM selects one response as the goal variable and the other responses as constrained variables; moreover, the deterministic input variables may also be subjected to constraints. GRSM generalizes the steepest descent search direction through the affine scaling search direction, borrowing ideas from interior point methods (a variation on Karmarkar's algorithm) in mathematical programming; see [3]. This search moves faster to the optimum than steepest descent, because the former avoids creeping along the boundary of the feasible area, which is determined by the constraints on the random outputs and the deterministic inputs. Moreover, this search tries to stay inside the feasible area, so the simulation program does not crash. Finally, this search is scale independent; see the proof in [2].

Formally, GRSM extends the classic RSM problem in (4.1) to the following constrained nonlinear random optimization problem:

$$\min_{\mathbf{z}} \mathsf{E}[w_0|\mathbf{z}] \tag{4.7}$$

such that the other (say) $(r-1)$ random outputs satisfy the constraints

$$\mathsf{E}[w_{h'}|\mathbf{z}] \geq a_{h'},\ h' = 1, \ldots, r-1, \tag{4.8}$$

and the k deterministic inputs z_j satisfy the box constraints

$$l_j \leq z_j \leq u_j,\ j = 1, \ldots, k. \tag{4.9}$$

An example is an inventory simulation, in which the sum of the expected inventory carrying costs and ordering costs should be minimized while the expected service percentage should be at least (say) 90 % so $a_1 = 0.9$ in (4.8); both the reorder quantity $z_1 = Q$ and the reorder level $z_2 = s$ should be non-negative so $z_1 \geq 0$ and $z_2 \geq 0$ in (4.9). Note that more complicated input constraints than (4.9)—namely, linear budget constraints—feature in the call-center simulation in [23]; input constraints resulting from output constraints are discussed in [35].

Analogously to the first steps of classic RSM based on (4.2), GRSM locally approximates the multivariate I/O function by r univariate first-order polynomials augmented with white noise:

$$\mathbf{y}_h = \mathbf{Z}\boldsymbol{\beta}_h + e_h,\ h = 0, \ldots r-1. \tag{4.10}$$

Like RSM, GRSM assumes that locally the white noise assumption holds for (4.10), so the BLUEs are the following LS estimators:

$$\hat{\boldsymbol{\beta}}_h = (\mathbf{Z}^{\mathsf{T}}\mathbf{Z})^{-1}\mathbf{Z}^{\mathsf{T}}\mathbf{w}_h,\ h = 0, \ldots, r-1. \tag{4.11}$$

The vector $\hat{\boldsymbol{\beta}}_0$ (LS estimator of first-order polynomial approximation of goal function) and the goal function (4.7) result in

$$\min_{\mathbf{z}} \hat{\boldsymbol{\beta}}_{0;-0}\mathbf{z}, \tag{4.12}$$

where $\hat{\boldsymbol{\beta}}_{0;-0} = (\hat{\beta}_{0;1}, \ldots, \hat{\beta}_{0,k})^{\mathsf{T}}$ is the LS estimate of the local gradient of the goal function. The $(r-1)$ estimates $\hat{\beta}_{h'}$ in (4.11) combined with the original output constraints (4.8) give

$$\hat{\boldsymbol{\beta}}_{h';-0}^{\mathsf{T}}\mathbf{z} \geq c_{h'},\ h' = 1, \ldots, r-1, \tag{4.13}$$

where $\hat{\boldsymbol{\beta}}_{h';-0} = (\hat{\beta}_{h';1}, \ldots, \hat{\beta}_{h',k})^{\mathsf{T}}$ is the estimated local gradient of constraint function h', and $c_{h'} = a_{h'} - \hat{\beta}_{h';0}$ is the modified right-hand side of this constraint function. The box constraints (4.9) remain unchanged.

Now the k-dimensional vectors $\hat{\boldsymbol{\beta}}_{h';-0}$ $(h' = 1, \ldots, r-1)$ in (4.13) are collected in the $(r-1) \times k$ matrix (say) $\mathbf{B}$. Likewise, we collect the $(r-1)$ elements $c_{h'}$ in the vector $\mathbf{c}$. We define $\mathbf{l}$ as the vector with the k elements l_j, and $\mathbf{u}$ as the vector with the elements u_j. Finally, we introduce the k-dimensional vectors with the non-negative slack variables $\mathbf{s}$, $\mathbf{r}$, and $\mathbf{v}$, to get the following equivalent problem formulation:

$$\begin{array}{ll} \text{minimize} & \hat{\boldsymbol{\beta}}_{0;-0}^{\mathsf{T}}\mathbf{z} \\ \text{subjectto} & \mathbf{Bz} - \mathbf{s} = \mathbf{c} \\ & \mathbf{z} + \mathbf{r} = \mathbf{u} \\ & \mathbf{z} - \mathbf{v} = \mathbf{l}. \end{array} \tag{4.14}$$

This optimization problem is linear in the inputs $\mathbf{z}$. GRSM uses this problem to derive the following search direction:

$$\mathbf{d} = -\left(\mathbf{B}^{\mathsf{T}}\mathbf{S}^{-2}\mathbf{B} + \mathbf{R}^{-2} + \mathbf{V}^{-2}\right)^{-1} \hat{\boldsymbol{\beta}}_{0;-0}, \tag{4.15}$$

where $\mathbf{S}$, $\mathbf{R}$, and $\mathbf{V}$ are diagonal matrixes with as main-diagonal elements the current estimated slack vectors $\mathbf{s}$, $\mathbf{r}$, and $\mathbf{v}$ in (4.14). Note that $\hat{\boldsymbol{\beta}}_{0;-0}$ in (4.15) is the estimated classic steepest ascent direction. As the value of a slack variable in (4.15) decreases so the corresponding constraint gets tighter, the GRSM search direction deviates more from the steepest descent direction. Possible singularity of the various matrices in (4.15) is discussed in [1].

Following the path defined by (4.15), GRSM must decide on the step size (say) λ along this path. GRSM chooses

$$\lambda = 0.8 \min_{h'} \left[\frac{c_{h'} - \hat{\boldsymbol{\beta}}_{h';-0}^{\mathsf{T}}\mathbf{z}_c}{\hat{\boldsymbol{\beta}}_{h';-0}^{\mathsf{T}}\mathbf{d}} \right], \tag{4.16}$$

where the factor 0.8 is chosen to decrease the probability that the local metamodel (4.13) is misleading when applied globally; $\mathbf{z}_c$ denotes the current input combination, so the new combination becomes $\mathbf{z}_c + \lambda\mathbf{d}$. The box constraints (4.9) for the deterministic inputs hold globally, so it is easy to check whether the new combination $\mathbf{z}_c + \lambda\mathbf{d}$ violates these constraints.

Analogously to classic RSM, GRSM proceeds stepwise. After each step along the search path, GRSM tests the following two hypotheses denoted by the superscripts (1) and (2):

1. Pessimistic null-hypothesis: $w_0(\mathbf{z}_c + \lambda\mathbf{d})$ (output of new combination) is no improvement over $w_0(\mathbf{z}_c)$ (output of old combination):

$$H_0^{(1)} : \mathsf{E}[w_0(\mathbf{z}_c + \lambda\mathbf{d})] \geq \mathsf{E}[w_0(\mathbf{z}_c)]. \tag{4.17}$$

2. This step is feasible, i.e., $w_{h'}(\mathbf{z}_c + \lambda\mathbf{d})$ satisfies the $(r-1)$ constraints in (4.8):

$$H_0^{(2)} : \mathsf{E}[w_{h'}(\mathbf{z}_c + \lambda\mathbf{d})] \geq a_{h'},\ h' = 1, \ldots, r-1. \tag{4.18}$$

To test these hypotheses, we may apply the following simple statistical procedures (more complicated parametric bootstrapping is used in [1], permitting non-normality and testing the relative improvement $w_0(\mathbf{z}_c + \lambda\mathbf{d})/w_0(\mathbf{z}_c)$ and the relative slacks $s_{h'}(\mathbf{z}_c + \lambda\mathbf{d})/s_{h'}(\mathbf{z}_c)$). To test (4.17), we apply the paired Student t_{m-1} statistic if CRN are used; we reject the hypothesis if significant improvement is observed. To test (4.18), we again apply a t_{m-1} statistic; because we test multiple hypotheses, we apply Bonferroni's inequality so we divide the classic α value by $(r-1)$ (number of tests).

Actually, a better solution may lie somewhere between $\mathbf{z}_c$ (old combination) and $\mathbf{z}_c + \lambda\mathbf{d}$ (new combination at maximum step size). Therefore, GRSM uses binary search, i.e., it simulates a combination that lies halfway between these two combinations—and is still on the search path. This halving of the stepsize may be applied several times.

Next, GRSM proceeds analogously to classic RSM, i.e., around the best combination found so far, GRSM selects a new local area. Again, a R-III design specifies the new simulation input combinations, and r first-order polynomials are fitted, which gives a new search direction. Note that we might use the m replications $\hat{\boldsymbol{\beta}}_r$ to estimate the accuracy of the search direction; to test the accuracy of the estimated optimum, we present a test in the next section.

Figure 4.1 illustrates GRSM for a problem with simple known test functions; obviously, in practice, we use simulation to estimate the true outputs of the various implicit I/O functions of the simulation model. This figure shows two inputs; see the two axes labeled z_1 and z_2. The goal function is to be minimized; the figure also shows two contour plots or iso-costs functions: $\mathsf{E}(w_0) = a_{0;1}$ and $\mathsf{E}(w_0) = a_{0;2}$ with $a_{0;2} < a_{0;1}$. There are two constrained random outputs; see $\mathsf{E}(w_1) = a_1$ and $\mathsf{E}(w_2) = a_2$. The search starts in the lower-right local area with a 2^2 design; see the four elongated points. Together with the replications that are not shown, the I/O data give a search direction; see the arrow leaving from point (0). The maximum step size

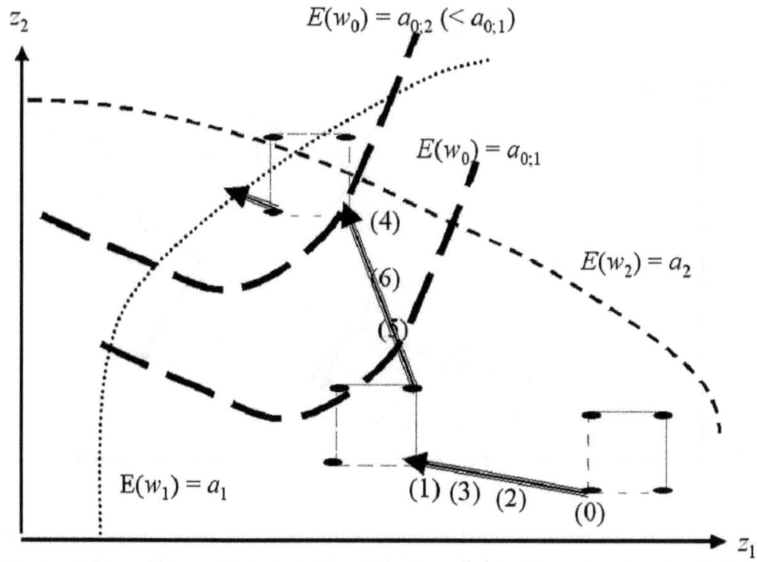

Fig. 4.1 GRSM illustration with two inputs, two contour plots for the goal output, two constraints for the other outputs, three local areas, three search directions, and six steps in these directions

along this path takes the search from point (0) to point (1). The binary search takes the search back to point (2), and next to point (3). Because the best point so far turns out to be point (1), the 2^2 design is used to select four points to be simulated in this local area; this point is one of the four points. This design gives a new search direction, which indeed avoids the boundary. The maximum step-size now takes the search to point (4). The binary search takes the search back to point (5), and next to point (6). Because the best point so far is now point (4), the 2^2 design is simulated in the local area with this point as one of its points. A new search direction is estimated, and the procedure continues.

Two GRSM examples are detailed in [1], illustrating and evaluating GRSM. One example is an inventory simulation specified in [7], which has a service-level constraint; no analytical solution is known. The other example is a test function with a known solution. The results for these examples are encouraging, as GRSM finds solutions that are both feasible and give low values for the goal functions.

4.6 Testing an Estimated Optimum in GRSM: KKT Conditions

Obviously, it is uncertain whether the optimum estimated by a heuristic such as GRSM is close enough to the true optimum. In deterministic nonlinear mathematical programming, the first-order necessary optimality conditions are known as the

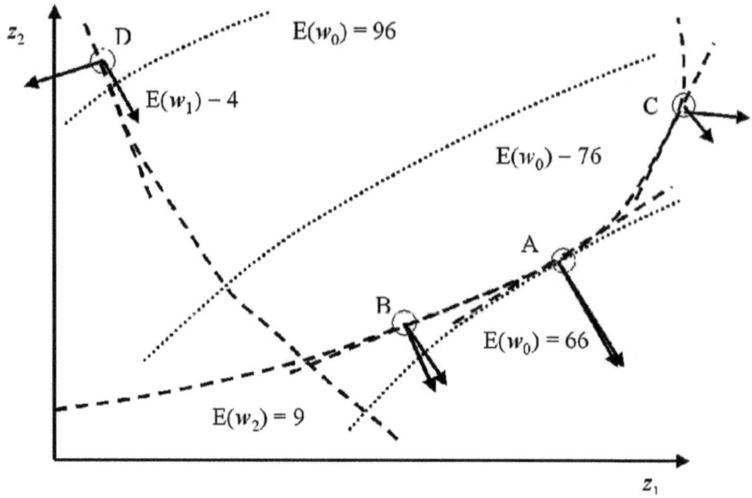

Fig. 4.2 A constrained nonlinear random optimization problem: three contour plots with goal values 66, 76, and 96; two other outputs with lower bounds 4 and 9; optimal point A; points B and C on bound 9; point D on bound 4; local gradients at A through D for goal function and binding constraint, perpendicular to local tangent lines for binding constraint

Karush–Kuhn–Tucker (KKT) conditions; see [22]. First we present the basic idea behind these conditions; next, we explain how to test these conditions in simulation.

Figure 4.2 illustrates the same type of problem as the one in Fig. 4.1. More precisely, Fig. 4.2 shows a goal function $\mathrm{E}(w_0)$ with the three contour plots that correspond with the values 66, 76, and 96; also see (4.7). Furthermore, the figure shows two constrained simulation outputs, namely $\mathrm{E}(w_1) \geq 4$ and $\mathrm{E}(w_2) \geq 9$; also see (4.8). The figure displays the boundaries of the feasible area, which is determined by $\mathrm{E}(w_1) = 4$ and $\mathrm{E}(w_2) = 9$. The optimum combination is point A. The points B and C lie on the boundary $\mathrm{E}(w_2) = 9$; point D lies on the boundary $\mathrm{E}(w_1) = 4$. Obviously, the points A and D lie far away from each other. The figure also displays the local gradients at these four points for the goal function and for the binding constraint, i.e., the constraint with a zero slack value in (4.8). These gradients are perpendicular to the local tangent lines; those lines are shown only for the binding constraint—not for the goal function.

Let $\mathbf{z}^0$ denote a local minimizer for the deterministic variant of our problem. The KKT conditions for $\mathbf{z}^0$ are then

$$\begin{aligned} &\boldsymbol{\beta}_{0;-0} = \sum_{h \in A\left(\mathbf{z}^0\right)} \lambda_h^0 \boldsymbol{\beta}_{h;-0}, \\ &\lambda_h^0 \geq 0, \\ &h \in A\left(\mathbf{z}^0\right), \end{aligned} \tag{4.19}$$

where $\boldsymbol{\beta}_{0;-0}$ denotes the k-dimensional vector with the gradient of the goal function, as we saw in (4.12); $A\left(\mathbf{z}^0\right)$ is the index set with the indices of the constraints that are binding at $\mathbf{z}^0$; λ_h^0 is the Lagrangian multiplier for binding constraint h; $\boldsymbol{\beta}_{h;-0}$ is the gradient of the output in that binding constraint; Fig. 4.2 has only one binding constraint at the points A through D. Altogether, (4.19) implies that the gradient of the objective is a nonnegative linear combination of the gradients of the binding constraints, at $\mathbf{z}^0$. Figure 4.2 shows that A satisfies (4.19); B has two gradients that point in different but similar directions—and so does C—whereas D has two gradients that point in completely different directions.

Note: If the optimum occurs inside the feasible area, then there are no binding constraints so the KKT conditions reduce to the condition that the goal gradient is zero. Classic RSM includes tests for a zero gradient estimated from a second-order polynomial; see again Sect. 4.2.

In simulation we must estimate the gradients; moreover, to check which constraints are binding, we must estimate the slacks of the constraints. This estimation changes the KKT conditions into a problem of nonlinear statistics. An asymptotic test is presented in [1] , but a small-sample bootstrap test is presented in [10]. We present the latter test, because it is simpler and it allows expensive simulation. Readers not interested in the technical details of this bootstrapping should skip the remainder of this section.

As in classic RSM, we assume locally constant (co)variances for each of the r simulation outputs (when moving to a new local area, the (co)variances may change, e.g., the points A through D in Fig. 4.2 do not have the same variance for the goal output). LS per univariate simulation output gives $\hat{\boldsymbol{\beta}}_h$ ($h = 0, 1, \ldots, r-1$) defined in (4.11). These estimators have the following estimated covariance matrix:

$$\widehat{\mathrm{Cov}}(\hat{\boldsymbol{\beta}}_h, \hat{\boldsymbol{\beta}}_{h'}) = \widehat{\mathrm{Cov}}(w_h, w_{h'}) \otimes (\mathbf{Z}^{\mathsf{T}}\mathbf{Z})^{-1} (h, h' = 0, \ldots, r-1), \tag{4.20}$$

where $\otimes$ denotes the Kronecker product and $\widehat{\mathrm{Cov}}(w_h, w_{h'})$ is an $r \times r$ matrix with the classic estimators of the (co)variances based on the m replications at the local center:

$$\widehat{\mathrm{Cov}}(w_h, w_{h'}) = (\hat{\sigma}_{h;h'}) = (\sum_{l=1}^{m} (w_{h;l} - \overline{w}_h)(w_{h';l} - \overline{w}_{h'}))\frac{1}{m-1}. \tag{4.21}$$

The Kronecker product implies that $\widehat{\mathrm{Cov}}(\hat{\boldsymbol{\beta}}_h, \hat{\boldsymbol{\beta}}_{h'})$ is an $rq \times rq$ matrix with q denoting the number of regression parameters (e.g., $q = 1 + k$ in a first-order polynomial); this matrix is formed from the $r \times r$ matrix $\widehat{\mathrm{Cov}}(w_h, w_{h'})$ by multiplying each of its elements by the entire $q \times q$ matrix $(\mathbf{Z}^{\mathsf{T}}\mathbf{Z})^{-1}$ (e.g., $\mathbf{Z}$ is an $N \times (1+k)$ matrix in Eq. (4.5)). The matrix $\widehat{\mathrm{Cov}}(w_h, w_{h'})$ is singular if $m \leq r$, e.g., the case study in [26] has $r = 2$ response types and $k = 14$ inputs so $m \geq 3$ replications of the center point are required. Of course, the higher m is, the higher is the power of the tests that use these replications. Cases with all n local points replicated or with CRN are not considered in [10]; these cases require further research.

In classic RSM we assume that the output is Gaussian, and now we assume that the r-variate simulation output is multivariate Gaussian. We use the center point to test whether a constraint is binding in the current local area, because this point is more representative of the local behavior than the points of the R-III design. (To save simulation runs, we should start a local experiment at its center point including replications; if it turns out that either no constraint is binding or at least one constraint is violated, then we need not test the other hypotheses so we do not simulate the remainder of the local design.) Actually, we test the following three null-hypotheses, denoted by the superscripts (1) through (3):

1. The current solution is fcasiblc and at least one constraint is binding; see (4.8):

$$H_0^{(1)} : \mathsf{E}[w_{h'}|\mathbf{x}=\mathbf{0}] = a_{h'},\ h' = 1,\ldots,r-1, \tag{4.22}$$

   where $\mathbf{x} = \mathbf{0}$ corresponds with the center of the local area expressed in the standardized inputs.
2. The expected value of the estimated goal gradient may be expressed as the expected value of a *linear* combination of the estimated gradients of the simulation outputs in the binding constraints, i.e., in (4.19) we replace the deterministic quantities by their random estimators:

$$H_0^{(2)} : \mathsf{E}[\hat{\boldsymbol{\beta}}_{0;-0}] = \mathsf{E}\left[\sum_{h\in A(\mathbf{z}^0)} \hat{\boldsymbol{\lambda}}_h^0 \hat{\boldsymbol{\beta}}_h\right]. \tag{4.23}$$

3. The Lagrangian multipliers in (4.23) are non-negative:

$$H_0^{(3)} : \mathsf{E}[\hat{\boldsymbol{\lambda}}] \geq \mathbf{0}. \tag{4.24}$$

Each of these three hypotheses requires multiple tests, so we apply Bonferroni's inequality. Moreover, we test these three hypotheses sequentially, so it is hard to control the final type-I and type-II error probabilities (classic RSM has the same type of problem, but has nevertheless acquired a track record in practice).

Sub 1: To test (4.22), we use the classic t statistic:

$$t_{m-1}^{(h')} = \frac{\overline{w}_{h'}(\mathbf{x}=\mathbf{0}) - a_{h'}}{\sqrt{\hat{\sigma}_{h';h'}/m}},\ h' = 1,\ldots,r-1, \tag{4.25}$$

where both the numerator and the denominator use the m replications at the local center point; see (4.21). This t statistic may give the following three results:

1. The statistic is significantly positive, i.e., the constraint for output h' is not binding. If none of the $(r-1)$ constraints is binding, then the optimal solution

is not yet found, assuming that at the optimum at least one constraint is binding; otherwise, classic RSM applies. The search for better solutions continues (see again Sect. 4.5).

2. The statistic is significantly negative, i.e., the current local area does not give feasible solutions so the optimal solution is not yet found. The search should back-up into the feasible area.
3. The statistic is non-significant, i.e., the current local area gives feasible solutions, and the constraint for output h' is binding. The index of this gradient is then included in $A\left(\mathbf{z}^0\right)$; see (4.23). And the KKT test proceeds as follows.

Sub 2 and 3: To estimate the linear combination in (4.23), we apply LS with as explanatory variables the estimated gradients of the (say) J binding constraints; these explanatory variables are now random. We collect these gradients in the $k \times J$ matrix $\hat{\mathbf{B}}_{J;-0}$. The parameters estimated through LS are $\hat{\boldsymbol{\lambda}}$. Let $\hat{\hat{\boldsymbol{\beta}}}_{0;-0}$ denote the LS estimator of the goal gradient:

$$\hat{\hat{\boldsymbol{\beta}}}_{0;-0} = \hat{\mathbf{B}}_{J;-0}(\hat{\mathbf{B}}_{J;-0}^{\mathsf{T}}\hat{\mathbf{B}}_{J;-0})^{-1}\hat{\mathbf{B}}_{J;-0}^{\mathsf{T}}\hat{\boldsymbol{\beta}}_{0;-0} = \hat{\mathbf{B}}_{J;-0}\hat{\boldsymbol{\lambda}} \tag{4.26}$$

with $\hat{\boldsymbol{\lambda}} = (\hat{\mathbf{B}}_{J;-0}^{\mathsf{T}}\hat{\mathbf{B}}_{J;-0})^{-1}\hat{\mathbf{B}}_{J;-0}^{\mathsf{T}}\hat{\boldsymbol{\beta}}_{0;-0}$. To quantify the validity of this linear approximation, we use the k-dimensional vector of its residuals

$$\hat{\mathbf{e}}(\hat{\hat{\boldsymbol{\beta}}}_{0;-0}) = \hat{\hat{\boldsymbol{\beta}}}_{0;-0} - \hat{\boldsymbol{\beta}}_{0;-0}. \tag{4.27}$$

Hypothesis (4.23) implies $\mathsf{E}[\hat{\mathbf{e}}(\hat{\hat{\boldsymbol{\beta}}}_{0;-0})] = \mathbf{0}$. This hypothesis involves a product of multivariates, so standard tests do not apply and we use bootstrapping. We do not apply distribution-free bootstrapping, because in expensive simulation only the center point is replicated a few times. Instead, we apply parametric bootstrapping, i.e., we assume a Gaussian distribution (like in classic RSM), and we estimate its parameters from the simulation's I/O data. This bootstrapping consists of the following four steps, where the superscript $*$ denotes a bootstrapped value:

1. Use the Monte Carlo method to sample $vec(\hat{\boldsymbol{\beta}}_{0;-0}^{*}, \hat{\mathbf{B}}_{J;-0}^{*})$, which is a $(k+kJ)$-dimensional vector formed by stapling (stacking) the k-dimensional goal gradient vector and the J k-dimensional vectors of the $k \times J$ matrix $\hat{\mathbf{B}}_{J;-0}^{*}$:

$$vec(\hat{\boldsymbol{\beta}}_{0;-0}^{*}, \hat{\mathbf{B}}_{J;-0}^{*}) \sim \mathcal{N}(vec(\hat{\boldsymbol{\beta}}_{0;-0}, \hat{\mathbf{B}}_{J;-0}), \widehat{\mathrm{Cov}}(vec(\hat{\boldsymbol{\beta}}_{0;-0}, \hat{\mathbf{B}}_{J;-0}))), \tag{4.28}$$

where $\widehat{\mathrm{Cov}}(vec(\hat{\boldsymbol{\beta}}_{0;-0}, \hat{\mathbf{B}}_{J;-0}))$ is the $(k+kJ) \times (k+kJ)$ matrix computed through (4.20).

2. Use the bootstrap values resulting from Step 1 to compute the LS estimate of the bootstrapped goal gradient using the bootstrapped gradients of the binding constraints as explanatory variables, i.e., use (4.26) adding the superscript $*$ to all random variables resulting in $\hat{\hat{\boldsymbol{\beta}}}_{0;-0}^{*}$ and $\hat{\boldsymbol{\lambda}}^{*}$.

3. Use $\hat{\hat{\boldsymbol{\beta}}}^*_{0;-0}$ from Step 2 and $\hat{\boldsymbol{\beta}}^*_{0;-0}$ from Step 1 to compute the bootstrap residual $\hat{\mathbf{e}}(\hat{\hat{\boldsymbol{\beta}}}^*_{0;-0}) = \hat{\hat{\boldsymbol{\beta}}}^*_{0;-0} - \hat{\boldsymbol{\beta}}^*_{0;-0}$, analogous to (4.27). Determine whether any of the bootstrapped Lagrangian multipliers $\hat{\boldsymbol{\lambda}}^*$ found in Step 2 is negative, i.e., augment a counter (say) c^* with the value 1 if this event occurs.
4. *Repeat* the preceding three steps (say) 1,000 times. This bootstrap sample gives the EDF of $\hat{\mathbf{e}}(\hat{\hat{\boldsymbol{\beta}}}^*_{0;-0;j})$—which denotes the bootstrapped residuals per input j ($j = 1,\ldots,k$)—and the final value of the counter c^*. Reject the hypothesis in (4.23) if this EDF implies a two-sided $(1-\alpha/(2k))$ confidence interval that does not cover the value 0, where the factor k is explained by Bonferroni's inequality. Reject (4.24) if the fraction $c^*/1000$ is significantly higher than 50 %; if the true Lagrangian multiplier is only slightly larger than zero, then nearly 50 % of the bootstrapped values is negative. To test the latter fraction, we approximate the binomial distribution through the normal distribution with mean 0.50 and variance $(0.50 \times 0.50)/1000 = 0.00025$.

The numerical examples in [10] give encouraging results, i.e., the classic t test for zero slacks performs as expected and the new bootstrap tests give observed type-I error rates close to the prespecified (nominal) rates, while the type-II error rate decreases as the tested input combination is farther away from the true optimum (see A through D in Fig. 4.2).

4.7 Robust Optimization

Taguchi emphasizes that in practice some inputs of a product are under complete control, whereas other inputs are not, e.g., the design of a car engine is completely controlled by the engineers, but the driving style is not (see again [41]). Consequently, a car design that allows some flexibility in its use is better, e.g., a car optimized only for the race circuit does not perform well in the city streets. Likewise, in simulation the perceived optimal solution may be far from optimal due to ignoring uncertainties in some inputs, e.g., the nominally optimal decision on the inventory control limits s (reorder level) and S (order-up-to level) may perform poorly if the solution ignores the uncertainty in the parameters assumed for the random demand and delivery time distributions. We first explain Taguchi's approach, updating and extending [17]; also see [18]. Then we briefly discuss the approach pioneered by Ben-Tal et al., which is detailed in [8, 9].

We see the following major differences between the two approaches. Originally, Ben-Tal et al. assumed static deterministic linear problems solved by linear programming (LP), whereas we assume dynamic random nonlinear problems solved by simulation. Ben-Tal et al. assume that uncertainty implies that the coefficients of the LP problem lie in a mathematical set called the *uncertainty set*, whereas we assume that the parameters of some variables of the simulation model (e.g., the demand distribution) have a given statistical distribution (e.g., demand has a Poisson distribution with a parameter λ estimated from historical data, so the estimated parameter $\hat{\lambda}$ has a Gaussian distribution with mean $\hat{\lambda}$ and standard deviation $\hat{\sigma}_{\hat{\lambda}}$). Currently, Ben-Tal et al. also consider multi-stage nonlinear problems and uncertainty sets based on historical data.

Taguchi's Robust Optimization

We use Taguchi's view of the world, distinguishing between two types of inputs:

(a) decision variables, denoted by (say) d_j $(j = 1, \ldots, k)$, so $\mathbf{d} = (d_1, \ldots, d_k)^{\mathsf{T}}$, and
(b) environmental or noise factors e_g $(g = 1, \ldots, c)$ so $\mathbf{e} = (e_1, \ldots, e_c)^{\mathsf{T}}$.

Taguchi assumes a single output, which we denote by w. He focuses on the mean μ_w and the variance σ_w^2 of this output, caused by the noise factors $\mathbf{e}$. However, we do not use Taguchi's scalar loss function such as the signal-to-noise or mean-to-variance ratio μ_w/σ_w^2; see [33, pp. 486–488]. Instead, we use both μ_w and σ_w^2 to characterize the statistical distribution of the output, and we try to solve the following problem:

$$\min \mu_w \text{ such that } \sigma_w^2 \leq T, \tag{4.29}$$

where T is some threshold; also see [33, pp. 488–495]. We also refer to the surveys on robust optimization in [11, 36].

Taguchi's worldview has been very successful in production engineering, but statisticians have seriously criticized his statistical techniques; see the panel report in [34]. Therefore, in [33, pp. 502–506], Taguchi's worldview is combined with RSM, which we adapt. We assume that $\mathbf{e}$ has the mean $\boldsymbol{\mu}_{\mathbf{e}}$ and the covariance matrix $\boldsymbol{\Omega}_{\mathbf{e}}$, whereas [33] assumes a constant variance σ_e^2, so $\boldsymbol{\Omega}_{\mathbf{e}} = \sigma_e^2\mathbf{I}$. To find a robust solution, [33] superimposes contour plots for the mean and variance of the output, whereas we use more general and flexible mathematical programming. This mathematical programming, however, requires specification of threshold values like T in (4.29); managers may find it hard to select specific values, so we may try different values and estimate the corresponding Pareto-optimal efficiency frontier. To estimate the variability of this frontier resulting from the various estimators, we may use bootstrapping (also see our bootstrapping in the preceding section). For details on our adaptation of the approach in [33], we refer to [17].

More precisely, [33] fits a second-order polynomial for $\mathbf{d}$ that is to be optimized. To model possible effects of $\mathbf{e}$, [33] fits a first-order polynomial in $\mathbf{e}$. Moreover, "control-by-noise" two-factor interactions are also estimated. Altogether, [33] fits the following "incomplete" second-order polynomial:

$$y = \beta_0 + \sum_{j=1}^{k} \beta_j d_j + \sum_{j=1}^{k} \sum_{j'=j}^{k} \beta_{j;j'} d_j d_{j'} + \sum_{g=1}^{c} \gamma_g e_g + \sum_{j=1}^{k} \sum_{g=1}^{c} \delta_{j;g} d_j e_g + \varepsilon$$

$$= \beta_0 + \boldsymbol{\beta}^\mathsf{T}\mathbf{d} + \mathbf{d}^\mathsf{T}\mathbf{B}\mathbf{d} + \boldsymbol{\gamma}^\mathsf{T}\mathbf{e} + \mathbf{d}^\mathsf{T}\boldsymbol{\Delta}\mathbf{e} + \varepsilon, \tag{4.30}$$

where we now use the symbol ε (instead of e) to denote the regression residual ε with $\mu_\varepsilon = 0$ if this metamodel has no lack-of-fit and with constant variance σ_ε^2; furthermore, $\boldsymbol{\beta} = (\beta_1, \ldots, \beta_k)^\mathsf{T}$, $\mathbf{B}$ denotes the $k \times k$ symmetric matrix with main-diagonal elements $\beta_{j;j}$ and off-diagonal elements $\beta_{j;j'}/2$, $\boldsymbol{\gamma} = (\gamma_1, \ldots, \gamma_c)^\mathsf{T}$, and $\boldsymbol{\Delta}$ denotes the $k \times c$ matrix with interactions $\delta_{j;g}$.

Clearly, (4.30) implies the following regression predictor for the true mean μ_w:

$$\mu_y = \beta_0 + \boldsymbol{\beta}^\mathsf{T}\mathbf{d} + \mathbf{d}^\mathsf{T}\mathbf{B}\mathbf{d} + \boldsymbol{\gamma}^\mathsf{T}\boldsymbol{\mu}_\mathbf{e} + \mathbf{d}^\mathsf{T}\boldsymbol{\Delta}\boldsymbol{\mu}_\mathbf{e}. \tag{4.31}$$

And the regression predictor for the true variance σ_w^2 is

$$\sigma_y^2 = (\boldsymbol{\gamma}^\mathsf{T} + \mathbf{d}^\mathsf{T}\boldsymbol{\Delta})\boldsymbol{\Omega}_\mathbf{e}(\boldsymbol{\gamma} + \boldsymbol{\Delta}^\mathsf{T}\mathbf{d}) + \sigma_\varepsilon^2 = \mathbf{l}^\mathsf{T}\boldsymbol{\Omega}_\mathbf{e}\mathbf{l} + \sigma_\varepsilon^2, \tag{4.32}$$

where $\mathbf{l} = (\boldsymbol{\gamma} + \boldsymbol{\Delta}^\mathsf{T}\mathbf{d}) = (\partial y/\partial e_1, \ldots, \partial y/\partial e_c)^\mathsf{T}$, so $\mathbf{l}$ is the gradient with respect to $\mathbf{e}$. Consequently, the larger the gradient's elements are, the larger σ_y^2. Furthermore, if there are no control-by-noise interactions so that $\boldsymbol{\Delta} = \mathbf{0}$, then we cannot control σ_y^2 through $\mathbf{d}$.

To enable estimation of the regression parameters in (4.30), we use a crossed design, i.e., we combine the design for $\mathbf{d}$ and the design for $\mathbf{e}$—as is usual in a Taguchian approach. To estimate the optimal $\mathbf{d}$, we use a CCD; see the discussion below (4.4). For the first-order polynomial in $\mathbf{e}$, we use a R-III design; see the example in Table 4.1. The combination of these two designs obviously enables the estimation of the two-factor interactions $\delta_{j;g}$. Note that designs that are more efficient than crossed designs are discussed in [17, 33].

To use linear regression analysis for the estimation of the q parameters in (say) $\boldsymbol{\zeta} = (\beta_0, \ldots, \delta_{k;c})^\mathsf{T}$ in (4.30), we reformulate (4.30) as

$$y = \boldsymbol{\zeta}^\mathsf{T}\mathbf{x} + \varepsilon, \tag{4.33}$$

where $\mathbf{x}$ is defined in the obvious way, e.g., the element corresponding with $\beta_{1;2}$ (interaction between d_1 and d_2) is $d_1 d_2$. Note that (4.33) is linear in $\boldsymbol{\zeta}$, but (4.30) is not linear in $\mathbf{d}$. Then (4.33) gives the LS estimator

$$\hat{\zeta} = (\mathbf{X}^\mathsf{T}\mathbf{X})^{-1}\mathbf{X}^\mathsf{T}\mathbf{w}, \tag{4.34}$$

where $\mathbf{X}$ is the $N \times q$ matrix of explanatory variables with $N = \sum_{i=1}^n m_i$ and n denoting the number of different simulated combinations of $\mathbf{d}$ and $\mathbf{e}$; $\mathbf{w}$ consists of the N simulation outputs. Obviously, the covariance matrix of $\hat{\zeta}$ is

$$\mathbf{cov}(\hat{\zeta}) = (\mathbf{X}^\mathsf{T}\mathbf{X})^{-1}\sigma_w^2. \tag{4.35}$$

The RSM metamodel (4.30) implies that σ_w^2 equals σ_ε^2. This variance is estimated by

$$MSR = \frac{(\hat{\mathbf{y}} - \mathbf{w})^{\mathsf{T}}(\hat{\mathbf{y}} - \mathbf{w})}{N - q}, \tag{4.36}$$

where $\hat{\mathbf{y}} = \hat{\boldsymbol{\zeta}}^{\mathsf{T}}\mathbf{x}$; also see (4.6).

To estimate μ_y, we simply plug $\hat{\zeta}$ defined by (4.34) into the right-hand side of (4.31), where $\mathbf{d}$ and $\boldsymbol{\mu}_{\mathbf{e}}$ are known. To estimate σ_y^2, we also plug $\hat{\zeta}$ into (4.32), where $\boldsymbol{\Omega}_{\mathbf{e}}$ is known. Note that (4.32) involves products of unknown parameters, so it implies a nonlinear estimator $\hat{\sigma}_y^2$; plugged-in estimators certainly create bias, but we ignore this bias.

Our final goal is to solve (4.29). We solve this constrained minimization problem through a mathematical programming solver, e.g., Matlab's "fmincon" [22]. This gives estimates of the robust decision variables and the corresponding mean and variance.

An example is detailed in [17], considering the economic order quantity (EOQ) for an environment with a demand rate that is uncertain but has a known distribution. This example demonstrates that if management prefers low variability of inventory costs, then they must pay a price, i.e., the expected costs increases. Furthermore, the classic EOQ assuming a known fixed demand rate and the robust EOQ do differ. More examples are referenced in [43].

Ben-Tal et al.'s Robust Optimization

Ben-Tal et al. emphasize that the nominal solution—which ignores the uncertainty in **e**—may easily violate the constraints in the given mathematical programming problem, so they derive a robust solution that gives a slightly worse value for the goal variable but increases the probability of satisfying the constraints. The mathematical challenge is to develop a computationally tractable "robust counterpart" of the original mathematical programming problem. Therefore, they propose a robust solution that is "immune" to variations within the *uncertainty set*. Given historical data on **e**, [43] derives a specific uncertainty set for **p**, which is the unknown density function of **e** that is compatible with the historical data on **e**. Technically, **p** belongs to this set with confidence $1 - \alpha$, assuming some phi-divergence measure such as the well-known chi-square distance. In this chapter we do not present the mathematical details of the derivation of tractable robust counterparts, but refer to [43].

Note that Taguchians assume a specific distribution for **e**, which implies $\boldsymbol{\mu}_{\mathbf{e}}$ and $\boldsymbol{\Omega}_{\mathbf{e}}$ in (4.31) and (4.32). This distribution may be estimated from historical data; however, an approach that uses only the original observed data on **e** is developed in [43], and several numerical examples demonstrate the effectiveness of this novel combination of the two approaches originated by Taguchi and Ben-Tal et al.

The examples in [43] include a deterministic simulation of the television example in [33, p. 512], and a random simulation of a distribution-center example in [40]. In this chapter we focus on simulation, so we discuss only the latter example. This example has five decision variables (e.g., number of forklifts) and two environmental variables (e.g., delay probabilities of suppliers); the response is the total throughput. An incomplete second-order polynomial like (4.30) is fitted. In [43], (4.29) is replaced by the following related problem:

$$\min \sigma_w^2 \text{ such that } \mu_w \leq T, \tag{4.37}$$

where the statistical parameters μ_w and σ_w^2 are based on the historical data using the phi-divergence criterion. These two examples in [43] again demonstrate that robust solutions may have better worst-case performance and also better average performance than "nominally" optimal solutions that ignore uncertainty.

4.8 Conclusions

In this chapter, we started with the basics of classic RSM, which minimizes the expected value of a single response variable in real-life experiments. Next we considered simulation experiments. We then added the adapted steepest descent search direction, which improves classic steepest descent. We also summarized GRSM for simulation with multivariate responses, assuming that one response is to be minimized while all the other responses and deterministic inputs must meet given constraints. Furthermore, we presented a bootstrap procedure for testing whether the KKT conditions hold for the estimated optimum. Finally, we considered robust optimization.

Future research may study the selection of the required number of replications and the use of replications to estimate the accuracy of the resulting estimated search direction or optimum. Bootstrapping may solve this problem, but more research is needed. Numerical evaluation of the adapted steepest descent method would benefit from more applications in practice. There is also a need for more research on the KKT testing procedure when all local points (not only the center) are replicated and CRN are used; more practical applications are also needed. In Taguchian robust optimization through RSM, we may vary the threshold values, which gives different optimal solutions and a corresponding Pareto frontier; bootstrapping this frontier might enable management to make the final compromise decision—but more research and applications are needed.

References

1. M. E. Angün. *Black Box Simulation Optimization: Generalized Response Surface Methodology*. CentER Dissertation Series, Tilburg University, Tilburg, the Netherlands, 2004.
2. E. Angün, D. den Hertog, G. Gürkan, and J. P. C. Kleijnen. Response surface methodology with stochastic constraints for expensive simulation. *Journal of the Operational Research Society*, 60(6):735–746, 2009.

3. E. R. Barnes. A variation on Karmarkar's algorithm for solving linear programming problems. *Mathematical Programming*, 36:174–182, 1986.
4. R. R. Barton. Response surface methodology. In S. I. Gass and M. C. Fu, editors, *Encyclopedia of Operations Research and Management Science*, pages 1307–1313. Springer, New York, 3rd edition, 2013.
5. R. R. Barton and M. Meckesheimer. Metamodel-based simulation optimization. In *Handbooks in Operations Research and Management Science*, Elsevier/North Holland, 13:535–574, 2006.
6. T. Bartz-Beielstein. *Experimental Research in Evolutionary Computation: The New Experimentalism*. Springer, Berlin, 2006.
7. S. Bashyam and M. C. Fu. Optimization of (s,S) inventory systems with random lead times and a service level constraint. *Management Science*, 44:243–256, 1998.
8. A. Ben-Tal and A. Nemirovski. Robust convex optimization. *Mathematics of Operations Research*, 23(4):769–805, 1998.
9. A. Ben-Tal and A. Nemirovski. Selected topics in robust convex optimization. *Mathematical Programming*, 112(1):125–158, 2008.
10. B. W. M. Bettonvil, E. del Castillo, and J. P. C. Kleijnen. Statistical testing of optimality conditions in multiresponse simulation-based optimization. *European Journal of Operational Research*, 199(2):448–458, 2009.
11. H. Beyer and B. Sendhoff. Robust optimization—a comprehensive survey. *Computer Methods in Applied Mechanics and Engineering,* 196:33–34, pp. 3190–3218, 2007.
12. G. E. P. Box and K. B. Wilson. On the experimental attainment of optimum conditions. *Journal of the Royal Statistical Society, Series B*, 13(1):1–38, 1951.
13. K. H. Chang, L. J. Hong, and H. Wan. Stochastic trust-region response-surface method (STRONG) – a new response-surface framework for simulation optimization, *INFORMS Journal on Computing*, 25(2):230–243, 2013.
14. K. H. Chang, M. K. Li, and H. Wan. Combining STRONG with screening designs for large-scale simulation optimization, *IIE Transactions*, 46:357–373, 2014.
15. M. Chih. A more accurate second-order polynomial metamodel using a pseudo-random number assignment strategy. *Journal of the Operational Research Society*, 64:198–207, 2013.
16. A. R. Conn, N. L. M. Gould, and P. L. Toint. *Trust-Region Methods*. SIAM, 2000.
17. G. Dellino, J. P. C. Kleijnen, and C. Meloni. Robust optimization in simulation: Taguchi and Response Surface Methodology. *International Journal of Production Economics*, 125(1):52–59, 2010.
18. G. Dellino, J. P. C. Kleijnen, and C. Meloni. Robust optimization in simulation: Taguchi and Krige combined. *INFORMS Journal on Computing*, 24(3):471–484, 2012.
19. R. L. Dykstra. Establishing the positive definiteness of the sample covariance matrix. The *Annals of Mathematical Statistics*, 41(6):2153–2154, 1970.
20. B. Efron and R. J. Tibshirani. *An Introduction to the Bootstrap*. Chapman & Hall, New York, 1993.
21. M. C. Fu and H. Qu. Regression models augmented with direct stochastic gradient estimators. *INFORMS Journal on Computing*, 26(3):484–499, 2014.
22. P. E. Gill, W. Murray, and M. H. Wright. *Practical Optimization*. Academic Press, London, 12th edition, 2000.
23. W. D. Kelton, R. P. Sadowski, and D. T. Sturrock. *Simulation with Arena*. McGraw-Hill, Boston, MA, 4th edition, 2007.
24. A. I. Khuri. Multiresponse surface methodology. In *Handbook of Statistics, volume 13*, edited by S. Ghosh and C. R. Rao, Elsevier, Amsterdam, 1996.
25. J. P. C. Kleijnen. *Statistical Techniques in Simulation, Part II*. Dekker, New York, 1975.
26. J. P. C. Kleijnen. Simulation and optimization in production planning: a case study. *Decision Support Systems*, 9:269–280, 1993.
27. J. P. C. Kleijnen. Response surface methodology for constrained simulation optimization: an overview. *Simulation Modelling Practice and Theory*, 16:50–64, 2008.
28. J. P. C. Kleijnen. *Design and Analysis of Simulation Experiments*. Springer, 2nd edition, 2015.

29. J. P. C. Kleijnen, D. den Hertog, and E. Angün. Response surface methodology's steepest ascent and step size revisited. *European Journal of Operational Research*, 159:121–131, 2004.
30. J. P. C. Kleijnen, D. den Hertog and E. Angün. Response surface methodology's steepest ascent and step size revisited: correction. *European Journal of Operational Research*, 170:664–666, 2006.
31. A. M. Law. *Simulation Modeling and Analysis*. McGraw-Hill, Boston, MA, 5th edition, 2014.
32. R. H. Myers, A. I. Khuri, and W. H. Carter. Response surface methodology: 1966–1988. *Technometrics*, 31(2):137–157, 1989.
33. R. H. Myers, D. C. Montgomery, and C. M. Anderson-Cook. *Response Surface Methodology: Process and Product Optimization using Designed Experiments*. Wiley, New York, 3rd edition, 2009.
34. V. N. Nair, editor. Taguchi's parameter design: a panel discussion. *Technometrics*, 34(2):127–161, 1992.
35. S. H. Ng, K. Xu, and W. K. Wong. Optimization of multiple response surfaces with secondary constraints for improving a radiography inspection process. *Quality Engineering*, 19(1):53–65, 2007.
36. G. J. Park, T. H. Lee, K. H. Lee, and K. H. Hwang. Robust design: an overview. *AIAA Journal*, 44(1):181–191, 2006.
37. H. Qu and M. C. Fu. On direct gradient enhanced simulation metamodels. In C. Laroque, J. Himmelspach, R. Pasupathy, O. Rose, and A. M. Uhrmacher, editors, *Proceedings of the 2012 Winter Simulation Conference*, pages 478–489, 2012.
38. R. Rikards and J. Auzins. Response surface method for solution of structural identification problems. *Fourth International Conference on Inverse Problems in Engineering*, Rio de Janeiro, Brazil, 2002.
39. S. C. Rosen, C. M. Harmonosky, and M. T. Traband. Optimization of systems with multiple performance measures via simulation: survey and recommendations. *Computers & Industrial Engineering*, 54(2):327–339, 2008.
40. W. Shi, J. P. C. Kleijnen, and Z. Liu. Factor screening for simulation with multiple responses: sequential bifurcation. *European Journal of Operational Research*, 237(1):136–147, 2014.
41. G. Taguchi. *System of Experimental Designs, Volumes 1 and 2*. UNIPUB/ Krauss International, White Plains, New York, 1987.
42. W. Van den Bogaard and J. P. C. Kleijnen. Minimizing waiting times using priority classes: A case study in response surface methodology. Discussion Paper FEW 77.056, (http://arno.uvt.nl/show.cgi?fid=105001 accessed 12 March 2014), 1977.
43. I. Yanikoglu, D. den Hertog, and J. P. C. Kleijnen. Adjustable robust parameter design with unknown distributions. CentER Discussion Paper (http://arno.uvt.nl/show.cgi?fid=129316 accessed 12 March 2014), 2013.

Chapter 5
Stochastic Gradient Estimation

Michael C. Fu

Abstract This chapter reviews simulation-based methods for estimating gradients, which are central to gradient-based simulation optimization algorithms such as stochastic approximation and sample average approximation. We begin by describing approaches based on finite differences, including the simultaneous perturbation method. The remainder of the chapter then focuses on the direct gradient estimation techniques of perturbation analysis, the likelihood ratio/score function method, and the use of weak derivatives (also known as measure-valued differentiation). Various examples are provided to illustrate the different estimators—for a single random variable, a stochastic activity network, and a single-server queue. Recent work on quantile sensitivity estimation is summarized, and several newly proposed approaches for using stochastic gradients in simulation optimization are discussed.

5.1 Introduction

For optimization problems with *continuous-valued* decision variables, the availability of gradients can dramatically improve the effectiveness of solution algorithms, but in the stochastic setting, since the outputs are themselves random, finding or deriving *stochastic* gradient estimators can itself be a challenging problem, which constitutes the subject of this chapter. The three succeeding chapters on stochastic approximation and sample average approximation—Chaps. 6, 7, and 8—highlight the central role that stochastic gradients play in simulation optimization. In addition to their use in gradient-based simulation optimization, these estimators have other important applications in simulation, most notably *sensitivity analysis*, e.g., factor screening to decide which factors are the most critical, and hedging of financial instruments and portfolios.

M.C. Fu (✉)
University of Maryland, College Park, MD, USA
e-mail: mfu@umd.edu

M.C. Fu (ed.), *Handbook of Simulation Optimization*, International Series in Operations Research & Management Science 216, DOI 10.1007/978-1-4939-1384-8_5

Consider the general optimization problem

$$\min_{x\in\Theta} f(x), \tag{5.1}$$

where $x \in \Theta \subseteq \mathbb{R}^d$. In the context of simulation optimization considered here, f is not directly available but instead the simulation model returns a noisy output $Y(x,\xi)$, where ξ represents the randomness. We consider two forms of the objective function, the commonly used expected value performance

$$f(x) = \mathsf{E}[Y(x,\xi)], \tag{5.2}$$

and the quantile

$$f(x) = q_\alpha(x) = \sup\{y : \mathsf{P}(Y(x,\xi) \le y) \le \alpha,\ 0 < \alpha < 1, \tag{5.3}$$

where $\alpha = \mathsf{P}(Y(x,\xi) \le q_\alpha(x))$ when Y is a continuous random variable.

We introduce two examples that will also be used later to illustrate the various direct gradient estimators:

- A *stochastic activity network* is a directed acyclic graph where the arcs have random activity times. The decision variables influence the distribution of these activity times. The output performance to be considered is the total time to go from a designated source to a designated sink in the network. We will specifically consider the longest path performance where the decision variables (input parameters) are in the individual activity time probability distributions.
- A first-come, first-served (FCFS) *single-server queue*, where the customer arrival process and the customer service times are both stochastic and independent of each other. The output performance to be considered is the average time spent in the system by a customer, denoted by T, and the input parameters will be in the service time distribution(s). When the arrival process is renewal, and the service times are independent and identically distributed (i.i.d.), this is known as a $G/G/1$ (or sometimes written $GI/GI/1$) queue. A simple optimization problem could be to choose the mean service time $x > 0$ to minimize

  $$f(x) = \mathsf{E}[T(x,\xi)] + c/x,$$

  where c can be viewed as the cost of having a faster server.

To solve (5.1) for either setting (5.2) or (5.3), a natural adaptation of steepest descent in deterministic nonlinear optimization is stochastic approximation (SA), which is an iterative update scheme on the parameter that takes the following general form for finding a zero of the objective function gradient:

$$x_{n+1} = \Pi_\Theta\left(x_n - a_n \hat{\nabla} f(x_n)\right), \tag{5.4}$$

where $\hat{\nabla} f$ is an estimate of the gradient ∇f, $\{a_n\}$ is the so-called gain (also known as step-size) sequence, and Π_Θ denotes a projection back into the feasible region Θ when the update (5.4) would otherwise take x_{n+1} out of Θ. Guaranteeing with probability 1 (w.p.1) convergence of x_n requires $a_n \to 0$, but at a rate that cannot be too quick, with a common set of conditions being

$$\sum_n a_n = \infty, \quad \sum_n a_n^2 < \infty.$$

In practice, a_n is often adjusted to a constant value after some number of iterations, which theoretically only guarantees weak convergence (in distribution). The gain sequence need not be deterministic, i.e., it could depend on the output that is generated, e.g., Kesten's rule [31], which decreases a_n only when the sign of the derivative estimate changes. Under appropriate conditions, there are also Central Limit Theorem results that characterize the asymptotic behavior of x_n (cf. [34]).

When $\hat{\nabla} f$ is an unbiased estimator of ∇f, the SA algorithm is generally referred to as being of the Robbins–Monro (RM) [38] type, whereas if $\hat{\nabla} f$ is only asymptotically unbiased, e.g., using a finite difference estimate with the difference going to zero at an appropriate rate, then the algorithm is referred to being of the Kiefer–Wolfowitz (KW) [32] type; see Chap. 6 for details. The Robbins–Monro SA algorithm generally has a canonical asymptotic convergence rate of $n^{-1/2}$, in contrast to $n^{-1/3}$ for the Kiefer–Wolfowitz SA algorithm.

A key challenge for using an SA algorithm for simulation optimization, is the sensitivity of the early transient finite-time behavior of (5.4) to the sequence $\{a_n\}$; for KW-type algorithms, there is the additional choice of the difference sequence. For example, the behavior of SA for the commonly used sequence $a_n = a/n$ $(a > 0)$ is very sensitive to the choice of a. If a is too small, then the algorithm will "crawl" towards the optimum, even at the $1/\sqrt{n}$ asymptotic rate. On the other hand, if a is chosen too large, then extreme oscillations may occur, resulting in an "unstable" progression. Iterate averaging, whereby the estimated optimum is not the latest value of x_n but an average of a window of most recent values, can reduce the sensitivity. Robust SA is a further generalization involving a weighted (based on $\{a_n\}$) average. Addressing the choice of $\{a_n\}$—as well as other issues such as how to project onto the feasible region Θ, which might be specified indirectly (e.g., in a mathematical programming formulation) and possibly involve "noisy" constraints that also have to be estimated along with the objective function—is one of the main topics of Chap. 6. Robust SA and other generalizations and extensions of iterate averaging, along with finite-time analysis of the resulting algorithms, are described in more detail in Chap. 7.

The rest of this chapter is organized as follows. Section 5.2 summarizes the finite difference approaches, including the simultaneous perturbation method that is especially useful in high-dimensional problems. Section 5.3 describes the direct gradient estimation techniques of perturbation analysis, the likelihood ratio/score function method, and the weak derivatives method (also known as measure-valued differentiation) in detail, including illustrative examples (Sects. 5.3.3 and 5.3.4), a summary of some basic theoretical tools (Sect. 5.3.5), and guidelines for the practitioner (Sect. 5.3.6). Section 5.4 treats the more recent work on quantile sensitivity estimation. Section 5.5 describes some new developments in using direct stochastic gradients in simulation optimization. Section 5.6 concludes by briefly describing the main application areas in historical context and providing some primary reference material for further reading. The content in Sects. 5.2 and 5.3 draws heavily from Fu [14], whereas the exposition on quantile sensitivities (Sect. 5.4) and recently proposed approaches to using stochastic gradients in simulation optimization (Sect. 5.5) is new.

An important note on notation: f will later be used to denote a probability density function (p.d.f.) rather than the objective function, so it will be replaced by J henceforth; also, what has been referred to as the decision variable(s) x in the optimization problem (5.1) and in the general SA recursion (5.4) will become the parameter (vector) θ. Specifically, the goal of the rest of the chapter will be to estimate either

$$\nabla_\theta J(\theta) = \nabla_\theta \mathsf{E}[Y(\theta,\xi)],$$

or in Sect. 5.4

$$q'_\alpha(\theta),$$

where q_α is defined by (5.3) and θ is scalar.

5.2 Indirect Gradient Estimators

We divide the approaches to stochastic gradient estimation into two main categories—indirect and direct—which we now more specifically define. An indirect gradient estimator usually has two characteristics: (a) it only estimates an *approximation* of the true gradient value, e.g., via a secant approximation in the scalar case; and (b) it uses only function evaluations (performance measure output samples) from the original (unmodified) system of interest. When used in SA, the resulting algorithms are commonly referred to as *gradient-free* or *stochastic zeroth-order* methods. A direct gradient estimator tries to estimate the true gradient using some additional analysis of the underlying stochastics of the model. More specifically, we will refer to the indirect gradient estimation approach as one in which the simulation output is treated as coming out of a given black box, by which we mean it satisfies two assumptions: (a) no knowledge of the underlying mechanics of the simulation model is used in deriving the estimators, such as knowing the input probability distributions; and (b) no changes are made in the execution of the simulation model itself, such as changing the input distribution for importance sampling. Note that this entails satisfying *both* assumptions; many of the direct gradient estimation techniques can be *implemented* without changing anything in the underlying simulation, but they may require some knowledge of the simulation model, such as the input distributions or some of the system dynamics. In the case of stochastic simulation, as opposed to online estimation based on an actual system, it could be argued that to carry out the simulation most of these mechanics need to be known, i.e., one cannot carry out a stochastic simulation without specifying the input distributions. Here, we simply use the two assumptions to distinguish between the two categories of approaches and not to debate whether an estimator is "model" dependent or not. In terms of stochastic approximation algorithms, indirect and direct gradient estimators generally correspond to Kiefer–Wolfowitz and Robbins–Monro algorithms, respectively.

We describe two indirect gradient estimators: finite differences and simultaneous perturbation. Following our definition, these approaches require no knowledge of the workings of the simulation model, which is treated as a black box.

5.2.1 *Finite Differences*

The straightforward brute-force method for estimating a gradient is simply to use finite differences, i.e., perturbing the value of each component of θ separately while holding the other components at the nominal value. If the value of the perturbation is too small, the resulting difference estimator could be extremely noisy, because the output is stochastic; hence, there is a trade-off between bias and variance in making this selection, and unless all components of the parameter vector are suitably "standardized" a priori, this choice must be done for each component separately, which could be a burdensome task for high-dimensional problems.

The simplest finite difference estimator is the one-sided forward difference gradient estimator, with ith component given by

$$\frac{Y(\theta + c_i e_i, \xi_{2,i}) - Y(\theta, \xi_{1,i})}{c_i}, \tag{5.5}$$

where c is the vector of differences (c_i the perturbation in the ith direction) and e_i denotes the unit vector in the ith direction.

A more accurate estimator is the two-sided symmetric (or central) difference gradient estimator, with ith component given by

$$\frac{Y(\theta + c_i e_i, \xi_{2,i}) - Y(\theta - c_i e_i, \xi_{1,i})}{2c_i}, \tag{5.6}$$

which corresponds to the estimator used in the original Kiefer–Wolfowitz SA algorithm. The variance reduction technique of common random numbers (CRN) can be thought of being the case where $\xi_{1,i} = \xi_{2,i} = \xi_i$. In stochastic simulation, using CRN can reduce the variance of the gradient estimators substantially, although in practice synchronization is an important issue, since merely using the same random number seeds is typically not effective. The symmetric difference estimator given by (5.6) is more accurate, but it requires $2d$ objective function estimates (simulation replications) per gradient estimate, as opposed to $d+1$ function estimates (simulation replications) for the one-sided estimator given by (5.5).

5.2.2 Simultaneous Perturbation

Introduced by Spall in 1992 [41], simultaneous perturbation stochastic approximation (SPSA) is targeted at high-dimensional problems, due to the property that the number of simulation replications needed to form an estimator of the gradient is independent of the dimension of the parameter vector. Specifically, the ith component of the simultaneous perturbation (SP) gradient estimator is given by

$$\frac{Y(\theta + \mathbf{C}\Delta, \xi_2) - Y(\theta - \mathbf{C}\Delta, \xi_1)}{2c_i\Delta_i}, \tag{5.7}$$

where $\Delta = (\Delta_1, \ldots, \Delta_d)$ is a d-dimensional vector of perturbations, which are generally assumed i.i.d. as a function of iteration and independent across components. In this case, $\mathbf{C}$ contains the set of differences for each component as a diagonal matrix with the differences $\{c_i\}$ on the diagonal. The key difference between this estimator and a finite difference estimator is that the *numerator* of (5.7)—corresponding to a difference in the function estimates—is the *same* for all components (i.e., independent of i), whereas the numerator in the symmetric difference estimator given by (5.6) involves a different pair of function estimates for each component (i.e., is a function of i). Thus, the full gradient estimator requires only *two* function estimates, *regardless of the size of the dimension* d. On the other hand, since d random numbers must be generated to produce the perturbation sequence Δ at each iteration, if generating function estimates is relatively inexpensive in terms of computation, then this procedure may not be computationally superior to the previous finite difference approaches. In most simulation optimization settings, however, generating simulation output responses $Y(\theta, \xi)$ is relatively quite expensive. SPSA has also been applied in situations where the output $J(\theta)$ is actually deterministic (no random Y) but expensive to generate, e.g., requires computationally intensive finite-element method calculations.

The key requirement on the perturbation sequence to guarantee w.p.1 convergence of SPSA is that each term have mean zero and finite inverse second moments. Thus, the normal (Gaussian) distribution is prohibited, and the most commonly used distribution is the symmetric Bernoulli, whereby the perturbation takes the positive and negative (equal in magnitude, e.g., ± 1) value w.p. 0.5. Intuitively, convergence comes about from the averaging property of the random directions selected at each iteration, i.e., in the long-run, each component will converge to the correct gradient even if at any particular iteration the estimator may appear odd. Thus, an interesting alternative to using random perturbation sequences $\{\Delta\}$ is to use deterministic sequences [3, 46], analogous to the use of quasi-Monte Carlo.

A very similar gradient estimator for use in SA algorithms is the random directions gradient estimator [33], whose ith component is given by

$$\frac{[Y(\theta + c\Delta, \xi_2) - Y(\theta - c\Delta, \xi_1)]\Delta_i}{2c_i}. \tag{5.8}$$

Instead of dividing by the perturbation component, the difference term multiplies the component. Thus, normal distributions can be used for the perturbation sequence, and convergence requirements translate the moment condition to a bound on the second moment, as well as zero mean. Of course, a correspondence to the SP estimator can be made by simply taking the componentwise inverse, but in practice the performance of the two resulting SA algorithms differs substantially.

An extensive and frequently updated annotated bibliography for SPSA can be found on the World Wide Web at http://www.jhuapl.edu/SPSA/.

5.3 Direct Gradient Estimators

When available, direct gradient estimators offer the following advantages:

- They are generally unbiased, which results in faster convergence rates when implemented in a simulation optimization algorithm, whether stochastic approximation, sample average approximation, or response surface methodology.
- They eliminate the need to determine appropriate values for the finite difference sequences, which influence the accuracy of the estimator. Smaller values of c in (5.5)–(5.8) lead to lower bias but usually at the cost of increased variance, to the point of possibly giving the wrong sign for small enough values.
- They are generally more computationally efficient.

We begin with the case where the output is an expectation, and write the output Y in terms of all the input random variables $\mathbf{X} \equiv \{X_i\}$:

$$J(\theta) = \mathsf{E}[Y(\mathbf{X})] = \mathsf{E}[Y(X_1,\ldots,X_N)], \tag{5.9}$$

where N is a fixed finite number, and for notational brevity, the display of the randomness ξ and the dependence on the parameter θ will often be suppressed in the following derivations. The various direct gradient estimation techniques are distinguished by their treatment of the dependence on θ in (5.9):

sample (pathwise) vs. *measure (distributional).*

As illustrated in the examples that follow, many settings allow either dependence, leading to different gradient estimators.

To derive direct gradient estimators, we write the expectation using what is sometimes called the law of the unconscious statistician:

$$\mathsf{E}[Y(\mathbf{X})] = \int y dF_Y(y) = \int Y(\mathbf{x}) dF_{\mathbf{X}}(\mathbf{x}), \tag{5.10}$$

where F_Y and $F_{\mathbf{X}}$ denote the distributions of Y and $\mathbf{X}$, respectively. In fact, when estimating expected value performance, stochastic simulation can be viewed as a way of implicitly carrying out this relationship, i.e., the simulation model is given input random variables with known distributions, and produces samples of output random variables, for which we would like to characterize the distributions.

For simplicity in discussion, we will assume henceforth that the parameter θ is scalar, because the vector case can be handled by taking each component individually. In view of the right-hand side of (5.10), we revisit the question as to the location of the parameter in a stochastic setting. Putting it in the sample performance $Y(\cdot;\theta)$ corresponds to the view of perturbation analysis (PA), whereas if it is absorbed in the distribution $F(\cdot;\theta)$, then the approach follows that of the likelihood ratio (LR) method (also known as the score function (SF) method) or weak derivatives (WD) (also known as measure-valued differentiation). In the general setting where the parameter is a vector, it is possible that some of the components would be most naturally located in the sample performance, while others would be easily retained in the distributions, giving rise to a mixed approach. For example, in an (s,S) inventory control system, it might be most effective to use PA for the control parameters (decision variables) s and S, and WD or LR/SF for the demand distribution parameters.

Let f_X denote the p.d.f. of all of the input random variables (not to be confused with the original objective function f as defined by (5.1)). Differentiating (5.10), and assuming an interchange of integration and differentiation is permissible, we write two cases:

$$\frac{d\mathsf{E}[Y(X)]}{d\theta} = \begin{cases} \displaystyle\int_{-\infty}^{\infty} Y(x)\frac{df_X(x;\theta)}{d\theta}dx & (5.11) \\[2ex] \displaystyle\int_0^1 \frac{dY(X(\theta;u))}{d\theta}du, & (5.12) \end{cases}$$

where x and u, as well as the integrals, are N-dimensional. For notational simplicity, these N-dimensional multiple integrals are written as a single integral throughout, and we also assume one random number u produces one random variate x. In (5.11), the parameter appears in the distribution directly, whereas in (5.12), the underlying uncertainty is considered the uniform random numbers; this dichotomy corresponds to the respective distributional (measure) and pathwise (sample) dependencies.

For expositional ease in introducing the approaches, we begin by assuming that the parameter only appears in X_1, which is generated *independently* of the other input random variables. So for the case of (5.12), we use the chain rule to write

$$\begin{aligned}\frac{d\mathsf{E}[Y(X)]}{d\theta} &= \int_0^1 \frac{dY(X_1(\theta;u_1),X_2,\ldots)}{d\theta}du \\ &= \int_0^1 \frac{\partial Y}{\partial X_1}\frac{dX_1(\theta;u_1)}{d\theta}du. \end{aligned} \qquad (5.13)$$

In other words, the estimator takes the form

$$\frac{\partial Y(X)}{\partial X_1}\frac{dX_1}{d\theta}, \tag{5.14}$$

where the parameter appears in the transformation from random number to random variate, and the derivative is expressed as the product of a sample path derivative and derivative of a random variable. The issue of what constitutes the latter will be taken up shortly, but this approach is called infinitesimal perturbation analysis (IPA). For the $M/M/1$ queue, the sample path derivative could be derived using Lindley's equation, relating the time in system of a customer to the service times (and interarrival times, which are not a function of the parameter).

Assume that X_1 has marginal p.d.f. $f_1(\cdot;\theta)$ and that the joint p.d.f. for the remaining input random variables $(X_2,\ldots)$ is given by f_{-1}, which has no (functional) dependence on θ. Then the assumed independence gives $f_X = f_1 f_{-1}$, and the expression (5.11) involving differentiation of a density (measure) can be further manipulated using the product rule of differentiation to yield the following:

$$\frac{d\mathrm{E}[Y(X)]}{d\theta} = \int_{-\infty}^{\infty} Y(x)\frac{\partial f_1(x_1;\theta)}{\partial\theta} f_{-1}(x_2,\ldots)dx \tag{5.15}$$

$$= \int_{-\infty}^{\infty} Y(x)\frac{\partial \ln f_1(x_1;\theta)}{\partial\theta} f_X(x)dx. \tag{5.16}$$

In other words, the estimator takes the form

$$Y(X)\frac{\partial \ln f_1(X_1;\theta)}{\partial\theta}. \tag{5.17}$$

Since the term $\frac{\partial \ln f_1(\cdot;\theta)}{\partial\theta}$ is the well-known (efficient) score function in statistics, this approach has been called the score function (SF) method. The other name given to this approach—the likelihood ratio (LR) method—comes from the closely related likelihood ratio function given by

$$\frac{f_1(\cdot;\theta)}{f_1(\cdot;\theta_0)},$$

which when differentiated with respect to θ gives

$$\frac{\partial f_1(\cdot;\theta)/\partial\theta}{f_1(\cdot;\theta_0)},$$

which is equal to the score function upon setting $\theta_0 = \theta$.

On the other hand, if instead of expressing the right-hand side of (5.15) as (5.16), the density derivative is written as

$$\frac{\partial f_1(x_1;\theta)}{\partial\theta} = c_1(\theta)\left(f_1^{(+)}(x_1;\theta) - f_1^{(-)}(x_1;\theta)\right),$$

it leads to the following relationship:

$$\frac{d\mathsf{E}[Y(X)]}{d\theta} = \int_{-\infty}^{\infty} Y(x) \frac{\partial f_X(x;\theta)}{\partial \theta} dx$$
$$= c_1(\theta)\left(\int_{-\infty}^{\infty} Y(x) f_1^{(+)}(x_1;\theta) f_{-1}(x_2,\ldots)dx - \int_{-\infty}^{\infty} Y(x) f_1^{(-)}(x_1;\theta) f_{-1}(x_2,\ldots)dx\right).$$

The triple $\left(c_1(\theta), f_1^{(+)}, f_1^{(-)}\right)$ constitutes a weak derivative (WD) for f_1, which is in general not unique. The corresponding WD estimator is of the form

$$c_1(\theta)\left(Y(X_1^{(+)}, X_2,\ldots) - Y(X_1^{(-)}, X_2,\ldots)\right), \tag{5.18}$$

where $X_1^{(-)} \sim f_1^{(-)}$ and $X_1^{(+)} \sim f_1^{(+)}$, henceforth often abbreviated $X^{(\pm)} \sim f^{(\pm)}$. The term *weak derivative* comes about from the possibility that $\frac{\partial f_1(\cdot;\theta)}{\partial \theta}$ in (5.15) may not be proper, but its *integral* may be well-defined, e.g., it might involve delta-functions (impulses), corresponding to probability mass functions (p.m.f.s) of discrete distributions. Note that even for a given WD representation, only the *marginal* distributions for the two random variables $X^{(\pm)}$ are specified, i.e., their joint distribution is not constrained, so the "estimator" given by (5.18) is not really completely specified. Since (5.18) is a difference of two terms that appear similar, one might expect that generating the two random variables using CRN rather than independently would be beneficial, and it is indeed true in many situations, but such a conclusion is problem dependent. However, for the Hahn–Jordan WD representation (to be described later in Sect. 5.3.2), independent generation turns out to be the method that minimizes the variance of the WD estimator [44].

If in the expression (5.12), the interchange of expectation and differentiation does not hold (e.g., if Y is an indicator function), then as long as there is more than one input random variable, appropriate conditioning will often allow the interchange as follows:

$$\frac{d\mathsf{E}[Y(X)]}{d\theta} = \int_0^1 \frac{d\mathsf{E}[Y(X(\theta;u))|Z(u)]}{d\theta} du, \tag{5.19}$$

where $Z \subseteq \{X_1,\ldots,X_N\}$. This conditioning is known as smoothed perturbation analysis (SPA), because it seeks to "smooth" out a discontinuous function. SPA leads to an estimator of the following form:

$$\frac{\partial \mathsf{E}[Y(X)|Z]}{\partial X_1} \frac{dX_1}{d\theta}. \tag{5.20}$$

Note that taking Z as the entire set leads back to (5.14).

Remark. For SPA, the conditioning in (5.19) was done with respect to a subset of the input random variables only. Further conditioning can be done on events in the system, which leads to an estimator of the following general form:

$$\frac{dY}{d\theta} + \mathsf{E}_Z[\delta Y|\mathscr{B}]\frac{d\mathsf{P}_Z(\mathscr{B})}{d\theta}, \tag{5.21}$$

where the subscript indicates a corresponding conditional expectation/probability, $\mathscr{B}$ is an appropriately selected event, and δY represents a change in the performance measure under the conditioned (usually called "critical") event. In this case, if the probability rate $\frac{d\mathsf{P}_Z(\mathscr{B})}{d\theta}$ is 0, the estimator (5.21) also reduces to IPA. On the other hand, if the IPA term $\frac{dY}{d\theta}$ is zero, the estimator may coincide with the WD estimator in certain cases, with correspondences between $c(\theta)$ and the probability rate, and between the difference term in (5.18) and the conditional expectation in (5.21).

5.3.1 Derivatives of Random Variables

PA estimators—e.g., those shown in (5.14), (5.20), (5.21)—require the notion of derivatives of random variables. The mathematical problem for defining such derivatives consists of constructing a family of random variables parameterized by θ on a *common* probability space, with the point of departure being a set of parameterized distribution functions $\{F(\cdot;\theta)\}$. We wish to construct $X(\theta) \sim F(\cdot;\theta)$ s.t. $\forall\theta \in \Theta$, $X(\theta)$ is differentiable w.p.1. The sample derivative is then defined in the intuitive manner as

$$\frac{dX(\theta,\omega)}{d\theta} = \lim_{\Delta\theta\to 0}\frac{X(\theta+\Delta\theta,\omega)-X(\theta,\omega)}{\Delta\theta},$$

where ω denotes a sample point in the underlying probability space. If the distribution of X is known, we have [21, 42]

$$\frac{dX(\theta)}{d\theta} = -\frac{\partial F(X;\theta)/\partial\theta}{\partial F(X;\theta)/\partial X}, \tag{5.22}$$

where we use the (slightly abusive) notation $\frac{\partial F(X;\theta)}{\partial X} = \left.\frac{\partial F(x;\theta)}{\partial x}\right|_{x=X}$.

Definition. For a distribution function $F(x;\theta)$, θ is said to be a *location* parameter if $F(x+\theta;\theta)$ does not depend on θ; θ is said to be a *scale* parameter if $F(x\theta;\theta)$ does not depend on θ; and θ is said to be a *generalized scale* parameter if $F(\bar{\theta}+x\theta;\theta)$ does not depend on θ, for some fixed $\bar{\theta}$ (usually a location parameter) not dependent on θ.

In these special cases, one can use the following sample derivatives for the three respective cases (location, scale, generalized scale):

$$\frac{dX}{d\theta} = 1, \quad \frac{dX}{d\theta} = \frac{X}{\theta}, \quad \frac{dX}{d\theta} = \frac{X - \bar{\theta}}{\theta}.$$

The most well-known example is the normal distribution, with the mean being a location parameter and the standard deviation a generalized scale parameter. Similarly, the two parameters in the Cauchy, Gumbel (extreme value), and logistic distributions also correspond to location and generalized scale parameters. Other examples include the mean of the exponential being a scale parameter; and the mean of the uniform distribution being a location parameter, with the half-width being a generalized scale parameter. In the special case $U(0,\theta)$, the single parameter is an ordinary scale parameter. Also, for $\mathscr{N}(\theta,(\theta\sigma)^2)$, θ is an ordinary scale parameter. See Table 5.1 in Sect. 5.3.3 for more examples.

Lastly, note that for a given distribution, there may be multiple ways to generate a random variate, which leads to different derivatives, some of which may be unbiased and some of which may not. This is called the role of representations, and is illustrated with a simple example (Example 5.5) in Sect. 5.3.3.

5.3.2 *Derivatives of Measures*

As we have seen already, both the LR/SF and WD estimators rely on differentiation of the underlying measure, so the parameters of interest should appear in the underlying (input) distributions. If this is not the case, then one approach is to try to "push" the parameter out of the performance measure into the distribution, so that the usual differentiation can be carried out. This is achieved by a change of variables, which is problem dependent.

Recall that we introduced the idea of a weak derivative by expressing the derivative of a density (p.d.f.) as an appropriately normalized difference of two p.d.f.s, i.e., the triple $(c(\theta), f^{(+)}, f^{(-)})$ satisfying

$$\frac{\partial f(x;\theta)}{\partial \theta} = c(\theta)\left(f^{(+)}(x;\theta) - f^{(-)}(x;\theta)\right).$$

This idea can be generalized without the need for a differentiable density, as long as the integral exists with respect to a certain set of (integrable) "test" functions, say $\mathscr{L}$, e.g., the set of continuous bounded functions.

Definition. The triple $(c(\theta), F^{(+)}, F^{(-)})$ is called a *weak derivative* for distribution (c.d.f.) F if for all functions $g \in \mathscr{L}$ (not a function of θ),

$$\frac{d}{d\theta}\int g(x)dF(x;\theta) = c(\theta)\left(\int g(x)dF^{(+)}(x;\theta) - \int g(x)dF^{(-)}(x;\theta)\right).$$

Remark. As mentioned earlier, the derivative is "weak" in the sense that the density derivative may not be defined in the usual sense, but in terms of generalized functions integrable with respect to the functions in $\mathscr{L}$, as in the "definition" of a delta function in terms of its integral. The concept of a weak derivative need not be restricted to probability measures, but any finite signed measures. Lastly, note that a WD gradient estimate may require as many as $2d$ additional simulations for the vector case (a pair for each component), unlike LR/SF and IPA estimators, which will always require just a single simulation.

One choice for the weak derivative (density) that is readily available is

$$\frac{\partial f}{\partial \theta} = c\left(f^{(+)} - f^{(-)}\right), \tag{5.23}$$

where

$$f^{(-)} = \frac{1}{c}\left(\frac{\partial f}{\partial \theta}\right)^{-}, \quad f^{(+)} = \frac{1}{c}\left(\frac{\partial f}{\partial \theta}\right)^{+}, \tag{5.24}$$

$(x)^{+} \equiv \max\{x,0\}$, $(x)^{-} \equiv \max\{-x,0\}$, and $c = \int\left(\frac{\partial f}{\partial \theta}\right)^{+} dx = \int\left(\frac{\partial f}{\partial \theta}\right)^{-} dx$, using the fact that

$$\int f(x)dx = 1 \quad \Longrightarrow \quad \int \frac{\partial f}{\partial \theta} dx = 0.$$

The representation given by (5.23) and (5.24) is the Hahn–Jordan decomposition, which will always exist for probability measures, and results in a decomposition involving two measures with complementary support. It can be shown in this case that generating the two random variables according to $f^{(+)}$ and $f^{(-)}$ independently minimizes variance for the WD estimator [44].

Remark. The representation is clearly not unique. In fact, for any non-negative integrable function h, we have

$$\frac{\partial f}{\partial \theta} = c\Big([f^{(-)} + h] - [f^{(+)} + h]\Big) = \tilde{c}\Big([f^{(-)} + h]/(1 + \int h) - [f^{(+)} + h]/(1 + \int h)\Big),$$

where $\tilde{c} = c(1 + \int h)$. Thus, one way to obtain the estimator using the original simulation is to choose a representation in which both measures have the same support as the original measure. Then importance sampling can be applied, so that the original simulation can be used to generate the estimator without the need for simulating the system under alternative input distributions. Perhaps the most direct way to achieve this is to add the original measure itself to both $f^{(-)}$ and $f^{(+)}$ and renormalize appropriately, i.e., choose $h = f$ above:

$$\frac{\partial f}{\partial \theta} = 2c\left([f^{(-)} + f]/2 - [f^{(+)} + f]/2\right).$$

5.3.3 Input Distribution Examples

We now demonstrate some of these concepts on a single random variable. Section 5.3.4 then considers the two examples introduced at the beginning of the chapter (stochastic activity network and single-server queue).

Example 5.1. Let $X \sim \exp(\theta)$, an exponential random variable with mean θ and p.d.f. given by

$$f(x;\theta) = \frac{1}{\theta} e^{-x/\theta} \mathbf{1}\{x > 0\},$$

where $\mathbf{1}\{\cdot\}$ denotes the indicator function. The usual construction of the random variable is

$$X(\theta;u) = -\theta \ln u,$$

where u represents a random number. Differentiating both expressions, we get

$$\begin{aligned}
\frac{\partial f(x;\theta)}{\partial \theta} &= \left[\frac{x}{\theta^2}\frac{1}{\theta}e^{-x/\theta} - \frac{1}{\theta^2}e^{-x/\theta}\right]\mathbf{1}\{x>0\} \\
&= f(x;\theta)\left[\frac{x}{\theta^2} - \frac{1}{\theta}\right] \\
&= \frac{1}{\theta}\left[\frac{x}{\theta^2}e^{-x/\theta}\mathbf{1}\{x>0\} - f(x;\theta)\right] \\
&= \frac{1}{\theta e}\left[\frac{e}{\theta}\left(\frac{x}{\theta}-1\right)e^{-x/\theta}\mathbf{1}\{x>\theta\} - \frac{e}{\theta}\left(1-\frac{x}{\theta}\right)e^{-x/\theta}\mathbf{1}\{0<x\le\theta\}\right], \\
\frac{dX(\theta;u)}{d\theta} &= -\ln u = \frac{X(\theta;u)}{\theta}.
\end{aligned}$$

In the third and fourth lines above, the density derivative (which is itself *not* a density) has been expressed as the difference of two densities multiplied by a constant. This demonstrates that the weak derivative representation is not unique, with the last decomposition being the Hahn–Jordan decomposition, noting that $x = \theta$ is the point at which $df(x;\theta)/d\theta$ changes sign. The following correspond to the LR/SF, WD (a) & (b), and IPA estimators, respectively:

$$Y(X)\frac{1}{\theta}\left(\frac{X_1}{\theta} - 1\right),$$

$$\frac{1}{\theta}\left[Y(X_{1a}^{(+)},\ldots) - Y(X_{1a}^{(-)},\ldots)\right], \quad \frac{1}{\theta e}\left[Y(X_{1b}^{(+)},\ldots) - Y(X_{1b}^{(-)},\ldots)\right],$$

$$\frac{dY}{dX_1}\frac{X_1}{\theta},$$

where $X_{1a}^{(-)} \sim exp(\theta)$ and $X_{1a}^{(+)} \sim Erl(2,\theta)$, where "$Erl$" is an Erlang distribution (see Table 5.1), and $X_{1b}^{(-)} \sim \theta - \text{trunc}(Erl(2,\theta),[0,\theta])$ and $X_{1b}^{(+)} \sim \theta + Erl(2,\theta)$, where "trunc($F,[a,b]$)" represents a distribution (c.d.f.) F truncated to the range $[a,b]$. Since an $Erl(2,\theta)$ distribution can be generated by the sum of two i.i.d. exponentially distributed random variables, one way to realize the first WD estimator would be to use $X_{1a}^{(-)} = X_1$ and then generate another $\tilde{X}_1 \sim exp(\theta)$ independent of the original X_1, giving the WD estimator

$$\frac{1}{\theta}\left[Y(X_1+\tilde{X}_1,\ldots) - Y(X_1,\ldots)\right].$$

The following is a simple example that demonstrates that the WD estimator is more broadly applicable than the LR/SF estimator.

Example 5.2. Let $X \sim U(0,\theta)$. Then we have the following:

$$\begin{aligned}
f(x;\theta) &= \frac{1}{\theta}\mathbf{1}\{0<x<\theta\}, \\
X(\theta;u) &= u\theta, \\
\frac{\partial f(x;\theta)}{\partial\theta} &= \frac{1}{\theta}\left[\delta(\theta-x) - \frac{1}{\theta}\mathbf{1}\{0<x<\theta\}\right] \\
&= \frac{1}{\theta}\left[\delta(\theta-x) - f(x;\theta)\right], \\
\frac{dX(\theta;u)}{d\theta} &= u \;=\; \frac{X(\theta;u)}{\theta},
\end{aligned} \tag{5.25}$$

where we define the Dirac δ-function as the "derivative" of a step function by

$$\mathbf{1}\{x \geq \theta\} = \int_{-\infty}^{x} \delta(y-\theta)dy. \tag{5.26}$$

On the right-hand side of Eq. (5.25), we have the difference of densities for a mass at θ and the original $U(0,\theta)$ distribution, respectively, i.e., the weak derivative representation $(1/\theta,\theta,F)$, where θ indicates a deterministic distribution with mass at θ. So, for example, the estimator in (5.18) would be given by

$$\frac{1}{\theta}\left(Y(\theta,X_2,\ldots) - Y(X_1,X_2,\ldots)\right).$$

This is a case where the LR/SF estimator is ill-defined, due to the δ-function. Another example is the following.

Example 5.3. Let $X \sim Par(\alpha,\theta)$, which represents the Pareto distribution with shape parameter $\alpha > 0$ and scale parameter $\theta > 0$, and p.d.f. given by

$$f(x) = \theta^{\alpha}\alpha x^{-(\alpha+1)}\mathbf{1}\{x \geq \theta\}.$$

Once again the LR/SF estimator does not exist (for θ), due to the appearance of the parameter in the indicator function that controls the support of the distribution, whereas WD estimators can be derived (see Table 5.1 at the end of the section).

However, the very general exponential family of distributions leads to a nice form for the LR/SF estimator.

Example 5.4. Let θ denote the vector of parameters in a p.d.f. that can be written in the following form:

$$f(x;\theta) = k(\theta)\exp\left(\sum_i v_i(\theta)t_i(x)\right)h(x),$$

where the functions h and $\{t_i\}$ are independent of θ, and the functions k and $\{v_i\}$ do not involve the argument. Then it is straightforward to derive

$$\frac{\partial \ln f(x;\theta)}{\partial \theta} = \frac{\nabla k(\theta)}{k(\theta)} + \sum_i \nabla v_i(\theta)t_i(x).$$

Examples include the normal, gamma, Weibull, and exponential, for the continuous case, and the binomial, Poisson, and geometric for the discrete case.

As mentioned in Sect. 5.3.1, the application of PA (both IPA or SPA) depends on the way the stochastic processes in the system are represented. We illustrate this through a simple random variable example. In terms of simulation, this means that a different representation used to generate the random variable could lead to a different PA estimator. For instance, in Example 5.1, an alternative equivalent representation is $X = -\theta\ln(1-u)$, which in this case leads to the same IPA estimator X/θ. Since the underlying distribution is identical for the different representations, the LR/SF and WD estimators are not dependent on the process representation, but as noted earlier, the same distribution has infinitely many possible WD estimators.

Example 5.5. For $\theta \in (0,1)$, let

$$X \sim \begin{cases} U(0,1) \text{ w.p. } \theta, \\ U(1,2) \text{ w.p. } 1-\theta, \end{cases}$$

a mixture of two uniform distributions, with $\mathsf{E}[X] = 1.5-\theta$ and $d\mathsf{E}[X]/d\theta = -1$. A straightforward construction/representation using two random numbers is

$$X = \mathbf{1}\{U_1 \leq \theta\}U_2 + \mathbf{1}\{U_1 > \theta\}(1+U_2), \tag{5.27}$$

where U_1, $U_2 \sim U(0,1)$ are *independent*. However, since

$$\frac{dX}{d\theta} = 0 \text{ w.p.1},$$

this clearly leads to a biased estimator. Note that viewed as a function of θ, X jumps from $1+U_2$ down to U_2 at $\theta = U_1$. However, an unbiased estimator can be obtained by using the following construction in which the "coin flipping" and uniform generation are correlated:

$$X = \mathbf{1}\{U \le \theta\}\frac{U}{\theta} + \mathbf{1}\{U > \theta\}\left(1 + \frac{U-\theta}{1-\theta}\right), \text{ where } U \sim U(0,1),$$

$$\Longrightarrow \frac{dX}{d\theta} = -\frac{U}{\theta^2}\mathbf{1}\{U \le \theta\} + \frac{U-1}{(1-\theta)^2}\mathbf{1}\{U > \theta\},$$

which is unbiased (has expectation equal to $d\mathsf{E}[X]/d\theta = -1$). This construction is based on the property that the distributions of the random variable U/θ under the condition $\{U < \theta\}$ and the random variable $(U-\theta)/(1-\theta)$ under the condition $\{U \ge \theta\}$ are both $U(0,1)$. From a simulation perspective, this representation has the additional advantage of requiring only a single random number to generate X instead of two as in the previous construction. In this case, the construction also corresponds to the inverse transform representation. In terms of the derivative, the crucial property of the representation is that X is continuous across $\theta = U$. One can easily construct other single random number representations that do not have this desirable characteristic, e.g.,

$$X = \mathbf{1}\{U \le \theta\}(1 - \frac{U}{\theta}) + \mathbf{1}\{U > \theta\}\left(1 + \frac{1-U}{1-\theta}\right), \text{ where } U \sim U(0,1),$$

$$\Longrightarrow \frac{dX}{d\theta} = \mathbf{1}\{U \le \theta\}\frac{U}{\theta^2} + \mathbf{1}\{U > \theta\}\frac{1-U}{(1-\theta)^2},$$

which is biased (has expectation +1), the intuitive reason being the discontinuity of X at $U = \theta$, where it jumps from 0 to 2.

For the first representation given by (5.27), which used two random numbers and led to a biased IPA estimator, SPA can be applied by conditioning on U_2 as follows:

$$X = \mathsf{E}[X_1|U_2] = U_2\theta + (1+U_2)(1-\theta) = 1 + U_2 - \theta,$$

leading to the trivially unbiased "estimator" $dX/d\theta = -1$.

To derive the WD and LR/SF estimators, the p.d.f. is given by

$$f(x;\theta) = \theta\mathbf{1}\{x \in (0,1]\} + (1-\theta)\mathbf{1}\{x \in (1,2]\},$$

so

$$\frac{\partial f(x;\theta)}{\partial\theta} = \mathbf{1}\{x\in(0,1]\} - \mathbf{1}\{x\in(1,2]\}, \tag{5.28}$$

$$\frac{\partial \ln f(x;\theta)}{\partial\theta} = \frac{1}{\theta}\mathbf{1}\{x\in(0,1]\} - \frac{1}{1-\theta}\mathbf{1}\{x\in(1,2]\}, \tag{5.29}$$

and the obvious WD from (5.28) is simply $(1, U(0,1), U(1,2))$, corresponding to the Hahn–Jordan decomposition, whereas the LR/SF estimator from (5.29) is given by

$$Y(X)\cdot\begin{cases}\frac{1}{\theta} & \text{if } X\le 1,\\ -\frac{1}{1-\theta} & \text{otherwise.}\end{cases}$$

However, as noted in the remark at the end of Sect. 5.3.2, the WD representation is not unique, so for example, one can add and subtract a $U(0,1)$ density in (5.28) to get

$$\mathbf{1}\{x\in(0,1]\} + \mathbf{1}\{x\in(0,1]\} - \mathbf{1}\{x\in(1,2]\} - \mathbf{1}\{x\in(0,1]\}$$

$$= 2\left[\mathbf{1}\{x\in(0,1]\} - \frac{1}{2}\mathbf{1}\{x\in(0,2]\}\right],$$

yielding the alternative WD representation $(2, U(0,1), U(0,2))$.

Discrete distributions present separate challenges for the different approaches. Basically, when the parameter appears in the support *probabilities*, then LR/SF and WD can be easily applied, whereas IPA is in general not applicable. The reverse is true, however, if the parameter appears instead in the support *values*. The next two examples demonstrate this dichotomy, where we work directly with the probability mass function (p.m.f.) $p(x;\theta) = \mathsf{P}(X = x)$, instead of densities with δ-functions. Let $Ber(p;a,b)$ denote a Bernoulli distribution that takes value a w.p. p and b w.p. $1-p$. We start with an example where the parameter θ is the Bernoulli probability.

Example 5.6. Let $X\sim Ber(\theta;a,b)$, $a\neq b$, which has p.m.f.

$$p(x;\theta) = \theta\mathbf{1}\{x=a\} + (1-\theta)\mathbf{1}\{x=b\},$$

so

$$\frac{\partial p}{\partial\theta} = \mathbf{1}\{x=a\} - \mathbf{1}\{x=b\},$$

which can be viewed as the difference of two (deterministic) masses at a and b (with $c(\theta)=1$), and is the Hahn–Jordan decomposition in this case. For the LR/SF estimator, we have

$$\frac{\partial \ln p}{\partial\theta} = \frac{\mathbf{1}\{x=a\} - \mathbf{1}\{x=b\}}{\theta\mathbf{1}\{x=a\} + (1-\theta)\mathbf{1}\{x=b\}} = \frac{1}{\theta}\mathbf{1}\{x=a\} - \frac{1}{1-\theta}\mathbf{1}\{x=b\}.$$

Note the similarity of both the WD and LR/SF estimators to the previous example. In this case, there is no way to construct X such that it will be differentiable w.p.1. For example, the natural construction/representation

$$X = a\mathbf{1}\{U \leq \theta\} + b\mathbf{1}\{U > \theta\}$$

yields $dX/d\theta = 0$ w.p.1, so IPA is not applicable.

In contrast, now consider an example where the parameter θ is one of the support values.

Example 5.7. Let $X \sim Ber(p;\theta;0)$, $\theta \neq 0$, $E[X] = p\theta$, $dE[X]/d\theta = p$, which has p.m.f.

$$p(x;\theta) = p\mathbf{1}\{x = \theta\} + (1-p)\mathbf{1}\{x = 0\},$$

which is not differentiable with respect to θ, so LR/SF and WD estimators cannot be derived. The natural random variable construction

$$X = \theta\mathbf{1}\{U \leq p\}$$

leads to the unbiased

$$\frac{dX}{d\theta} = \mathbf{1}\{U \leq p\} = \mathbf{1}\{X = \theta\} = \frac{X}{\theta}.$$

The IPA estimator $dX/d\theta = \mathbf{1}\{X = \theta\}$ holds even if additional values are added to the underlying support, as long as the additional values do not involve θ. If θ enters into them, then the estimator can be easily modified to reflect the additional dependence.

For many common input distributions, Table 5.1 provides the necessary derivatives needed to implement each of the three methods (IPA, LR/SF, WD). Recall also that the two parameters in the Cauchy, Gumbel, and logistic distributions (not given in the table) are location and (generalized) scale parameters, so the IPA expressions would be the same as for the normal distribution. The entry for the mean of the normal has an interesting implementation for the WD estimator, based on the observation that a normally distributed random variable $\mathcal{N}(\mu, \sigma^2)$ can be generated via the product of a uniform $U(0,1)$ random number and a double-sided Maxwell $Mxw(\mu, \sigma^2)$ random variate (generated independently of each other, cf. [25], which also provides a method for generating from this distribution). Implementation using such pairs of independent $U(0,1)$ and $Mxw(\mu, \sigma^2)$ distributed random variates results in a WD derivative estimator with provably the lowest variance for any polynomial output function. Of course, in most settings the output is not polynomial; furthermore, the WD estimator requires an additional simulation replication per partial derivative.

Table 5.1 Derivatives for common input distributions (NA = not applicable)

Distribution	Parameterization	p.d.f./p.m.f. $f(x)/p(x)$	Support
Bernoulli	$Ber(p;a,b)$	$p\mathbf{1}\{x=a\}+(1-p)\mathbf{1}\{x=b\}$	$\{a,b\}$
geometric	$geo(p)$	$(1-p)^{x-1}p$	$\mathbb{Z}^+$
negative binomial	$negbin(n,p)$	$\binom{x-1}{n-1}(1-p)^{x-n}p^n$	$\{n,n+1,\ldots\}$
binomial	$bin(n,p)$	$\binom{n}{x}p^x(1-p)^{n-x}$	$\mathbb{N}$
Poisson	$Poi(\lambda)$	$\frac{e^{-\lambda}\lambda^x}{x!}$	$\mathbb{N}$
normal (Gaussian)	$\mathcal{N}(\mu,\sigma^2)$	$\frac{1}{\sqrt{2\pi}\sigma}e^{-\frac{(x-\mu)^2}{2\sigma^2}}$	$\mathbb{R}$
Maxwell (2-sided)	$Mxw(\mu,\sigma^2)$	$\frac{(x-\mu)^2}{\sqrt{2\pi}\sigma^3}e^{-\frac{(x-\mu)^2}{2\sigma^2}}$	$\mathbb{R}$
uniform	$U(a,b)$	$\frac{1}{b-a}$	$[a,b]$
exponential	$exp(\beta)$	$\beta^{-1}e^{-x/\beta}$	$x\geq 0$
Weibull	$Wei(\alpha,\beta)$	$\alpha\beta^{-\alpha}x^{\alpha-1}e^{-(x/\beta)^\alpha}$	$x\geq 0$
gamma	$gam(\alpha,\beta)$	$\frac{\beta^{-\alpha}x^{\alpha-1}e^{-x/\beta}}{\Gamma(\alpha)}$	$x\geq 0$
Erlang	$Erl(\alpha,\beta)$	$\frac{\beta^{-n}x^{n-1}e^{-x/\beta}}{(n-1)!}$	$x\geq 0$
Pareto	$Par(\alpha,\beta)$	$\alpha\beta^\alpha x^{-(\alpha+1)}$	$x\geq\beta$

$\alpha>0$ is generally the shape parameter, μ is a location parameter, and $\beta>0$ is a scale parameter; $\Gamma(\alpha)=\int_0^\infty t^{\alpha-1}e^{-t}dt$; $\Gamma(\alpha)=(\alpha-1)\Gamma(\alpha-1)$; $geo(p)=negbin(1,p)$, $Erl(n,\beta)=gam(n,\beta)$ for $n\in\mathbb{Z}^+$; $Wei(1,\beta)=gam(1,\beta)=exp(\beta)$. The definition of the parameter β used here in the exponential, Weibull, and gamma distributions is the inverse of what is often found in the literature ([35] being a notable exception), but makes β a scale parameter. This is why the WD expressions below for the exponential, Weibull, and gamma distributions differ slightly from the table in [24].

input dist $X\sim F$	IPA $\frac{dX}{d\theta}$	LR/SF $\frac{\partial\ln f(X;\theta)}{\partial\theta}$	WD $(c(\theta),F^{(+)},F^{(-)})$
$Ber(\theta;a,b)$	NA	$\frac{1}{\theta}\mathbf{1}\{X=a\}$ $-\frac{1}{1-\theta}\mathbf{1}\{X=b\}$	$(1,a,b)$
$Ber(p;\theta,b)$	$\mathbf{1}\{X=\theta\}$	NA	NA
$geo(\theta)$	NA	$\frac{1}{\theta}+\frac{1-X}{1-\theta}$	$(\frac{1}{\theta},geo(\theta),negbin(2,\theta))$
$bin(n,\theta)$	NA	$\frac{X}{\theta}-\frac{n-X}{1-\theta}$	$(n,1+bin(n-1,\theta),bin(n-1,\theta))$
$Poi(\theta)$	NA	$\frac{X}{\theta}-1$	$(1,1+Poi(\theta),Poi(\theta))$
$\mathcal{N}(\theta,\sigma^2)$	1	$\frac{X-\theta}{\sigma^2}$	$\left(\frac{1}{\sqrt{2\pi}\sigma},\theta+Wei(2,\frac{1}{2\sigma^2}),\ \theta-Wei(2,\frac{1}{2\sigma^2})\right)$
$\mathcal{N}(\mu,\theta^2)$	$\frac{X-\mu}{\theta}$	$\frac{1}{\theta}\left[\left(\frac{X-\mu}{\theta}\right)^2-1\right]$	$(\frac{1}{\theta},Mxw(\mu,\theta^2),\mathcal{N}(\mu,\theta^2))$
$U(0,\theta)$	$\frac{X}{\theta}$	NA	$(\frac{1}{\theta},\theta,U(0,\theta))$
$U(\theta-\gamma,\theta+\gamma)$	1	NA	$(\frac{1}{2\gamma},\theta+\gamma,\theta-\gamma)$
$U(\mu-\theta,\mu+\theta)$	$\frac{X-\mu}{\theta}$	NA	$(\frac{1}{\theta},Ber(0.5;\mu-\theta,\mu+\theta),U(\mu-\theta,\mu+\theta))$
$exp(\theta)$	$\frac{X}{\theta}$	$\frac{1}{\theta}\left(\frac{X}{\theta}-1\right)$	$(\frac{1}{\theta},Erl(2,\theta),exp(\theta))$
$Wei(\alpha,\theta)$	$\frac{X}{\theta}$	$\frac{1}{\theta}\left[\left(\frac{X}{\theta}\right)^\alpha-\alpha\right]$	$(\frac{\alpha}{\theta},[Erl(2,\theta^\alpha)]^{1/\alpha},Wei(\alpha,\theta))$
$gam(\alpha,\theta)$	$\frac{X}{\theta}$	$\frac{1}{\theta}\left(\frac{X}{\theta}-\alpha\right)$	$(\frac{\alpha}{\theta},gam(\alpha+1,\theta),gam(\alpha,\theta))$
$Par(\alpha,\theta)$	$\frac{X}{\theta}$	NA	$(\frac{\alpha}{\theta},Par(\alpha,\theta),\theta)$

5.3.4 Output Examples

We consider the two examples introduced at the beginning of the chapter: stochastic activity network and single-server queue.

Stochastic Activity Network

A stochastic activity network will be given by a directed acyclic graph, defined by M nodes and N directed arcs representing activities. The activity times are given by random variables $X_i,\ i=1,\ldots,N$. Without loss of generality, we take node 1 as the source (origin) and node M as the sink (destination). A path P is a set of directed arcs going from source to sink. Let $\mathscr{P}$ denote the set of all paths from source to sink, and P^* denote the set of arcs on the optimal path corresponding to the project duration given by Y (e.g., shortest or longest path, depending on the problem), i.e.,

$$Y = \sum_{j \in P^*} X_j,$$

where P^* itself is a random variable. We wish to estimate $d\mathsf{E}[Y]/d\theta$, where θ is a parameter in the distribution(s) of the activity times $\{X_i\}$.

Example 5.8. An example of a five-node network with six arcs is shown in Fig. 5.1, in which there are three paths: $\mathscr{P} = \{(1,4,6),(1,3,5,6),(2,5,6)\}$. If the longest path is the performance measure of interest, then

$$\begin{aligned} Y &= \max\{X_1+X_4+X_6, X_1+X_3+X_5+X_6, X_2+X_5+X_6\} \\ &= X_6 + \max\{X_1+X_4, X_1+X_3+X_5, X_2+X_5\}. \end{aligned}$$

For a specific realization, $\{X_1=9,\ X_2=15,\ X_3=8,\ X_4=16,\ X_5=11,\ X_6=12\}$, $Y = 12+\max\{9+16, 9+8+11, 15+11\} = 40$ and $P^* = (1,3,5,6)$.

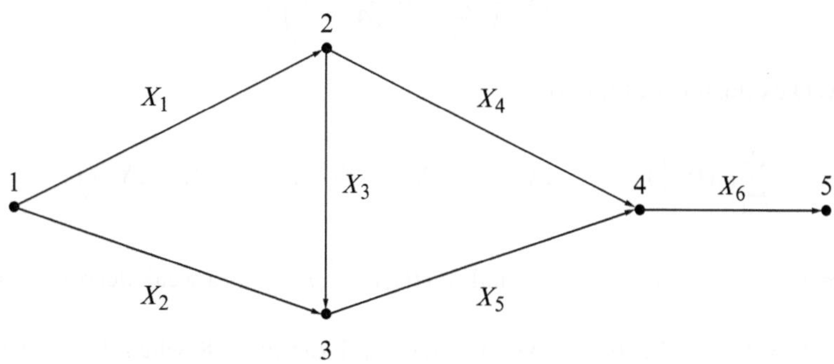

Fig. 5.1 Example stochastic activity network

Denote the c.d.f. and p.d.f. of X_i by F_i and f_i, respectively. For simplicity, assume all of the activity times are *independent*. Even so, it should be clear that duration of paths in $\mathscr{P}$ will not in general be independent, e.g., Example 5.8, where all three of the paths include arc 6, so clearly the durations are not independent.

Let θ be a parameter in the distribution of a single X_i, i.e., in f_i and F_i only. Then the IPA estimator is given by

$$\frac{dY}{d\theta} = \frac{dX_i}{d\theta}\mathbf{1}\{i \in P^*\}.$$

The LR/SF estimator is given by

$$Y\frac{\partial \ln f_i(X_i;\theta)}{\partial\theta}.$$

The WD estimator is of the form

$$c(\theta)\left(Y(X_1,\ldots,X_i^{(+)},\ldots,X_N) - Y(X_1,\ldots,X_i^{(-)},\ldots,X_N)\right)$$

where $X_i^{(\pm)} \sim F_i^{(\pm)}$, and $(c(\theta),F_i^{(+)},F_i^{(-)})$ is a weak derivative for F_i.

If we allow the parameter to possibly appear in all of the distributions, then the IPA estimator is found by applying the chain rule:

$$\frac{dY}{d\theta} = \sum_{i\in P^*}\frac{dX_i}{d\theta},$$

whereas the LR/SF and WD estimators are derived by applying the product rule of differentiation to the underlying input distribution, applying the independence that allows the joint distribution to be expressed as a product of marginals. In particular, the LR/SF estimator is given by

$$Y(X)\left(\sum_{i=1}^{N}\frac{\partial \ln f_i(X_i;\theta)}{\partial\theta}\right).$$

The WD estimator is of the form

$$\sum_{i=1}^{N} c_i(\theta)\left(Y(X_1,\ldots,X_i^{(+)},\ldots,X_N) - Y(X_1,\ldots,X_i^{(-)},\ldots,X_N)\right),$$

where $X_i^{(\pm)} \sim F_i^{(\pm)}, i = 1,\ldots,N$, and $(c_i(\theta),F_i^{(+)},F_i^{(-)})$ is a weak derivative for F_i.

Example 5.9. We illustrate several cases for Example 5.8 when $\theta = 10$ is the mean of the exponential distribution for one or all of the activity times. For the WD estimator, assume that the WD used is the entry from Table 5.1, i.e., $(1/\theta, Erl(2,\theta), exp(\theta))$, so that for the distribution(s) in which θ enters, $X_i^{(+)} \sim$

$Erl(2,10)$ and $X_i^{(-)} = X_i$. Assume the same values for X_i as in Example 5.8, and the following outputs for $X_i^{(+)}$: $X_1^{(+)} = 17$, $X_2^{(+)} = 33$, $X_3^{(+)} = 15$, $X_4^{(+)} = 40$, $X_5^{(+)} = 20$, $X_6^{(+)} = 25$.

Case 1: θ is the mean of the first activity time, i.e., $X_1 \sim exp(\theta)$.
The IPA estimate is simply $X_1/\theta = 9/10 = 0.9$, since X_1 is on the critical path in Example 5.8. The LR/SF estimate is given by $(40)(1/10)(9/10-1) = -0.4$. For WD, in the "+" network, the longest path remains the same as in the original network, and the longest path length simply increases by the difference in X_1, so the WD estimate is given by (1/10)(48-40) = 0.8.

Case 2: θ is the mean of the second activity time, i.e., $X_2 \sim exp(\theta)$.
The IPA estimate is 0, since X_2 is not on the critical path for Example 5.8. The LR/SF estimate is given by $(40)(1/10)(15/10-1) = 2$. For WD, in the "+" network, the longest path becomes (2,5,6), and the WD estimate is given by (1/10)(56-40) = 1.6.

Case 3: θ is the mean of the sixth activity time, i.e., $X_6 \sim exp(\theta)$.
The IPA estimate is $X_6/\theta = 12/10 = 1.2$, since X_6 is always on the critical path. The LR/SF estimate is given by $(40)(1/10)(12/10-1) = 0.8$, and the WD estimate is given by (1/10)(53-40) = 1.3.

Case 4: θ is the mean of *all* of the activity times, i.e., $X_i \sim exp(\theta)$ i.i.d.
The IPA estimate is $(X_1+X_3+X_5+X_6)/\theta = 40/10 = 4.0$. The LR/SF estimate is given by $(40)(1/10)(-0.1+0.5-0.2+0.6+0.1+0.2) = 4.4$. For WD, the longest path has to be computed separately for six different network realizations; and the WD estimate is the sum of the six differences: (1/10)(8+16+7+21+9+13) = 7.4.

If instead we were interested in estimating $\mathsf{P}(Y > y)$ for some fixed y, the WD and LR/SF estimators would simply replace Y with the indicator function $\mathbf{1}\{Y > y\}$. For example, in Case 1 of Example 5.9, for any $y < 40$, the LR/SF estimate is given by (1/10)(9/10-1) = -0.01, and the WD estimate is (1/10)(1-1) = 0; for $y \geq 40$, the LR/SF estimate is 0; for $y \geq 48$, the WD estimate is (1/10)(0-0) =0, whereas for $40 \leq y < 48$, the WD estimate is (1/10)(1-0)=0.1. However, IPA would not apply, since the indicator function is inherently discontinuous, so an extension of IPA such as SPA is required. On the other hand, if the performance measure were $\mathsf{P}(Y > \theta)$, then since the parameter does not appear in the distribution of the input random variables, WD and LR/SF estimators cannot be derived without first carrying out an appropriate change of variables. These cases are addressed in [15].

Single-Server Queue

We illustrate each of the three direct gradient estimators for the FCFS $G/G/1$ queue. Let A_i be the interarrival time between the $(i-1)$st and ith customer, X_i the service time of the ith customer, and T_i the system time (in queue plus in service) of the

ith customer. The sample performance of interest is the average system time over the first N customers $\overline{T}_N = \frac{1}{N}\sum_{i=1}^{N} T_i$, and we take θ as a parameter in the service time distribution(s). Assume that the system starts empty, so that the times of the first N service completions are completely determined by the first N interarrival times and first N service times. Also assume that the arrival process is independent of the service times, which are also independent of each other but not necessarily identically distributed, with the p.d.f. and c.d.f. for X_i given by f_i and F_i, respectively.

The system time of a customer for a FCFS single-server queue satisfies the recursive Lindley equation:

$$T_{i+1} = X_{i+1} + (T_i - A_{i+1})^+. \tag{5.30}$$

The IPA estimator is obtained by differentiating (5.30):

$$\frac{dT_{i+1}}{d\theta} = \frac{dX_{i+1}}{d\theta} + \frac{dT_i}{d\theta}\mathbf{1}\{T_i \geq A_{i+1}\}, \tag{5.31}$$

so that the IPA estimator for the derivative of average system time is given by

$$\frac{d\overline{T}_N}{d\theta} = \frac{1}{N}\sum_{i=1}^{N}\frac{dT_i}{d\theta} = \frac{1}{N}\sum_{m=1}^{M}\sum_{i=n_{m-1}+1}^{n_m}\sum_{j=n_{m-1}+1}^{i}\frac{dX_j}{d\theta}, \tag{5.32}$$

where M is the number of busy periods observed and n_m is the index of the last customer served in the mth busy period ($n_0 = 0$ and $n_M = N$ for M complete busy periods), and expressions for $dX/d\theta$ for many input distributions can be found in Table 5.1. Implementation of the estimator involves keeping track of two running quantities, one for (5.31) and another for the summation in (5.32); thus, the additional computational overhead is minimal, and *no alteration of the underlying simulation is required.*

To derive an LR/SF estimator, we use the fact that the interarrival times and service times are independently generated, so the joint p.d.f. on the input random variables will simply be the product of the p.d.f.s of the joint interarrival time distribution and the individual service time distributions given by

$$g(A_1,\ldots,A_N)\prod_{i=1}^{N} f_i(X_i;\theta),$$

where g denotes the joint p.d.f. of the interarrival times. Thus, the straightforward LR/SF estimator would be given by

$$\left(\frac{d\overline{T}_N}{d\theta}\right)_{LR} = \overline{T}_N \sum_{i=1}^{N}\frac{\partial \ln f_i(X_i;\theta)}{\partial\theta}, \tag{5.33}$$

where expressions for some common input distributions can be found in Table 5.1.

The WD estimator is also relatively straightforward, just incorporating the product rule of differentiation as before:

$$\left(\frac{d\overline{T}_N}{d\theta}\right)_{WD} = \sum_{i=1}^{N} c_i(\theta)\left[\overline{T}_N(A_1,\ldots,A_N,\ldots,X_i^{(+)},\ldots) - \overline{T}_N(A_1,\ldots,A_N,\ldots,X_i^{(-)},\ldots)\right],$$

where $X_i^{(\pm)} \sim F_i^{(\pm)}$, $i = 1,\ldots,N$ for $(c_i(\theta), F_i^{(+)}, F_i^{(-)})$ a weak derivative of F_i (again, see Table 5.1). Note that in general, implementation of the estimator requires $2N$ separate sample paths and resulting sample performance estimates whenever the parameter appears in N input random variables.

Example 5.10. We illustrate the numerical calculation for the three estimators when $\theta = 10$ is the mean of the exponential distribution for two cases: the first service time only or all of the service times. Again, for the WD estimator, assume that the WD used is the entry from Table 5.1, i.e., $(1/\theta, Erl(2,\theta), exp(\theta))$, so that for the distribution(s) in which θ enters, $X_i^{(+)} \sim Erl(2,10)$ and $X_i^{(-)} = X_i$. Take $N = 5$, with the first five arrivals occurring at $t = 10, 20, 30, 40, 50$, i.e., $A_i = 10,\ i = 1,2,3,4,5$, and the following service times generated:

$$X_1 = 15,\ X_2 = 7,\ X_3 = 11,\ X_4 = 9,\ X_5 = 6.$$

For these values, it turns out that all five customers are in the same busy period, i.e., all except the first customer have to wait, and we get the following outputs:

$$T_1 = 15,\ T_2 = 12,\ T_3 = 13,\ T_4 = 12,\ T_5 = 8;\ \overline{T}_5 = 12.0.$$

For the WD estimate, we also need the following (only first entry for the 1st case):

$$X_1^{(+)} = 25,\ X_2^{(+)} = 12,\ X_3^{(+)} = 21,\ X_4^{(+)} = 19,\ X_5^{(+)} = 11.$$

Letting $T_j^{(+i)} \equiv T_j(\ldots,X_i^{(+)},\ldots)$ and $\overline{T}_N^{(+i)} \equiv \overline{T}_N(\ldots,X_i^{(+)},\ldots) = \frac{1}{N}\sum_{j=1}^{N} T_j^{(+i)}$, we compute the following values for $T_j^{(+i)}$ and $\overline{T}_5^{(+i)}$:

i \ j	1	2	3	4	5	$\overline{T}_5^{(+i)}$
1	25	22	23	22	18	22.0
2	15	17	18	17	13	16.0
3	15	12	23	22	18	18.0
4	15	12	13	22	18	16.0
5	15	12	13	12	13	13.0

Note that since all the service times $X^{(+)}$ are longer than the original service times, all five customers remained in a single busy period on the "+" path.

Case 1: θ is the mean of the first service time only, i.e., $X_1 \sim exp(\theta)$.
The IPA estimate is simply $[5(X_1/\theta)]/5 = 15/10 = 1.5$; the LR/SF estimate is $(12)(1/10)(15/10 - 1) = 0.6$; and the WD estimate is $(1/10)(22-12) = 1.0$.

Case 2: θ is the mean of *all* of the service times, i.e., $X_i \sim exp(\theta)$ i.i.d.
The IPA estimate is $[(5X_1 + 4X_2 + 3X_3 + 2X_4 + X_5)/\theta]/5 = 3.2$; the LR/SF estimate is $(12)(1/10)(0.5 - 0.3 + 0.1 - 0.1 - 0.4) = -0.24$; and the WD estimate is $(1/10)(10+4+6+4+1) = 2.5$.

Variance Reduction

Both the LR/SF and WD estimators may have variance problems if the parameter appears in all of the distributions, e.g., if it is the common mean when the service times are i.i.d. The variance of the LR/SF estimator given by (5.33) increases linearly with N, so some sort of truncation is generally necessary. For the single-server queue example, the regenerative structure provides such a mechanism, to be described shortly. For the WD estimator, although the variance of the estimator may not increase with N, implementation may not be practical for large N. However, in many cases, the expression can be simplified, making the computation more acceptable. As discussed earlier, the variance properties of a WD estimator depend heavily on the particular weak derivative(s) used and the coupling (correlation) between $X_i^{(+)}$ and $X_i^{(-)}$.

Using regenerative theory, the mean steady-state system time can be written as a ratio of expectations:

$$\mathsf{E}[T] = \frac{\mathsf{E}[Q]}{\mathsf{E}[\eta]},$$

where η is the number of customers served in a busy period and Q is the sum of the system times of customers served in a busy period. Differentiation yields

$$\frac{d\mathsf{E}[T]}{d\theta} = \frac{d\mathsf{E}[Q]/d\theta}{\mathsf{E}[\eta]} - \frac{d\mathsf{E}[\eta]/d\theta}{\mathsf{E}[\eta]}\mathsf{E}[T].$$

Applying the natural LR/SF estimators for each of the terms separately leads to the following regenerative estimator over M busy periods, for the i.i.d. case where θ appears in the common service time p.d.f. f_X:

$$\left(\frac{d\overline{T}_N}{d\theta}\right)_{LR} = \frac{1}{N}\sum_{m=1}^{M}\left\{\left(\sum_{i=n_{m-1}+1}^{n_m} T_i\right)\sum_{i=n_{m-1}+1}^{n_m}\frac{\partial \ln f_X(X_i;\theta)}{\partial\theta}\right\}$$
$$-\frac{1}{N}\sum_{m=1}^{M}\left\{(n_m - n_{m-1})\sum_{i=n_{m-1}+1}^{n_m}\frac{\partial \ln f_X(X_i;\theta)}{\partial\theta}\right\}\overline{T}_N.$$

The advantage of this estimator is that the summations are bounded by the length of the busy periods, so provided the busy periods are relatively short, the variance of the estimators should be tolerable.

Higher Derivatives

For the WD estimator, a second derivative estimator would take exactly the same form as before, the only difference being that $(c_i(\theta), F_i^{(+)}, F_i^{(-)})$ should be a weak *second* derivative of F_i.

Using the regenerative method as before, the second derivative LR/SF estimator is also relatively easy to derive:

$$\left(\frac{d^2\overline{T}_N}{d\theta^2}\right)_{LR} = \frac{1}{N}\sum_{m=1}^{M}\left\{\left(\sum_{i=1}^{n_m} T_i\right)\sum_{i=n_{m-1}+1}^{n_m}\left[\frac{\partial^2 \ln f_X(X_i;\theta)}{\partial\theta^2} + \left(\frac{\partial \ln f_X(X_i;\theta)}{\partial\theta}\right)^2\right]\right\}$$
$$-\frac{1}{N}\sum_{m=1}^{M}\left\{(n_m - n_{m-1})\sum_{i=n_{m-1}+1}^{n_m}\left[\frac{\partial^2 \ln f_X(X_i;\theta)}{\partial\theta^2} + \left(\frac{\partial \ln f_X(X_i;\theta)}{\partial\theta}\right)^2\right]\right\}\frac{1}{N}\sum_{j=1}^{N} T_j.$$

On the other hand, IPA will not work for higher derivatives for the single-server queue example. The implicit assumption used in deriving an IPA estimator is that small changes in the parameter results in small changes in the sample performance, which translates to the boundary condition in (5.31) being unchanged by differentiation. In general, the interchange (5.11) will typically hold if the sample performance is continuous with respect to the parameter. For the Lindley equation, although T_{n+1} in (5.30) has a "kink" at $T_n = A_{n+1}$, it is still continuous at that point, which is the intuition behind why IPA works. Unfortunately, the "kink" means that the derivative given by (5.31) has a discontinuity at $T_n = A_{n+1}$, so that IPA will fail for the second derivative.

An unbiased SPA second derivative estimator can be derived under the additional assumption that the arrival process has independent interarrival times, by conditioning on all *previous* interarrival and service times at each departure, which determines the system time, say T_n, with the corresponding next interarrival time, A_{n+1}, unconditioned. We provide a brief informal derivation based on sample path intuition (refer to Fig. 5.2). For the right-hand estimator, in which we assume $\Delta T_n > 0$ (technically it should refer to $\Delta\theta$), the only "critical" events are those departures that terminate a busy period, with the possibility that two busy periods coalesce (idle period disappears) due to a perturbation. Letting g_n and G_n denote the

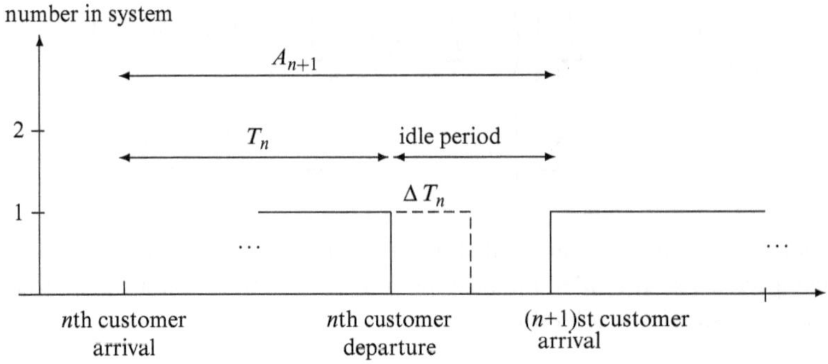

Fig. 5.2 Quantities used in deriving FCFS single-server queue SPA estimator

respective p.d.f. and c.d.f. of A_n, the corresponding probability rate (conditional on T_n) is then calculated as follows:

$$\lim_{\Delta\theta\to 0} \frac{\mathsf{P}(T_n + \Delta T_n \geq A_{n+1} | T_n < A_{n+1})}{\Delta T_n} = \frac{g_{n+1}(T_n)}{1 - G_{n+1}(T_n)} \frac{dT_n}{d\theta},$$

and the corresponding effect would be that the ΔT_n perturbation would be propagated to the next busy period. The complete SPA estimator is given by

$$\left(\frac{d^2\overline{T}_N}{d\theta^2}\right)_{SPA} = \frac{1}{N} \sum_{m=1}^{M} \sum_{i=n_{m-1}+1}^{n_m} \sum_{j=n_{m-1}+1}^{i} \frac{d^2 X_j}{d\theta^2} + \frac{1}{M} \sum_{m=1}^{M} \frac{g_{n_m+1}(T_{n_m})}{1 - G_{n_m+1}(T_{n_m})} \left(\frac{dT_{n_m}}{d\theta}\right)^2,$$

where $\frac{d^2X}{d\theta^2}$ is well-defined when $F_X(X;\theta)$ is twice differentiable, and, in particular, $\frac{d^2X}{d\theta^2} = 0$ for location, scale, and generalized scale parameters.

5.3.5 Rudimentary Theory

A basic requirement for the stochastic gradient estimator is that it be unbiased.

Definition. The gradient estimator $\hat{\nabla}_\theta J(\theta)$ is *unbiased* if $\mathsf{E}[\hat{\nabla} J(\theta)] = \nabla_\theta J(\theta)$.

Basically, unbiasedness requires the exchange of the operations of differentiation (limit) and integration (expectation), as was assumed in deriving (5.11) and (5.12). Although in theory *uniform integrability* is both a necessary and sufficient condition allowing the desired exchange of limit and expectation operators, in practice the key result used in the theoretical proofs of unbiasedness is the (Lebesgue) dominated convergence theorem. In the case of PA, the bounding involves properties of the performance measure, whereas in LR/SF and WD, the bounding involves the distribution functions (probability measures).

Theorem 5.1 (Dominated Convergence Theorem). If $\lim_{n\to\infty} g_n = g$ w.p.1 and $|g_n| \leq M \; \forall n$ w.p.1 with $\mathsf{E}[M] < \infty$, then $\lim_{n\to\infty} \mathsf{E}[g_n] = \mathsf{E}[g]$.

Take $\Delta\theta \to 0$ instead of $n \to \infty$, and g is the gradient estimator, so $g_{\Delta\theta}$ is the limiting sequence that defines the sample (path) gradient. Verifying that an actual bound exists is often a non-trivial task in applications, especially in the case of perturbation analysis.

Considering the two equations in (5.11), we translate these conditions to

$$g_{\Delta\theta} = \frac{Y(\theta+\Delta\theta) - Y(\theta)}{\Delta\theta}, \tag{5.34}$$

$$g_{\Delta\theta} = Y(x)\frac{f(x;\theta+\Delta\theta) - f(x;\theta)}{\Delta\theta}, \tag{5.35}$$

for IPA and LR/SF, respectively.

For IPA, the dominated convergence theorem bound implied by (5.34) corresponds to Lipschitz continuity on the sample performance function Y, so that the usual conditions required are piecewise differentiability and Lipschitz continuity of Y, where the Lipschitz modulus is integrable, i.e., $\exists M > 0$ with $\mathsf{E}[M] < \infty$ s.t.

$$|Y(\theta+\Delta\theta) - Y(\theta)| \leq M|\Delta\theta|.$$

In practice, the following generalization of the mean value theorem is useful.

Theorem 5.2 (Generalized Mean Value Theorem). Let Y be a continuous function that is differentiable on a compact set $\tilde{\Theta} = \Theta \backslash \tilde{D}$, where $\tilde{D}$ is a set of countably many points. Then, $\forall \theta, \theta+\Delta\theta \in \Theta$,

$$\left| \frac{Y(\theta+\Delta\theta) - Y(\theta)}{\Delta\theta} \right| \leq \sup_{\theta\in\tilde{\Theta}} \left| \frac{dY}{d\theta} \right|.$$

If $Y(\theta)$ can be shown to be continuous and piecewise differentiable on Θ w.p.1, then it just remains to show

$$\mathsf{E}\left[\sup_{\theta\in\tilde{\Theta}} \left| \frac{dY}{d\theta} \right| \right] < \infty,$$

to satisfy the conditions required for unbiasedness via the dominated convergence theorem. Basically, in order for the chain rule to be applicable, the sample performance function needs to be continuous with respect to the underlying random variable(s). This translates into requirements on the form of the performance measure and on the dynamics of the underlying stochastic system. The applicability of IPA may depend on how the input processes are constructed/generated, as was illustrated in Example 5.5, where one representation led to a biased estimator while another led to an unbiased estimator. In applying SPA, there is the choice of conditioning quantities (cf. (5.19)/(5.20)), which affects how easily the resulting conditional expectation can be estimated from sample paths. In Example 5.5, the representation that led to a biased IPA estimator only had two random variables, so there was a limited choice on what to condition, and the obvious choice led immediately to an unbiased SPA estimator.

For the LR/SF method, the bound is applied to the (joint) p.d.f. (or p.m.f.). *Note that the bound on $f(x;\theta)$ is with respect to the parameter θ and not its usual argument.* For WD, the required interchange is guaranteed by the definition of the weak derivative, but the sample performance must be in the set of "test" functions $\mathscr{L}$ in the definition, which again generally requires the dominated convergence theorem.

The previous examples can be used to show in very simple cases where difficulties arise. Consider the p.d.f.

$$f(x;\theta) = \frac{1}{\theta}\mathbf{1}\{0 < x < \theta\},$$

where the LR/SF method does not apply. In this case, f viewed as a function of θ for fixed x has a discontinuity at $\theta = x$. Similarly, consider the function

$$\mathsf{P}(Y > y) = \mathsf{E}[\mathbf{1}\{Y > y\}],$$

for which IPA will not work. In this case, the performance measure is an indicator function, which is discontinuous in its argument. In both of these simple examples, the dominated convergence theorem cannot be applied, because the required quantity cannot be bounded. However, since the dominated convergence theorem provides only *sufficient* conditions, it is possible in some very special situations (neither of which these two examples satisfy), unbiasedness may still hold.

In addition to the basic requirement for an unbiased estimator, it is important for the estimators to have low variance. There are also a multitude of choices of WD triples for a given input distribution, and this determines both the amount of additional simulation required and the variance of the resulting WD output gradient estimator. For LR/SF estimators, the variance of the estimator could also be a problem if care is not taken in implementation, e.g., a naïve estimator may lead to a linear increase in variance with respect to the simulation horizon, as in the single-server queue example.

5.3.6 Guidelines for the Practitioner

Here we summarize some key considerations in applying the three direct gradient estimation methods (PA, LR/SF, WD) (cf. [14]):

- IPA is generally inapplicable if there is a *discontinuity* in the sample performance or the underlying system dynamics; the *commuting condition* (see [21]) can be used to check the latter by considering possible event sequences in the system. Smoothness may depend on system representation, as mentioned in Sect. 5.3.1 and illustrated in Example 5.5.
- SPA uses *conditional Monte Carlo*, so just as in its use for variance reduction, the chief challenges of applying this approach include choosing what to condition on and being able to compute (or estimate) the resulting conditional expectation. The derived estimator may require additional simulations; see [18] for a comprehensive treatment of SPA.
- LR/SF and WD are more difficult to apply when the parameter does not appear explicitly in a probability distribution (so-called "structural" parameters), in which case an appropriate change of variables needs to be found.
- When the parameter of interest is known to be a *location* or (generalized) *scale* parameter of the input distribution, then IPA is particularly easy to apply, regardless of how complicated the actual distribution may be.
- For the LR/SF method, if the parameter appears in an input distribution that is reused frequently such as in an i.i.d. sequence of random variables, e.g., interarrival and service times in a queueing system, truncation of some sort will usually be required to mitigate the linear increase in variance.
- Application of the WD method generally requires two selections to be made: (a) which (non-unique) weak derivative $(c, F^{(+)}, F^{(-)})$ representation to use; and (b) how to correlate (or couple) the random variables generated from $F^{(+)}$ and $F^{(-)}$. Table 5.1 provides recommendations for many common distributions, and the Hahn–Jordan decomposition given by (5.23) and (5.24) always provides a fallback option. For the continuous distribution WD representations in Table 5.1, the use of common random numbers can often reduce variance, whereas for the Hahn–Jordan WD representation, it is best to generate the random variables independently [44]. High-dimensional vectors may require many additional simulations.
- For discrete distributions, IPA can usually be applied if the parameter occurs in the possible *values* of the input random variable, whereas LR/SF and WD can be applied if the parameter occurs in the *probabilities*.
- *Higher derivative* estimators are generally easy to derive using the LR/SF or WD method, but the former often leads to estimators with large variance and the latter may require a large number of additional simulations.

5.4 Quantile Sensitivity Estimation

We begin by considering the sensitivities of the order statistics, which naturally leads to quantile sensitivity estimation. This is the approach taken in [44]; see [16, 29, 30] for alternative approaches.

For a sample size n of random variables $Y_j,\ j=1,\ldots,n$, we consider the ith order statistic $Y_{(i)}$, where the order statistics are defined by

$$Y_{(1)} \leq Y_{(2)} \leq \cdots \leq Y_{(i)} \leq \cdots \leq Y_{(n)}.$$

For simplicity, assume the $\{Y_j\}$ are all continuous random variables, so that equality occurs w.p.0. The order-statistics definition assumes neither independence nor identical distributions for the underlying $\{Y_j\}$. When $i = \lceil \alpha n \rceil$, where $\lceil x \rceil$ denotes the ceiling function that returns the next integer greater than or equal to x, $Y_{(i)}$ will correspond to the quantile estimator for q_α, which we denote by $\hat{q}_\alpha^n(\theta) \equiv Y_{(\lceil \alpha n \rceil)}$. We write $Y_{(i)}(Y_1,\ldots,Y_n)$ as necessary to show explicit dependence on the $\{Y_j\}$, which will be the case for the WD estimator.

Under the setting that Y_i are i.i.d. and θ is a (scalar) parameter in the (common) distribution of $\{Y_j\}$, the first objective will be to estimate

$$J'(\theta) \equiv \frac{d\mathsf{E}[Y_{(i)}]}{d\theta}.$$

Then the respective IPA, LR/SF, and WD estimators are given by

$$\frac{dY_{(i)}}{d\theta}, \tag{5.36}$$

$$Y_{(i)} \sum_{j=1}^{n} \frac{\partial \ln f_{Y_j}(Y_j;\theta)}{\partial \theta}, \tag{5.37}$$

$$c(\theta) \sum_{j=1}^{n} \left[Y_{(i)}(Y_1,\ldots,Y_j^{(+)},\ldots,Y_n) - Y_{(i)}(Y_1,\ldots,Y_j^{(-)},\ldots,Y_n) \right], \tag{5.38}$$

where f_{Y_j} denotes the p.d.f. of Y_j. Note that for the IPA estimator, $dY_{(i)}/d\theta$ corresponds to the $dY/d\theta$ for $Y_{(i)}$ and NOT the ith order statistic of $\{dY_i/d\theta\}$, i.e., if we write the order statistics for the IPA estimators of $\{dY_i/d\theta\}$ as

$$\left[\frac{dY}{d\theta}\right]_{(1)} \leq \cdots \left[\frac{dY}{d\theta}\right]_{(j)} \leq \cdots \left[\frac{dY}{d\theta}\right]_{(n)},$$

then in general,

$$\frac{dY_{(j)}}{d\theta} \neq \left[\frac{dY}{d\theta}\right]_{(j)}.$$

For the setting where Y_i are i.i.d., since the dependence of Y on its arguments doesn't depend on the order,

$$\mathsf{E}\left[Y_{(j)}\frac{\partial \ln f_Y(Y_j)}{\partial \theta}\right] = \mathsf{E}\left[Y_{(k)}\frac{\partial \ln f_Y(Y_k)}{\partial \theta}\right] \;\forall j,k,$$

where f_Y denotes the common p.d.f., so that the following LR/SF estimator

$$nY_{(j)}\frac{\partial \ln f_Y(Y_j)}{\partial \theta} \;\forall j,$$

has the same expectation as the original LR/SF estimator given by (5.37), but now the linearly (with n) increasing variance becomes quadratic in n. The other problem with these estimators, whether this one or the original (5.37), is that they depend on f_Y, which is in general unknown in the simulation setting, where the Y_i denote the output of i.i.d. simulation replications, a function of input random variables, say $X_1,\ldots,X_n$, whose distributions are known.

Similarly, we can eliminate the linearly (with n) increasing number of replications for the WD estimator by noting that if Y^* is independent of all Y_i (which are i.i.d.) but not necessarily having the same distribution, then the following is true: $\forall j,k \in \{1,\ldots,n\}$,

$$Y_{(i)}(Y_1,\ldots,Y_{j-1},Y^*,Y_{j+1},\ldots,Y_n) \stackrel{d}{=} Y_{(i)}(Y_1,\ldots,Y_{k-1},Y^*,Y_{k+1},\ldots,Y_n).$$

As a result, the following estimator with the same expectation and order $n-1$ less pairs of simulations can be used:

$$nc(\theta)\left[Y_{(i)}(Y_1,\ldots,Y_j^{(+)},\ldots,Y_n) - Y_{(i)}(Y_1,\ldots,Y_j^{(-)},\ldots,Y_n)\right]$$

for any j. Again, as in the LR/SF case, the $Y_j^{(+)}$ and $Y_j^{(-)}$ need to be derived as a function of the (common) distribution of the $\{Y_i\}$.

It turns out that although all of the order statistics sensitivity estimators are unbiased, going to the quantile estimation setting by increasing the sample size only leads to asymptotic unbiasedness and not consistency in the general case, so that batching is required to obtain a consistent estimator of the quantile sensitivity q'_α, where q_α is defined by (5.3). Specifically, although the usual quantile estimator $\hat{q}^n_\alpha \equiv Y_{(\lceil \alpha n\rceil)}$ is strongly consistent, i.e.,

$$\lim_{n\to\infty} \hat{q}^n_\alpha = q_\alpha \text{ w.p.1},$$

for the quantile sensitivity estimator $\hat{q}'^n_\alpha(\theta) \equiv dY_{(\lceil \alpha n\rceil)}/d\theta$, *it does* **not** *follow in general that*

$$\lim_{n\to\infty} \hat{q}'^n_\alpha(\theta) = q'_\alpha(\theta)$$

in any sense (strong or weak), except that the mean converges correctly, i.e.,

$$\lim_{n\to\infty} \mathsf{E}[\hat{q}'^{n}_{\alpha}(\theta)] = q'_{\alpha}(\theta).$$

To obtain consistency requires batching, i.e., for k batches each of sample size n, the estimator is

$$\bar{q}'^{n,k}_{\alpha}(\theta) = \frac{1}{k}\sum_{i=1}^{k} \hat{q}'^{n,i}_{\alpha}(\theta),$$

where $\hat{q}'^{n,i}_{\alpha}(\theta)$ is the ith estimate out of k batches for whichever estimator is used—IPA, LR/SF, or WD, given by (5.36), (5.37), or (5.38), respectively—for $\hat{q}'^{n,i}_{\alpha}(\theta)$. Then it can be established that

$$\lim_{\substack{k\to\infty\\ n\to\infty}} \hat{q}'^{n,k}_{\alpha}(\theta) = q'_{\alpha}(\theta).$$

The unbatched IPA quantile sensitivity estimator (5.36) does turn out to be provably consistent if the following (very restrictive) condition is satisfied [30]:

There exists a function ϕ s.t. $\frac{dY}{d\theta} = \phi(Y)$.

We illustrate this condition with two simple examples, the first of which satisfies this condition, and the second of which does not. In both of these toy examples, the distribution of the output Y is known, so the LR/SF and WD quantile sensitivity estimators can also be written down explicitly.

Example 5.11. Take Example 5.1 with $Y = X\theta$ where X is exponentially distributed with mean 1, i.e., $X \sim exp(1)$, so $Y \sim exp(\theta)$. Since $dY/d\theta = X = Y/\theta$, the condition is satisfied with $\phi(y) = y/\theta$, and the unbatched IPA estimator $Y_{\lceil \alpha n\rceil}/\theta$ is consistent.

Example 5.12. Take $Y = \theta X_1 + X_2$, where X_1 and X_2 are both $\mathcal{N}(0,1)$ and independent, so $Y \sim \mathcal{N}(0,\theta^2+1)$. Then $dY/d\theta = X_1 = (Y - X_2)/\theta$, which still involves the input random variable X_2, and there is no way to write it in terms of Y only.

However, $\mathsf{E}[dY/d\theta|Y] = \mathsf{E}[X_1|Y] = \theta Y/(\theta^2+1) = -\partial_2 F_Y(Y;\theta)/\partial_1 F_Y(Y;\theta)$, where ∂_i denotes the partial derivative with respect to the ith argument ($i = 1,2$), and the relationship

$$\mathsf{E}\left[\frac{dY}{d\theta}\middle| Y\right] = -\frac{\partial_2 F_Y(Y;\theta)}{\partial_1 F_Y(Y;\theta)}$$

can be shown to hold in general [30].

5.5 New Approaches for Using Direct Stochastic Gradients in Simulation Optimization

This section provides an overview of several new developments in using direct stochastic gradient estimators for simulation optimization: three approaches to enhancing metamodels for response surface methodology (RSM), and combining indirect and direct estimators when used in stochastic approximation (SA), a simulation optimization approach introduced earlier in the introduction and treated in-depth in Chaps. 6 and 7. The new estimator is called a Secant-Tangents AveRaged (STAR) gradient, because it averages two direct (tangent) gradient estimators and one indirect (finite-difference secant) gradient estimator.

Before describing the other three approaches, we briefly summarize RSM; see Chap. 4 for details. Like SA, RSM is a sequential search procedure. The central component of RSM is the fitting of the local response surface using a metamodel, and the most common procedure used is regression. Specifically, after the preliminary scaling and screening of the input variables (called factors in the experimental design terminology), there are two main phases to RSM. In Phase I, which is iterative, a linear regression model is generally used to estimate a search direction to explore. Once a relatively flat area is found, RSM proceeds to Phase II where a higher-order—usually quadratic—model is fitted, which is used to estimate the optimum.

Because regression analysis arose from physically observed processes, it assumes that the only data points generated are measurements of the value of the dependent variable for each combination of independent variable values. In the simulation setting, the availability of direct gradient estimates opens up new possibilities that have just recently begun to be exploited. We begin by discussing a promising new approach that generalizes traditional regression, which is called Direct Gradient Augmented Regression (DiGAR).

Another metamodeling technique that can be used for RSM is kriging, which also arose from physical measurements. In the simulation setting, a generalization called stochastic kriging, is apropos. We discuss two enhancements to stochastic kriging that exploit the availability of direct gradient estimates: Stochastic Kriging with Gradients (SKG), which is analogous to DiGAR, and Gradient Extrapolated Stochastic Kriging (GESK), which uses the gradients in a totally different manner by generating new output data.

For these three approaches, we follow the notation of statisticians in using y for the output and x for the input.

5.5.1 *Direct Gradient Augmented Regression (DiGAR)*

Consider the usual regression setting with independent variable x and dependent variable y, where $n > 1$ data points $(x_1, y_1), \ldots, (x_n, y_n)$ are given. Both independent and dependent variables take values from a continuous domain. For expositional ease, here we describe only the one-dimensional setting, for which the basic DiGAR model is the following:

$$y_i = \beta_0 + \beta_1 x_i + \varepsilon_i, \tag{5.39}$$

$$g_i = \beta_1 + \varepsilon_i', \tag{5.40}$$

where $g_i, i = 1,2,\ldots,n$ are the gradient estimates with residuals $\{\varepsilon_i'\}$, and $\{\varepsilon_i\}$ denote the residuals of the outputs. The first line corresponds to traditional regression, whereas the second line adds the direct gradient estimates in the natural way. A weighted least-squares objective function is defined by

$$L = \alpha \sum_{i=1}^{n} (y_i - \beta_0 - \beta_1 x_i)^2 + (1-\alpha) \sum_{i=1}^{n} (g_i - \beta_1)^2, \tag{5.41}$$

where $\alpha \in [0,1]$. Note that $\alpha = 1$ corresponds to standard regression, whereas $\alpha = 0.5$ corresponds to the ordinary least squares (OLS) DiGAR model where the function estimates and derivative estimates are equally weighted, and $\alpha = 0$ corresponds to using only the gradient information. Solving the least-squares problem by minimizing (5.41) yields the following α–DiGAR estimators for the parameters in the regression model given by (5.39)/(5.40):

$$\hat{\beta}_0 = \bar{y} - \hat{\beta}_1 \bar{x}, \;\; \hat{\beta}_1 = \frac{\frac{1}{n}\sum_{i=1}^{n}(x_i - \bar{x})(y_i - \bar{y}) + \frac{1-\alpha}{\alpha}\bar{g}}{\frac{1}{n}\sum_{i=1}^{n}(x_i - \bar{x})^2 + \frac{1-\alpha}{\alpha}}, \tag{5.42}$$

where the bars indicate the sample means of the respective quantities. If the estimators for the gradients are also unbiased, then the α–DiGAR estimators are also unbiased. In the homogeneous setting where the variances of the function and gradient estimates are given by σ^2 and σ_g^2, respectively, the following result (Proposition 4 in [19]) provides conditions under which the new DiGAR estimator guarantees variance reduction for the slope estimator in the regression model, assuming unbiased estimators in the uncorrelated setting.

Theorem 5.3. *For* $\mathsf{E}[\varepsilon_i] = \mathsf{E}[\varepsilon_i'] = 0 \;\forall i,\; \mathrm{Cov}(\varepsilon_i, \varepsilon_j) = \mathrm{Cov}(\varepsilon_i', \varepsilon_j') = 0, i \neq j,$ $\mathrm{Cov}(\varepsilon_i, \varepsilon_j') = 0 \;\forall\, i, j,$

$$\frac{\sigma_g^2}{\sigma^2} \le \frac{2\alpha}{1-\alpha} + \frac{1}{\frac{1}{n}\sum_{i=1}^{n}(x_i - \bar{x})^2} \Longleftrightarrow \mathrm{Var}(\hat{\beta}_1^{DiGAR}) \le \mathrm{Var}(\hat{\beta}_1^{standard}),$$

where $\hat{\beta}_1^{DiGAR}$ *and* $\hat{\beta}_1^{standard}$ *denote the* α*-DiGAR slope estimator (5.42) and the standard slope estimator, respectively.*

Even stronger guarantees are available in the maximum likelihood estimator (MLE) setting (Proposition 7 in [19]). However, all of these results are for the one-dimensional uncorrelated setting. For the multivariate setting $\mathbf{x}_i = \{x_{ij}, j = 1,\ldots,d\}$ with corresponding output $\{y_i\}$ and partial derivatives $\{g_{ij}\}$, the extension of the least-squares objective function (5.41) for the α-DiGAR model

$$y_i = \beta_0 + \sum_{j=1}^{d} \beta_j x_{ij} + \varepsilon_i,$$

$$g_{ij} = \beta_j + \varepsilon'_{ij},$$

with non-negative weights that sum to 1, $\{\alpha_j, j = 0,1,\ldots,d\}$, is

$$\alpha_0 \sum_{i=1}^{n}(y_i - \beta_0 - \sum_{j=1}^{d} \beta_j x_{ij})^2 + \sum_{j=1}^{d} \alpha_j \sum_{i=1}^{n}(g_{ij} - \beta_j)^2,$$

which when minimized yields the following slope estimators:

$$\hat{\beta}_j = \frac{\sum_{i=1}^{n}(x_{ij} - \bar{x}_j)(y_i - \bar{y}) - \sum_{k \neq j} \beta_k \sum_{i=1}^{n}(x_{ik} - \bar{x}_k)(x_{ij} - \bar{x}_j) + n\frac{\alpha_j}{\alpha_0}\bar{g}_j}{\sum_{i=1}^{n}(x_{ij} - \bar{x}_j)^2 + n\frac{\alpha_j}{\alpha_0}}, \quad j = 1,\ldots,d,$$

which reduces to the previous expression (5.42) with $\alpha_0 = \alpha, \alpha_1 = 1 - \alpha$, when there is just a single input.

In the simulation optimization setting, simulation replications are often computationally expensive, so it is desirable to use as few of them as possible for each value of the input variable. When applying RSM for sequential search, the direction of improved performance is perhaps the most critical output of the fitted metamodel in Phase I. In several numerical experiments reported in [19] for an $M/M/1$ queue with relatively small number of simulation replications per design point, the slope of the standard linear regression model often gave the wrong sign, whereas the DiGAR model always estimated the sign correctly. Thus, where applicable, DiGAR should provide a better metamodel for simulation optimization using RSM.

5.5.2 *Stochastic Kriging with Gradients (SKG)*

The stochastic kriging (SK) model, introduced by [1], takes multivariate input $\{(\mathbf{x}_i, n_i)\}$, $i = 1,2,\ldots,k$, which generates $y_j(\mathbf{x}_i)$ as the simulation output from replication j at design point $\mathbf{x}_i$, where $\mathbf{x} = (x_1, x_2, \ldots, x_d)^T \in \mathbb{R}^d$. Stochastic kriging models $y_j(\mathbf{x}_i)$ as

$$y_j(\mathbf{x}_i) = \mathbf{f}(\mathbf{x}_i)^T \boldsymbol{\beta} + \mathsf{M}(\mathbf{x}_i) + \varepsilon_j(\mathbf{x}_i), \tag{5.43}$$

where $\mathbf{f}(\mathbf{x}_i) \in \mathbb{R}^p$ is a vector with known functions of $\mathbf{x}_i$, and $\boldsymbol{\beta} \in \mathbb{R}^p$ is a vector with unknown parameters to be estimated. It is assumed that M is a realization of a mean zero stationary random process (or random field). The simulation noise for replication j taken at $\mathbf{x}_i$ is denoted as $\varepsilon_j(\mathbf{x}_i)$. The trend term $\mathbf{f}(\mathbf{x}_i)^T\boldsymbol{\beta}$ represents the overall surface mean. The stochastic nature in M is sometimes referred to as *extrinsic* uncertainty. The uncertainty in ε_j comes from the nature of stochastic simulation, and it is sometimes referred to as *intrinsic* uncertainty.

Given the simulation response outputs $\{y_j(\mathbf{x}_i)\}_{j=1}^{n_i}$, $i = 1,2,\ldots,k$, denote the sample mean of response output and simulation noise at $\mathbf{x}_i$ as

$$\bar{y}(\mathbf{x}_i) = \frac{1}{n_i}\sum_{j=1}^{n_i} y_j(\mathbf{x}_i), \quad \bar{\varepsilon}(\mathbf{x}_i) = \frac{1}{n_i}\sum_{j=1}^{n_i} \varepsilon_j(\mathbf{x}_i), \tag{5.44}$$

and model the averaged response output as

$$\bar{y}(\mathbf{x}_i) = \mathbf{f}(\mathbf{x}_i)^T\boldsymbol{\beta} + \mathsf{M}(\mathbf{x}_i) + \bar{\varepsilon}(\mathbf{x}_i).$$

The SKG framework, introduced in [10, 11], parallels DiGAR in the stochastic kriging setting by modeling the added gradient information analogously:

$$\bar{g}^r(\mathbf{x}_i) = \left(\frac{\partial \mathbf{f}(\mathbf{x}_i)}{\partial x_r}\right)^T \boldsymbol{\beta} + \frac{\partial \mathsf{M}(\mathbf{x}_i)}{\partial x_r} + \bar{\delta}^r(\mathbf{x}_i).$$

5.5.3 Gradient Extrapolated Stochastic Kriging (GESK)

Rather than modeling the gradient directly as in DiGAR and SKG, Gradient Extrapolated Stochastic Kriging (GESK) *extrapolates* in the neighborhood of the original design points $\{\mathbf{x}_i\}, i = 1,2,\ldots,k$, i.e., additional response data are generated via linear extrapolations using the gradient estimates as follows:

$$\mathbf{x}_i^+ = \mathbf{x}_i + \Delta\mathbf{x}_i, \quad y_j(\mathbf{x}_i^+) = y_j(\mathbf{x}_i) + \mathbf{g}_j(\mathbf{x}_i)^T\Delta\mathbf{x}_i, \tag{5.45}$$

where $\Delta\mathbf{x} = (\Delta x_1, \Delta x_2, \ldots, \Delta x_d)^T$, and $\bar{y}(\mathbf{x}_i^+)$ is defined similarly as $\bar{y}(\mathbf{x}_i)$ as in (5.44). Different extrapolation techniques can be applied in (5.45), and multiple points can also be added to the neighborhood of $\mathbf{x}_i$. For simplicity, here we assume that the same step size is used for all design points, i.e., $\Delta\mathbf{x}_i = \Delta\mathbf{x}$, $i = 1,2,\ldots,k$, and that only a single additional point is added in the neighborhood of each point. Figure 5.3 depicts the idea where the gradient is indicated by the arrow and the extrapolated point by the cross.

The GESK model requires the choice of step sizes for the extrapolated points; large step sizes allow better coverage but at the cost of additional bias since the linearity is less likely to hold further from the original point. Thus, there is a basic

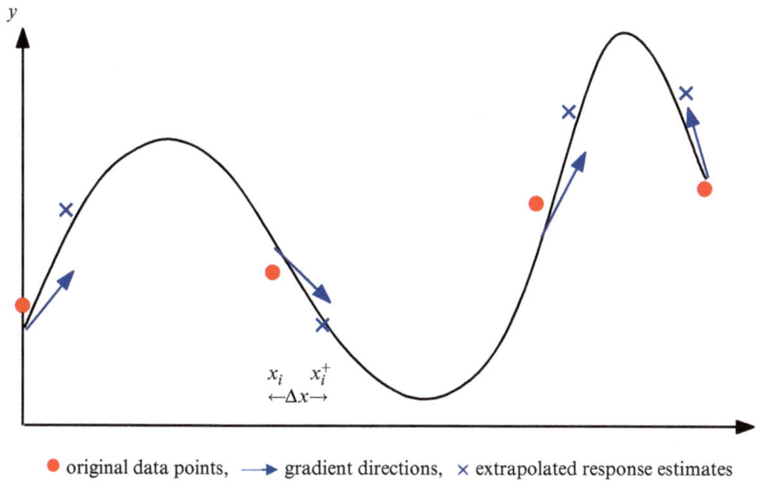

Fig. 5.3 Illustration of gradient-extrapolated response points

bias-variance tradeoff to consider. In [37], this tradeoff is analyzed in a simplified setting, leading to conditions under which improvement can be guaranteed. To illustrate the potential improvements in performance that SKG and GESK offer over ordinary stochastic kriging, we present a simple stylized numerical example from [37] for a highly nonlinear function with added noise.

Numerical Example

The output is $y_j(x) = f(x) + \varepsilon_j(x)$, where $f(x) = \exp(-1.4x)\cos(7\pi x/2)$ and $\varepsilon_j(x) \sim \mathcal{N}(0,1)$, and the gradient estimate is given by $g_j(x) = f'(x) + \delta_j(x)$, where $\delta_j(x) \sim \mathcal{N}(0,25)$. Note that the variance of the direct gradient estimates are higher than those of the response outputs, generally the situation found in stochastic simulation settings. The Gaussian correlation function $R_{\mathrm{M}}(x,x') = \exp\{-\theta(x-x')^2\}$ is used for the stochastic kriging models. The number of design points is six, and the number of replications per design point is 50. Predictions are made at $N = 200$ equally spaced points in $[-2,0]$. Figure 5.4 shows the results in the form of graphs for a typical macro-replication, where both SKG and GESK are considerably better than SK as a result of incorporating gradient estimates, and both SKG and GESK better capture the trend of the response surface.

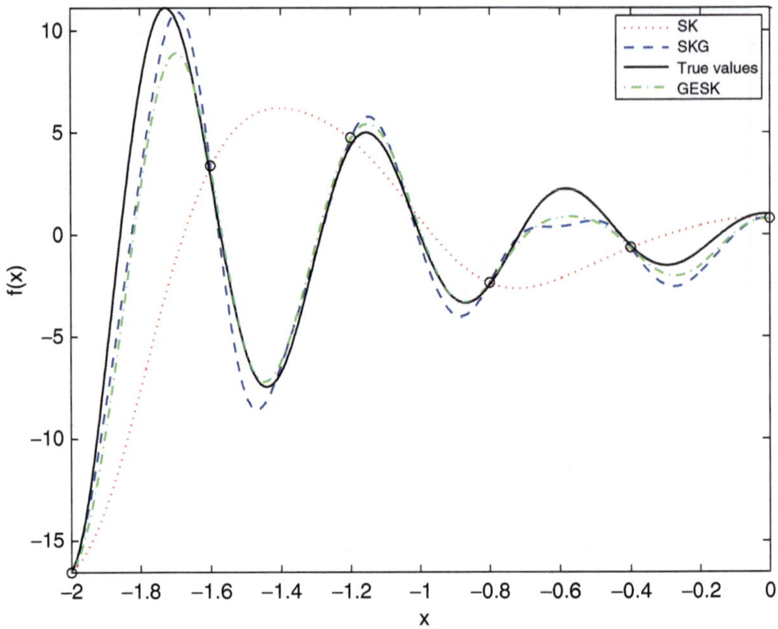

Fig. 5.4 Fitted curves for a representative macro-replication (6 points, 50 replications per point)

5.5.4 Secant-Tangents AveRaged Stochastic Approximation (STAR-SA)

Another proposal for the use of direct stochastic gradients is combining them with indirect gradients obtained from function estimates. In the deterministic optimization setting, there is nothing gained by the use of the function values themselves if exact gradients are available, but in the simulation setting the direct gradient estimates are noisy, so averaging them with indirect (finite difference) gradient estimates could potentially reduce the variance at the cost of adding some bias. This is the idea of the Secant-Tangents AveRaged (STAR) gradient estimator used in the STAR-SA algorithm introduced in [8, 9]:

$$g^{\mathrm{STAR}}(x,\xi) = \alpha \frac{Y(x+c,\xi) - Y(x-c,\xi)}{2c} + (1-\alpha)\frac{g(x+c,\xi)+g(x-c,\xi)}{2}, \tag{5.46}$$

where $\alpha \in [0,1]$, and for notational convenience, the same noise is assumed for both points, e.g., through common random numbers, which is a convex combination of a symmetric finite difference (secant) and an average of two direct gradient (tangent) estimators. More details are provided in the following chapter on stochastic approximation, Chap. 6, including the extension to higher dimensions in the form of STAR-SPSA [8].

5.6 Concluding Remarks

The direct estimation techniques PA, LR/SF, and WD have been applied to a wide variety of application domains, the dominant ones being queueing, inventory, and finance. Historically, the early research was all on queueing systems, motivated originally by problems in manufacturing and communication networks. The first work in the inventory control setting was [13, 23]. The first work in finance was [5, 17], which considered IPA, SPA, and LR/SF estimators (cf. the book [22]). More recent research in the finance setting includes the first work on quantile sensitivity estimation [29]; the first combined IPA/LR estimator for options pricing [45]; and the development of various WD estimators, as described in the dissertation [44], which also includes an extensive treatment of IPA and LR/SF. Other areas of application include preventive maintenance, statistical process control, and traffic light signal control. Stochastic gradient estimation approaches not covered in this chapter include frequency domain experimentation and Malliavin calculus, the latter primarily used in continuous-time finance settings; see [14] for references on these various applications.

More details on IPA can be found in the books by Glasserman [21], Ho and Cao [26], and Cao [6], whereas a comprehensive treatment of SPA can be found in the book by Fu and Hu [18]. Although IPA and SPA are the best known forms of perturbation analysis, other versions include rare perturbation analysis [4], structural IPA [12], discontinuous perturbation analysis [40], and augmented IPA [20]. This chapter has treated gradient estimation, hence the focus on *infinitesimal* PA, but perturbation analysis originally arose from investigating the effects of a finite perturbation [27]; see also [7,28,43]. LR/SF is discussed in the books by Rubinstein and Shapiro [39], Glasserman [22], and Asmussen and Glynn [2], all of which also include discussion of IPA. The weak derivative method was introduced by Pflug [36], and many of the WD entry derivations in Table 5.1 can be found in [24].

Acknowledgements This work was supported in part by the National Science Foundation under Grants CMMI-0856256 and ECCS-0901543, and by the Air Force Office of Scientific Research under Grant FA9550-10-10340.

References

1. B. Ankenman, B. L. Nelson, and J. Staum. Stochastic kriging for simulation metamodeling. *Operations Research*, 58(2):371–382, 2010.
2. S. Asmussen and P. Glynn. *Stochastic Simulation: Algorithms and Analysis*. Springer, New York, 2007.
3. S. Bhatnagar, M. C. Fu, S. I. Marcus, and I. J. Wang. Two-timescale simultaneous perturbation stochastic approximation using deterministic perturbation sequences. *ACM Transactions on Modeling and Computer Simulation*, 13(2):180–209, 2003.
4. P. Brémaud and F. J. Vázquez-Abad. On the pathwise computation of derivatives with respect to the rate of a point process: The phantom RPA method. *Queueing Systems: Theory and Applications*, 10:249–270, 1992.

5. M. Broadie and P. Glasserman. Estimating security price derivatives using simulation. *Management Science*, 42(2):269–285, 1996.
6. X.-R. Cao. *Realization Probabilities: The Dynamics of Queuing Systems*. Springer-Verlag, Boston, Massachusetts, 1994.
7. C. G. Cassandras and S. G. Strickland. On-line sensitivity analysis of Markov chains. *IEEE Transactions on Automatic Control*, AC-34:76–86, 1989.
8. M. Chau, M. C. Fu, and H. Qu. Multivariate stochastic approximation using a Secant-Tangents AveRaged (STAR) gradient estimator. Technical report, Working paper, University of Maryland, College Park, 2014.
9. M. Chau, H. Qu, and M. C. Fu. A new hybrid stochastic approximation algorithm. In *Proceedings of the 12th International Workshop on Discrete Event Systems*, 2014.
10. X. Chen. *Enhancing Stochastic Kriging Metamodels for Computer Simulation*. PhD thesis, Northwestern University, 2012.
11. X. Chen, B. Ankenman, and B. L. Nelson. Enhancing stochastic kriging metamodels with gradient estimators. *Operations Research*, 61(2):512–528, 2013.
12. L. Y. Dai and Y. C. Ho. Structural infinitesimal perturbation analysis for derivative estimation in discrete event dynamic systems. *IEEE Transactions on Automatic Control*, 40:1154–1166, 1995.
13. M. C. Fu. Sample path derivatives for (s,S) inventory systems. *Operations Research*, 42(2):351–364, 1994.
14. M. C. Fu. Gradient estimation. In S. G. Henderson and B. L. Nelson, editors, *Handbooks in Operations Research and Management Science: Simulation*, chapter 19, pages 575–616. Elsevier, 2006.
15. M. C. Fu. Sensitivity analysis in Monte Carlo simulation of stochastic activity networks. In F. B. Alt, M. C. Fu, and B. L. Golden, editors, *Perspectives in Operations Research: Papers in Honor of Saul Gass' 80th Birthday*, pages 351–366. Springer, 2006.
16. M. C. Fu, L. J. Hong, and J. Q. Hu. Conditional Monte Carlo estimation of quantile sensitivities. *Management Science*, 55(12):2019–2027, 2009.
17. M. C. Fu and J. Q. Hu. Sensitivity analysis for Monte Carlo simulation of option pricing. *Probability in the Engineering and Informational Sciences*, 9(3):417–446, 1995.
18. M. C. Fu and J. Q. Hu. *Conditional Monte Carlo: Gradient Estimation and Optimization Applications*. Kluwer Academic Publishers, 1997.
19. M. C. Fu and H. Qu. Regression models augmented with direct stochastic gradient estimators. *INFORMS Journal on Computing*, 26(3):484–499, 2014.
20. A. Gaivoronski, L. Y. Shi, and R. S. Sreenivas. Augmented infinitesimal perturbation analysis: An alternate explanation. *Discrete Event Dynamic Systems: Theory and Applications*, 2:121–138, 1992.
21. P. Glasserman. *Gradient Estimation via Perturbation Analysis*. Kluwer Academic Publishers, Boston, Massachusetts, 1991.
22. P. Glasserman. *Monte Carlo Methods in Financial Engineering*. Springer, New York, 2004.
23. P. Glasserman and S. Tayur. Sensitivity analysis for base-stock levels in multi-echelon production-inventory systems. *Management Science*, 41:263–281, 1995.
24. B. Heidergott, F. Vazquez-Abad, G. Pflug, and T. Farenhorst-Yuan. Gradient estimation for discrete-event systems by measure-valued differentiation. *ACM Transactions on Modeling and Computer Simulation*, 20(1):5.1–5.28, 2010.
25. B. Heidergott, F. Vazquez-Abad, and W. Volk-Makarewicz. Sensitivity estimation for Gaussian systems. *European Journal of Operational Research*, 187(1):193–207, 2008.
26. Y. C. Ho and X. R. Cao. *Perturbation Analysis and Discrete Event Dynamic Systems*. Kluwer Academic, 1991.
27. Y. C. Ho, M. A. Eyler, and T. T. Chien. A gradient technique for general buffer storage design in a serial production line. *International Journal of Production Research*, 17:557–580, 1979.
28. Y. C. Ho and S. Li. Extensions of infinitesimal perturbation analysis. *IEEE Transactions on Automatic Control*, AC-33:827–838, 1988.
29. L. J. Hong. Estimating quantile sensitivities. *Operations Research*, 57(1):118–130, 2009.

30. G. Jiang and M. C. Fu. On estimating quantile sensitivities via infinitesimal perturbation analysis. *Operations Research*, under review, 2014.
31. H. Kesten. Accelerated stochastic approximation. *The Annals of Mathematical Statistics*, 29(1):41–59, 1958.
32. J. Kiefer and J. Wolfowitz. Stochastic estimation of the maximum of a regression function. *Annals of Mathematical Statistics*, 23:462–266, 1952.
33. H. J. Kushner and D. Clark. *Stochastic Approximation Methods for Constrained and Unconstrained Systems*. Springer-Verlag, 1978.
34. H. J. Kushner and G. G. Yin. *Stochastic Approximation Algorithms and Applications*. Springer-Verlag, 2nd edition, 2003.
35. A. M. Law and W. D. Kelton. *Simulation Modeling and Analysis*. McGraw-Hill, New York, 3rd edition, 2000.
36. G. C. Pflug. Sampling derivatives of probabilities. *Computing*, 42:315–328, 1989.
37. H. Qu and M. C. Fu. Gradient extrapolated stochastic kriging. *ACM Transactions on Modeling and Computer Simulation*, 24(4): forthcoming, 2014.
38. H. Robbins and S. Monro. A stochastic approximation method. *The Annals of Mathematical Statistics*, 22:400–407, 1951.
39. R. Y. Rubinstein and A. Shapiro. *Discrete Event Systems: Sensitivity Analysis and Stochastic Optimization by the Score Function Method*. John Wiley & Sons, 1993.
40. L. Y. Shi. Discontinuous perturbation analysis of discrete event dynamic systems. *IEEE Transactions on Automatic Control*, 41:1676–1681, 1996.
41. J. C. Spall. Multivariate stochastic approximation using simultaneous perturbation gradient approximation. *IEEE Transactions on Automatic Control*, 37:332–341, 1992.
42. R. Suri and M. A. Zazanis. Perturbation analysis gives strongly consistent sensitivity estimates for the $M/G/1$ queue. *Management Science*, 34:39–64, 1988.
43. P. Vakili. Using a standard clock technique for efficient simulation. *Operations Research Letters*, 10(8):445–452, 1991.
44. W. Volk-Makarewicz. *Advances in Derivative Estimation: Ranked Data, Quantiles, and Options*. PhD thesis, Vrije (Free) University of Amsterdam, 2014.
45. Y. Wang, M. C. Fu, and S. I. Marcus. A new stochastic derivative estimator for discontinuous payoff functions with application to financial derivatives. *Operations Research*, 60(2):447–460, 2012.
46. X. Xiong, I. J. Wang, and M. C. Fu. An asymptotic analysis of stochastic approximation with deterministic perturbation sequences. In *Proceedings of the 2002 Winter Simulation Conference*, pages 285–291. IEEE, Piscataway, NJ, 2002.

Chapter 6
An Overview of Stochastic Approximation

Marie Chau and Michael C. Fu

Abstract This chapter provides an overview of stochastic approximation (SA) methods in the context of simulation optimization. SA is an iterative search algorithm that can be viewed as the stochastic counterpart to steepest descent in deterministic optimization. We begin with the classical methods of Robbins–Monro (RM) and Kiefer–Wolfowitz (KW). We discuss the challenges in implementing SA algorithms and present some of the most well-known variants such as Kesten's rule, iterate averaging, varying bounds, and simultaneous perturbation stochastic approximation (SPSA), as well as recently proposed versions including scaled-and-shifted Kiefer–Wolfowitz (SSKW), robust stochastic approximation (RSA), accelerated stochastic approximation (AC-SA) for convex and strongly convex functions, and Secant-Tangents AveRaged stochastic approximation (STAR-SA). We investigate the empirical performance of several of the recent algorithms by comparing them on a set of numerical examples.

6.1 Introduction

Stochastic approximation (SA) is a recursive algorithm that can be viewed as the stochastic counterpart to steepest descent in deterministic optimization. SA was introduced by Robbins and Monro in 1951 [32] to solve noisy root-finding problems and was later applied to the setting of stochastic optimization by solving for the zero of the gradient. The gradient-free setting was addressed by Kiefer and Wolfowitz in 1952 [24]. SA is currently one of the most widely applicable and most useful methods for simulation optimization.

Consider the stochastic optimization problem

$$\min_{x\in\Theta} f(x), \tag{6.1}$$

M. Chau • M.C. Fu (✉)
University of Maryland, College Park, MD, USA
e-mail: mchau@math.umd.edu; mfu@umd.edu

M.C. Fu (ed.), *Handbook of Simulation Optimization*, International Series in Operations Research & Management Science 216, DOI 10.1007/978-1-4939-1384-8_6

where $f(x) = \mathsf{E}[Y(x,\xi)]$ is a performance measure, $Y(x,\xi)$ is a sample performance, ξ denotes the stochastic effects, and $\Theta \subseteq \mathbb{R}^d$ is a continuous parameter space. In this case, the objective is to find a sequence $\{x_n\}$ that converges to a unique (local) optimum

$$x^* = \arg\min_{x\in\Theta} f(x), \tag{6.2}$$

by using the recursion

$$x_{n+1} = \Pi_\Theta\left(x_n - a_n\hat{\nabla}f(x_n)\right), \tag{6.3}$$

where $\Pi_\Theta(x)$ is a projection of x back into the feasible region Θ if $x \notin \Theta$, $a_n > 0$ is the step size or gain size, $\hat{\nabla}f(x_n)$ is an estimate of the gradient $\nabla f(x_n)$, and x_N is the output, where N is the stopping time, which we denote by x_N^*. The projection operator is only required in the constrained setting. Moreover, the minimization problem in (6.1) and (6.2) could easily be changed to maximization by changing the sign of a_n in (6.3). The two classical methods, Robbins–Monro (RM) and Kiefer–Wolfowitz (KW), estimate $\nabla f(x_n)$ using unbiased direct gradient estimates and finite difference gradient estimates, respectively. Under certain conditions, RM and KW have the respective asymptotic convergence rates $O(n^{-1/2})$ and $O(n^{-1/3})$.

Advances in SA have included the development of new algorithms, modifications to existing ones, and new asymptotic and finite-time theory. Asymptotic convergence properties of KW, RM and their variations have been a major research focus (cf. [12, 14, 15, 28, 31, 41]). The original RM and KW algorithms apply to one-dimensional problems, but they were later extended to the multidimensional case [2]. Furthermore, the earlier conditions used to prove convergence for RM and KW were relaxed to obtain almost sure convergence [2]. The estimate x_n in (6.3) was shown to be asymptotically normal [15] with the optimal convergence rate of $O(n^{-1/2})$ [8]. More recently, researchers have placed greater emphasis on finite-time theory as well as error bounds on the difference between objective value at the current estimate and the optimal objective value (i.e., $\mathsf{E}[f(x_N^*) - f(x^*)]$), which the next chapter treats in more detail.

Although recursion (6.3) is quite simple, the choice of step-size sequence $\{a_n\}$, gradient estimator $\hat{\nabla}f(x_n)$, projection operator Π_Θ, and output x_N^* each has a significant impact on the performance of the algorithm.

We first discuss the influence of step-size sequence $\{a_n\}$ on the finite-time performance. It is widely known that the practical performance of the classical RM and KW algorithms is highly dependent on the choice of $\{a_n\}$ and often performs poorly without tuning. The algorithm can experience a long oscillatory period if the gain sequence $\{a_n\}$ is "too large," where the iterates jump back and forth without approaching the optimum x^*, which can be seen in Fig. 6.1 (left graph), or a degraded convergence rate if $\{a_n\}$ is "too small" relative to the magnitude of the gradient, where the iterates barely move, which can be seen in Fig. 6.1 (right graph) (only x_1 is labeled since the other iterates are in the same position).

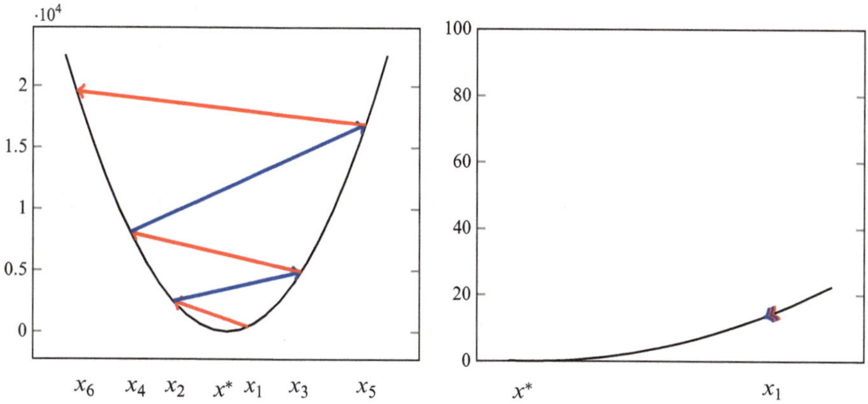

Fig. 6.1 Sensitivity of SA to step size a_n when a_n is "too large" relative to the gradient (left graph) and when a_n is "too small" relative to the gradient (right graph)

One approach to tackle the sensitivity is to develop robust step-size sequences, i.e., an adaptive step-size rule. The earliest attempt at adaptive step sizes was Kesten's rule, which can be applied to both RM and KW [23]. This step size only decreases when there is a directional change in the iterates, i.e., $(x_{n+1} - x_n)(x_n - x_{n-1}) < 0$. The idea behind this adaptive step size is that, if the iterates move in the same direction, there is reason to believe they are not in close proximity of the optimum, so the momentum should not be decreased. Later, this idea was extended by increasing the step size, as opposed to keeping it constant when the consecutive errors in the estimates are of the same sign, to increase the speed to convergence [37]. A recent attempt, called scaled-and-shifted Kiefer–Wolfowitz (SSKW) and described in more detail in Sect. 6.4.1, adaptively adjusts the step-size sequence $\{a_n\}$ finitely many times during the course of a modified version of KW [4]. The rationale behind the procedure is to increase the gain size so the iterates are able to move from one endpoint to the other in the one-dimensional case (ideas are analogous in the multidimensional case [3]), which ensures the step sizes are large enough to make noticeable progress towards the optimum in finite-time. If the step sizes $\{a_n\}$ are too large, then they decrease at a faster pace during the recourse stage. Another method used to select an adaptive step size is generating an approximation of the inverse of the Hessian, which is the stochastic analogue to the deterministic Newton–Raphson method [42]. According to Yakowitz et al. [45], "...the optimal choice [of step-size sequence] involves the Hessian of the risk [objective] function, which is typically unknown and hard to estimate." Hessians have been estimated using a set of finite difference gradient approximations [16], heuristics based on quasi-Newton methods [22], and finite difference approximations using gradient estimates [33, 42]. The choice of $\{a_n\}$ has a significant impact on the finite-time performance of the algorithm, which is quite difficult to characterize theoretically.

Another approach to reduce the sensitivity of the estimated optimum to $\{a_n\}$ is to modify the output so that the algorithm puts less emphasis on the last iterate. The underlying idea behind methods of this type is to take longer steps and incorporate a subset of the iterates into the output to decrease the reliance on the last iterate. Iterate averaging, which takes the average of all iterates as the output as opposed to the final iterate, was the earliest proposal [31, 34]. This algorithm can be easily implemented, is "robust," since it is less sensitive to the initial step size choice, and exhibits $O(n^{-1/2})$ asymptotic convergence rate under appropriate conditions. A generalization of iterate averaging, called robust stochastic approximation (RSA) algorithm [30], uses the step-size sequence $\{a_n\}$ for weighting the iterates, which will be described in more detail in Sect. 6.4. Another generalization introduces a proximity function into the objective function, which acts as a regularization term to prevent the next iterate update x_{n+1} from being too far from x_n. An example from this class of algorithms called the accelerated stochastic approximation (AC-SA) algorithm [20] is detailed in Sect. 6.4.

Asymptotic theory for SA initially only considered functions satisfying specific global conditions; however, it is only necessary for the requirements to hold for a compact set as long as it contains the optimum, so the projection operator is particularly important in the constrained optimization setting. The feasible region Θ must be large enough to increase the likelihood that $x^* \in \Theta$, but enlarging the search space may deteriorate the performance of the algorithm. Adaptively increasing the search space still leads to provable convergence with an appropriate projection operator, such as adaptively projecting the iterates onto an increasing compact set [1].

The gradient estimate is also clearly central to any SA algorithm, and thus the subject of stochastic gradients is treated in depth in Chap. 5. The most common gradient estimate is obtained using finite differences, because it only requires performance measures and no other additional information from the system. For each dimension, the finite difference gradient estimate requires two performance measures, and if the measurements are highly volatile, then the gradient estimates can be noisy. Furthermore, finite difference estimates become computationally expensive in high dimensions, since the cost grows linearly with the parameter dimension [2]. Simultaneous perturbation stochastic approximation (SPSA) [38] only requires two estimates of the objective function to approximate the gradient, and the computational cost is independent of the dimension of the parameter space. Recently, a new SA algorithm called Secant-Tangents AveRaged stochastic approximation (STAR-SA) has been proposed, and it employs a hybrid gradient estimator that combines direct gradient estimates with a symmetric finite difference gradient estimate [5, 6].

The remainder of the chapter is organized as follows. We introduce the classical stochastic approximation methods, Robbins–Monro (RM) and Kiefer–Wolfowitz (KW) in Sect. 6.2. In Sect. 6.3, we present some of the most useful enhancements for simulation optimization: Kesten's rule, Ruppert–Polyak averaging iterates, varying bounds, and SPSA. We describe four recent developments, scaled-and-shifted

Kiefer–Wolfowitz (SSKW), robust stochastic approximation (RSA), accelerated stochastic approximation (AC-SA) for convex and strongly convex functions, and Secant-Tangents AveRaged (STAR) stochastic approximation in Sect. 6.4. In Sect. 6.5, we present numerical experiments comparing KW-type algorithms (original KW, KW with Kesten's rule, SSKW), RM-type methods (original RM, RM with iterate averaging, RSA, AC-SA), and single versus mixed gradients (RM, KW, STAR-SA). Finally, we close with concluding remarks in Sect. 6.6.

6.2 Classical Methods

The classical RM and KW algorithms address unconstrained stochastic optimizations problems, so we consider the recursive scheme

$$x_{n+1} = x_n - a_n \hat{\nabla} f(x_n), \tag{6.4}$$

which is identical to (6.3) with the exception of the projection operator.

6.2.1 *Robbins–Monro (RM) Algorithm*

The RM algorithm was introduced to solve the root-finding problem

$$M(x) = \alpha$$

for $x \in \mathbb{R}$, where $M(x)$ is a monotone function and $\alpha \in \mathbb{R}$. However, it was later applied to a specific case of root-finding in the stochastic optimization setting, where the objective is to optimize a stochastic objective function $f(x)$ by setting $M(x) = \nabla f(x)$ and $\alpha = 0$. RM solves this problem iteratively as in (6.3) by replacing $\hat{\nabla} f(x_n)$ with an unbiased estimator, and the output is taken as the last iterate, x_N^*, where N is the stopping time. However in RM, the direct gradient measurements are still approximations to the actual gradient because of the presence of noise ($\hat{\nabla} f(x_n) = \nabla f(x_n) + \varepsilon_n$, where ε_n is noise with zero mean). Given the appropriate parameters, this algorithm converges asymptotically at a rate of $O(n^{-1/2})$ [35].

Theorem 6.1 (Theorem 2 [32]). *Assume $\nabla f(x)$ has a unique root x^* and suppose $\hat{\nabla} f(x)$ is an unbiased gradient estimator, i.e., $\nabla f(x) = \mathsf{E}[\hat{\nabla} f(x)]$. If the sequence $\{x_n\}$ is generated from (6.4) and the following conditions hold:*

1. *$\{a_n\}$ is a sequence of positive constants such that $\sum_{n=1}^{\infty} a_n = \infty$ and $\sum_{n=1}^{\infty} a_n^2 < \infty$.*
2. *$\nabla f(x) \geq 0$ for $x > x^*$ and $\nabla f(x) \leq 0$ for $x < x^*$.*
3. *There exists a positive constant C such that $\mathsf{P}(|\hat{\nabla} f(x)| \leq C) = 1\ \forall x$.*

Then $x_n \xrightarrow{L^2} x^*$ as $n \to \infty$, where $\xrightarrow{L^2}$ denotes mean-squared convergence.

The most well-known conditions are restrictions on the gain sequence $\{a_n\}$. Generally, $a_n \to 0$ but $\sum_{n=1}^{\infty} a_n = \infty$, which prevents the step size from converging to zero too quickly, so the iterates are able to make progress to x^* and do not get stuck at a poor estimate. The usual form is $a_n = \frac{\theta_a}{(n+A)^\alpha}$, where $\theta_a > 0$, $A \geq 0$ and $\frac{1}{2} < \alpha \leq 1$, with $A=0$ and $\alpha = 1$ as a commonly used choice. The objective function f is assumed to have a global minimum with a bounded derivative.

6.2.2 *Kiefer–Wolfowitz (KW) Algorithm*

The KW stochastic approximation algorithm is referred to as a gradient-free or stochastic zeroth-order method in the following chapter, since it only requires noisy measurements of the function and does not require additional information on the system dynamics or input distributions. The original KW iterative scheme

$$x_{n+1} = x_n - a_n \frac{Y(x_n + c_n, \xi_n^+) - Y(x_n - c_n, \xi_n^-)}{2c_n}, \tag{6.5}$$

estimates the gradient using a symmetric finite difference gradient estimate, and under certain conditions, KW can achieve an asymptotic convergence rate of $O(n^{-1/3})$. In addition, common random numbers (CRN) can be employed to decrease the variance of estimates, and KW can achieve an asymptotic convergence rate of $O(n^{-1/2})$ in certain settings [25].

Theorem 6.2 (Theorem in [24]). *Assume* $f(x) = \mathsf{E}[Y(x,\xi)]$. *If the sequence* $\{x_n\}$ *is generated from (6.5) and the following conditions hold:*

1. *Let* $\{a_n\}$ *and* $\{c_n\}$ *be positive tuning sequences satisfying the conditions*

$$c_n \to 0, \ \sum_{n=1}^{\infty} a_n = \infty, \ \sum_{n=1}^{\infty} a_n c_n < \infty, \ \sum_{n=1}^{\infty} a_n^2 c_n^{-2} < \infty.$$

2. $f(x)$ *is strictly decreasing for* $x < x^*$, *strictly increasing for* $x > x^*$.
3. $\mathrm{Var}[Y(x,\xi)] < \infty$ *and the following regularity conditions hold:*

   1) *There exist positive constants* β *and* B *such that*

$$|x' - x^*| + |x'' - x^*| < \beta \Longrightarrow |f(x') - f(x'')| < B|x' - x''|.$$

   2) *There exist positive* ρ *and* R *such that*

$$|x' - x''| < \rho \Longrightarrow |f(x') - f(x'')| < R.$$

   3) *For every* $\delta > 0$ *there exists a positive* $\pi(\delta)$ *such that*

$$|x - x^*| > \delta \Longrightarrow \inf_{\frac{\delta}{2} > \varepsilon > 0} \frac{|f(x+\varepsilon) - f(x-\varepsilon)|}{\varepsilon} > \pi(\delta).$$

Then $x_n \xrightarrow{p} x^*$ *as* $n \to \infty$, *where* $\xrightarrow{p}$ *denotes convergence in probability.*

Condition 1 assures that the step size a_n does not converge to zero too fast, so the iterates do not get stuck at a poor estimate. In addition, the condition restricts the finite difference step size c_n from decreasing too quickly, which constrains the noise of the gradients. The second condition insures that there is a global optimum. The first regularity condition requires $f(x)$ to be locally Lipschitz in a neighborhood of x^*; the second one prevents $f(x)$ from changing drastically in the feasible region; and the last one prohibits the function from being very flat outside a neighborhood of x^* so that the iterates approach the optimum. Although the KW algorithm converges asymptotically, its finite-time performance is dependent on the choice of tuning sequences, $\{a_n\}$ and $\{c_n\}$. If the current x_n is in a relatively flat region of the function and the a_n is small, then the convergence will be slow. On the other hand, if the x_n is located in a very steep region of the function and $\{a_n\}$ is large, then the iterates will experience a long oscillation period. If $\{c_n\}$ is too small, the gradient estimates using finite differences could be extremely noisy.

KW has been extended to higher dimensions, and two common gradients considered are symmetric differences and forward differences whose ith component is given by

$$\hat{\nabla} f_i(x_n) = \begin{cases} \dfrac{Y(x_n + c_n e_i, \xi_{n,i}^+) - Y(x_n - c_n e_i, \xi_{n,i}^-)}{2c_n} & \text{symmetric difference,} \\ \dfrac{Y(x_n + c_n e_i, \xi_{n,i}^+) - Y(x_n, \xi_{n,i})}{c_n} & \text{one-sided forward difference,} \end{cases}$$

where e_i denotes a d-dimensional ith unit basis vector, $c_n \in \mathbb{R}^+$, and $Y(x, \xi)$ is an unbiased estimate of $f(x)$. This method perturbs each component of x_n (i.e., $x_{n,i}$ for $i = 1, \ldots, d$) one at a time while holding all others constant and returns a corresponding function value estimate. For instance, symmetric differences requires the estimate of two function values $f(x_n + c_n e_i)$ and $f(x_n - c_n e_i)$ for $i = 1, \ldots, d$, and forward differences requires $f(x_n)$ and $f(x_n + c_n e_i)$ for $i = 1, \ldots, d$; therefore, using symmetric and one-sided forward difference estimates involves $2d$ and $d + 1$ simulation replications, respectively. Although using the symmetric difference scheme is computationally more expensive, it has the potential to reach an asymptotic convergence rate of $O(n^{-1/3})$ compared to $O(n^{-1/4})$ for forward differences. For $d = 1$, the computational cost is identical for both the symmetric difference and one-sided forward difference. Compared with the RM algorithm, however, KW convergence rates are typically inferior, although under certain conditions with CRN $\xi_{n,i}^+ = \xi_{n,i}^-$, KW algorithms also can achieve the $O(n^{-1/2})$ asymptotic convergence rate. For simulation optimization, RM is not always applicable since additional information is needed, which may not be readily available or is difficult to obtain. For KW, there is an additional task of appropriately choosing the difference sequence $\{c_n\}$. In general, KW is a simple algorithm to implement for simulation optimization applications, albeit costly in high-dimensional settings.

6.3 Well-Known Variants

In this section, we elaborate on Kesten's rule, iterate averaging, adaptively varying bounds, and SPSA.

6.3.1 *Kesten's Rule*

It is well-known that the classical SA algorithms are extremely sensitive to the step-size sequence $\{a_n\}$. Therefore, it could be advantageous to consider adaptive step sizes that adjust based on the ongoing performance of the algorithm, in hopes of adapting them to the characteristics of the function at the current location of the iterate and proximity of the current iterate to the optimum. Kesten's rule [23] decreases the step size only when there is a directional change in the iterates. The notion behind this adaptive step size is that, if the iterates continue in the same direction, there is reason to believe they are approaching the optimum and the pace should not be decreased in order to accelerate the convergence. If the errors in the estimate values change signs, it is an indication that either the step size is too large and the iterates are experiencing long oscillation periods or the iterates are in the vicinity of the true optimum; either way, the step size should be reduced to a more appropriate step size or to hone in on x^*. The following algorithm is for the one-dimensional case $d = 1$.

SA Algorithm Using Kesten's Rule

- Input. Choose $x_1 \in \Theta, \{a_n\}$, Π_Θ, and stopping time N.
- Initialize.
  - Let $n = 2$ and $k = 1$.
  - Generate an estimate $\hat{\nabla} f(x_1)$ of $\nabla f(x_1)$.
  - Compute $x_2 = \Pi_\Theta(x_1 - a_1 \hat{\nabla} f(x_1))$.
- While $n < N$,
  - Step 1. Generate an estimate $\hat{\nabla} f(x_n)$ of $\nabla f(x_n)$.
  - Step 2. Compute $x_{n+1} = \Pi_\Theta(x_n - a_k \hat{\nabla} f(x_n))$. If $(x_{n+1} - x_n)(x_n - x_{n-1}) < 0$, go to Step 3. Otherwise, go to Step 4.
  - Step 3. Let $n = n+1$ and $k = k+1$. Go to Step 1.
  - Step 4. Let $n = n+1$. Go to Step 1.
- Output. $x_N^* = x_N$.

Kesten's rule can be applied to both RM and KW and still guarantee convergence in probability, as long as $\{a_n\}$ satisfies condition (1) in Theorems 1 and 2 for RM and KW, respectively [23]. An extension of Kesten's rule to higher dimensions is discussed in [11]. See [18] for an extensive review of both deterministic and stochastic step sizes.

6.3.2 Averaging Iterates

Iterate averaging approaches SA from a different angle. Instead of fine-tuning the step sizes to adapt to the function characteristics, iterate averaging takes bigger steps (i.e., a_n larger than $O(n^{-1})$) for the estimates to oscillate around the optimum, so the average of the iterates will result in a good approximation to the true optimum. The idea is simple, and yet can be very effective. It is easy to see that for this method to be successful, it is essential for the iterates to surround the optimum in a balanced manner, and the domain for which the iterates oscillate shrinks as n increases. Averaging trajectories reduces the sensitivity to the initial step size choice. The algorithm follows recursion (6.3) for the RM case; however, instead of the taking the last iterate x_N as the output, the optimum is estimated by

$$x_N^* = \frac{1}{N}\sum_{n=1}^{N} x_n,$$

which is an average of N iterates, where N is the stopping time. Under "classic" assumptions, iterate averaging achieves the same convergence rate as the RM method. Furthermore, $\sqrt{n}(x_n^* - x^*)$ is asymptotically normal with mean zero and the smallest covariance matrix, which is the inverse of the average Fisher information matrix. (cf. [31]). A constant step size can be applied and yields convergence in distribution [28].

A variation of this method is called the "sliding window" average, which is based on the last m iterates:

$$x_N^* = \frac{1}{m}\sum_{n=N-m+1}^{N} x_n. \tag{6.6}$$

An advantage of (6.6) is it ignores the first $N-m$ iterates, which may be poor estimates, since the first iterate is arbitrary, and averages only the last m, which are assumed to be closer to x^*. Asymptotic normality for a growing window is shown in [26, 28], which also includes constant step sizes. Another modification of the original method incorporates x_N^* with x_N in the components being averaged, which is known as the feedback approach [27]. These methods are suited for problems where the iterates hover around the optimum. In an empirical study, iterate averaging was applied to SPSA [29]. The results suggest that if the Hessian of $f(x)$ is large, averaging is considered ideal, since it is associated with a high variability in $f(x)$, which indicates the iterates are moving around the optimum. In general, averaging iterates leads to more robustness with respect to step-size sequence because of the reduced sensitivity, while converging at the same optimal asymptotic rate as RM. Inspired by iterate averaging, weighted averages for KW was presented to achieve the optimal asymptotic convergence rate $O(n^{-1/2})$ under certain conditions [13]. Under certain parameter settings, iterate averaging and weighted averaging produce the same estimator.

6.3.3 Varying Bounds

Initially, the asymptotic theory for SA only considered functions satisfying specific global conditions; however, subsequently it was shown the requirements need only hold on a compact set $\Theta \in \mathbb{R}^d$ containing the optimum. Therefore, the projection operator is particularly important in the constrained optimization setting. Since the optimum is unknown, the compact set should be large enough so that $x^* \in \Theta$ with high probability; however, this may increase the potential of an algorithm to perform poorly due to the size of the parameter search space [1] . For instance, if the compact set is very large, the step size is extremely small, and the current iterate is extremely far from the optimum, then the convergence is likely to be slow; however, if the compact set is small and contains the optimum, then the iterates will never be too far from the optimum. Even if the step sizes are small, the convergence will be much faster in comparison to the algorithm restricted to a much larger set.

One of the first ideas was to project the iterates onto a predetermined fixed point once the magnitude of the iterate surpassed an arbitrarily specified threshold, with the threshold increasing after it is exceeded [10]. This method converges asymptotically, but in practice, it has its pitfalls. When an iterate is projected onto an arbitrary fixed point, in a sense, the algorithm restarts from this "initial" value with a smaller sequence of step sizes. Not only does it lose all of the progress gained from the iterations prior to the projection, but the reduction in step size could hinder the convergence by moving even slower towards the optimum. To circumvent this issue, it was shown that it suffices to project the iterates onto a predetermined bounded set [46]. This is a slight improvement, since the iterates do not start from the same position with an even smaller step size. However, it still has its limitations, since the initial start values are restricted to the predetermined compact set. Later, an algorithm defined over a growing feasible region by writing Θ as an increasing sequence of compact sets (i.e., $\Theta_m \subseteq \Theta_{m+1}$, where $\Theta = \cup_m \Theta_m$) was introduced [1]. The orthogonal projection operator changes from Π_{Θ_m} to $\Pi_{\Theta_{m+1}}$ if $x_n \notin \Theta_m$. The idea is to start with a smaller feasible region Θ_1 and only increase when there is reason to believe the optimum $x^* \notin \Theta_1$ (i.e., when the $x_n \notin \Theta_1$). Since the projection is made onto the current compact set Θ_m, the progress gained up to that point is not lost. The feasible region Θ is written as $\cup_{m=1}^{\infty} \Theta_m$, so it is impossible for $x^* \notin \Theta_m$ for some m. If x^* is contained in one of the earlier compact sets and if they grow slowly, the empirical results could improve significantly. The key in the performance is to choose the sequence $\{\Theta_m\}$ appropriately. If it grows too quickly, the results might be very similar to that of the original SA algorithm. The following algorithm and convergence result are for the RM multidimensional case $d \geq 1$, where $||\cdot||$ denotes the Euclidean norm.

SA with Varying Bounds

- Input. Choose $x_1 \in \Theta_1$, $\{a_n\}$ and $\{\Theta_m\}$.
- Initialize. Let $n = 1$ and $m = 1$.
- While $n < N$,
  - Step 1. Generate an estimate $\hat{\nabla} f(x_n)$ of $\nabla f(x_n)$.
  - Step 2. Compute $x'_{n+1} = x_n - a_n \hat{\nabla} f(x_n)$. If $x'_{n+1} \in \Theta_m$, go to Step 3. Otherwise, go to Step 4.
  - Step 3. Let $x_{n+1} = x'_{n+1}$, $n = n+1$ and go to Step 1.
  - Step 4. Let $x_{n+1} = \Pi_{\Theta_m}(x'_{n+1})$, $n = n+1$, $m = m+1$ and go to Step 1.
- Output. $x^*_N = x_N$.

Theorem 6.3 (Theorem 2 [1]). *Let the sequence $\{x_n\}$ be generated using the above algorithm, $\varepsilon_n = \hat{\nabla} f(x) - \mathsf{E}[\hat{\nabla} f(x)|\mathscr{F}_n]$, and $\beta_n = \mathsf{E}[\hat{\nabla} f(x)|\mathscr{F}_n] - \nabla f(x)$, where $\mathscr{F}_n$ is the smallest σ-algebra used to generate x_{n+1}. If the following conditions hold:*

1. *The sequence $\{\Theta_m\}$ is a set of compact convex sets such that $\Theta_m \subseteq \Theta_{m+1}$ for all m and $\cup_{m=1}^{\infty} \Theta_m = \Theta$.*
2. *The positive sequences of real numbers $\{a_n\}$ and $\{c_n\}$ converge to zero such that $\sum_{n=1}^{\infty} a_n = \infty$, $\sum_{n=1}^{\infty} a_n c_n < \infty$, and $\sum_{n=1}^{\infty} a_n^2 c_n^{-2} < \infty$.*
3. *There exists $\kappa \geq 0$ such that $\mathsf{E}[||\varepsilon_n||^2|\mathscr{F}_n] \leq \frac{\kappa}{c_n^2}(1 + ||x_n - x^*||^2)$ a.s. for all n.*
4. *$||\beta_n||$ is bounded a.s. for all n, and $\sum_{n=1}^{\infty} a_n ||\beta_n|| < \infty$ a.s.*
5. *There exist a positive sequence of real numbers $\{M_n\}$ and integer $N \geq 1$ such that $\sum_{n=1}^{\infty} a_n^2 M_n^2 < \infty$ and for all $n \geq N$, $\sup_{x \in \Theta_{n-1}} ||f(x)|| \leq M_n$.*
6. *There exists a unique $x^* \in \Theta$ such that $\nabla f(x^*) = 0$, and for all $0 < \delta \leq 1$, $\inf_{x \in \Theta: \delta \leq ||x - x^*|| \leq \delta^{-1}} f(x)^{\mathsf{T}}(x - x^*) > 0$.*

Then $x_n \to x^*$ a.s. as $n \to \infty$.

If an appropriate increasing sequence of compact sets is chosen, the finite-time performance can improve significantly, but this optimal choice is still an open problem.

6.3.4 Simultaneous Perturbation Stochastic Approximation (SPSA)

Simultaneous perturbation stochastic approximation (SPSA) specifically addresses multivariate optimization problems [38]. Similar to KW-type algorithms, SPSA only requires the objective function values to approximate the underlying gradient and is therefore easy to implement. However, SPSA only requires two functional evaluations at each iteration regardless of the dimension of the parameter space Θ, which could potentially reduce the computational cost significantly in high-dimensional problems. SPSA perturbs the vector x randomly in all directions simultaneously (hence, the name of the method) and the ith component of the gradient estimate has the form

$$\hat{\nabla} f_i(x_n) = \frac{Y(x_n + c_n\Delta_n, \xi_n^+) - Y(x_n - c_n\Delta_n, \xi_n^-)}{2c_n\Delta_{n,i}}, \tag{6.7}$$

where $\Delta_n = (\Delta_{n,1}, \ldots, \Delta_{n,d}) \in \mathbb{R}^d$ and generally assumed to be i.i.d. and independent across components, $c_n \in \mathbb{R}^+$ is the finite difference step size, and $\xi_n^{\pm}$ denotes the randomness. Observe that the numerator in (6.7) involves two function estimates and is identical for all i; therefore, the cost of the full gradient (aside from generating Δ_n) is independent of dimension.

SPSA Algorithm

- Input. Choose $x_1 \in \Theta$, $\{a_n\}$, $\{c_n\}$, and stopping time N.
- Initialize. Let $n = 1$.
- While $n < N$,
  - Step 1. Generate a d-dimensional random perturbation vector Δ_n.
  - Step 2. Generate an estimate of $\nabla f(x_n)$:

$$\hat{\nabla} f(x_n) = \frac{Y(x_n + c_n\Delta_n, \xi_n^+) - Y(x_n - c_n\Delta_n, \xi_n^-)}{2c_n} \begin{bmatrix} \Delta_{n,1}^{-1} \\ \vdots \\ \Delta_{n,d}^{-1} \end{bmatrix}$$

  - Step 3. Compute $x_{n+1} = x_n - a_n \hat{\nabla} f(x_n)$.
  - Step 4. Let $n = n + 1$. Go to Step 1.
- Output. $x_N^* = x_N$.

Theorem 6.4 (Theorem 7.1 [40]). *Suppose f has a unique minimum $x^* \in \Theta$ and $\{x_n\}$ is generated using SPSA. If the following conditions hold:*

1. *The positive sequences of real numbers $\{a_n\}$ and $\{c_n\}$ converge to zero such that $\sum_{n=1}^{\infty} a_n = \infty$ and $\sum_{n=1}^{\infty} a_n^2 c_n^{-2} < \infty$.*
2. *The function $f(x) \in C^3$ and bounded on $\mathbb{R}^d$.*
3. *$||x_n|| < \infty$ for all n.*
4. *$\mathsf{E}[\varepsilon_n^+ - \varepsilon_n^- | \Delta_n, \mathscr{F}_n] = 0$ and $\mathsf{E}[(Y(x_n \pm c_n \Delta_n, \xi_n^{\pm})/\Delta_{n,i})^2]$ is uniformly bounded for all n, i.*
5. *x^* is an asymptotically stable solution of the differential equation $\partial x(t)/\partial t = -\nabla f(x(t))$.*
6. *For each n, $\{\Delta_{n,i}\}_{i=1}^{d}$ are identically distributed, $\{\Delta_{n,i}\}$ are independent and symmetrically distributed with zero mean and uniformly bounded in magnitude for all n, i.*

Then $x_n \to x^*$ a.s. as $n \to \infty$.

The optimal convergence rate for SPSA is $O(n^{-1/3})$ [38]. Various convergence proofs have been presented with slight modifications to the conditions (cf. [9,13,19, 38,43]). The perturbation sequence $\{\Delta_n\}$, where $\Delta_n = (\Delta_{n,1}, \ldots, \Delta_{n,d})$ with $\{\Delta_{n,i}\}$ independent, must have mean zero (i.e., $\mathsf{E}[\Delta_n] = 0$), and finite inverse moments (i.e., $\mathsf{E}[|\Delta_{n,i}|^{-1}] < \infty$ for $i = 1, \ldots, d$). As a result, the Gaussian distribution is not applicable. Instead, the most common distribution used is the symmetric Bernoulli taking a positive and negative value (i.e., ± 1) with probability 0.5. In addition, an appropriately scaled x_n is approximately normal for large n, and the relative efficiency of SPSA depends on the geometric shape of $f(x)$, choice of $\{a_n\}$ and $\{c_n\}$, distribution of $\{\Delta_{n,i}\}$, and noise level.

Many extensions to the original SPSA algorithm have been developed, e.g., the constrained setting using projection operators [17, 36]. A slight modification is the averaging of the SPSA gradient estimators. Instead of generating one gradient estimate at each iteration, multiple gradient estimates can be generated at additional computational cost and averaged to reduce the noise. An accelerated form of SPSA approximates the second-order Hessian $\nabla^2 f(x)$ to accelerate the convergence [40], analogous to the Newton–Raphson method. Iterate averaging in the SPSA setting has also been explored, but performs relatively poor in finite-time [13,39]. All in all, SPSA has been shown to be an effective SA method for tackling high-dimensional problems, with ease of implementation and the asymptotic theory to support it.

6.4 Recent Modifications

This section presents several recently proposed modifications that focus on improving the finite-time performance of SA: the scaled-and-shifted Kiefer–Wolfowitz (SSKW) algorithm, the robust SA (RSA) algorithm, the accelerated SA (AC-SA) algorithm, and the Secant-Tangents AveRaged stochastic approximation (STAR-SA) algorithm. The theoretical results for RSA and AC-SA focus on an alternative way to analyze the performance of the estimates through $f(x_N^*) - f(x^*)$. The inequality in (6.15) is an alternative way to view the performance of SA, which focuses on the distance between the function evaluated at the estimate and the optimal function value (i.e., $\mathsf{E}[f(x_N^*) - f(x^*)]$) as opposed to a distance between the estimate and optimum (e.g., $\mathsf{E}(x_N^* - x^*)^2$). To illustrate the difference, consider an extremely flat function on the entire feasible region. The alternative performance measure will indicate that almost any iterate in the feasible region will be a good estimate, whereas performance based on the mean-squared error (MSE) of the estimate and optimum will be more sensitive to the estimate x_N^*. Further details on these two algorithms are provided in the next chapter.

6.4.1 Scaled-and-Shifted Kiefer–Wolfowitz (SSKW)

The scaled-and-shifted Kiefer–Wolfowitz (SSKW) algorithm [4] adaptively adjusts $\{a_n\}$ and $\{c_n\}$ finitely many times during the course of the algorithm to adapt to the characteristics of the function and noise level in hopes of preventing slow convergence in finite-time. The idea is to increase $\{a_n\}$ so the iterates are able to make noticeable progress towards the optimum with the option of decreasing $\{a_n\}$ later if it is too large. Furthermore, if the direction of the gradient is classified as incorrect, then $\{c_n\}$ is increased to reduce the noise. Note that KW only requires two parameter choices $\{a_n\}$ and $\{c_n\}$, whereas SSKW requires eleven, as seen in the algorithm below.

SSKW Algorithm

Scaling Phase

- Input. $\{a_n\}$, $\{c_n\}$, $[l,u]$, Π_Θ, stopping time N, and
  - h_0 = number of forced boundary hits,
  - γ_0 = scale up factor for $\{c_n\}$,
  - k_a = maximum number of shifts of $\{a_n\}$,
  - v_a = initial upper bound of shift,
  - ϕ_a = maximum scale up factor for $\{a_n\}$,
  - k_c = maximum number of scale ups for $\{c_n\}$,
  - c_0 = maximum value of $\{c_n\}$ after scale ups (i.e., $c_n \le c^{max} = c_0(u-l)$),
  - g_0 = maximum number of gradient estimates in scaling phase,
  - m_{max} = maximum number of adaptive iterations ($m_{max} \le N$).
- Initialize.
  - Choose $x_1 \in [l + c_1, u - c_1]$.
  - Let $n = 1$, $m = 1$, $g = 1$, $sh = 0$, and $sc = 0$.
- Do while $m \le h_0$ and $g \le g_0$.
  - Step 1.
    - Generate an estimate $\hat{\nabla} f(x_n)$ using symmetric differences.

- Compute x_{n+1} using recursion (6.3).
  - If $x_{n+1} \in (l + c_n, x_n)$, go to Step 2.
  - If $x_{n+1} \in (x_n, u - c_n,)$, go to Step 3.
  - If $x_{n+1} > u - c_{n+1}$ and $x_n = u - c_n$ or if $x_{n+1} < l + c_{n+1}$ and $x_n = l + c_n$, go to Step 4, if $sc \leq k_c$.
  - If $x_{n+1} > u - c_{n+1}$ and $x_n = l + c_n$ or if $x_{n+1} < l + c_{n+1}$ and $x_n = u - c_n$, go to Step 5.

– Step 2.
  - Scale $\{a_n\}$ up by $\alpha = \min\{\phi_a, (u - c_{n+1} - x_n)/(x_{n+1} - x_n)\}$ and use $\{\alpha a_n\}$ for the remaining iterations.
  - Set $x_{n+1} = l + c_{n+1}$. Let $n = n+1$, $m = m+1$, $g = g+1$ and go to Step 1.

– Step 3.
  - Scale $\{a_n\}$ up by $\alpha = \min\{\phi_a, (l + c_{n+1} - x_n)/(x_{n+1} - x_n)\}$ and use $\{\alpha a_n\}$ for the remaining iterations.
  - Set $x_{n+1} = u - c_{n+1}$. Let $n = n+1$, $m = m+1$, $g = g+1$ and go to Step 1.

– Step 4.
  - Scale $\{c_n\}$ up by $\gamma = \min\{\gamma_0, c^{max}/c_n\}$ and use $\{\gamma c_n\}$ for the remaining iterations.
  - Let $sc = sc + 1$ and go to Step 5.

– Step 5.
  - Set $x_{n+1} = \min\{u - c_{n+1}, \max\{x_{n+1}, l + c_{n+1}\}\}$.
  - Let $n = n+1$, $g = g+1$ and go to Step 1.

Shifting Phase

- While $n \leq m_{max}$ and $n \leq N$,

– Step 1.
  - Generate an estimate $\hat{\nabla} f(x_n)$ using symmetric differences.
  - Compute x_{n+1} using (6.3).
    - If $x_{n+1} > u - c_{n+1}$ and $x_n = l + c_n$ or if $x_{n+1} < l + c_{n+1}$ and $x_n = u - c_n$, go to Step 2, if $sh < k_a$.
    - If $x_{n+1} > u - c_{n+1}$ and $x_n = u - c_n$ or if $x_{n+1} < l + c_{n+1}$ and $x_n = l + c_n$, go to Step 3, if $sc < k_c$.
    - Otherwise, go to Step 4.

– Step 2.
  - Find smallest integer β' such that $x_{n+1} \in (l + c_n, u - c_n)$ with $a_{n+\beta'}$.
  - Set $\beta = \min(\nu_a, \beta')$ and shift $\{a_n\}$ to $\{a_{n+\beta}\}$. If $\beta = \nu_a$, set $\nu_a = 2\nu_a$.
  - Let $sh = sh + 1$ and go to Step 4.

– Step 3.

  · Scale $\{c_n\}$ up by $\gamma = \min\{\gamma_0, c^{max}/c_n\}$ and use $\{\gamma c_n\}$ for the remaining iterations.
  · Let $sc = sc + 1$ and go to Step 4.

– Step 4.

  · Set $x_{n+1} = \min\{u - c_{n+1}, \max\{x_{n+1}, l + c_{n+1}\}\}$.
  · Let $n = n + 1$ and go to Step 1.

KW Algorithm

- If $n > m_{max}$ and $n < N$, then SSKW reverts back to KW and stop when $n = N$.
- Output. $x_N^* = x_N$.

The SSKW algorithm has two pre-processing phases, scaling and shifting, which adjust the tuning sequences in order to improve the finite-time performance, before reverting back to the original KW algorithm. In the scaling phase, the $\{a_n\}$ is scaled up by a factor α, i.e., $\{a_n\}$ to $\{\alpha a_n\}$, so the iterates can move from one boundary to the other to ensure the step sizes are not too small relative to the gradient. In the shifting phase, the sequence $\{a_n\}$ is decreased by shifting or "skipping" a finite number (β) of terms from $\{a_n\}$ to $\{a_{n+\beta}\}$, when the iterates fall outside of the feasible region when the sign of the gradient is correct. This acts as a recourse stage and reduces the step size faster in case the step-size sequence $\{a_n\}$ is too large. During both phases, $\{c_n\}$ is scaled up by γ, i.e., $\{c_n\}$ to $\{\gamma c_n\}$, if the previous iterate is at the boundary and the update falls outside the feasible region but is moving in the wrong direction. This increase is an attempt to reduce the noise of the gradient estimate. These adjustments do not affect the asymptotic convergence, since the scaling phase only scales $\{a_n\}$ up by a constant, the shifting phase only skips a finite number of terms in $\{a_n\}$, and the perturbation sequence $\{c_n\}$ is only scaled up by a constant, all of which occur finitely many times.

6.4.2 *Robust Stochastic Approximation (RSA)*

The robust SA (RSA) method is intended to be relatively insensitive to the choice of the step-size sequence, similar to Polyak–Ruppert iterate averaging. The form of RSA is identical to (6.3) with the exception of the output. Instead of $x_N^* = x_N$, where x_N is the last iterate, x_N^* is calculated as

$$x_N^* = \frac{\sum_{n=1}^N a_n x_n}{\sum_{n=1}^N a_n},$$

where $a_n > 0$ for all n. It is clear that if $a_n = a$, where $a \in \mathbb{R}^+$ for all n, then $x_N^* = \frac{1}{N}\sum_{n=1}^N x_n$, giving the uniformly weighted average of Polyak–Ruppert. As mentioned earlier, iterate averaging under a constant step size for a moving window is asymptotically normal [28]. A finite-time bound was derived for $\mathsf{E}[f(x_n^*) - f(x^*)]$ under RSA when f is assumed convex [30]. Assume there exists $C > 0$ such that $\mathsf{E}[||\nabla f(x)||^2] \leq C^2$ for all $x \in \Theta$. Then for an N-step iteration policy,

$$\mathsf{E}[f(x_N^*) - f(x^*)] \leq \frac{||x_0 - x^*||^2 + C^2\sum_{n=1}^N a_n^2}{2\sum_{n=1}^N a_n}. \tag{6.8}$$

For equal weights or iterate averaging, the bound on the right hand side of (6.8) can be minimized if

$$a_n = a := \frac{D_\Theta}{C\sqrt{N}},$$

where $D_\Theta = \max_{x,y\in\Theta} ||x-y||$. The distance $||x_0 - x^*||$ in the place of D_Θ tightens the bound in (6.8), but x^* is unknown so the improvement may not be practically meaningful. This step size requires the number of iterations N to be fixed. Similar to iterate averaging, a sliding window average can also be employed in RSA. The estimate consists of the last $N-K+1$ estimates and has the form

$$x_{N,K}^* = \frac{\sum_{n=K}^N a_n x_n}{\sum_{n=K}^N a_n}. \tag{6.9}$$

If we consider the varying step size

$$a_n = \frac{\theta D_\Theta}{C\sqrt{n}}, \tag{6.10}$$

for $\theta > 0$, then we have the bound

$$\mathsf{E}[f(x_{N,K}^*) - f(x^*)] \leq \frac{D_\Theta C}{\sqrt{N}}\left[\frac{2}{\theta}\left(\frac{N}{N-K+1}\right) + \frac{\theta}{2}\sqrt{\frac{N}{K}}\right], \tag{6.11}$$

for $1 \leq K \leq N$.

6.4.3 *Accelerated Stochastic Approximation (AC-SA) for Strongly Convex Functions*

The accelerated SA (AC-SA) algorithm [21] takes a similar approach to iterate averaging and RSA by taking long strides and incorporating each of the iterates into the output. The next two algorithms, accelerated SA for strongly convex and convex functions, take advantage of the smoothness factor of the function if it exists. AC-SA for convex functions is a special case of AC-SA for strongly convex functions, so we first introduce AC-SA for strongly convex functions and then restrict the strong convexity parameter for the convex case.

AC-SA is an example of a proximal method, which introduce a proximity function into the objective function. The prox-function acts as a regularization term to prevent the next iterate update x_{n+1} from being too far from x_n and is comprised of a distance generating function or Bregman function $\omega : \Theta \to \mathbb{R}$, which is continuously differentiable and strongly convex with modulus $\nu > 0$ satisfying

$$\langle x-y, \nabla\omega(x) - \nabla\omega(y)\rangle \geq \nu||x-y||^2 \; \forall x,y \in \Theta,$$

where $\langle\cdot,\cdot\rangle$ denotes the inner product. A prox-function with the given distance generating function is

$$V(x,y) = V_\omega(x,y) = \omega(y) - [\omega(x) + \langle\nabla\omega(x), y-x\rangle].$$

As $x_n \to x^*$, the regularization term disappears, so minimizing $f(x)$ plus a regularizer is equivalent to minimizing the function $f(x)$.

Consider a strongly convex function $f(\cdot)$ satisfying

$$\frac{\mu}{2}||y-x||^2 \leq f(y) - f(x) - \langle\nabla f(x), y-x\rangle \leq \frac{L}{2}||y-x||^2 + M||y-x||, \quad (6.12)$$

for all $x,y \in \Theta$ where $\mu > 0$ is the strong convexity parameter. Notice that if f is Lipschitz continuous with Lipschitz constant $M/2$, then (6.12) holds with $M > 0$, $L = 0$, and $\mu = 0$, and if f has Lipschitz continuous gradients with Lipschitz constant L, then (6.12) holds with $M = 0$, $L > 0$, and $\mu = 0$.

The AC-SA algorithm updates three sequences, $\{x_n^{md}\}, \{x_n^{ag}\}$, and $\{x_n\}$. Here, "*md*" and "*ag*" are abbreviations for median and aggregate, respectively, and median is used in a loose sense.

Accelerated SA Method for Strongly Convex Functions

- Input.
  - Specify $V(x,y)$, $\{\alpha_n\}$ and $\{\gamma_n\}$ be given such that $\alpha_1 = 1$, $\alpha_n \in (0,1)$ for $n \geq 2$, and $\gamma_n > 0$ for $n \geq 1$ and a stopping time N.
- Initialize. Choose $x_0^{ag} = x_0 \in \Theta$ and let $n = 1$.
- While $n < N$,
  - Step 1. Generate an estimate $\hat{\nabla} f(x_n)$ of $\nabla f(x_n)$.
  - Step 2. Compute

$$x_n^{md} = \frac{\alpha_n[(1-\alpha_n)\mu+\gamma_n]}{\gamma_n+(1-\alpha_n^2)\mu}x_{n-1} + (1-\alpha_n)\frac{(1-\alpha_n)(\mu+\gamma_n)}{\gamma_n+(1-\alpha_n^2)\mu}x_{n-1}^{ag}$$

$$x_n = \arg\min_{x\in\Theta}\{\alpha_n[\langle\nabla f(x_n^{md}),x\rangle + \mu V(x_n^{md},x)] + [(1-\alpha_n)\mu+\gamma_n]V(x_{n-1},x)\}$$

$$x_n^{ag} = \alpha_n x_n + (1-\alpha_n)x_{n-1}^{ag}$$

  - Step 3. Let $n=n+1$ and go to Step 1.
- Output. $x_N^* = x_N^{ag}$.

Note: $V(x,y) = \frac{1}{2}||x-y||^2$ using the Euclidean norm with $\nu = 1$ is a common prox-function. Refer to [20] for details.

Theorem 6.5 (Theorem 1 [20]). *Assume* $V(x,y) \le \frac{1}{2}||x-y||^2$ for all $x,y\in\Theta$ *when* $\mu < 0$ *and* $\mathsf{E}[(\hat{\nabla} f(x) - \nabla f(x))^2] \le \sigma^2$ $\forall x\in\Theta$. *Choose* $\{\alpha_n\}$ *and* $\{\gamma_n\}$ *such that*

$$\nu(\mu+\gamma_n) > L\alpha_n^2, \tag{6.13}$$

$$\gamma_n/\Gamma_n = \gamma_{n+1}/\Gamma_{n+1} \text{ for } n\ge 1, \tag{6.14}$$

where

$$\Gamma_n = \begin{cases} 1 & \text{if } n=1; \\ (1-\alpha_n)\Gamma_{n-1} & \text{if } n\ge 2. \end{cases}$$

Then,

$$\mathsf{E}[f(x_N^{ag}) - f(x^*)] \le \Gamma_N\left(\gamma_1 V(x_0,x^*) + \sum_{n=1}^{N}\frac{2(M^2+\sigma^2)\alpha_n^2}{\Gamma_n[\nu(\mu+\gamma_n)-L\alpha_n^2]}\right). \tag{6.15}$$

Consider $\alpha_n = 2/(n+1)$, $\gamma_n = 4L/[\nu n(n+1)]$, and $\Gamma_n = 2/[n(n+1)]$. It can be easily checked that these choices satisfy conditions (6.13) and (6.14). Under these conditions, the right hand side of (6.15) can be bounded by

$$\frac{4LV(x_0,x^*)}{\nu N(N+1)} + \frac{8(M^2+\sigma^2)}{\nu\mu(N+1)}, \tag{6.16}$$

for $\mu > 0$. The bounds in (6.15) and (6.16) rely on additional information of the function and gradient, which are unknown, so they must be approximated.

6.4.4 *Accelerated Stochastic Approximation (AC-SA) for Convex Functions*

AC-SA for convex functions is a special case of AC-SA for strongly convex functions with $\mu = 0$. The algorithm is identical to AC-SA for strongly convex function with the exception of the x_n^{md} and x_n update since $\mu = 0$. The resulting updates are

$$x_n^{md} = \alpha_n x_{n-1} + (1-\alpha_n)x_{n-1}^{ag},$$
$$x_n = \arg\min_{x\in\Theta}\{\alpha_n\langle\nabla f(x_n^{md}),x\rangle + \gamma_n V(x_{n-1},x)\}.$$

Interestingly, if $V(x,y) = \frac{1}{2}||x-y||^2$, then the update for x_n simplifies to

$$x_n = \Pi_\Theta\left(x_{n-1} - \frac{\alpha_n}{\gamma_n}\hat{\nabla} f(x_n^{md})\right), \tag{6.17}$$

which has a similar form to the standard SA algorithm. Notice in the update for x_n in (6.17), α_n/γ_n takes the place of the step size a_n in (6.3) and the gradient estimate $\hat{\nabla} f$ is evaluated at x_n^{md} as opposed to x_{n-1}. If we consider the same parameter setting as in the strongly convex case, the "step size" α_n/γ_n increases with n. Furthermore, the lower and upper bounds for the optimal objective function can be computed online and the difference converges to 0 as the number of iterations goes to infinity [20].

Theorem 6.6 (Proposition 7 [20]). *Assume that the assumptions in Theorem 6.5 hold for $\mu = 0$ and the sequences $\alpha_n = 2/(n+1)$ and $\gamma_n = 4\gamma/[\nu n(n+1)]$ for $\gamma \geq 2L$. Then*

$$\mathrm{E}[f(x_N^{ag}) - f(x^*)] \leq \frac{4\gamma V(x_0,x^*)}{\nu N(N+1)} + \frac{4(M^2+\sigma^2)(N+2)}{3\gamma}, \tag{6.18}$$

where

$$\gamma = \max\left\{2L, \left[\frac{\nu(M^2+\sigma^2)N(N+1)(N+2)}{3V(x_0,x^*)}\right]^{1/2}\right\}$$

minimizes the bound in (6.18).

6.4.5 Secant-Tangents AveRaged Stochastic Approximation (STAR-SA)

The Secant-Tangents AveRaged (STAR) stochastic approximation algorithm estimates the gradient using a hybrid estimator, which is a convex combination of a symmetric finite difference and an average of two direct gradient estimators:

$$\hat{\nabla} f(x_n) = \alpha_n \frac{Y(x_n + c_n, \varepsilon_n^+) - Y(x_n - c_n, \varepsilon_n^-)}{2c_n} + (1-\alpha_n)\left(\frac{Y'(x_n + c_n, \delta_n^+) + Y'(x_n - c_n, \delta_n^-)}{2}\right), \quad (6.19)$$

where $\varepsilon_n^{\pm}$ and $\delta_n^{\pm}$ denote the randomness (i.e., $f(x_n \pm c_n) = \mathsf{E}[Y(x_n \pm c_n, \varepsilon_n^{\pm})]$ and $f'(x_n \pm c_n) = \mathsf{E}[Y'(x_n \pm c_n, \xi_n^{\pm})]$), $\alpha_n \in [0,1]$ for all n, $c_n \to 0$ and $\alpha_n \to 0$ as $n \to \infty$. The STAR gradient estimate requires function and gradient estimates on two points, $x_n \pm c_n$ for each $\hat{\nabla} f(x_n)$. In a setting where direct gradients are available, if the direct gradient is very noisy relative to the function estimates, it is difficult to decide between implementing RM or KW, even though RM converges faster asymptotically. Since the performance of neither algorithm is always superior to the other, the STAR gradient incorporates both. The weights of the convex combination play a critical role in the performance of STAR-SA and can be chosen to minimize the variance of the gradient estimate such that it is less than the variance of both the symmetric finite difference gradient estimate and direct gradient estimate. If

$$\alpha_n^* = \frac{\sigma_g^2 c_n^2 + \rho \sigma_f \sigma_g c_n^2}{\sigma_f^2 + \sigma_g^2 c_n^2 + 2\rho \sigma_f \sigma_g c_n},$$

where $\mathrm{Var}[Y(x,\varepsilon)] = \sigma_f^2$, $\mathrm{Var}[Y'(x,\xi)] = \sigma_g^2$, and $\mathrm{Corr}(Y(x,\varepsilon), Y'(x,\xi)) = \rho$, then STAR-SA is theoretically optimal in terms of MSE compared to RM and KW for simple quadratic functions, and the variance of the STAR gradient is less than that of RM and KW under certain conditions.

Theorem 6.7 (Theorem 3 [6]). *Let $\{x_n\}$ be a sequence generated using recursion (6.3) and gradient estimate (6.19). Assume*

1. *There exist positive sequences $\{a_n\}$, $\{c_n\}$, and $\{\alpha_n\}$ such that $\alpha_n \in [0,1]$ for all n, $\sum_{n=1}^{\infty} a_n \alpha_n = \infty$, $\sum_{n=1}^{\infty} a_n c_n < \infty$, $\sum_{n=1}^{\infty} a_n^2 < \infty$, and $\sum_{n=1}^{\infty} a_n^2 c_n^{-2} < \infty$.*
2. *There exist $B, C > 0$ such that $\mathsf{P}(|f''(x)| \le B) = 1$ and $\mathsf{P}(|f'(x)| \le C) = 1$ for all $x \in \Theta$.*
3. *There exist $K_0, K_1 > 0$ such that $K_0|x - x^*| \le |f'(x)| \le K_1|x - x^*|$ for all $x \in \Theta$.*
4. *$f'(x)(x - x^*) > 0$ for all $x \in \mathbb{R}\backslash\{x^*\}$.*
5. *For $c > 0$, $\sigma^2 = \sup_{x\in\mathbb{R}} \mathrm{Var}[Y(x+c,\xi^+) - Y(x-c,\xi^-)|x] < \infty$ for all $x \in \Theta$.*
6. *ε_n^+, ε_n^-, δ_n^+, δ_n^- are i.i.d. with mean zero for all n.*

Then $x_n \xrightarrow{L^2} x^*$ as $n \to \infty$.

Numerical experiments show that STAR-SA is competitive against RM and KW, even when the number of iterations for RM are doubled due to the increase in computational cost of STAR-SA [6]. In the experimental results, the STAR-SA algorithm either performs significantly better than RM and KW or the MSE is close to that of the algorithm with the lower MSE. STAR-SA has been extended to higher dimensions by considering simultaneous perturbation gradient estimates, instead of using a symmetric finite difference gradient estimate, to take advantage of its potential efficiency and robustness [5].

6.5 Numerical Experiments

We present three sets of numerical experiments comparing the mean-squared error (MSE) of various SA algorithms on several contrasting functions. The first set of experiments illustrates the sensitivity of KW and two variants to the choice of the two step-size sequence parameters, taken from [7]. The second set compares three SA algorithms, the robust SA (RSA) method, the accelerated SA (AC-SA) method, and the original RM algorithm, under various initial settings and step-size parameters for RM (i.e., starting values, compact intervals, noise levels, and step sizes). The last set of numerical experiments explores the potential gains from using the STAR gradient estimate, which utilizes both direct and indirect gradient estimates, as opposed to using them separately, as in RM and KW, respectively. Since the numerical experiments consider maximization problems, the sign of a_n and α_n/γ_n in recursion (6.3) and (6.17), respectively, must be adjusted accordingly.

Sensitivity Analysis of KW and Its Variants

We perform a sensitivity analysis of KW and KW using Kesten's rule (denoted henceforth by KWK) with symmetric finite difference gradient estimates, and we compare the results with SSKW. Using the parameter settings $a_n = \theta_a/n$, $c_n = \theta_c/n^{1/4}$, $\theta_a > 0$, $\theta_c > 0$ arbitrary but fixed, $N =$ 10,000 iterations, and 1,000 sample paths, our analysis replicates the results of [4] for $f(x) = -0.001x^2$ on the interval $[-50, 50]$, where SSKW performs significantly better than KW in terms of MSE and oscillatory period; however, this result is obtained using what seem to be nearly worst-case parameter setting for KW. In our experiments, we consider a wide range of parameters and initial settings for KW and KWK: 19 initial starting values uniformly spaced within the truncated interval $x_1 \in \{-50+5k \mid k = 1,2,\ldots 19\}$, 45 different θ_a values parametrized by $\theta_a \in \{10^s k \mid k = 1,2,\ldots,9, s = 0,1,\ldots,4\}$, and 10 different θ_c values parametrized by $\theta_c \in \{10^s k \mid k = 1,2,\ldots,5, s = 0,1\}$. In total, there are 8,550 combinations.

The numerical results illustrate the sensitivity of the classical SA methods to the parameters. In fact, near optimal performance can be obtained with fine-tuning.

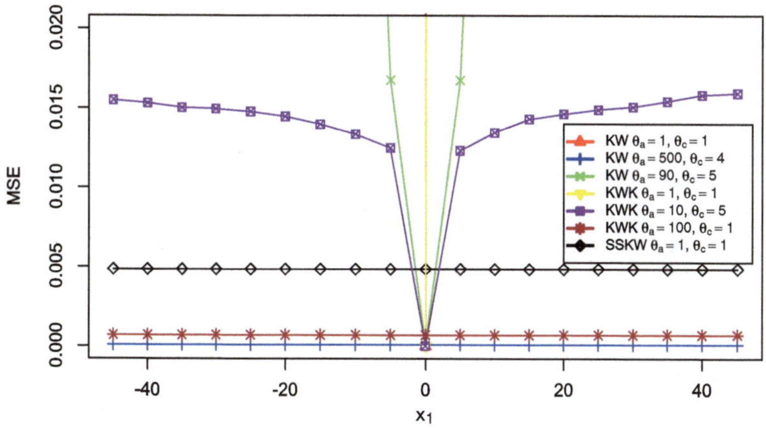

Fig. 6.2 MSE of the 10000th iterate of KW and KWK for three parameter settings and SSKW for $f(x) = -.001x^2$, $\sigma = 0.001$, $a_n = \theta_a/n$, $c_n = \theta_c/n^{1/4}$

Out of the 8,550 combinations, KW outperforms SSKW in half of the cases, which indicates that with some tuning, KW yields good performance for a fairly wide range of tunable parameters. Figure 6.2 plots the MSE of KW, KWK, and SSKW for $f(x) = -.001x^2, \sigma = 0.001$ against the initial starting values x_1 for several parameter choices that are a good representation of a majority of the results. The parameter value $\theta_a = \theta_c = 1$, identical to the settings in [4], is among the worst for KW and KWK, represented by a nearly vertical orange line for both algorithms, as a result of the overlapping red and yellow lines for KW and KWK, respectively. For this parameter setting, SSKW beats KW and KWK significantly for all initial values with the exception of $x_1 = 0$. The first column in Table 6.1 compares the MSE all three algorithms with $x_1 = 0.01$, and clearly, KW outperforms SSKW in almost all cases. Of course, a practitioner would have no way of knowing whether or not the starting iterate was close to the true optimum, so these results do not indicate that KW will always perform well. They do indicate, however, that KW exhibits substantial variation in performance. In the case where $\theta_a = 90, \theta_c = 5$ and $\theta_a = 10, \theta_c = 5$, KW and KWK, respectively, outperform SSKW in a neighborhood around the optimum. There are also well-tuned parameters such as $\theta_a = 500$, $\theta_c = 4$ for KW and $\theta_a = 100$, $\theta_c = 1$ for KWK that outperform SSKW for all initial start values. When KW and KWK perform better than SSKW, the difference is not as pronounced as when SSKW outperforms KW, but careful tuning can partially mitigate the sensitivity of KW to parameters such as the initial iterate.

In addition, we implement KW and its variants using the same parameters ($a_n = 1/n$, $c_n = 1/n^{1/4}$, $x_1 = 30$) as in [4] on $f(x) = 100e^{-.006x^2}$ to test the algorithms under the same setting for a different function. Figure 6.3 plots the MSE of the 10000th iterate as a function of the initial start value. The horizontal line for all noise levels indicates that SSKW is insensitive the initial start value. KW and KWK outperform SSKW within certain intervals around the optimum for

Table 6.1 MSE of the 100th, 1000th, and 10000th iteration for KW and its variants with $a_n = 1/n$, $c_n = 1/n^{1/4}$

		$f(x) = -0.001x^2$ [-50, 50] $x_1 = .01$			$f(x) = 100e^{-0.006x^2}$ [-50, 50] $x_1 = 30$		
σ	Algorithm	100	1000	10000	100	1000	10000
0.001	SSKW	$5.10x10^{-2}$	$1.70x10^{-2}$	$5.00x10^{-3}$	$5.07x10^{-2}$	$1.68x10^{-2}$	$4.84x10^{-3}$
	KW	10^{-4}	10^{-4}	10^{-4}	763.8	653.3	431.4
	KWK	$1.12x10^{-4}$	$1.08x10^{-4}$	$1.04x10^{-4}$	10^{-7}	$3x10^{-8}$	10^{-8}
0.01	SSKW	$5.10x10^{-2}$	$1.70x10^{-2}$	$5.00x10^{-3}$	5.07	1.68	$4.90x10^{-1}$
	KW	10^{-4}	10^{-4}	10^{-4}	763.8	653.3	431.2
	KWK	$2.10x10^{-3}$	$2.11x10^{-3}$	$2.05x10^{-3}$	$9.54x10^{-6}$	$2.76x10^{-6}$	$8.41x10^{-7}$
0.1	SSKW	$5.10x10^{-2}$	$1.70x10^{-2}$	$5.00x10^{-3}$	165.8	57.4	16.0
	KW	10^{-4}	10^{-4}	10^{-4}	763.4	651.4	418.2
	KWK	$2.01x10^{-1}$	$2.03x10^{-1}$	$1.97x10^{-1}$	$5.65x10^{-2}$	$2.76x10^{-4}$	$8.41x10^{-5}$
1.0	SSKW	$5.10x10^{-2}$	$1.70x10^{-2}$	$5.00x10^{-3}$	187.2	57.8	18.7
	KW	10^{-4}	10^{-4}	10^{-4}	722.5	562.5	415.7
	KWK	20.1	20.3	19.7	456.9	315.1	239.7

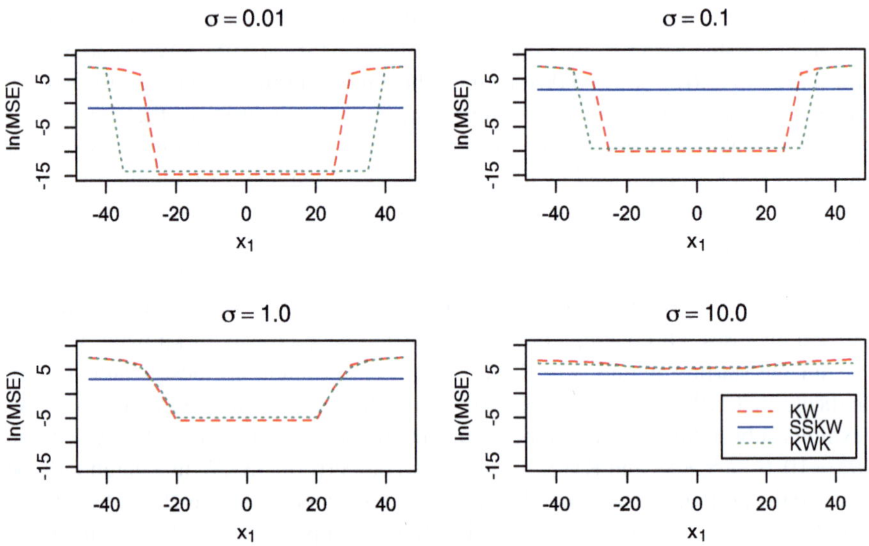

Fig. 6.3 MSE Comparison of KW and its variants for $f(x) = 100e^{-.006x^2}$ with values $a_n = 1/n$, $c_n = 1/n^{1/4}$, $N = 10000$

$\sigma \in \{0.001, 0.01, 0.1, 1.0\}$ and KWK's better performance intervals overlap those of KW. However, KW using the deterministic step size $1/n$ performs better than KWK where the intervals overlap, which can be seen in Fig. 6.3. Unfortunately, outside of those intervals, KW and KWK have a tendency to perform poorly.

RM, RM with Iterate Averaging, Robust SA and Accelerated SA

We investigate the MSE performance of RSA and AC-SA using direct gradient estimates and compare the results against the classical RM algorithm and RM with iterate averaging. We consider the optimal parameter settings for RSA and AC-SA, which require additional knowledge of the function, its gradient, and the optimum, so in practice, they must be approximated.

We consider a simple quadratic function, $f(x) = -\frac{1}{3}x^2$, on the truncated intervals $[-50,50]$ and $[-5,95]$ with $x_1 = 30.0$, $\sigma = 1.0$, and 1,000 sample paths. For the RM and RM with iterate averaging algorithm, we employ a common step size $a_n = \theta_a/n$, where $\theta_a = 10.0$. RM performed relatively well for a wide range of multiplicative constants. We chose to use $\theta_a = 10.0$, although it did not yield the lowest MSE at the 1000th or 10000th iteration from preliminary numerical tests. For RSA, we adopt a constant step size that minimizes the finite-time bound in (6.8), where $C = 100/3, 190/3$ for the intervals [-50, 50] and [-5, 95], respectively, and $D_\Theta = 100$. For the AC-SA algorithm, we consider $\alpha_n = 2/(n+1)$ and $\gamma_n = 4\gamma/[n(n+1)]$, where γ is given in (6.19) with $\nu = 1$, $L = 2/3$ and $M = 0$.

Figure 6.4 plots the MSE as a function of the number of iterations from 1 to 10000 on a log scale. The results for both the centered and skew truncated intervals appear to have the same behavior across all four algorithms. RM performs well with a good parameter choice, although it is not the best, but averaging the iterates improves the performance, resulting in a smoother monotonically decreasing MSE curve as the number of iterations increase. Compared to a decently/reasonably tuned RM and RM with iterate averaging algorithm, RSA appears to be inferior, at least in this simple numerical experiment. The most interesting curve is from the AC-SA algorithm, where one can observe periodic oscillations, which decrease in magnitude as the number of iterations increase. We further investigated this behavior by analyzing individual sample paths, and the estimates $\{x_n^{ag}\}$ appear to have the same behavior, following a smooth oscillating path/curve. From Fig. 6.4, the AC-SA curve appears to level off and hover slightly over the RSA curve. The stopping time dictates the relative performance of AC-SA when there are a smaller number of iterations because of the oscillations. For the case of the skewed interval, there is a small range of iterations where AC-SA outperforms RSA, RM, and RM with iterate averaging, as well as other small ranges where it outperforms RSA. Keep in mind that these experiments are for a simple quadratic function for a particular setting, so the relative performance will most likely change in a different setting.

From our numerical experiments, one can conclude that RM and RM with iterate averaging has the potential to outperform RSA and AC-SA if the step-size parameter is chosen appropriately for a wide range of choices. In this case, iterate averaging improves the performance of RM for all 10,000 iterations. Both the AC-SA and RSA algorithms require additional knowledge to choose the optimal step size that minimizes the bound in (6.15) and (6.8) for AC-SA and RSA, respectively.

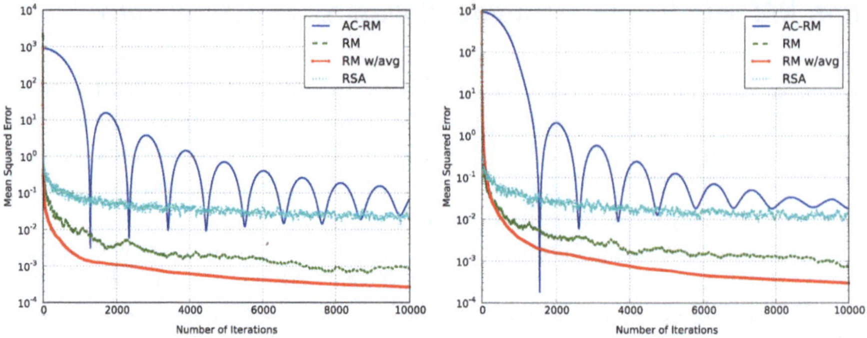

Fig. 6.4 MSE under RM, RM with averaging, RSA, and AC-SA for $f(x) = -\frac{1}{3}x^2$, $x_1 = 30.0$, $\sigma = 1.0$, for symmetrical $[-50, 50]$ (*left graph*) and skewed $[-5, 95]$ (*right graph*) intervals

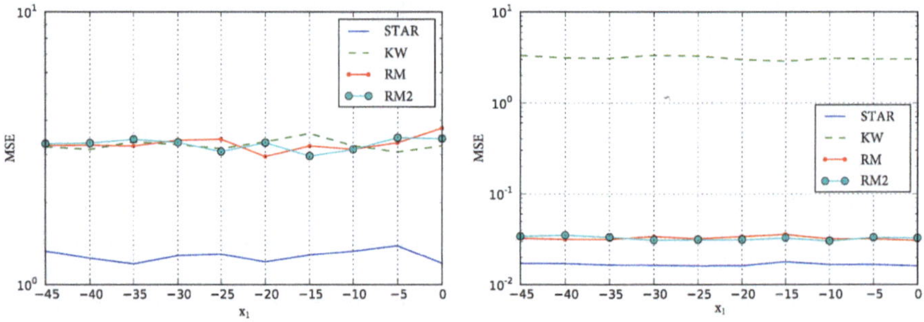

Fig. 6.5 MSE of 1000th iterate under STAR-SA, RM, and KW for $f(x) = -0.1x^2$, $\sigma_g = 1.0$, and two levels of σ_f: 0.1 (*left graph*) and 1.0 (*right graph*)

STAR-SA, RM, and KW

We implement STAR-SA, RM, and KW under various combinations of noise levels σ_f and σ_g, for $f(x) = -ax^2$, $a > 0$ and $\Theta = [-50, 50]$. The gain sequence and finite difference step sizes are θ_a/n and $\theta_c/n^{1/4}$, respectively, and the MSE results are based on 1,000 sample paths. For a fairer comparison, the number of iterations for RM is doubled, since STAR-SA and KW require twice the number of sample path runs. We consider the following values for the parameter and initial settings: steepness level $a \in \{10^k | k = -3, -2.5, \ldots, 1.5, 2\}$, $x_1 \in \{-50 + 5k | k = 1, \ldots, 10\}$, $\theta_a \in \{1, 10, 100\}$, $\theta_c \in \{0.1, 1.0\}$, $\sigma_f \in \{10^k | k = -3, \ldots, 1\}$, $\sigma_g \in \{10^k | k = -3, \ldots, 1\}$, and $N \in \{100, 1000, 10000\}$. Although STAR-SA, RM, and KW were implemented for all settings, only the case $f(x) = -0.1x^2$, $a_n = 10n^{-1}$, $c_n = 0.1n^{-1/4}$, $\sigma_f \in \{0.001, 0.1, 1.0\}$, $\sigma_g \in \{0.001, 0.1, 1.0\}$ and $N = 1000$ will be described in detail.

The STAR-SA algorithm outperforms KW and RM for 6 out of the 9 combinations for all initial start values considered. For $\sigma_f = 0.001$ and $\sigma_f < \sigma_g$, the MSE of STAR-SA is lower than that of RM, but is approximately equal to that of KW. The only case where the MSE of the STAR-SA algorithm is not approximately less than or equal to the MSE of KW and RM is when both noise levels are very low, i.e., $\sigma_f = \sigma_g = 0.001$, which is not shown. In this case, RM performs better than STAR-SA except when the start value is close to the optimum $x^* = 0$. In fact, the MSE of STAR-SA decreases as x_1 approaches x^*. In addition, the MSE of KW and RM are close to the optimum when $\sigma_f = 0.001$ and $\sigma_g = 0.001, 0.1$, respectively, whereas the MSE of STAR-SA is close to 0 for $(\sigma_f, \sigma_g) \in \{(0.001, 0.001), (0.001, 0.1), (0.001, 1.0), (0.1, 0.001), (0.1, 0.1), (1.0, 0.001), (1.0, 0.1)\}$. Figure 6.5 illustrates the MSE results when $\sigma_f = 1.0$. When the noise of the function is high, KW performs poorly, RM outperforms KW, and STAR-SA has the lowest MSE. Figure 6.5 shows a case where the performance of KW and RM are similar, but the MSE of the STAR-SA algorithm is lower. Overall, from the numerical experiments conducted, STAR-SA either performs significantly better than both RM and KW in terms of MSE or the MSE is approximately equal to that of the algorithm with the lower MSE.

6.6 Concluding Remarks

Stochastic approximation has an enormous body of literature in all aspects of theory, algorithms, and applications. From its origins in statistics, it has now reached many disciplines in engineering and the social sciences, with well-known successes in such areas as signal processing, pattern recognition, and machine learning. Clearly, simulation optimization is another fertile area for its application.

This chapter introduced the two main versions of SA:

- KW-like methods that rely only on function estimates, known as *gradient-free* or *stochastic zeroth-order* algorithms; and
- RM-like methods that make use of direct estimates of first-order derivative information, known as *stochastic gradient* or *stochastic first-order* algorithms.

The latter methods generally perform better in practice, but they require information that is not always available. Asymptotically, they can obtain an $O(n^{-1/2})$ convergence rate, whereas the former are generally limited to a $O(n^{-1/3})$ convergence rate. Among the gradient-free methods, SPSA has been particularly successful for high-dimensional problems.

The finite-time behavior of any SA algorithm depends heavily on the choice of the step-size or gain sequence, and various approaches to handling this challenge have been presented, from Kesten's rule to iterate averaging, with the latter procedure highly recommended.

Classical notions of convergence in SA address the iterates $\{x_n\}$, whereas recent finite-time analysis has turned to the properties of the function values $\{f(x_n)\}$. Chapter 7 focuses on some recent SA algorithms mainly tailored to convex stochastic programming problems and provides such convergence properties.

Finally, SA methods are aimed at continuous-valued optimization problems, but there is some work attempting to apply SA to discrete optimization problems. A recent Ph.D. dissertation [44] addresses this setting and includes a summary of previous work in the area.

Acknowledgements This work was supported in part by the National Science Foundation under Grants CMMI 0856256 and ECCS 0901543, and by the Air Force Office of Scientific Research under Grant FA9550-10-10340.

References

1. S. Andradóttir. A stochastic approximation algorithm with varying bounds. *Operations Research*, 43(6):1037–1048, 1995.
2. J. R. Blum. Multidimensional stochastic approximation methods. *The Annals of Mathematical Statistics*, 25(4):737–744–200, 1954.
3. M. Broadie, D. Cicek, and A. Zeevi. An adaptive multidimensional version of the Kiefer-Wolfowitz stochastic approximation algorithm. In M. D. Rossetti, R. R. Hill, B. Johansson, A. Dunkin, and R. G. Ingalls, editors, *Proceedings of the Winter Simulation Conference*, pages 601–612. IEEE, Piscataway, NJ, 2009.
4. M. Broadie, D. Cicek, and A. Zeevi. General bounds and finite-time improvement for the Kiefer-Wolfowitz stochastic approximation algorithm. *Operations Research*, 59(5):1211–1224, 2011.
5. M. Chau, M. C. Fu, and H. Qu. Multivariate stochastic approximation using a Secant-Tangents AveRaged (STAR) gradient estimator. Technical report, Working paper, University of Maryland, College Park, 2014.
6. M. Chau, H. Qu, and M. C. Fu. A new hybrid stochastic approximation algorithm. In *Proceedings of the 12th International Workshop on Discrete Event Systems*, 2014.
7. M. Chau, H. Qu, M. C. Fu, and I. O. Ryzhov. An empirical sensitivity analysis of the Kiefer-Wolfowitz algorithm and its variants. In R. Pasupathy, S.-H. Kim, A. Tolk, R. Hill, and M. E. Kuhl, editors, *Proceedings of the 2013 Winter Simulation Conference*, pages 945–965. IEEE, Piscataway, NJ, 2013.
8. H. F. Chen. Convergence rate of stochastic approximation algorithms in the degenerate case. *SIAM Journal on Control and Optimization*, 36(1):100–114, 1998.
9. H. F. Chen, T. E. Duncan, and B. Pasik-Duncan. A Kiefer-Wolfowitz algorithm with randomized differences. *IEEE Transactions on Automatic Control*, 44(3):442–453, 1999.
10. H. F. Chen and Y. M. Zhu. Stochastic approximation procedure with randomly varying truncations. *Scientia Sinica Series A*, 29:914–926, 1986.
11. B. Delyon and A. B. Juditsky. Accelerated stochastic approximation. *SIAM Journal on Optimization*, 3(4):868–881, 1993.
12. C. Derman. An application of Chung's lemma to the Kiefer-Wolfowitz stochastic approximation procedure. *The Annals of Mathematical Statistics*, 27(2):532–536, 1956.
13. J. Dippon and J. Renz. Weighted means in stochastic approximation of minima. *SIAM Journal on Control Optimization*, 35(5):1811–1827, 1997.
14. V. Dupač. On the Kiefer-Wolfowitz approximation method. *Časopis pro pěstování Matematiky*, 82(1):47–75, 1957.
15. V. Fabian. Stochastic approximation of minima with improved asymptotic speed. *The Annals of Mathematical Statistics*, 38(1):191–200, 1967.
16. V. Fabian. Stochastic approximation. In Rustagi, editor, *Optimizing Methods in Statistics*, pages 439–470. Academic Press, 1997.

17. M. C. Fu and S. D. Hill. Optimization of discrete event systems via simultaneous perturbation stochastic approximation. *IIE Transactions*, 29(3):233–243, 1997.
18. A. P. George and W. B. Powell. Adaptive stepsizes for recursive estimation with applications in approximate dynamic programming. *Machine Learning*, 65(1):167–198, 2006.
19. L. Gerencsér. Convergence rates of moments in stochastic approximation with simultaneous perturbation gradient approximation and resetting. *IEEE Transactions on Automatic Control*, 44(5):894–905, 1998.
20. S. Ghadimi and G. Lan. Optimal stochastic approximation algorithms for strongly convex stochastic composite optimization I: a generic algorithmic framework. *SIAM Journal on Optimization*, 22(4):1469–1492, 2012.
21. S. Ghadimi and G. Lan. Optimal stochastic approximation algorithms for strongly convex stochastic composite optimization II: shrinking procedures and optimal algorithms. *SIAM Journal on Optimization*, 23(4):2061–2089, 2013.
22. C. Kao, W. T. Song, and S. P. Chen. A modified quasi-Newton method for optimization in simulation. *International Transactions in Operations Research*, 4(3):223–233, 1997.
23. H. Kesten. Accelerated stochastic approximation. *The Annals of Mathematical Statistics*, 29(1):41–59, 1958.
24. K. Kiefer and J. Wolfowitz. Stochastic estimation of the maximum of a regression function. *The Annals of Mathematical Statistics*, 23(3):462–466, 1952.
25. N. L. Kleinman, J. C. Spall, and D. Q. Naiman. Simulation-based optimization with stochastic approximation using common random numbers. *Management Science*, 45(11):1570–1578, 1999.
26. H. J. Kushner and J. Yang. Stochastic approximation with averaging of the iterates: optimal asymptotic rates of convergence for general processes. *SIAM Journal on Control and Optimization*, 31:1045–1062, 1993.
27. H. J. Kushner and J. Yang. Stochastic approximation with averaging and feedback: rapidly convergent "on-line" algorithms. *IEEE Transactions on Automatic Control*, 40:24–34, 1995.
28. H. J. Kushner and G. G. Yin. *Stochastic Approximation and Recursive Algorithms and Applications*. Springer, New York, NY, 2003.
29. J. L. Maryak. Some guidelines for using iterate averaging in stochastic approximation. *Proceedings of the 36th IEEE Conference on Decision and Control*, 3:2287–2290, 1997.
30. A. Nemirovski, A. Juditsky, G. Lan, and A. Shapiro. Robust stochastic approximation approach to stochastic programming. *SIAM Journal on Optimization*, 19(4):1574–1609, 2009.
31. B. T. Polyak and A. B. Juditsky. Acceleration of stochastic approximation by averaging. *SIAM Journal on Control and Optimization*, 30(4):838–855, 1992.
32. H. Robbins and S. Monro. A stochastic approximation method. *The Annals of Mathematical Statistics*, 22:400–407, 1951.
33. D. Ruppert. A Newton-Raphson version of the multivariate Robbins-Monro procedure. *The Annals of Statistics*, 13(1):236–245, 1985.
34. D. Ruppert. Efficient estimations from a slowly convergent Robbins-Monro process. Technical report, Cornell University Operations Research and Industrial Engineering, Ithaca, NY, February 1988.
35. J. Sacks. Asymptotic distribution of stochastic approximation procedures. *The Annals of Mathematical Statistics*, 29(2):351–634, 1958.
36. P. Sadegh. Constrained optimization via stochastic approximation with a simultaneous perturbation gradient approximation. *Automatica*, 33(5):889–892, 1997.
37. G. Sardis. Learning applied to successive approximation algorithms. *IEEE Transactions on Systems, Science and Cynbernetics*, SSC(6):97–103, 1970.
38. J. C. Spall. Multivariate stochastic approximation using a simultaneous perturbation gradient approximation. *IEEE Transactions on Automatic Control*, 37(3):332–341, 1992.
39. J. C. Spall. Adaptive stochastic approximation by the simultaneous perturbation method. *IEEE Transactions on Automatic Control*, 45(10):1839–1853, 2000.
40. J. C. Spall. *Introduction to Stochastic Search and Optimization: Estimation, Simulation, and Control*. John Wiley, Hoboken, NJ, 2003.

41. A. B. Tsybakov and B. T. Polyak. Optimal order of accuracy of search algorithms in stochastic optimization. *Problemy Peredachi Informatsii*, 26(2):45–63, 1990.
42. J. H. Venter. An extension of the Robbins-Monro procedure. *The Annals of Mathematical Statistics*, 38(1):181–190, 1967.
43. I. J. Wang and E. K. P. Chong. A deterministic analysis of stochastic approximation with randomized differences. *IEEE Transactions on Automatic Control*, 43(12):1745–1749, 1998.
44. Q. Wang. *Optimization with Discrete Simultaneous Perturbation Stochastic Approximation Using Noisy Loss Function Measurements*. PhD thesis, Johns Hopkins University, 2013.
45. S. Yakowitz, P. L'Ecuyer, and F. Vazquez-Abad. Global stochastic optimization with low-dispersion point sets. *Operations Research*, 48(6):939–950, 2000.
46. G. Yin and Y. M. Zhu. Almost sure convergence of stochastic approximation algorithms with nonadditive noise. *International Journal of Control*, 49(4):1361–1376, 1989.

Chapter 7
Stochastic Approximation Methods and Their Finite-Time Convergence Properties

Saeed Ghadimi and Guanghui Lan

Abstract This chapter surveys some recent advances in the design and analysis of two classes of stochastic approximation methods: stochastic first- and zeroth-order methods for stochastic optimization. We focus on the finite-time convergence properties (i.e., iteration complexity) of these algorithms by providing bounds on the number of iterations required to achieve a certain accuracy. We point out that many of these complexity bounds are theoretically optimal for solving different classes of stochastic optimization problems.

7.1 Introduction

The problem of interest in this chapter is given in the form

$$f^* := \min_{x \in \Theta} f(x) \tag{7.1}$$

where $f(x) = \mathsf{E}[F(x,\xi)]$, $\Theta \subseteq \mathbb{R}^n$ is a nonempty closed convex set, ξ is a random vector with probability distribution P supported on set $\Xi \subseteq \mathbb{R}^d$, and $F : \Theta \times \Xi \to \mathbb{R}$. We assume that the expectation

$$\mathsf{E}[F(x,\xi)] = \int_{\Xi} F(x,\xi) dP(\xi) \tag{7.2}$$

is well-defined and finite-valued for every $x \in \Theta$. Moreover, we assume that $f(\cdot)$ is continuous on Θ.

A basic difficulty of solving stochastic optimization problem (7.1) is that the multidimensional integral (expectation) (7.2) cannot be computed with high accuracy for dimension $d > 5$. One popular method to solve (7.1) is the sample average approximation (SAA) approach of Chap. 8 (cf. [31, 32]). In SAA, problem (7.1) is first approximated by

S. Ghadimi • G. Lan (✉)
University of Florida, Gainesville, FL, USA
e-mail: sghadimi@ufl.edu; glan@ise.ufl.edu

M.C. Fu (ed.), *Handbook of Simulation Optimization*, International Series in Operations Research & Management Science 216, DOI 10.1007/978-1-4939-1384-8_7

$$\min_{x\in\Theta}\left\{\frac{1}{N}\sum_{k=1}^{N}F(x,\xi_k)\right\}, \tag{7.3}$$

where $\xi_k, k=1,\ldots,N$ are given i.i.d. samples of ξ. Note that the SAA method is not an algorithm; the obtained SAA problem (7.3) still has to be solved by an appropriate iterative numerical procedure. Indeed, one possible drawback for SAA is the high iteration cost (at least linear with respect to N) for numerical algorithms to solve (7.3). In addition, this approach cannot be used for online optimization where the decision vector x needs to be updated whenever a new sample ξ_k is generated.

In this chapter, we focus on a different approach, stochastic approximation, aimed at solving problem (7.1) directly. These methods are online approaches, updating the decision vector x_k whenever a new sample ξ_k is generated at iteration k. More specifically, depending on the information available, we consider two different types of stochastic approximation methods.

1. *Stochastic first-order (SFO) methods*: Given $x_k \in \Theta$ and a random sample ξ_k, we have access to the stochastic function value $F(x_k,\xi_k)$ and stochastic (sub)gradient $G(x_k,\xi_k)$ such that

$$\mathsf{E}_{\xi_k}[G(x_k,\xi_k)] \in \partial f(x_k),$$

   where $\partial f(x)$ denotes the subdifferential[1] of f at x. Observe that the above condition will be satisfied by a measurable selection $G(x,\xi) \in \partial_x F(x,\xi)$ under some mild regularity assumptions (cf. Strassen [34]).
2. *Stochastic zeroth-order (SZO) methods*: Given $x_k \in \Theta$ and a random sample ξ_k, we only have access to the stochastic function value $F(x_k,\xi_k)$, and intend to approximate stochastic (sub)gradients by using available stochastic function values.

Clearly, SZO methods are built upon SFO methods and their analysis will be slightly more complicated than that of the latter, due to the error associated with approximating stochastic (sub)gradients by stochastic function values.

The development of SFO methods goes back to the pioneering paper by Robbins and Monro [27], whose stochastic approximation (SA) algorithm solves problem (7.1) by mimicking the simplest gradient descent method, i.e., for chosen $x_1 \in \Theta$ and a sequence $\gamma_k > 0$, $k \geq 1$, of step sizes, it generates the iterates by the formula

$$x_{k+1} := \Pi_\Theta\big(x_k - \gamma_k G(x_k,\xi_k)\big), \tag{7.4}$$

[1] A subgradient of a function f at x_0 is a vector $y \in \mathbb{R}^n$ such that $f(x) \geq f(x_0) + y^T(x-x_0),\ \forall x \in \Theta$. The set of all such subgradients is called the subdifferential of f at the point x_0.

where Π_Θ denotes projection onto the set Θ given by $\Pi_\Theta(x) = \arg\min_{z\in\Theta} \|x-z\|_2$. It has been shown that the classical SA algorithm possesses the "asymptotically optimal" rate of convergence for solving strongly convex stochastic programming (SP) problems [3, 29]. Motivated by Robbins and Monro [27], Kiefer and Wolfowitz [15] introduced the first SZO method where the stochastic gradient $G(x_k, \xi_k)$ in (7.4) is approximated by using finite differences. Spall [33] improved Kiefer and Wolfowitz's algorithm by introducing simultaneous perturbation stochastic approximation (SPSA). These SA algorithms became widely used in stochastic optimization (see, e.g., [1, 5, 6, 17, 24, 28, 33] and references therein) and, due to especially low demand for computer memory, in signal processing (cf. [1] and references therein). However, these SA algorithms are very sensitive to the choice of step sizes and typically perform poorly in practice (e.g., [33, Sect. 4.5.3]). An important improvement of the classical SA, based on iterates averaging, was developed by Polyak [25] and Polyak and Juditsky [26]. Their methods were shown to be more robust with respect to the selection of step sizes than the classical SA and also exhibit the "asymptotically optimal" rate of convergence for solving strongly convex SP problems. Please see [17, 20] for an account of the earlier history of SA methods.

This chapter focuses on recent development of SFO SA methods for SP. New SA-type methods have been introduced to solve SP problems that are not necessarily strongly convex. Furthermore, motivated by complexity theory in convex optimization [21], the finite-time convergence properties of these methods have been analyzed. For example, Nemirovski et al. [20] presented a modified SA approach, robust SA, for solving general nonsmooth convex SP problems. They demonstrated that the robust SA exhibits an optimal (unimprovable) $O(1/\varepsilon^2)$ iteration complexity for solving these problems, where ε represents a bound on the distance of the estimated optimal value from f^*. In other words, one can identify a worst-case convex SP instance that prevents any algorithm from performing better than this complexity bound. This method has been shown in [19, 20] to be competitive to the aforementioned SAA approach (see, e.g., [16, 30]) and even significantly outperforms it for solving a class of convex stochastic optimization problems. Similar techniques, based on subgradient averaging, have been proposed in [13, 14, 22]. While these techniques dealt with nonsmooth convex programming problems, Lan [18] presented an accelerated SA algorithm for nonsmooth and smooth stochastic optimization. Lan demonstrated that by properly choosing the step-size policy, the accelerated SA algorithm exhibits the optimal iteration complexity for solving different classes of convex stochastic optimization problems (see also [8, 9] for discussions about strong convexity). While convexity has played an important role in establishing the convergence of all these SA algorithms, Ghadimi and Lan [10] also developed randomized stochastic gradient methods, along with its accelerated version in [11] for solving general smooth (not necessarily convex) stochastic optimization problems.

While there exists a somewhat long history for the development of zeroth-order (or derivative-free) methods in nonlinear programming (see the monograph by

Conn et al. [4] and references therein), there exist only a few complexity results for these types of methods, mostly for convex programming (e.g., [21, 23]) and deterministic nonconvex programming problems (e.g., [2,7,23,35]). The complexity studies on SZO methods appear in the literature only recently. In particular, Nesterov [23] derived some provably tight bounds for approximating first-order information by zeroth-order information using the Gaussian smoothing technique. Based on this technique, he established the $O(n^2/\varepsilon^2)$ complexity for solving general nonsmooth convex SP problems (see p. 17 of [23]). By incorporating the Gaussian smoothing technique [23] into the randomized stochastic gradient method, Ghadimi and Lan [10] present a randomized stochastic gradient free (RSGF) method for solving a class of simulation-based optimization problems without requiring the convexity assumption. In addition, when applied to smooth stochastic convex optimization problems, the bound in [10] has a much weaker dependence on n than the one previously established by Nesterov for solving general nonsmooth convex SP problems. Such an improvement is obtained by explicitly making use of the smoothness properties of the objective function and carefully choosing the step sizes and smoothing parameters.

In this chapter, we provide an overview of several SFO and SZO methods [8–10, 18, 20, 23]. We focus on the description of these algorithms, the selection on the step-size policy and the associated complexity results, rather than the more involved convergence analysis of these algorithms, which can be found in the original papers. This chapter is structured as follows. In Sect. 7.2, we present different SFO methods including robust and accelerated stochastic approximation methods, which are optimal for solving different classes of stochastic convex optimization problems. We then present a new stochastic approximation type method, the randomized stochastic gradient method in Sect. 7.3. Section 7.4 is devoted to a randomized stochastic gradient-free method for solving a certain class of simulation-based optimization problems. Concluding remarks are made in Sect. 7.5.

7.2 Convex Stochastic Optimization

In this section, we discuss several stochastic first-order methods for solving convex stochastic optimization problems where the objective function f in (7.1) is convex, but not necessarily differentiable. We first show that the robust SA algorithm can achieve the optimal complexity for solving these general nonsmooth convex stochastic optimization problems. We then present the accelerated SA algorithm, which can achieve the optimal complexity for solving both nonsmooth and smooth convex optimization problems. We then show that the accelerated SA algorithm can achieve the optimal or nearly optimal complexity bounds for solving nonsmooth and smooth strongly convex problems.

7.2.1 *Robust SA Method for Nonsmooth Stochastic Convex Optimization*

In this subsection, we present an optimal method, robust SA, for solving general nonsmooth stochastic convex optimization problems. We start by formally describing this algorithm.

Algorithm 1 The robust SA (RSA) method

Let $x_0 \in \Theta$ be given.
for $k = 0, \ldots, N-1$ **do**
  Set $x_{k+1} := \Pi_\Theta \left(x_k - \gamma_k G(x_k, \xi_k)\right)$ for some $\gamma_k \in (0, +\infty)$.
end for
Output $\bar{x}_N = \frac{\sum_{k=1}^N \gamma_k x_k}{\sum_{k=1}^N \gamma_k}$.

Observe that the updating step of Algorithm 1 differs from that of the subgradient descent method in that one uses the stochastic subgradient $G(x_k, \xi_k)$ in place of an exact subgradient $g(x_k) \in \partial f(x_k)$. Also, note that the generalization of the above robust SA algorithm, the mirror-descent SA in [20], can use possibly non-Euclidean projections (prox-mapping) instead of simple Euclidean projections used in Algorithm 1. Moreover, the robust SA method described above differs from the classical SA in the selection of the output solution and the step-size policy. Specifically, we choose the averaging solution $\bar{x}_N$ as the output solution of the robust SA method, rather than the last iterate x_N as in the classic SA method. Also, while the step-size policy γ_k for the above robust SA method is $O(1/\sqrt{N})$ as mentioned after Theorem 7.1, it is $O(1/k)$ for the classic SA method.

To establish the convergence of the robust SA algorithm, we assume that the stochastic subgradients $G(x, \xi)$ have bounded "variance," i.e.,

$$\mathsf{E}\left[\|G(x,\xi)\|^2\right] \le M^2, \quad \forall x \in \Theta. \tag{7.5}$$

Under this assumption, the quality of the output solution $\bar{x}_N$ of the robust SA method can be quantified as follows (cf. [20, p. 1583]).

Theorem 7.1. *Suppose that f is convex and condition (7.5) holds. Then for the N-step of robust SA algorithm, we have*

$$\mathsf{E}\left[f(\bar{x}_N) - f^*\right] \le \frac{\|x_0 - x^*\|^2 + M^2 \sum_{t=1}^N \gamma_t^2}{2\sum_{t=1}^N \gamma_t} \tag{7.6}$$

for any $N \ge 0$, where x^* is an arbitrary optimal solution of (7.1).

In implementations of the robust SA algorithm, different policies to specify γ_k can be applied (see [20]). We discuss only the *constant step-size* policy, where the number N of iterations is fixed in advance, and $\gamma_k = \gamma$, $k = 1,\ldots,N$. In that case,

$$\bar{x}_N = \frac{1}{N}\sum_{k=1}^{N} x_k. \tag{7.7}$$

Also assume for the sake of simplicity that Θ is bounded [2] and define

$$D_\Theta := \max_{x,z\in\Theta} \|x - z\|. \tag{7.8}$$

By choosing the step sizes as

$$\gamma_k = \gamma := \frac{\sqrt{2}D_\Theta}{M\sqrt{N}}, \quad k = 1,\ldots,N, \tag{7.9}$$

we have in view of (7.6) that

$$\mathsf{E}\left[f(\bar{x}_N) - f^*\right] \le \frac{MD_\Theta}{\sqrt{2N}}. \tag{7.10}$$

Observe that the choice of γ in (7.9) is obtained by minimizing the right-hand side (RHS) of (7.6) subject to $\gamma_1 = \ldots = \gamma_N = \gamma$. Also, note that by choosing the constant step-size policy, the output solution (7.7) of Algorithm 1 is the same as that of the improved classical SA method by Polyak and Juditsky [26]. However, the choice of the step sizes in the latter is different from the one in (7.9).

By (7.10), the robust SA method needs at most

$$\frac{M^2 D_\Theta^2}{2\varepsilon^2}$$

iterations to find an ε-solution of (7.1), i.e., a point $\bar{x} \in \Theta$ s.t. $\mathsf{E}[f(\bar{x}) - f^*] \le \varepsilon$. This bound is unimprovable for the SFO methods to solve general nonsmooth convex optimization problems, in view of the complexity theory by Nemirovski and Yudin [21, Sect. 5.3]. By Markov's inequality it follows from (7.10) that for any $\varepsilon > 0$,

$$\mathsf{P}\left(f(\bar{x}_N) - f^* > \varepsilon\right) \le \frac{D_\Theta M}{\varepsilon\sqrt{2N}}. \tag{7.11}$$

[2]This assumption can be relaxed, e.g., by simply setting $\gamma_k = \frac{\sqrt{2}}{M\sqrt{N}}$.

It is possible to obtain finer bounds for (7.11) when imposing the following more restrictive condition:

$$\mathsf{E}\left[\exp\left\{\|G(x,\xi)\|^2/M^2\right\}\right] \le \exp\{1\}. \tag{7.12}$$

Note that condition (7.12) is stronger than (7.5). Indeed, if a random variable F satisfies $\mathsf{E}[\exp\{F/a\}] \le \exp\{1\}$ for some $a > 0$, then by Jensen's inequality $\exp\{\mathsf{E}[F/a]\} \le \mathsf{E}[\exp\{F/a\}] \le \exp\{1\}$, and therefore $\mathsf{E}[F] \le a$. Of course, condition (7.12) holds if for all $(x,\xi) \in \Theta \times \Xi$:

$$\|G(x,\xi)\| \le M.$$

The following result has been established in [20, Proposition 2.2].

Proposition 7.1. *Suppose that condition (7.12) holds. Then for the constant stepsize policy* (7.9), *the following inequality holds for any* $\Omega \ge 1$:

$$\mathsf{P}\left(f(\bar{x}_N) - f^* > (12+2\Omega)D_\Theta M N^{-1/2}\right) \le 2\exp\{-\Omega\}. \tag{7.13}$$

It follows from (7.13) that the number N of steps required by the algorithm to solve the problem with accuracy $\varepsilon > 0$, and a (probabilistic) confidence $1-\beta$, is of order $O\left(\varepsilon^{-2}\log^2(1/\beta)\right)$. Note that in practice one can modify the robust SA algorithm so that the approximate solution $\bar{x}_N$ is obtained by averaging over a part of the trajectory (see [20] for details).

7.2.2 *Accelerated SA Methods for Nonsmooth and Smooth Stochastic Optimization*

While the robust SA method is theoretically optimal for solving general nonsmooth convex optimization problems, it does not make use of the smoothness properties that the objective function f might have. As a result, it does not achieve the best possible complexity bounds for solving smooth stochastic optimization problems. In this subsection, we discuss a new stochastic optimization method, the accelerated SA, which is optimal for solving not only nonsmooth but also smooth stochastic optimization problems. Below, we present the framework of this method for convex problems.

Algorithm 2 The accelerated SA (AC-SA) method for convex problems

Let $x_0^{ag} = x_0 \in \Theta$, step-size parameters $\{\alpha_k\}_{k\geq 1}$ and $\{\gamma_k\}_{k\geq 1}$ be given s.t. $\alpha_1 = 1$, $\alpha_k \in (0,1)$ for any $k \geq 2$, and $\gamma_k > 0$ for any $k \geq 1$.
for $k = 1,\ldots,N$ **do**
Set

$$x_k^{md} = \alpha_k x_{k-1} + (1-\alpha_k)x_{k-1}^{ag}, \tag{7.14}$$

$$x_k = \Pi_\Theta\left(x_{k-1} - \alpha_k/\gamma_k G(x_k^{md},\xi_k)\right), \tag{7.15}$$

$$x_k^{ag} = \alpha_k x_k + (1-\alpha_k)x_{k-1}^{ag}. \tag{7.16}$$

end for
Output x_N^{ag}.

In this subsection, we assume that the objective function $f(\cdot)$ in (7.1) can be either smooth or nonsmooth. In the former, we assume that $f(\cdot)$ has Lipschitz continuous gradients, i.e.,

$$\|\nabla f(y) - \nabla f(x)\| \leq L\|y-x\|, \quad \forall x,y \in \Theta. \tag{7.17}$$

In the latter, we assume that $f(\cdot)$ is a general Lipschitz continuous function, i.e.,

$$|f(y) - f(x)| \leq K\|y-x\|, \quad \forall x,y \in \Theta. \tag{7.18}$$

Moreover, to establish the convergence properties of the AC-SA method, we make the following assumption regarding the error in estimating the true (sub)gradient of $f(\cdot)$ by the stochastic gradient:

$$\mathsf{E}\left[\|G(x,\xi_k) - g(x)\|^2\right] \leq \sigma^2, \quad \forall x \in \Theta, \quad \forall k \geq 1. \tag{7.19}$$

where $g(x) \in \partial f(x)$. The following result shows the main convergence properties of the AC-SA method, which was first shown in [18, p. 378] (see also [8, p. 1475]).

Theorem 7.2. *Suppose that condition (7.19) holds and the step sizes $\{\alpha_k\}_{k\geq 1}$ and $\{\gamma_k\}_{k\geq 1}$ in the AC-SA method are chosen such that*

$$\gamma_1/\Gamma_1 = \gamma_2/\Gamma_2 = \ldots, \tag{7.20}$$

where

$$\Gamma_k := \begin{cases} 1, & k = 1, \\ (1-\alpha_k)\Gamma_{k-1}, & k \geq 2. \end{cases} \tag{7.21}$$

Then,

a) if condition (7.17) holds and $\gamma_k > L\alpha_k^2$*, then for any* $N \geq 1$*, we have*

$$\mathsf{E}[f(x_N^{ag}) - f^*] \leq \Gamma_N \left(\frac{\gamma_1 \|x_0 - x^*\|^2}{2} + \sum_{k=1}^{N} \frac{\sigma^2 \alpha_k^2}{\Gamma_k[\gamma_k - L\alpha_k^2]} \right). \tag{7.22}$$

b) if condition (7.18) holds, then for any $N \geq 1$*, we have*

$$\mathsf{E}[f(x_N^{ag}) - f^*] \leq \Gamma_N \left(\frac{\gamma_1 \|x_0 - x^*\|^2}{2} + 2(4K^2 + \sigma^2) \sum_{k=1}^{N} \frac{\alpha_k^2}{\Gamma_k \gamma_k} \right). \tag{7.23}$$

We now discuss the step-size policies for the above result. By choosing $\alpha_k = 2/(k+1)$ and $\gamma_k = 4\gamma/[k(k+1)]$, we have $\Gamma_k = 2/[k(k+1)]$, which consequently implies that condition (7.20) holds. Now, if condition (7.17) holds, by choosing

$$\gamma = \gamma_N^* = \max \left\{ 2L, \left[\frac{\sigma^2 N(N+1)(N+2)}{3\|x_0 - x^*\|^2} \right]^{\frac{1}{2}} \right\}, \tag{7.24}$$

we have $\gamma_k > L\alpha_k^2$, and in the view of (7.22), we obtain

$$\mathsf{E}[f(x_N^{ag}) - f^*] \leq \frac{4L\|x_0 - x^*\|^2}{N(N+1)} + \frac{4\|x_0 - x^*\|\sigma}{\sqrt{N+1}}. \tag{7.25}$$

Observe that by choosing α_k, γ_k, and Γ_k as mentioned above, γ_N^* in (7.24) is obtained by minimizing the RHS of (7.22) with respect to γ over the interval $[2L, +\infty)$. Also, note that the bound in (7.25) implies that the AC-SA method can find an ε-solution of (7.1) in at most

$$2\|x_0 - x^*\| \sqrt{\frac{2L}{\varepsilon}} + \frac{64\|x_0 - x^*\|^2 \sigma^2}{\varepsilon^2}$$

iterations, which is unimprovable for the SFO methods to solve smooth convex optimization problems, in view of the complexity theory by Nemirovski and Yudin [21, Sect. 5.3].

Moreover, if condition (7.18) holds, by choosing

$$\gamma = \gamma_N^* = \left[\frac{(4K^2 + \sigma^2) N(N+1)(2N+1)}{6\|x_0 - x^*\|^2} \right]^{\frac{1}{2}}, \tag{7.26}$$

and in the view of (7.23), we have

$$\mathsf{E}[f(x_N^{ag}) - f^*] \le \frac{3\|x_0 - x^*\|(2K+\sigma)}{\sqrt{N}}. \tag{7.27}$$

Note that if conditions (7.19) and (7.18) hold, then condition (7.5) holds for any $M \ge K+\sigma$. Thus, according to our discussion in the previous subsection, (7.27) is also unimprovable for the SFO methods to solve general nonsmooth convex optimization problems.

Applying Markov's inequality to (7.25) and (7.27), for any $\varepsilon > 0$, we obtain

$$\mathsf{P}\left(f(x_N^{ag}) - f^* > \varepsilon\right) \le \frac{1}{\varepsilon}\left(\frac{4L\|x_0 - x^*\|^2}{N(N+1)} + \frac{4\|x_0 - x^*\|\sigma}{\sqrt{N+1}}\right),$$

and

$$\mathsf{P}\left(f(x_N^{ag}) - f^* > \varepsilon\right) \le \frac{3\|x_0 - x^*\|(2K+\sigma)}{\varepsilon\sqrt{N}},$$

respectively, for smooth and nonsmooth convex stochastic optimization problems. To improve the above probability bounds, we need to modify the light-tail assumption (7.12) for the error associated with estimating the true (sub)gradient by the stochastic (sub)gradient, i.e.,

$$\mathsf{E}\left[\exp\{\|G(x,\xi_k) - g(x)\|^2/\sigma^2\}\right] \le \exp\{1\}, \ \ \forall x \in \Theta, \ \ \forall k \ge 1. \tag{7.28}$$

Define

$$R_\Theta(x^*) := \max_{x\in\Theta} \|x - x^*\|.$$

The following result is from [18, Corollary 1] (see also [8, Proposition 7]).

Proposition 7.2. *Suppose that condition (7.28) holds and λ be any positive number.*

a) If condition (7.17) holds and γ is set to (7.24), then,

$$\mathsf{P}\left(f(x_N^{ag}) - f^* > \frac{4L\|x_0 - x^*\|^2}{N(N+1)} + \frac{4(1+2\lambda)\sigma R_\Theta(x^*)}{\sqrt{N}}\right)$$
$$\le \exp\{-\lambda^2/3\} + \exp\{-\lambda\}. \tag{7.29}$$

b) If condition (7.18) holds and γ is set to (7.26), then,

$$\mathsf{P}\left(f(x_N^{ag}) - f^* > 3\left[2K\|x_0 - x^*\| + (1+3\lambda)\sigma R_\Theta(x^*)\right] N^{-1/2}\right)$$
$$\le \exp\{-\lambda^2/3\} + \exp\{-\lambda\}. \tag{7.30}$$

In view of (7.29) and (7.30), we conclude that the number N of steps required by the AC-SA algorithm to solve smooth or nonsmooth problems with accuracy $\varepsilon > 0$, and a confidence $1-\beta$, is of order $\mathcal{O}\left(\varepsilon^{-1/2}+\varepsilon^{-2}\log^2(1/\beta)\right)$ and $\mathcal{O}\left(\varepsilon^{-2}\log^2(1/\beta)\right)$, respectively.

One critical problem associated with SA-type methods is that it is difficult to check the accuracy of the generated solutions. In the remaining part of this subsection, we show that one can compute, with little additional computational effort, certain stochastic lower bounds of the optimal value of (7.1) during the execution of the AC-SA algorithms. These stochastic lower bounds, when grouped with certain stochastic upper bounds on the optimal value, can provide online accuracy certificates for the generated solutions. Note that similar validation procedures have also been developed by Lan et al. [19] for the robust SA algorithm discussed in Sect. 7.2.1.

Define

$$\widetilde{\mathrm{lb}}_N = \min_{x\in\Theta} \Gamma_N \sum_{k=1}^{N} \frac{\alpha_k}{\Gamma_k} \tilde{l}_f(x_k^{md},\xi_k,x), \tag{7.31}$$

where

$$\tilde{l}_f(z,\xi,x) := F(z,\xi) + \langle G(z,\xi), x-z\rangle.$$

Observe that by definition of Γ_k in (7.21), we have

$$\sum_{k=1}^{N}\frac{\alpha_k}{\Gamma_k} = \frac{\alpha_1}{\Gamma_1} + \sum_{k=2}^{N}\frac{1}{\Gamma_k}\left(1-\frac{\Gamma_k}{\Gamma_{k-1}}\right) = \frac{1}{\Gamma_1} + \sum_{k=2}^{N}\left(\frac{1}{\Gamma_k}-\frac{1}{\Gamma_{k-1}}\right) = \frac{1}{\Gamma_N}. \tag{7.32}$$

Noting this observation, convexity of f and the fact that x_k^{md} is a function of $\xi_{[k-1]} = (\xi_1,\ldots,\xi_{k-1})$, and ξ_k is independent of $\xi_{[k-1]}$, we have

$$\begin{aligned}
\mathsf{E}\left[\widetilde{\mathrm{lb}}_N\right] &= \mathsf{E}\left[\mathsf{E}_{\xi_{[N-1]}}\left[\min_{x\in\Theta}\left(\Gamma_N\sum_{k=1}^{N}\frac{\alpha_k}{\Gamma_k}\tilde{l}_f(x_k^{md},\xi_k,x)\right)\right]\right] \\
&\le \mathsf{E}\left[\min_{x\in\Theta}\mathsf{E}_{\xi_{[k-1]}}\left[\left(\Gamma_N\sum_{k=1}^{N}\frac{\alpha_k}{\Gamma_k}\tilde{l}_f(x_k^{md},\xi_k,x)\right)\right]\right] \\
&= \mathsf{E}\left[\min_{x\in\Theta}\Gamma_N\sum_{k=1}^{N}\frac{\alpha_k}{\Gamma_k}\left\{f(x_k^{md})+\langle g(x_k^{md}),x-x_k^{md}\rangle\right\}\right] \le f^*,
\end{aligned}$$

i.e., on average, $\widetilde{\mathrm{lb}}_N$ gives a lower bound for the optimal value of (7.1). To see how good the lower bound $\widetilde{\mathrm{lb}}_N$ is, we estimate the expectations and probabilities of the corresponding errors in Proposition 7.3 (cf. [8, p. 1489]). To establish the large-deviation results for $\widetilde{\text{lb}}_k$, we also need the following light-tail condition about the stochastic function values $F(x,\xi)$.

$$\mathsf{E}\left[\exp\{\|F(x,\xi_k)-f(x)\|^2/Q^2\}\right] \le \exp\{1\}, \ \forall x\in\Theta, \ \forall k\ge 1, \tag{7.33}$$

for some $Q>0$.

Proposition 7.3. *Suppose that condition (7.19) holds and the step sizes in the AC-SA method are given such that* $\alpha_k = 2/(k+1)$ *and* $\gamma_k = 4\gamma/[k(k+1)]$. *Then,*

a) if condition (7.17) holds and γ *is given by (7.24) with replacing* $\|x_0-x^*\|$ *by* $R_\Theta(x_0)$, *then for any* $N\ge 1$, *we have*

$$\mathsf{E}[f(x_N^{ag})-\widetilde{\text{lb}}_N] \le \frac{4LR_\Theta(x_0)^2}{N(N+1)} + \frac{4R_\Theta(x_0)\sigma}{\sqrt{N+1}}. \tag{7.34}$$

If in addition, conditions (7.28) and (7.33) hold, then we have

$$\mathsf{P}\left(f(x_N^{ag})-f^* > \frac{4LR_\Theta(x_0)^2}{N(N+1)} + \frac{4R_\Theta(x_0)\sigma + \lambda[4\sigma(2R_\Theta(x_0)+R_\Theta(x^*))+2Q]}{\sqrt{N}}\right)$$
$$\le 2\exp\{-\lambda^2/3\}+\exp\{-\lambda\}, \tag{7.35}$$

b) if condition (7.18) holds and γ *is given by (7.26) with replacing* $\|x_0-x^*\|$ *by* $R_\Theta(x_0)$, *then for any* $N\ge 1$, *we have*

$$\mathsf{E}[f(x_N^{ag})-\widetilde{\text{lb}}_N] \le \frac{3R_\Theta(x_0)(2K+\sigma)}{\sqrt{N}}. \tag{7.36}$$

If in addition, conditions (7.28) and (7.33) hold, then we have

$$\mathsf{P}\left(f(x_N^{ag})-f^* > 3R_\Theta(x_0)(2K+\sigma)+\lambda[4\sigma(2R_\Theta(x_0)+R_\Theta(x^*))+2Q]N^{-1/2}\right)$$
$$\le 2\exp\{-\lambda^2/3\}+\exp\{-\lambda\}. \tag{7.37}$$

We now describe a way to enhance the above lower bounds with efficiently computable upper bounds on the optimal value f^* so that one can assess the quality of the generated solutions in an online manner. Define

$$\overline{\text{ub}}_N = \beta_N^{-1} \sum_{k=\lceil N/2\rceil}^{N} kF(x_k^{ag},\xi_k), \ \forall N\ge 1, \tag{7.38}$$

where $\beta_N := \sum_{k=\lceil N/2\rceil}^N k$ and $\lceil\cdot\rceil$ denotes the ceiling function that returns the next integer above the argument. Since $\mathsf{E}_{\xi_k}[F(x_k^{ag},\xi_k)] = f(x_k^{ag})$, we have $\mathsf{E}[\overline{\text{ub}}_N]\ge f^*$, i.e., $\bar{\mathrm{ub}}_N$, $N \geq 1$, on average, provide online upper bounds on f^*. Accordingly, we define the new online lower bounds as

$$\overline{\mathrm{lb}}_N = \beta_N^{-1} \sum_{k=\lceil N/2 \rceil}^{N} k \widetilde{\mathrm{lb}}_k, \ \forall N \geq 1, \tag{7.39}$$

where $\widetilde{\mathrm{lb}}_k$ is defined in (7.31). Using this observation and Proposition 7.3, we have

$$\begin{aligned} \mathsf{E}[\overline{\mathrm{ub}}_N - \overline{\mathrm{lb}}_N] &= \beta_N^{-1} \sum_{k=\lceil N/2 \rceil}^{N} k[f(x_k^{ag}) - \widetilde{\mathrm{lb}}_k] \leq \beta_N^{-1} \sum_{k=\lceil N/2 \rceil}^{N} kO\left(\frac{1}{\sqrt{k}}\right) \\ &= O\left(\beta_N^{-1} \sum_{k=\lceil N/2 \rceil}^{N} \sqrt{k}\right) = O\left(\frac{1}{\sqrt{N}}\right), \ N \geq 3, \end{aligned}$$

where the last identity follows from the facts that $\sum_{k=\lceil N/2 \rceil}^{N} \sqrt{k} = O(N^{\frac{3}{2}})$ and that

$$\beta_N \geq \frac{1}{2}\left[N(N+1) - \left(\frac{N}{2}+1\right)\left(\frac{N}{2}+2\right)\right] \geq \frac{1}{8}\left(3N^2 - 2N - 8\right).$$

Therefore, the gap between online upper bound $\overline{\mathrm{ub}}_N$ and lower bound $\overline{\mathrm{lb}}_N$ converges to 0 in the same order of magnitude as the one between $f(x_N^{ag})$ and $\widetilde{\mathrm{lb}}_N$ for both smooth and nonsmooth convex SP problems (cf. [8, p. 1490]).

7.2.3 *Accelerated SA Methods for Strongly Convex Optimization*

In this subsection, we consider strongly convex problems with $f(\cdot)$ in (7.1) satisfying the following condition:

$$\frac{\mu}{2}\|y-x\|^2 \leq f(y) - f(x) - \langle g(x), y-x \rangle \leq \frac{L}{2}\|y-x\|^2 + 2K\|y-x\|, \ \ \forall x,y \in \Theta, \tag{7.40}$$

where $g(x) \in \partial f(x)$ and $\mu > 0$ is the strong convexity parameter. Note that if L or K is zero, the right hand side inequality in the above relation immediately follows from (7.18) or (7.17). Our goal is to show that the AC-SA algorithm, when employed with proper step-size parameters, is nearly optimal for solving strongly convex SP problems. Moreover, we show that the optimal complexity for solving this class of problems can be achieved by properly restarting the AC-SA algorithm with certain step-size policy.

Below, we generalize the AC-SA algorithm presented in the previous subsection for strongly convex problems.

Algorithm 3 The accelerated SA (AC-SA) method for strongly convex problems

---

Let $x_0^{ag} = x_0 \in \Theta$, step-size parameters $\{\alpha_k\}_{k\geq 1}$ and $\{\gamma_k\}_{k\geq 1}$ be given s.t. $\alpha_1 = 1$, $\alpha_k \in (0,1)$ for any $k \geq 2$, and $\gamma_k > 0$ for any $k \geq 1$.

for $k = 1,\ldots,N$ **do**

Set

$$x_k^{md} = \frac{(1-\alpha_k)(\mu+\gamma_k)}{\gamma_k + (1-\alpha_k^2)\mu} x_{k-1}^{ag} + \frac{\alpha_k[(1-\alpha_k)\mu+\gamma_k]}{\gamma_k+(1-\alpha_k^2)\mu} x_{k-1}, \tag{7.41}$$

$$x_k = \arg\min_{x\in\Theta} \left\{ \alpha_k[\langle G(x_k^{md},\xi_k),x\rangle + \frac{\mu}{2}\|x_t^{md}-x\|^2] + \frac{(1-\alpha_k)\mu+\gamma_k}{2}\|x_{k-1}-x\|^2 \right\}, \tag{7.42}$$

$$x_k^{ag} = \alpha_k x_k + (1-\alpha_k) x_{k-1}^{ag}. \tag{7.43}$$

end for

Output x_N^{ag}.

The next result is established in [8, Theorem 1] about the main convergence properties of the AC-SA method for solving strongly convex problems.

Theorem 7.3. *Suppose that condition (7.19) holds and the step sizes $\{\alpha_k\}_{k\geq 1}$ and $\{\gamma_k\}_{k\geq 1}$ in the AC-SA method are chosen such that condition (7.20) holds. If condition (7.40) holds and $\mu+\gamma_k > L\alpha_k^2$, then for any $N \geq 1$, we have*

$$\mathsf{E}[f(x_N^{ag}) - f^*] \leq \Gamma_N \left(\frac{\gamma_1\|x_0-x^*\|^2}{2} + \sum_{k=1}^{N} \frac{2(4K^2+\sigma^2)\alpha_k^2}{\Gamma_k(\mu+\gamma_k-L\alpha_k^2)} \right), \tag{7.44}$$

where Γ_k is defined in (7.21).

By choosing step sizes $\alpha_k = 2/(k+1)$ and $\gamma_k = 4L/[k(k+1)]$, and in the view of (7.44), for any $N \geq 1$, we have

$$\mathsf{E}[f(x_N^{ag}) - f^*] \leq \frac{2L\|x_0-x^*\|^2}{N(N+1)} + \frac{8(4K^2+\sigma^2)}{\mu(N+1)}, \tag{7.45}$$

which in the view of the complexity theory by Nemirovski and Yudin [21, Sect. 7.2], implies that the AC-SA algorithm is nearly optimal for solving strongly convex stochastic optimization problems. More specifically, while the second term in bound (7.45) is unimprovable, the first term can be much improved. Observe also that the above complexity bound is actually optimal for nonsmooth strongly convex problems, i.e., $L = 0$ and $K > 0$ in (7.40).

By (7.45) and Markov's inequality, we conclude that, if condition (7.19) holds, then the iteration complexity of the AC-SA algorithm for solving strongly convex SP problems to accuracy ε with confidence level $1-\beta$ is bounded by

$$O\left(\frac{1}{\beta}\left[\sqrt{\frac{L\|x_0-x^*\|^2}{\varepsilon}}+\frac{K^2+\sigma^2}{\mu\varepsilon}\right]\right). \tag{7.46}$$

Moreover, if condition (7.28) holds, then (7.46) can be improved to

$$O\left(\sqrt{\frac{L\|x_0-x^*\|^2}{\nu\varepsilon}}+\frac{K^2+\sigma^2}{\mu\varepsilon}+\frac{\sigma^2}{\mu\varepsilon}\log\frac{1}{\beta}+\left[\frac{\sigma R_\Theta(x^*)}{\varepsilon}\log\frac{1}{\beta}\right]^2\right). \tag{7.47}$$

Note that the above iteration-complexity bound has a significantly worse dependence on ε than the one in (7.46), although it depends only logarithmically on $1/\beta$. It can be shown that by properly shrinking the feasible region once a while during the execution of the AC-SA method, the bound in (7.47) can be significantly improved to

$$O\left(\frac{1}{\varepsilon}\left[\log\frac{1}{\varepsilon\beta}\right]^2\right),$$

in terms of its dependence on ε (see [9, Proposition 5] for more details). It is worth to point out that, similarly to the bounds (7.31), (7.38), and (7.39) in the previous subsection, we can define online lower and upper bounds on the optimal value of problem (7.1), which allow us to provide online accuracy certificates to evaluate quality of the generated solutions by the AC-SA algorithm for strongly convex problems (see subsection 5 in [8] for more details).

One problem associated with the above AC-SA method is that it is not optimal for solving strongly convex SP problems. In the remaining part of this subsection, we show that by properly restarting the AC-SA algorithm for convex problems with a certain step-size policy, it can achieve the optimal iteration complexity for solving strongly convex SP problems. Below we present a multi-stage AC-SA algorithm for strongly convex problems.

Algorithm 4 The multi-stage AC-SA method

Let $p_0 \in \Theta$, and a bound $\mathscr{V}_0$ such that $f(p_0) - f(x^*) \le \mathscr{V}_0$ be given.
for $s = 1, \ldots, S$ **do**
a) Run N_s iterations of the AC-SA algorithm for convex problems with input $x_0 = p_{s-1}$, $\{\alpha_k\}_{k\ge1} = 2/(k+1)$ and $\{\gamma_k\}_{k\ge1} = 4\gamma/k(k+1)$ with $\gamma = \gamma_s$, where

$$N_s = \left\lceil \max\left\{4\sqrt{\frac{2L}{\mu}}, \frac{128(4K^2+\sigma^2)}{3\mu\mathscr{V}_0 2^{-(s+1)}}\right\}\right\rceil, \tag{7.48}$$

$$\gamma_s = \max\left\{2L, \left[\frac{\mu(4K^2+\sigma^2)}{3\mathscr{V}_0 2^{-(s-1)}} N_s(N_s+1)(N_s+2)\right]^{\frac{1}{2}}\right\}, \tag{7.49}$$

b) Set $p_s = x^{ag}_{N_s}$, where $x^{ag}_{N_s}$ is the solution obtained in Step a).
end for
Output $p_S = x^{ag}_{N_S}$.

The following result about the convergence properties of the multi-stage AC-SA algorithm is presented in [9, Proposition 7].

Proposition 7.4. *Let $\{p_s\}_{s\ge1}$ be computed by the multi-stage AC-SA algorithm. If condition (7.19) holds, then*

$$\mathsf{E}[f(p_s) - f^*] \le \mathscr{V}_s \equiv \mathscr{V}_0 2^{-s}, \quad \forall s \ge 0. \tag{7.50}$$

As a consequence, the multi-stage AC-SA algorithm will find a solution $\bar{x} \in \Theta$ of (7.1) such that $\mathsf{E}[f(\bar{x}) - f^*] \le \varepsilon$ for any $\varepsilon \in (0, \mathscr{V}_0)$ in at most $S := \lceil \log \mathscr{V}_0/\varepsilon \rceil$ stages. Moreover, the total number of iterations performed by this algorithm to find such a solution is bounded by $O(\mathscr{T}(\varepsilon))$, where

$$\mathscr{T}(\varepsilon) := \sqrt{\frac{L}{\mu}} \max\left(1, \log\frac{\mathscr{V}_0}{\varepsilon}\right) + \frac{K^2+\sigma^2}{\mu\varepsilon}. \tag{7.51}$$

By (7.51) and in view of the complexity theory by Nemirovski and Yudin [21, Sect. 7.2], we conclude that the multi-stage algorithm exhibits optimal iteration complexity for solving strongly convex problems. Moreover, suppose that we run the multi-stage AC-SA algorithm for $S_\beta := \lceil \log \mathscr{V}_0/(\beta\varepsilon) \rceil$ stages for a given $\beta \in (0,1)$. Then by (7.50) and Markov's inequality, we have $\mathsf{P}(f(p_{S_\beta}) - f^* > \varepsilon) \le \beta$, which implies that the total number of iterations performed by the multi-stage AC-SA algorithm to solve strongly convex problems with accuracy $\varepsilon > 0$ and a confidence level $1-\beta$ is of order $O(\mathscr{T}(\beta\varepsilon))$.

As shown in [9, Proposition 9], by properly shrinking the feasible region of the problem (7.1) and under condition (7.28), we can improve the large-deviation properties of the multi-stage AC-SA algorithm to

$$\sqrt{\frac{L}{\mu}} \max\left(1, \log\frac{\mathscr{V}_0}{\varepsilon}\right) + \frac{K^2+\sigma^2}{\mu\varepsilon} + \left[\ln\frac{\log(\mathscr{V}_0/\varepsilon)}{\beta}\right]^2 \frac{\sigma^2}{\mu\varepsilon}.$$

7.3 Nonconvex Stochastic Optimization

The convergence of existing SA methods including the ones presented in the previous section, requires $f(\cdot)$ to be convex [8, 9, 18–20]. In this section, we study a class of nonconvex stochastic optimization problems. More specifically, we assume that $f(\cdot)$ in (7.1) is differentiable (but possibly nonconvex) with Lipschitz-continuous gradients (i.e., (7.17) holds). In addition, we assume that $\Theta \equiv \mathbb{R}^n$ and the function f bounded from below. Observe also that to guarantee the convexity of $f(\cdot)$, one often needs to assume the random variables ξ_k, $k \geq 1$, to be independent of the search sequence $\{x_k\}$. Below we present a new SA-type algorithm that can deal with both convex and nonconvex stochastic optimization problems, and allow random noises to be dependent on the search sequence. This algorithm is obtained by incorporating a certain randomization scheme into the classical SA method.

Algorithm 5 The randomized stochastic gradient (RSG) method

Let initial point x_1, iteration limit N, step sizes $\{\gamma_k\}_{k\geq 1}$ and random variable R with probability mass function $P_R(\cdot)$ supported on $\{1,\ldots,N\}$ be given.
for $k = 1,\ldots,R$ **do**
  Set $x_{k+1} = x_k - \gamma_k G(x_k, \xi_k)$, for some $\gamma_k \in (0, 2/L)$.
end for
Output x_R.

Note that, in comparison with the classical SA algorithm, we have used a random iteration count, R, to terminate the execution of the RSG algorithm. Equivalently, one can view such a randomization scheme from a slightly different perspective described as follows. Instead of terminating the algorithm at the R-th step, one can also run the RSG algorithm for N iterations but randomly choose a search point x_R (according to P_R) from its trajectory as the output of the algorithm. Clearly, using the latter scheme, we just need to run the algorithm for the first R iterations and the remaining $N-R$ iterations are surpluses. Note however, that the primary goal to introduce the random iteration count R is to derive new complexity results for nonconvex SP, rather than save the computational efforts in the last $N-R$ iterations of the algorithm. Indeed, if R is uniformly distributed, the computational gain from such a randomization scheme is simply a factor of two.

The following result established in [10, Theorem 2.1] describes the main convergence properties of the RSG method.

Theorem 7.4. *Suppose that the probability mass function $P_R(\cdot)$ in the RSG method is chosen such that*

$$P_R(k) := \mathsf{P}(R=k) = \frac{2\gamma_k - L\gamma_k^2}{\sum_{j=1}^N (2\gamma_j - L\gamma_j^2)}, \quad k=1,\ldots,N. \tag{7.52}$$

If condition (7.19) holds, then

a) for any $N \geq 1$, we have

$$\frac{1}{L}\mathsf{E}[\|\nabla f(x_R)\|^2] \leq \frac{D_f^2 + \sigma^2 \sum_{k=1}^N \gamma_k^2}{\sum_{k=1}^N (2\gamma_k - L\gamma_k^2)}, \tag{7.53}$$

where the expectation is taken with respect to R and $\xi_{[N]} := (\xi_1,\ldots,\xi_N)$,

$$D_f := \left[\frac{2(f(x_1)-f^*)}{L}\right]^{\frac{1}{2}}, \tag{7.54}$$

and f^* denotes the optimal value of problem (7.1);

b) if, in addition, problem (7.1) is convex with an optimal solution x^*, then, for any $N \geq 1$,

$$\mathsf{E}[f(x_R) - f^*] \leq \frac{\|x_1 - x^*\|^2 + \sigma^2 \sum_{k=1}^N \gamma_k^2}{\sum_{k=1}^N (2\gamma_k - L\gamma_k^2)}, \tag{7.55}$$

where the expectation is taken with respect to R and $\xi_{[N]}$.

We now describe a possible strategy for the selection of the step sizes $\{\gamma_k\}$ in the RSG method. For the sake of simplicity, let us assume that a constant step-size policy is used, i.e., $\gamma_k = \gamma$, $k=1,\ldots,N$, for some $\gamma \in (0, 2/L)$. Note that the assumption of constant step sizes does not hurt the efficiency estimate of the RSG method. By choosing the step sizes as

$$\gamma_k = \min\left\{\frac{1}{L}, \frac{D_f}{\sigma\sqrt{N}}\right\}, \quad k=1,\ldots,N, \tag{7.56}$$

and in the view of (7.53), we have

$$\frac{1}{L}\mathsf{E}[\|\nabla f(x_R)\|^2] \leq \mathscr{B}_N := \frac{LD_f^2}{N} + \frac{2D_f\sigma}{\sqrt{N}}. \tag{7.57}$$

If in addition, problem (7.1) is convex, then by choosing the step sizes as

$$\gamma_k = \min\left\{\frac{1}{L}, \frac{\|x_1 - x^*\|}{\sigma\sqrt{N}}\right\}, \quad k=1,\ldots,N, \tag{7.58}$$

and in the view of (7.55), we have

$$\mathsf{E}[f(x_R) - f^*] \leq \frac{L\|x_1 - x^*\|^2}{N} + \frac{2\|x_1 - x^*\|\sigma}{\sqrt{N}}. \tag{7.59}$$

Note that D_f and $\|x_1 - x^*\|$ in the definition of γ_k in (7.56) and (7.58), respectively, can be replaced by any constant $\tilde{D} > 0$. In this case, the second terms in (7.57) and (7.59) are just affected by some constants. Thus, the RSG method allows us to have unified treatment for both smooth convex and nonconvex SP problems.

Moreover, (7.57) implies that the number of iterations performed by the RSG method to find a solution $\bar{x}$ such that $\mathsf{E}[\|\nabla f(\bar{x})\|^2] \leq \varepsilon$ is order $O(\sigma^2/\varepsilon^2)$, which is, to the best of our knowledge, the first complexity result in the literature for smooth nonconvex SP problems. Also, (7.59) implies that the RSG method posses nearly optimal convergence rate for convex SP problems.

One possible drawback of the step-size policy in (7.56) and (7.58) is that we need to estimate L to obtain an upper bound on the step sizes $\{\gamma_k\}$. Note that similar requirements also exist in some other first-order methods (e.g., gradient descent methods and Nesterov's accelerated gradient methods). While under the deterministic setting, one can somehow relax such requirements by using certain line-search procedures to enhance the practical performance of these methods, it is more difficult to devise similar line-search procedures for the stochastic setting, since the exact values of $f(x_k)$ and $\nabla f(x_k)$ are not available. It should also be noted that we do not need very accurate estimate for L in the RSG method. Indeed, it can be easily checked that the RSG method exhibits an $O(1/\sqrt{N})$ rate of convergence if the step sizes $\{\gamma_k\}$ are set to

$$\min\left\{\frac{1}{qL}, \frac{\tilde{D}}{\sigma\sqrt{N}}\right\}, \quad k = 1, \ldots, N$$

for any $q \in (1, \sqrt{N}]$. In other words, we can overestimate the value of L by a factor up to $\sqrt{N}$ and the resulting RSG method still exhibits similar rate of convergence. A common practice in stochastic optimization is to estimate these parameters by using the stochastic gradients computed at a small number of trial points (see, e.g., [8,9,19,20]). It is also worth noting that, although in general the selection of P_R will depend on γ_k and hence on L, such a dependence is not necessary in some special cases. In particular, if the step sizes $\{\gamma_k\}$ are chosen according to a constant step-size policy (e.g., (7.56)), then R is uniformly distributed on $\{1, \ldots, N\}$. See [12, Sect. 6] for more details about the practical implementation of the RSG method.

By (7.57) and Markov's inequality, we have

$$\mathsf{P}\left(\|\nabla f(x_R)\|^2 \geq \varepsilon\right) \leq \frac{1}{\varepsilon}\left(\frac{L^2 D_f^2}{N} + \frac{2LD_f\sigma}{\sqrt{N}}\right). \tag{7.60}$$

It then follows that the number of iterations performed by the RSG method for finding an (ε,β)-solution of problem (7.1) such that $\mathrm{P}\left(\|\nabla f(x_R)\|^2 \geq \varepsilon\right) \leq \beta$, after disregarding a few constant factors, can be bounded by

$$O\left(\frac{1}{\beta\varepsilon} + \frac{\sigma^2}{\beta^2\varepsilon^2}\right). \tag{7.61}$$

The above complexity bound is rather pessimistic in terms of its dependence on β.

We now describe a variant of the RSG method that can considerably improve the complexity bound in (7.61). This procedure consists of two phases: an optimization phase used to generate a list of candidate solutions via a few independent runs of the RSG method and a post-optimization phase in which a solution is selected from this candidate list.

Algorithm 6 The two-phase RSG (2-RSG) method

Let initial point x_1, number of runs S, iteration limit N, and sample size T be given.
Optimization phase:
for $s = 1,\ldots,S$ **do**
Call the RSG method with input x_1, iteration limit N, step sizes $\{\gamma_k\}$ in (7.56) and probability mass function P_R in (7.52). Let $\bar{x}_s$ be the output of this procedure.
end for
Post-optimization phase:
Output $\bar{x}^*$ from the candidate list $\{\bar{x}_1,\ldots,\bar{x}_S\}$ such that

$$\|g(\bar{x}^*)\| = \min_{s=1,\ldots,S} \|g(\bar{x}_s)\|, \;\; g(\bar{x}_s) := \frac{1}{T}\sum_{k=1}^{T} G(\bar{x}_s,\xi_k). \tag{7.62}$$

Observe that in (7.62), we define the best solution $\bar{x}^*$ as the one with the smallest value of $\|g(\bar{x}_s)\|$, $s = 1,\ldots,S$. Alternatively, one can choose $\bar{x}^*$ from $\{\bar{x}_1,\ldots,\bar{x}_S\}$ such that

$$\tilde{f}(\bar{x}^*) = \min_{s=1,\ldots,S} \tilde{f}(\bar{x}_s), \;\; \tilde{f}(\bar{x}_s) = \frac{1}{T}\sum_{k=1}^{T} F(\bar{x}_s,\xi_k). \tag{7.63}$$

Also, in the 2-RSG method described above, the number of computations for the stochastic gradient is given by $S\times N$ and $S\times T$, respectively, for the optimization phase and post-optimization phase. The following result presented in [10, Theorem 2.4] summarizes the convergence properties of the 2-RSG method.

Theorem 7.5. *If condition (7.19) holds, the following statements hold for the 2-RSG method applied to problem (7.1).*

a) Let $\mathcal{B}_N$ be defined in (7.57). We have

$$\mathrm{P}\left(\|\nabla f(\bar{x}^*)\|^2 \geq 2\left[4L\mathcal{B}_N + \frac{3\lambda\sigma^2}{T}\right]\right) \leq \frac{S+1}{\lambda} + 2^{-S}, \ \forall \lambda > 0; \tag{7.64}$$

b) Let $\varepsilon > 0$ and $\beta \in (0,1)$ be given. If the parameters (S,N,T) are set to

$$S = S(\beta) := \lceil \log(2/\beta) \rceil, \tag{7.65}$$

$$N = N(\varepsilon) := \left\lceil \max\left\{ \frac{32L^2D_f^2}{\varepsilon}, \left[\frac{64LD_f\sigma}{\varepsilon}\right]^2 \right\} \right\rceil, \tag{7.66}$$

$$T = T(\varepsilon,\beta) := \left\lceil \frac{24(S+1)\sigma^2}{\beta\varepsilon} \right\rceil, \tag{7.67}$$

then the 2-RSG method can compute an (ε,β)-solution of problem (7.1) after taking at most

$$S(\beta)\,[N(\varepsilon) + T(\varepsilon,\beta)] \tag{7.68}$$

computations of the stochastic gradient.

In view of (7.65), (7.66) and (7.67), the complexity bound in (7.68), after disregarding a few constant factors, is equivalent to

$$O\left(\frac{\log(1/\beta)}{\varepsilon} + \frac{\sigma^2}{\varepsilon^2}\log\frac{1}{\beta} + \frac{\log^2(1/\beta)\sigma^2}{\beta\varepsilon}\right). \tag{7.69}$$

The above bound can be considerably smaller than the one in (7.61) up to a factor of $1/\left[\beta^2\log(1/\beta)\right]$, when the second terms in both bounds dominate.

The complexity bound (7.69) can be further improved. Specifically, if condition (7.28) holds and the sample size for estimating the gradients at post-optimization phase is set to

$$T = T'(\varepsilon,\beta) := \frac{24\sigma^2}{\varepsilon}\left[1 + \left(3\ln\frac{2(S+1)}{\beta}\right)^{\frac{1}{2}}\right]^2, \tag{7.70}$$

then the complexity of the 2-RSG method for computing an (ε,β)-solution of problem (7.1) is order

$$O\left(\frac{\log(1/\beta)}{\varepsilon} + \frac{\sigma^2}{\varepsilon^2}\log\frac{1}{\beta} + \frac{\log^2(1/\beta)\sigma^2}{\varepsilon}\right). \tag{7.71}$$

Clearly, the third term of the above bound is significantly smaller than the corresponding one in (7.69) by a factor of $1/\beta$ (c.f., [10, Corollary 2.5]).

7.4 Randomized Stochastic Zeroth-Order (SZO) Methods

In this section, we still consider the situation when we $f(\cdot)$ in (7.1) is differentiable (but possibly nonconvex) with Lipschitz-continuous gradients (i.e., (7.17) holds). In addition, we assume that $\Theta \equiv \mathbb{R}^n$ and the function $f(\cdot)$ bounded from below. However, throughout this section, we assume that only stochastic function values $F(x,\xi)$ of the objective function $f(\cdot)$ in (7.1) are given. Our goal is to present SZO methods for solving these types of simulation optimization problems.

To exploit zeroth-order information, we consider a smooth approximation of the objective function f given by a convolution with Gaussian distribution. Specifically, we consider a smooth approximation of f defined as

$$f_\mu(x) = \frac{1}{(2\pi)^{\frac{n}{2}}} \int f(x+\mu u) e^{-\frac{1}{2}\|u\|^2} du = \mathsf{E}_u[f(x+\mu u)], \tag{7.72}$$

where $\mu > 0$ is the smoothing parameter. The following result due to Nesterov [23] describes some properties of $f_\mu(\cdot)$.

Theorem 7.6. *The following statements hold for any f satisfying in (7.17).*

a) The gradient of f_μ given by

$$\nabla f_\mu(x) = \frac{1}{(2\pi)^{\frac{n}{2}}} \int \frac{f(x+\mu u) - f(x)}{\mu} u e^{-\frac{1}{2}\|u\|^2} du, \tag{7.73}$$

is Lipschitz continuous with constant L_μ such that $L_\mu \le L$.

b) For any $x \in \mathbb{R}^n$,

$$|f_\mu(x) - f(x)| \le \frac{\mu^2}{2} Ln, \tag{7.74}$$

$$\|\nabla f_\mu(x) - \nabla f(x)\| \le \frac{\mu}{2} L(n+3)^{\frac{3}{2}}. \tag{7.75}$$

c) For any $x \in \mathbb{R}^n$,

$$\frac{1}{\mu^2}\mathsf{E}_u[\{f(x+\mu u) - f(x)\}^2 \|u\|^2] \le \frac{\mu^2}{2} L^2 (n+6)^3 + 2(n+4)\|\nabla f(x)\|^2. \tag{7.76}$$

Below, we modify the RSG method in the previous section to use stochastic zeroth-order rather than first-order information for solving problem (7.1).

Algorithm 7 The randomized stochastic gradient free (RSGF) method

---

Let initial point x_1, iteration limit N, step sizes $\{\gamma_k\}_{k\geq 1}$ and random variable R with probability mass function $P_R(\cdot)$ supported on $\{1,\ldots,N\}$ be given.
for $k = 1,\ldots,R$ **do**
Generate u_k by Gaussian random vector generator and compute $G_\mu(x_k,\xi_k,u_k)$ given by

$$G_\mu(x_k,\xi_k,u_k) = \frac{F(x_k+\mu u_k,\xi_k) - F(x_k,\xi_k)}{\mu} u_k. \tag{7.77}$$

Set $x_{k+1} = x_k - \gamma_k G_\mu(x_k,\xi_k,u_k)$ for some $\gamma_k \in (0, 2/(n+4)L)$.
end for
Output x_R.

Note that by (7.73) and (7.77), we have

$$\mathsf{E}_{\xi,u}[G_\mu(x,\xi,u)] = \mathsf{E}_u\left[\mathsf{E}_\xi[G_\mu(x,\xi,u)|u]\right] = \nabla f_\mu(x), \tag{7.78}$$

which clearly implies that $G_\mu(x,\xi,u)$ is an unbiased estimate of $\nabla f_\mu(x)$. Hence, if the variance $\tilde{\sigma}^2 \equiv \mathsf{E}_{\xi,u}[\|G_\mu(x,\xi,u) - \nabla f_\mu(x)\|^2]$ is bounded, we can directly apply the convergence results in Theorem 7.4 to the above RSGF method. However, there still exist a few problems in this approach. First, we do not know an explicit expression of the bound $\tilde{\sigma}^2$. Secondly, this approach does not provide any information regarding how to appropriately specify the smoothing parameter μ. The latter issue is critical for the implementation of the RSGF method. The next result established in [10, Theorem 3.2], shows the main convergence properties of the RSGF method.

Theorem 7.7. *Suppose that the probability mass function $P_R(\cdot)$ in the RSGF method is given by*

$$P_R(k) := \mathsf{P}(R=k) = \frac{\gamma_k - 2L(n+4)\gamma_k^2}{\sum_{j=1}^N\left[\gamma_j - 2L(n+4)\gamma_j^2\right]}, \quad k=1,\ldots,N. \tag{7.79}$$

If condition (7.19) holds, then

a) for any $N \geq 1$, we have

$$\frac{1}{L}\mathsf{E}[\|\nabla f(x_R)\|^2] \leq \frac{1}{\sum_{k=1}^N\left[\gamma_k - 2L(n+4)\gamma_k^2\right]}$$

$$\left\{D_f^2+2\mu^2(n+4)\left[1+L(n+4)^2\sum_{k=1}^{N}(\frac{\gamma_k}{4}+L\gamma_k^2)\right]+2(n+4)\sigma^2\sum_{k=1}^{N}\gamma_k^2\right\}, \tag{7.80}$$

where the expectation is taken with respect to R, $\xi_{[N]}$ and $u_{[N]}$.

b) *if, in addition, problem (7.1) is convex with an optimal solution x^*, then, for any $N \geq 1$,*

$$\mathsf{E}[f(x_R)-f^*] \leq \frac{1}{2\sum_{k=1}^{N}\left[\gamma_k-2(n+4)L\gamma_k^2\right]} \tag{7.81}$$

$$\left\{\|x_1-x^*\|^2+2\mu^2L(n+4)\sum_{k=1}^{N}\left[\gamma_k+L(n+4)^2\gamma_k^2\right]+2(n+4)\sigma^2\sum_{k=1}^{N}\gamma_k^2\right\}, \tag{7.82}$$

where the expectation is taken with respect to R, $\xi_{[N]}$ and $u_{[N]}$.

Similarly to the RSG method, we can specialize the convergence results in the above theorem with a constant step-size policy. Specifically, if condition (7.19) holds, then by choosing the step size γ_k and the smoothing parameter μ as

$$\gamma_k = \frac{1}{\sqrt{n+4}}\min\left\{\frac{1}{4L\sqrt{n+4}},\frac{D_f}{\sigma\sqrt{N}}\right\}, \quad k=1,\ldots,N, \tag{7.83}$$

$$\mu \leq \frac{D_f}{(n+4)\sqrt{2N}}, \tag{7.84}$$

we have

$$\frac{1}{L}\mathsf{E}[\|\nabla f(x_R)\|^2] \leq \bar{\mathscr{B}}_N := \frac{12(n+4)LD_f^2}{N}+\frac{8D_f\sqrt{n+4}\sigma}{\sqrt{N}}. \tag{7.85}$$

If, in addition, problem (7.1) is convex with an optimal solution x^*, γ_k and μ are chosen such that

$$\gamma_k = \frac{1}{\sqrt{n+4}}\min\left\{\frac{1}{4L\sqrt{n+4}},\frac{\|x_1-x^*\|}{\sigma\sqrt{N}}\right\}, \quad k=1,\ldots,N, \tag{7.86}$$

$$\mu \leq \frac{\|x_1-x^*\|}{\sqrt{n+4}}, \tag{7.87}$$

then

$$\mathsf{E}[f(x_R)-f^*] \leq \frac{5L(n+4)\|x_1-x^*\|^2}{N}+\frac{4\|x_1-x^*\|\sqrt{n+4}\sigma}{\sqrt{N}}. \tag{7.88}$$

The above bound implies that the complexity of the RSGF method, when the problem is convex, to find a solution $\bar{x}$ such that $\mathsf{E}[f(\bar{x}) - f^*] \le \varepsilon$, is bounded by $O(n/\varepsilon^2)$, which has a weaker dependence (by a factor of n) than the one established by Nesterov for solving general nonsmooth convex SP problems (see page 17 of [23]).

By (7.85) and Markov's inequality, we have

$$\mathsf{P}\left(\|\nabla f(x_R)\|^2 \ge \varepsilon\right) \le \frac{1}{\varepsilon}\left(\frac{12(n+4)L^2D_f^2}{N} + \frac{8LD_f\sqrt{n+4}\sigma}{\sqrt{N}}\right), \tag{7.89}$$

which implies that the total number of computation of stochastic performed by the RSGF method for finding an (ε, β)-solution of (7.1) can be bounded by

$$O\left(\frac{n}{\beta\varepsilon} + \frac{n}{\beta^2}\frac{\sigma^2}{\varepsilon^2}\right). \tag{7.90}$$

Similarly to the RSG method, we can design a two-phase variant of the RSGF method to improve the dependence of the above bound on the confidence level β as follows.

Algorithm 8 The two-phase RSGF (2-RSGF) method

Let initial point x_1, number of runs S, iteration limit N, and sample size T be given.
Optimization phase:
for $s = 1, \ldots, S$ **do**
  Call the RSGF method with input x_1, iteration limit N, step sizes $\{\gamma_k\}$ in (7.86), probability mass function P_R in (7.79) and the smoothing parameter μ satisfying in (7.87). Let $\bar{x}_s$ be the output of this procedure.
end for
Post-optimization phase:
Output $\bar{x}^*$ from the candidate list $\{\bar{x}_1, \ldots, \bar{x}_S\}$ such that

$$\|g_\mu(\bar{x}^*)\| = \min_{s=1,\ldots,S} \|g_\mu(\bar{x}_s)\|, \;\; g_\mu(\bar{x}_s) := \frac{1}{T}\sum_{k=1}^{T} G_\mu(\bar{x}_s, \xi_k, u_k). \tag{7.91}$$

The following result presented in [10, Theorem 3.4] summarizes the convergence properties of the 2-RSGF method.

Theorem 7.8. *If condition (7.19) holds, the following statements hold for the 2-RSGF method applied to problem (7.1).*

a) Let $\bar{\mathscr{B}}_N$ be defined in (7.85). We have

$$\mathsf{P}\left(\|\nabla f(\bar{x}^*)\|^2 \geq 8L\bar{\mathscr{B}}_N + \frac{3(n+4)L^2D_f^2}{2N} + \frac{24(n+4)\lambda}{T}\left[L\bar{\mathscr{B}}_N + \frac{(n+4)L^2D_f^2}{N} + \sigma^2\right]\right) \leq \frac{S+1}{\lambda} + 2^{-S}, \ \forall \lambda > 0.$$

b) Let $\varepsilon > 0$ and $\beta \in (0,1)$ be given. If S is set to $S(\beta)$ as in (7.65), and the iteration limit N and sample size T, respectively, are set to

$$N = \hat{N}(\varepsilon) := \max\left\{\frac{12(n+4)(6LD_f)^2}{\varepsilon}, \left[\frac{144L\sqrt{n+4}D_f\sigma}{\varepsilon}\right]^2\right\}, \tag{7.92}$$

$$T = \hat{T}(\varepsilon,\beta) := \frac{24(n+4)(S+1)}{\beta}\max\left\{1, \frac{6\sigma^2}{\varepsilon}\right\}, \tag{7.93}$$

then the 2-RSGF method can compute an (ε,β)-solution of problem (7.1) after taking at most

$$2S(\beta)\left[\hat{N}(\varepsilon) + \hat{T}(\varepsilon,\beta)\right] \tag{7.94}$$

computations of the stochastic function values.

Observe that in the view of (7.65), (7.92) and (7.93), the complexity of the 2-RSGF method to find (ε,β)-solution of (7.1) is bounded by

$$O\left(\frac{n\log(1/\beta)}{\varepsilon} + \frac{n\sigma^2}{\varepsilon^2}\log\frac{1}{\beta} + \frac{n\log^2(1/\beta)}{\beta}\left[1 + \frac{\sigma^2}{\varepsilon}\right]\right). \tag{7.95}$$

The above bound is considerably smaller than the one in (7.90), up to a factor of $O\left(1/[\beta^2\log(1/\beta)]\right)$, when the second terms are the dominating ones in both bounds.

7.5 Summary

In this chapter, we surveyed some recent advances on the design and analysis of SA-type methods for solving different classes of stochastic optimization problems. We focused on the finite-time convergence properties of these methods, indicating the optimality of their associated complexity bounds when the problem is convex. We also emphasized the large-deviation properties of the rate of convergence for these new SA-type methods. Proofs and numerical experiments have not been included here, but can be found in the original papers [8–10, 18, 20, 23].

Acknowledgements This work was supported in part by the National Science Foundation under Grants CMMI-1000347, CMMI-1254446, and DMS-1319050, and by the Office of Naval Research under Grant N00014-13-1-0036.

References

1. A. Benveniste, M. Métivier, and P. Priouret. *Algorithmes adaptatifs et approximations stochastiques*. Masson, 1987. English translation: *Adaptive Algorithms and Stochastic Approximations*, Springer Verlag (1993).
2. C. Cartis, N. I. M. Gould, and P. L. Toint. On the oracle complexity of first-order and derivative-free algorithms for smooth nonconvex minimization. *SIAM Journal on Optimization*, 22:66–86, 2012.
3. K. Chung. On a stochastic approximation method. *Annals of Mathematical Statistics*, pages 463–483, 1954.
4. A. R. Conn, K. Scheinberg, and L. N. Vicente. *Introduction to Derivative-Free Optimization*. SIAM, Philadelphia, 2009.
5. Y. Ermoliev. Stochastic quasigradient methods and their application to system optimization. *Stochastics*, 9:1–36, 1983.
6. A. Gaivoronski. Nonstationary stochastic programming problems. *Kybernetika*, 4:89–92, 1978.
7. R. Garmanjani and L. N. Vicente. Smoothing and worst-case complexity for direct-search methods in nonsmooth optimization. *IMA Journal of Numerical Analysis*, 33:1008–1028, 2013.
8. S. Ghadimi and G. Lan. Optimal stochastic approximation algorithms for strongly convex stochastic composite optimization, I: a generic algorithmic framework. *SIAM Journal on Optimization*, 22:1469–1492, 2012.
9. S. Ghadimi and G. Lan. Optimal stochastic approximation algorithms for strongly convex stochastic composite optimization, II: shrinking procedures and optimal algorithms. *SIAM Journal on Optimization*, 23:2061–2089, 2013.
10. S. Ghadimi and G. Lan. Stochastic first- and zeroth-order methods for nonconvex stochastic programming. *SIAM Journal on Optimization*, 23:2341–2368, 2013.
11. S. Ghadimi and G. Lan. Accelerated gradient methods for nonconvex nonlinear and stochastic optimization. Technical report, Department of Industrial and Systems Engineering, University of Florida, Gainesville, FL 32611, USA, June 2013.
12. S. Ghadimi, G. Lan, and H. Zhang. Mini-batch stochastic approximation methods for constrained nonconvex stochastic programming. Manuscript, Department of Industrial and Systems Engineering, University of Florida, Gainesville, FL 32611, USA, August 2013.
13. A. Juditsky, A. Nazin, A. B. Tsybakov, and N. Vayatis. Recursive aggregation of estimators via the mirror descent algorithm with average. *Problems of Information Transmission*, 41:n.4, 2005.
14. A. Juditsky, P. Rigollet, and A. B. Tsybakov. Learning by mirror averaging. *Annals of Statistics*, 36:2183–2206, 2008.
15. J. Kiefer and J. Wolfowitz. Stochastic estimation of the maximum of a regression function. *Annals of Mathematical Statistics*, 23:462–466, 1952.
16. A. J. Kleywegt, A. Shapiro, and T. Homem-de-Mello. The sample average approximation method for stochastic discrete optimization. *SIAM Journal on Optimization*, 12:479–502, 2001.
17. H. J. Kushner and G. Yin. *Stochastic Approximation and Recursive Algorithms and Applications*, volume 35 of *Applications of Mathematics*. Springer-Verlag, New York, 2003.
18. G. Lan. An optimal method for stochastic composite optimization. *Mathematical Programming*, 133(1):365–397, 2012.

19. G. Lan, A. S. Nemirovski, and A. Shapiro. Validation analysis of mirror descent stochastic approximation method. *Mathematical Programming*, 134(2):425–458, 2012.
20. A. S. Nemirovski, A. Juditsky, G. Lan, and A. Shapiro. Robust stochastic approximation approach to stochastic programming. *SIAM Journal on Optimization*, 19:1574–1609, 2009.
21. A. S. Nemirovski and D. Yudin. *Problem complexity and method efficiency in optimization.* Wiley-Interscience Series in Discrete Mathematics. John Wiley, XV, 1983.
22. Y. E. Nesterov. Primal-dual subgradient methods for convex problems. *Mathematical Programming*, 120:221–259, 2006.
23. Y. E. Nesterov. Random gradient-free minimization of convex functions. Technical report, Center for Operations Research and Econometrics (CORE), Catholic University of Louvain, January 2010.
24. G. Pflug. Optimization of stochastic models. In *The Interface Between Simulation and Optimization*. Kluwer, Boston, 1996.
25. B. Polyak. New stochastic approximation type procedures. *Automat. i Telemekh.*, 7:98–107, 1990.
26. B. Polyak and A. Juditsky. Acceleration of stochastic approximation by averaging. *SIAM J. Control and Optimization*, 30:838–855, 1992.
27. H. Robbins and S. Monro. A stochastic approximation method. *Annals of Mathematical Statistics*, 22:400–407, 1951.
28. A. Ruszczyński and W. Sysk. A method of aggregate stochastic subgradients with on-line stepsize rules for convex stochastic programming problems. *Mathematical Programming Study*, 28:113–131, 1986.
29. J. Sacks. Asymptotic distribution of stochastic approximation. *Annals of Mathematical Statistics*, 29:373–409, 1958.
30. A. Shapiro. Monte Carlo sampling methods. In A. Ruszczyński and A. Shapiro, editors, *Stochastic Programming*. North-Holland Publishing Company, Amsterdam, 2003.
31. A. Shapiro. Sample average approximation. In S. I. Gass and M. C. Fu, editors, *Encyclopedia of Operations Research and Management Science*, pages 1350–1355. Springer, 3rd edition, 2013.
32. A. Shapiro, D. Dentcheva, and A. Ruszczyński. *Lectures on Stochastic Programming: Modeling and Theory*. SIAM, Philadelphia, 2009.
33. J. Spall. *Introduction to Stochastic Search and Optimization: Estimation, Simulation, and Control*. John Wiley, Hoboken, NJ, 2003.
34. V. Strassen. The existence of probability measures with given marginals. *Annals of Mathematical Statistics*, 38:423–439, 1965.
35. L. N. Vicente. Worst case complexity of direct search. *EURO Journal on Computational Optimization*, 1:143–153, 2013.

Chapter 8
A Guide to Sample Average Approximation

Sujin Kim, Raghu Pasupathy, and Shane G. Henderson

Abstract This chapter reviews the principles of sample average approximation (SAA) for solving simulation optimization problems. We provide an accessible overview of the area and survey interesting recent developments. We explain when one might want to use SAA and when one might expect it to provide good-quality solutions. We also review some of the key theoretical properties of the solutions obtained through SAA. We contrast SAA with stochastic approximation (SA) methods in terms of the computational effort required to obtain solutions of a given quality, explaining why SA "wins" asymptotically. However, an extension of SAA known as retrospective optimization can match the asymptotic convergence rate of SA, at least up to a multiplicative constant.

8.1 Introduction

How does one solve an optimization problem of the form

$$\min_{x \in \Theta} f(x), \tag{8.1}$$

where $\Theta \subseteq \mathbb{R}^d$ $(d < \infty)$ and the real-valued function $f(\cdot)$ cannot be computed exactly, but can be estimated through a (stochastic) simulation? The principle of Sample Average Approximation (SAA) allows one to tackle such problems through the use of sampling and optimization methods for deterministic problems. We introduce SAA, describe its properties through both examples and theory, and relate

---

S. Kim
National University of Singapore, Singapore
e-mail: iseks@nus.edu.sg

R. Pasupathy
Purdue University, West Lafayette, IN, USA
e-mail: pasupath@purdue.edu

S.G. Henderson (✉)
Cornell University, Ithaca, NY, USA
e-mail: sgh9@cornell.edu

M.C. Fu (ed.), *Handbook of Simulation Optimization*, International Series in Operations Research & Management Science 216, DOI 10.1007/978-1-4939-1384-8_8

SAA to established concepts in stochastic simulation. Our goal is to communicate the essence of the idea and the key results in the area, rather than to provide an exhaustive discussion of what is known about SAA. As such, this chapter is best viewed as a *guide* rather than a *survey*. Similar guides for the strongly related area of stochastic programming at a more introductory level can be found in [58].

Throughout, we assume that the function f cannot be observed or computed directly, but we know that $f(x) = \mathsf{E}[Y(x,\xi)]$, where ξ is a random element with a distribution that does not depend on x, and $Y(\cdot,\cdot)$ is a (deterministic) real-valued function. Implicit in this statement is that for each fixed $x \in \Theta$, $\mathsf{E}|Y(x,\xi)| < \infty$. We suppress measure-theoretic considerations unless they come into play at a practical level. Nevertheless, we attempt to state results precisely.

In SAA, we select and fix $\xi_1, \xi_2, \ldots, \xi_n$, all having the same distribution as ξ, and set

$$f_n(x) = \frac{1}{n}\sum_{i=1}^{n} Y(x,\xi_i).$$

Given the (fixed) sample $\xi_1, \xi_2, \ldots, \xi_n$, the function $f_n(\cdot)$ is deterministic, and so we can apply deterministic optimization algorithms to solve the problem

$$\min_{x\in\Theta} f_n(x). \tag{8.2}$$

We then take an optimizer, X_n^* say, of (8.2) as an estimator of an optimal solution of (8.1). Unless otherwise stated, we assume that $(\xi_1, \xi_2, \ldots, \xi_n)$ form an independent and identically distributed (i.i.d.) sample. The independence assumption is sometimes relaxed, mostly in variance reduction schemes or randomized quasi-Monte Carlo schemes where dependence is deliberately introduced.

A popular example to illustrate SAA is the continuous newsvendor problem where we buy x units of some commodity at a cost $c > 0$ per unit, observe demand ξ, and sell as many units as we can at price $s > c$; see, e.g., [57, p. 330]. The goal is to choose x so as to maximize profit. (Of course, one can convert this problem to a minimization problem as in (8.1) simply by multiplying by -1.) The profit for a given realization ξ is $Y(x,\xi) = s\min\{x,\xi\} - cx$. This function is concave in x, has slope $s-c>0$ for sufficiently small x and slope $-c<0$ for sufficiently large x. It follows that the same is true for the approximating function $f_n(\cdot)$, which therefore achieves its maximum. In fact, it is straightforward to show that an optimizer X_n^* of $f_n(\cdot)$ occurs at the $1-c/s$ quantile of the empirical distribution associated with the observed demands $\xi_1, \xi_2, \ldots, \xi_n$, i.e., the $\lceil n(1-c/s)\rceil$th smallest of the observed demands. If we assume that the distribution of ξ is continuous at its $1-c/s$ quantile, which is optimal for the true problem (e.g., [41, p. 353]), then X_n^* converges to this value as $n \to \infty$ almost surely (a.s.). So in this case, SAA is successful, in that the sequence of optimizers $\{X_n^*\}$ converges to a true optimizer.

In general, is SAA a reasonable approach? What kinds of problems are such that SAA works, in the sense that X_n^* can be expected to converge to the set of optimizers of (8.1) as $n \to \infty$ in some sense? What kinds of problems are such that (8.2) is amenable to deterministic optimization algorithms? Is this procedure competitive with alternative algorithms, in the sense that the solutions returned after a given computational effort are comparable in quality?

Most of these questions have been addressed in previous surveys of SAA, so what is different here? We emphasize the intuition behind SAA, developing concepts through a range of examples as well as through theory. Mostly we do not prove results here, but instead give references to complete proofs, and provide proof sketches where that helps build understanding. Many of those proofs can be found in the excellent surveys [55–57].

It is hard to pin down the origin of the SAA concept. Certainly there are strong ties to maximum likelihood estimation and M-estimation in statistics, but perhaps the strongest roots of the idea from an Operations Research perspective lie in variants called the stochastic counterpart method [47, 48] and sample-path optimization [20, 37, 43].

We focus on the unconstrained optimization problem (8.1), but SAA can also encompass constrained optimization problems, even when the constraints must also be evaluated using simulation; see [60, 61] and the next chapter. The SAA principle is very general, having been applied to settings including chance constraints [1], stochastic-dominance constraints [24] and complementarity constraints [18].

The rest of this chapter is organized as follows. Section 8.2 provides a set of examples that showcase when SAA is appropriate in the sense that the optimization problem (8.2) has reasonable structure that allows for numerical solution. Section 8.3 provides verifiable conditions under which one can expect the problems (8.1) and (8.2) to share important structural properties such as continuity and differentiability. This section also showcases the close connection between problems that are "SAA appropriate" and those that are amenable to infinitesimal perturbation analysis (IPA) [13, 17] for gradient estimation. This section can also be viewed as a review of IPA with particular emphasis on multidimensional problems. In Sect. 8.4, we review some key properties of SAA, particularly with regard to large-sample performance. Sects 8.5 and 8.6 delve into the selection of the sample size n in some detail. These sections relate the computational effort required to achieve a given solution quality in SAA to that of a competing method known as stochastic approximation. It turns out that SAA is not as efficient as stochastic approximation, at least in the asymptotic sense. (In the non-asymptotic world of small sample sizes, it is harder to make clear conclusions, although some results are known for idealized versions of both approaches; see, e.g., [54].) This leads one to the class of methods collectively known as retrospective optimization, which is an extension of SAA. We review some recent results on retrospective optimization that show that this class of methods can match stochastic approximation in terms of asymptotic efficiency.

8.2 When Is SAA Appropriate?

One hopes the X_n^* obtained by solving the SAA problem converges to a solution of the true problem x^*. The critical condition for convergence is a uniform version of the strong law of large numbers (ULLN), which takes the form

$$\sup_{x\in\Theta}|f_n(x)-f(x)| = \sup_{x\in\Theta}\left|\frac{1}{n}\sum_{i=1}^{n} Y(x,\xi_i) - \mathsf{E}[Y(x,\xi)]\right| \to 0$$

as $n \to \infty$ a.s. The ULLN ensures that the optimal objective value of the SAA problem converges to that of the true problem. With additional conditions, the optimal SAA solution converges to the true optimal solution. When the sample function is convex (concave for a maximization problem), the pointwise law of large numbers ensures that the ULLN holds on a compact set. We will further discuss conditions under which the ULLN holds in Sect. 8.4.

In many problems, sample functions are not smooth and may have discontinuities. However, the true problem may still exhibit nice structure, being smooth and even convex. In such a case, if the ULLN holds, we may still be able to use a deterministic optimization technique to effectively solve the nonsmooth sample average problem and thereby obtain a good approximate solution.

In this section, we provide examples that illustrate how SAA works, what issues may arise when SAA is applied, and how we may deal with them in various settings. Henceforth, all vectors are assumed to be column vectors, and x^T denotes the transpose of x.

Example 8.1 (Multi-Dimensional Newsvendor Problem). Consider a firm that manufactures p products from q resources. For given resource type $i = 1,\dots,q$ and product type $j = 1,\dots,p$, let a_{ij} be the amount of resource i required to produce one unit of product j, and v_j be the unit margin for product j, i.e., revenue minus processing cost. Suppose that a manager must decide on a resource vector $x = (x_1,\dots,x_q)$ before the product demand vector $\xi = (\xi_1,\dots,\xi_p)$ is observed. After the demand becomes known, the manager chooses a production vector $y = (y_1,\dots,y_p)$ so as to maximize the operating profit in the linear program

$$\begin{aligned} \mathscr{P}_{(x,\xi)}:\ \max\nolimits_{y\in\mathbb{R}^p_+}\ & v^\mathsf{T}y \\ \text{s.t.}\quad & Ay \le x \quad \text{(capacity constraints)} \\ & y \le \xi \quad \text{(demand constraints)}. \end{aligned}$$

Here, A is a $(q \times p)$ matrix and a_{ij} is the (i,j) element of A. Let $\Pi(x,\xi)$ denote the maximal operating profit function for a given resource level vector x and a given demand vector ξ. This is precisely the optimal objective value of the problem $\mathscr{P}_{(x,\xi)}$. Then $\Pi(x,\xi) = v^\mathsf{T}y^*(x,\xi)$, where $y^*(x,\xi)$ is an associated optimal production vector.

Suppose that the demand ξ can be viewed as a random vector and the probability distribution of ξ is known. Let $\pi(x)$ denote the expected maximal operating profit, where

$$\pi(x) = \mathsf{E}\left[\Pi(x,\xi)\right],$$

for all $x \in \mathbb{R}^q_+$. Let $c_i, i = 1,\ldots,q$, be the unit investment cost for resource i. By incorporating the investment cost into the operating profit, the value of the firm is defined as $\Pi(x,\xi) - c^\mathsf{T}x$, for a fixed (x,ξ). The manager's objective is now to choose the resource level x so as to maximize the expected firm value. This leads to the following stochastic optimization problem:

$$\max_{x\in\mathbb{R}^q_+} f(x) = \pi(x) - c^\mathsf{T}x. \tag{8.3}$$

This problem is known as the *multi-dimensional newsvendor problem* [59]. For simplicity, we focus our attention on the single-period newsvendor model, but the structure of the optimal policy in the single-period model can be extended to a dynamic setting under reasonable conditions [19]. In general, a closed-form solution for the multi-dimensional newsvendor problem is unattainable, unlike the single-dimensional problem. We illustrate how the SAA approach can be applied to this example and present some technical details.

From linear programming theory, we can show that both the sample path function $\Pi(\cdot,\xi)$ and the expected objective function π exhibit nice structural properties. First, $\Pi(\cdot,\xi)$ is concave for any fixed ξ, and so is $\pi(\cdot) = \mathsf{E}[\Pi(\cdot,\xi)]$. If ξ has a discrete probability distribution, then both $\Pi(\cdot,\xi)$ and $\pi(\cdot)$ are piecewise linear and concave. However, we focus on random demand with a continuous probability distribution, and we would like to determine conditions under which $\pi(\cdot)$ is differentiable everywhere. Assume that ξ is finite a.s. Consider the dual problem of the linear program $\mathscr{P}_{(x,\xi)}$:

$$\begin{aligned} \mathscr{D}_{(x,\xi)} : \min\nolimits_{(\mu,\lambda)\in\mathbb{R}^{p+q}_+} \quad & x^\mathsf{T}\lambda + \xi^\mathsf{T}\mu \\ \text{s.t.} \quad & A^\mathsf{T}\lambda + \mu \geq v. \end{aligned}$$

Since ξ is finite, the primal problem has a finite optimal solution and the optimal value of the primal problem is equal to that of the dual problem. Let $\lambda(x,\xi)$ denote the optimal shadow value of the capacity constraint in the primal problem $\mathscr{P}_{(x,\xi)}$. Using duality theory, it can be shown that

$$\Pi(x,\xi) \leq \Pi(x_0,\xi) + \lambda(x_0,\xi)^\mathsf{T}(x - x_0), \tag{8.4}$$

and hence $\lambda(\cdot,\xi)$ is a subgradient of $\Pi(\cdot,\xi)$. Since $\Pi(\cdot,\xi)$ is concave for a fixed ξ, it is differentiable except on a set A with Lebesgue measure zero. Since ξ is a continuous random variable, A is also negligible with respect to the probability measure. Thus, $\lambda(x,\xi)$ is unique and $\nabla_x\Pi(x,\xi) = \lambda(x,\xi)$ at a fixed x a.s. Taking the expectation in Eq. (8.4) yields that $\mathsf{E}[\lambda(\cdot,\xi)]$ is a subgradient of $\pi(\cdot)$. Therefore, $\mathsf{E}[\lambda(x,\xi)]$ is unique for all $x \in \mathbb{R}^q_+$ so that $\pi(\cdot)$ is differentiable and

$$\nabla\pi(\cdot) = \mathsf{E}[\lambda(\cdot,\xi)] = \mathsf{E}[\nabla_x\Pi(\cdot,\xi)]. \tag{8.5}$$

Note that $\Pi(\cdot,\xi)$ does not have to be differentiable everywhere, but expectation with respect to a continuous random variable ξ yields a smooth function $\pi(\cdot)$. Equation (8.5) establishes that one can interchange the expectation and differentiation operators. In Sect. 8.3 we will discuss how this interchange property basically ensures that SAA is appropriate for tackling an optimization problem.

The analysis above shows that the true function $\pi(\cdot)$ and the sample function $\Pi(\cdot,\xi)$ share the same nice structural properties; smoothness and concavity. This allows the multi-dimensional newsvendor problem to be effectively solved by the SAA method. The sample average approximation function

$$f_n(x) = \frac{1}{n}\sum_{k=1}^{n}\Pi(x,\xi_k) - c^{\mathsf{T}}x$$

is piecewise linear and concave, but not smooth everywhere. However, the sample average approximation function can be quickly smoothed out as the sample size n grows, so in practice, one can choose sufficiently large n, and then apply an algorithm for optimization of smooth concave functions to solve the sample average approximation problem using the gradient estimator $\frac{1}{n}\sum_{k=1}^{n}\lambda(x,\xi_k) - c$. If the sample average function is not smooth enough and any gradient-based algorithm is not appropriate to use, a subgradient method for convex optimization can be applied to $-f_n(\cdot)$. One can also apply two-stage stochastic linear programming algorithms to solve the sampled problem [8].

Example 8.2 (Multi-Mix Blending Problem). Consider a simple blending problem in which q products are made with p raw materials. The blend products have to satisfy certain pre-specified quality requirements. The total processing costs incurred depend on the product blending options used. Additionally, the production output has to meet minimum requirements. A complication arises when some materials are available in different quantities at different prices.

For the sake of illustration, we consider a problem with only one quality measure. For given raw material type $i = 1,\ldots,p$, and product type $j = 1,\ldots,q$, let

- Q_i be the value of the quality parameter for raw material i,
- b_j be the threshold acceptable level of quality per unit of product j,
- c_{ij} be the cost of processing one unit of raw material i for product j,
- x_{ij} be the amount of raw material i blended into product j,
- d_j be the minimum output level required of product j, and
- a_i be the available amount of raw material i.

The classical multi-mix blending problem is to determine the amount of raw material x_{ij} that minimizes the total processing cost subject to quality requirements. This problem can be formulated as a linear program. We modify this problem with the assumption that the raw material quality parameters Q_i are random with known probability distributions. In this case, the quality requirement constraints can only be satisfied with a certain probability. Thus, instead of minimizing the total processing cost, the manager chooses the amounts x of raw material to be blended in order to maximize the probability of achieving the target quality while keeping the total processing cost within a certain level. This leads to the following constrained stochastic optimization problem:

$$\min_{x\in\mathbb{R}_+^{p\times q}} f(x) = \mathsf{P}\left(\sum_{i=1}^{p}(b_j - Q_i)x_{ij} > 0, j = 1,\dots,q\right) \tag{8.6}$$

$$\text{s.t.} \quad \sum_{i=1}^{p}\sum_{j=1}^{q} c_{ij}x_{ij} \le \tau \quad \text{(total processing cost constraints)},$$

$$\sum_{i=1}^{p} x_{ij} \ge d_j, j = 1,\dots,q \quad \text{(demand constraints)},$$

$$\sum_{j=1}^{q} x_{ij} \le a_i, i = 1,\dots,p \quad \text{(resource constraints)}.$$

In general, analytic evaluation of the probability objective function is intractable, particularly when the quality parameters Q_i are correlated. In applying SAA, the random element ξ is taken to be all the quality parameters Q_i. The corresponding sample function $Y(x,\xi)$ is

$$Y(x,\xi) = \mathbf{1}\left\{\sum_{i=1}^{p}(b_j - Q_i)x_{ij} > 0, j = 1,\dots,q\right\},$$

where $\mathbf{1}\{\cdot\}$ denotes the indicator function.

Suppose that ξ is a nonnegative random vector with a continuous density function. Note that for any feasible solution x, $f(x)$ is an integral of a density function over a polyhedral set parameterized by x. By using a classical result in mathematical analysis, it can be shown that the true function f is differentiable and the gradient can be expressed as a surface integral [25]. By applying Propositions 8.2 and 8.5 in Sect. 8.3, we can show that the ULLN for f_n holds. Therefore, as long as we can obtain a solution to the sample problem, we can guarantee the convergence of the SAA optimal values. However, the sample function has a discontinuity whenever $\sum_{i=1}^{p}(b_j - Q_i)x_{ij} = 0$ for some $j = 1,\dots,q$, and $\nabla_x Y(x,\xi) = 0$ for any x except discontinuity points, i.e., the sample average function $f_n(x)$ is differentiable except on a set A of probability zero, and $\nabla f_n(x) = 0$ on A^c. This problem is ill-posed and any point $x \in A^c$ is a stationary point. Thus, any locally convergent algorithm that searches for a stationary point is not applicable.

One approach to this problem is to approximate the sample function by using a smooth (or piecewise linear) function. The indicator sample function has a very simple structure, only taking on values of zero or one. At any discontinuous point $x \in A$, we can obtain a smooth approximate function by smoothly connecting the sample function on an open neighborhood of x. The resulting approximate function can have a non-zero gradient that is selected to point in an uphill direction in this neighborhood. Then, we can develop an algorithm that employs the gradient to search for an approximate optimal solution. In the example above, we can use the smooth approximation $\phi(h(x,\xi),\varepsilon)$ of the sample function $Y(x,\xi)$, where $h(x,\xi) = \min\{\sum_{i=1}^{p}(b_j - Q_i)x_{ij}, j = 1,\ldots,q\}$ and $\phi : \mathbb{R} \times \mathbb{R}\backslash\{0\} \to \mathbb{R}$ is a continuously differentiable real-valued function such that for any $z \in \mathbb{R}$ and a given $\varepsilon > 0$,

(a) $\phi(z,0) = \mathbf{1}\{z \in (0,\infty)\}$, and
(b) $\mathbf{1}\{z \in (0,\infty)\} \le \phi(z,\varepsilon) \le \mathbf{1}\{z \in (-\varepsilon,\infty)\}$.

If the smoothing parameter ε goes to zero as n increases, then under a set of regularity conditions, the optimal solution of the smooth approximate problem converges to a stationary point of the true problem. This smoothing technique has been widely used, particularly for minimizing risk measures such as Value-at-Risk (VaR) and conditional Value-at-Risk (CVaR) [2, 16], as well as for handling chance constraints [23]. Xu and Zhang [63] provide simple examples of smoothing techniques and discuss the local convergence of the SAA method with a smoothed sample function.

The smoothing approach above changes the true objective function to be optimized, and while the magnitude of the change can be controlled through the parameter ε, one might wonder whether this approximation can be avoided. Sometimes a conditional expectation can be used to smooth jumps in the sample function $Y(\cdot,\xi)$. This is called smoothed perturbation analysis (SPA) [15]. SPA was developed to overcome difficulties in applying infinitesimal perturbation analysis (see Sect. 8.3) in nonsmooth settings, and has been applied to a large class of stochastic discrete event systems. To illustrate the SPA technique, we consider a company that produces only one type of product using two types of raw material. Then, for any $x = (x_1, x_2) > 0$, the corresponding objective function is

$$\begin{aligned} f(x) = \mathsf{E}[Y(x,\xi)] &= \mathsf{E}\left[\mathbf{1}\left\{b\sum_{i=1}^{2} x_i - Q_1x_1 - Q_2x_2 > 0\right\}\right] \\ &= \mathsf{E}\left[\mathsf{E}\left[\mathbf{1}\left\{b\sum_{i=1}^{2} x_i - Q_1x_1 - Q_2x_2 > 0\right\}\middle| Q_2\right]\right] \\ &= \mathsf{E}\left[F_{Q_1}\left(\frac{b\sum_{i=1}^{2} x_i - Q_2x_2}{x_1}\right)\right], \end{aligned} \tag{8.7}$$

where F_{Q_1} is the cumulative distribution function of Q_1 (under the assumption that Q_1 has a density). If the density of Q_1 is continuous, then at any fixed Q_2, the function inside the expectation in (8.7) is differentiable at x. Thus, if F_{Q_1} is known, $f(\cdot)$ can be approximated instead by taking the sample average of this smooth sample function, the expectation of which is the exact function we truly wish to minimize.

Smoothing techniques may not always be applicable, particularly when the sample function has a complex structure. However, even when not applicable, the gap between the sample function $f_n(\cdot)$ and the true function $f(\cdot)$ may still converge to 0 uniformly on the domain Θ as the sample size n grows, so provided that one can solve the SAA problem (8.2), the sequence of optimal objective function values will converge.

Example 8.3 (Bus Scheduling Problem). Passengers arrive at a bus stop according to a Poisson process with rate λ over the interval $[0,1]$. We wish to schedule the arrival times of d infinite-capacity buses over this time interval so as to minimize the expected sum of passenger wait times. We assume that an additional bus arrives at time 1 so that all waiting times are well defined and can be calculated.

Let $\tilde{x}_j$ denote the scheduled time of arrival of the jth bus, $j = 1,2,\ldots,d+1$ where $\tilde{x}_{d+1} = 1$, and let $x_j = \tilde{x}_j - \tilde{x}_{j-1}$ denote the length of the time interval between the arrivals of buses $j-1$ and j, $j = 1,2,\ldots,d+1$, where $\tilde{x}_0 = 0$. The random element ξ here may be taken to be N, the number of passengers to arrive over the time interval $[0,1]$, along with the times $T_1, T_2, \ldots, T_N$ of their arrival.

The sample function $Y(x,\xi)$ may be written

$$Y(x,\xi) = \sum_{i=1}^{N} \sum_{j=1}^{d+1} (\tilde{x}_j - T_i) \mathbf{1}\{T_i \in (\tilde{x}_{j-1}, \tilde{x}_j]\}.$$

The order-statistic property of Poisson processes may be used to show directly (e.g., [44, p. 68]) that

$$f(x) = \mathsf{E}[Y(x,\xi)] = \frac{\lambda}{2} \sum_{j=1}^{d+1} x_j^2,$$

so that $f(x)$ is convex and quadratic, and hence smooth. However, the sample function has a discontinuity whenever the arrival time of a bus coincides with the time of a passenger arrival.

Like the multi-mix blending problem, for any fixed ξ, $Y(\cdot,\xi)$ is differentiable except on a set A of probability zero. Unlike the multi-mix blending problem, the gradient of the sample function $\nabla_x Y(x,\xi)$ at $x \in A^c$ can take a non-zero value. For example, when $d = 1$, the derivative of $Y(x,\xi)$ for a fixed ξ is zero on $(0,T_1)$ and positive at any differentiable point $x \in (T_1,1)$. Note that the gradient of the sample function does not provide any useful information about the (quadratic) true function, and hence any gradient-based algorithm is highly unlikely to work well in this setting. Approximating the sample function with smoothing techniques is not a trivial problem in this case due to the complexity of the sample function. One could try to minimize the sample average approximation problem in this example using techniques such as metamodeling. If the difference between the sample average and the true function is small enough for sufficiently large sample size n, then the true function can be well approximated with a quadratic metamodel. Indeed, by applying Propositions 8.2 and 8.5, we can show that the gap between the sample and the true function does not persist and eventually converges to 0 uniformly on $[0,1]^d$. Thus, although SAA can in principle be applied to this example, it may not be the best approach, due to the lack of useful structure in the sample functions.

In some examples, sample functions can have appealing structural properties in some variables but not in others. For example, in [14], an (s,S) inventory system is considered. When the problem is reparameterized with decision variables $\Delta = S - s$ and S, then for each fixed Δ, the sample functions are piecewise linear and convex in S. It is then natural to solve the problem by searching only over Δ, using the value of S that minimizes the sample function at each fixed Δ. More precisely, for each fixed Δ, let $S_n^*(\Delta)$ minimize $f_n(\Delta,\cdot)$. Then one can reduce the two-dimensional optimization problem of minimizing $f_n(\cdot,\cdot)$ to a one-dimensional optimization problem where one minimizes $f_n(\Delta, S_n^*(\Delta))$ over Δ. As discussed in [14], this latter problem is not unimodal in Δ, so it is difficult to solve numerically.

8.3 Detecting When SAA Is Appropriate

The key principles exemplified in Sect. 8.2 are that

1. SAA is appropriate only when the approximating functions f_n have some structure that enables the application of an efficient deterministic optimization algorithm, and
2. the limiting function f that we actually want to minimize shares that structure, so that the properties of the limiting function such as the location of local minima are similar to those of the approximating function.

The term "some structure" is intentionally vague because it can mean different things in different problems. For example, by "some structure" we might mean that all sample functions are convex, in which case convex optimization techniques can be applied to the sample functions, and we can ensure convergence to a global minimum of both the sample functions and the limiting function. Alternatively, the sample functions might not be convex, but might be differentiable, in which case gradient-based methods could be applied. A weaker property that can be exploited by numerical optimization algorithms is Lipschitz continuity.

The approximating functions f_n are observable, because we can generate them in finite time, while the limiting function f is not directly observable. Nevertheless, one can often infer structural properties of f through the corresponding properties of the approximating functions f_n and regularity conditions that ensure that these properties persist in the limit as $n \to \infty$.

In this section, we give sufficient conditions involving only the sample functions $Y(\cdot,\cdot)$ (from which the approximating functions $f_n(\cdot)$ are built) for the *true* function $f(\cdot)$ to be continuous or differentiable at a fixed point x. If these conditions apply at each point x in the domain, then one can conclude that $f(\cdot)$ is continuous or differentiable over that domain. Perhaps surprisingly, one can often arrive at this conclusion even when the sample functions do not possess these same properties over the entire domain. Therefore, Principle 2 does not follow automatically from Principle 1.

These observations will not be surprising to those who are familiar with infinitesimal perturbation analysis (IPA), and indeed, the results presented here can be viewed as a recasting of those ideas in the SAA setting. If we take as given that SAA-appropriate problems are those for which both the approximating functions $f_n(\cdot)$ and $f(\cdot)$ are differentiable, and the derivatives of $f_n(\cdot)$ converge to those for $f(\cdot)$, then we arrive at an underlying theme of this section, which is the following meta-principle:

SAA-appropriate problems are almost exactly those in which IPA applies.

In contrast to much of the IPA literature, we explicitly treat the case where d, the dimension of the domain Θ, can be greater than one. The ideas involved are similar to the one-dimensional case, but some additional care is required. See [17, Chap. 1] for an excellent treatment of the one-dimensional case.

Our first result [27] gives sufficient conditions for $f(\cdot)$ to be continuous at a fixed point $x \in \Theta$. The result is disarmingly straightforward to state and prove. Throughout this chapter, $\|\cdot\|$ will denote the Euclidean norm. Let $B(x,\delta) = \{y : \|y-x\| \leq \delta\}$ denote the closed ball of radius δ around x.

Proposition 8.1. *Fix $x \in \Theta$. Suppose that $Y(\cdot,\xi)$ is continuous at x a.s., i.e., for all ξ in a set of probability 1, $Y(x+h,\xi) \to Y(x,\xi)$ as $h \to 0$. Suppose further that the family of random variables*

$$\{Y(x+h,\xi) : x+h \in B(x,\delta)\}$$

is uniformly integrable, for some $\delta > 0$. Then $f(\cdot)$ is continuous at x.

Proof. Continuity of $Y(\cdot,\xi)$ on a set A of probability 1 ensures that

$$\begin{aligned} f(x) &= \mathsf{E}[Y(x,\xi)\mathbf{1}\{\xi \in A\}] \\ &= \mathsf{E}\left[\lim_{h\to 0} Y(x+h,\xi)\mathbf{1}\{\xi \in A\}\right] \\ &= \lim_{h\to 0} \mathsf{E}[Y(x+h,\xi)\mathbf{1}\{\xi \in A\}] \qquad (8.8) \\ &= \lim_{h\to 0} f(x+h), \end{aligned}$$

where the interchange (8.8) is justified by the uniform integrability assumption. ∎

As an immediate corollary we have the following result on the global continuity of $f(\cdot)$.

Corollary 8.1. *Suppose that the conditions of Proposition 8.1 hold at each $x \in \Theta$. Then $f(\cdot)$ is continuous on Θ.*

What is perhaps surprising about this corollary is that we may be able to establish that $f(\cdot)$ is continuous on Θ even when the sample functions $Y(\cdot,\xi)$ are discontinuous on Θ almost surely! The apparent contradiction dissolves when one realizes that the assumption of Proposition 8.1 requires continuity of $Y(\cdot,\xi)$ only locally at x. There may be discontinuities of this function at points outside a neighbourhood of x, and this neighbourhood can depend on ξ.

As an example, let us revisit the bus-scheduling problem from Sect. 8.2. The sample functions $Y(\cdot,\xi)$ have discontinuities at all points x such that a bus arrival time coincides with a passenger arrival time in the interval $(0,1)$. Consequently, the sample functions $Y(\cdot,\xi)$ are discontinuous on Θ almost surely. However, for a fixed $x \in \Theta$, $Y(\cdot,\xi)$ is continuous at x unless a passenger arrival coincides with one of the bus arrival times encoded in x, which occurs with probability 0. Furthermore, the sum of the waiting times of the N arriving passengers is bounded by N, which has finite expectation, and so $\{Y(y,\xi) : y \in B(x,\delta)\}$ is uniformly integrable. Proposition 8.1 then ensures that f is continuous at x, and Corollary 8.1 allows us to conclude that f is continuous on Θ. We already knew that f is continuous on Θ, because it is a convex quadratic. However, this same argument can be used to show continuity in other examples where the form of $f(\cdot)$ is unknown. See [27] for an example involving locating multiple ambulances. This result for the bus-scheduling example is a special case of the following general result.

Proposition 8.2. *Suppose that*

(i) for any fixed ξ, $Y(\cdot,\xi)$ is a piecewise Lipschitz continuous function, i.e., there exists a countable partition of Θ such that the restriction of $Y(\cdot,\xi)$ to the interior of each component is Lipschitz continuous,
(ii) the Lipschitz constants in all components are bounded by an integrable random variable $L(\xi)$,
(iii) the jump size at any discontinuous point $x \in \Theta$ is bounded by a random variable $J(\xi)$ with $\mathsf{E}[J^2(\xi)] < \infty$, and
(iv) for any $x \in \Theta$, $m(x,x+h,\xi) \to 0$ a.s. as $\|h\| \to 0$, where $m(x,x+h,\xi)$ is the number of discontinuity points of the sample function $Y(\cdot,\xi)$ restricted to the line segment joining x and $x+h$ and satisfies

$$\sup_{x+h \in B(x,\delta)} m(x,x+h,\xi) \leq M(\xi),$$

for some $\delta > 0$ and a random variable $M(\xi)$ with $\mathsf{E}[M^2(\xi)] < \infty$.

Then the assumptions in Proposition 8.1 hold, and hence $f(\cdot)$ is continuous on Θ.

Proof. Fix $x \in \Theta$. We have

$$|Y(x+h,\xi) - Y(x,\xi)| \leq \|h\| L(\xi) + J(\xi) m(x, x+h, \xi). \tag{8.9}$$

By Assumption (*iv*), the right hand side of (8.9) converges to zero a.s. as $h \to 0$. Thus, $Y(\cdot,\xi)$ is continuous at x a.s. Since $m(x+h,x,\xi) \leq M(\xi)$ (over $x+h \in B(x,\delta)$), the right hand side of (8.9) is dominated by an integrable random variable $\|h\|L(\xi) + J(\xi)M(\xi)$. Thus, $\{|Y(x+h,\xi) - Y(x,\xi)| : x+h \in B(x,\delta)\}$ is uniformly integrable, and so is $\{Y(x+h,\xi) : x+h \in B(x,\delta)\}$. ∎

As with continuity, one can obtain differentiability results for $f(\cdot)$ based on local properties of the sample functions $Y(\cdot,\cdot)$.

Proposition 8.3. *Fix x in the interior of Θ. Suppose that $Y(\cdot,\xi)$ is differentiable at x w.p.1, and let $\nabla Y(x,\xi)$ be its gradient. Suppose further that the family of random variables*

$$\left\{ \frac{Y(x+h,\xi) - Y(x,\xi)}{\|h\|} : 0 < \|h\| \leq \delta \right\} \tag{8.10}$$

is uniformly integrable, for some $\delta > 0$. Then $f(\cdot)$ is differentiable at x, and $\nabla f(x) = \mathsf{E}[\nabla Y(x,\xi)]$.

Proof. We have that for all ξ in a set A of probability 1,

$$Y(x+h,\xi) = Y(x,\xi) + h^{\mathsf{T}} \nabla Y(x,\xi) + \|h\| R(x,\xi,h), \tag{8.11}$$

where the remainder term $R(x,\xi,h) \to 0$ as $h \to 0$. For $\xi \notin A$, define $\nabla Y(x,\xi) = 0$ and $R(x,\xi,h) = 0$. Taking $h = re_i$, i.e., the ith unit vector scaled by r, for each $i = 1,2,\ldots,d$, and letting $r \to 0$, the uniform integrability assumption implies that $\mathsf{E}[|\partial Y(x,\xi)/\partial x_i|] < \infty$. Hence, all d components of $\mathsf{E}[\nabla Y(x,\xi)]$ exist and are finite. Taking expectations in (8.11), we obtain

$$f(x+h) = f(x) + h^{\mathsf{T}} \mathsf{E}[\nabla Y(x,\xi)] + \|h\| \mathsf{E}[R(x,\xi,h)],$$

so the result will follow if we show that $\mathsf{E}[R(x,\xi,h)] \to 0$ as $h \to 0$. From (8.11), we have that for $\xi \in A$,

$$\frac{Y(x+h,\xi) - Y(x,\xi)}{\|h\|} = \frac{h^{\mathsf{T}}}{\|h\|} \nabla Y(x,\xi) + R(x,\xi,h),$$

and the left-hand side is uniformly integrable (over $\|h\| \in (0,\delta]$) by assumption. But each component of $\nabla Y(x,\xi)$ is integrable, and therefore, $h^{\mathsf{T}} \nabla Y(x,\xi)/\|h\|$ is uniformly integrable for $\|h\| \in (0,\delta]$. It follows that $R(x,\xi,h)$ is uniformly integrable for $h \in (0,\delta]$, and therefore, $\mathsf{E}[R(x,\xi,h)] \to 0$ as $h \to 0$ as required. ∎

Corollary 8.2. *Suppose that the conditions of Proposition 8.3 hold at each x in the interior of Θ. Then $f(\cdot)$ is differentiable on the interior of Θ with $\nabla f(x) = \mathsf{E}[\nabla Y(x,\xi)]$.*

It is again striking that under certain verifiable conditions, one can show that $f(\cdot)$ is differentiable throughout the interior of Θ, even if the sample functions $Y(\cdot,\xi)$ are not. In fact, this is the norm in applications arising in discrete-event simulation, in that the functions $Y(\cdot,\xi)$ typically fail to be differentiable on "seams" in Θ that have measure 0.

The uniform integrability assumption is almost always verified (either locally or on the interior of Θ) by showing that $Y(\cdot,\xi)$ is Lipschitz continuous with Lipschitz constant $L(\xi)$ on the appropriate set, where $\mathsf{E}[L(\xi)] < \infty$. Indeed, the Lipschitz condition ensures that $|Y(x+h,\xi) - Y(x,\xi)| \leq L(\xi)\|h\|$, and the uniform integrability requirement follows immediately. But how can this property be verified? In one dimension, one can appeal to the following result, known as the generalized mean value theorem, in which the Lipschitz constant for a sample function $Y(\cdot,\xi)$ arises from a bound on the (sample) derivative. For a proof, see [12, Sect. 8.5].

Theorem 8.1. *Let g be a continuous real-valued function on the closed interval $[a,b]$ that is differentiable everywhere except possibly on a set C of at most countably many points. Then for all x and $x+h$ in $[a,b]$ with $h \neq 0$,*

$$\left|\frac{g(x+h)-g(x)}{h}\right| \leq \sup_{y\in[a,b]\setminus C} |g'(y)|.$$

In higher dimensions, we can again apply this result. One difficulty is that real-valued (sample) functions arising in discrete-event simulation often fail to be differentiable along "seams," so the set of non-differentiable points can be uncountable. Fortunately, it is sufficient for our purposes to apply the generalized mean-value theorem along certain line segments only. So long as these line segments intersect the non-differentiable set in at most countably many places, we can apply the generalized mean-value theorem. The following proposition gives sufficient conditions for the uniform integrability condition (8.10) in Proposition 8.3.

Proposition 8.4. *For some $\delta > 0$ suppose that for all ξ in a set of probability 1,*

(i) $Y(\cdot,\xi)$ is continuous in $B(x,\delta)$;
(ii) $C(\xi) \cap [x,y]$ is countable for all $y \in B(x,\delta)$, where $C(\xi)$ denotes the points of non-differentiability of $Y(\cdot,\xi)$ in $B(x,\delta)$ and $[x,y]$ denotes the line segment joining x and y; and
(iii) $\sup_{y\in B(x,\delta)\setminus C(\xi)} \|\nabla Y(y,\xi)\| \leq L(\xi) < \infty$.

If $\mathsf{E}[L(\xi)] < \infty$, then the uniform integrability condition (8.10) holds.

Proof. For $\|h\| \leq \delta$ and ξ in the set of probability 1,

$$|Y(x+h,\xi) - Y(x,\xi)| \leq \sup_{y\in[x,x+h]\setminus C(\xi)} |h^{\mathsf{T}}\nabla Y(y,\xi)| \tag{8.12}$$

$$\leq \|h\| \sup_{y\in[x,x+h]\setminus C(\xi)} \|\nabla Y(y,\xi)\| \tag{8.13}$$

$$\leq \|h\| L(\xi),$$

where (8.12) and (8.13) follow from the generalized mean-value theorem and the Cauchy–Schwarz inequality, respectively. The result follows since $L(\xi)$ is integrable. ■

Sometimes one can verify the Lipschitz property directly, as in the following example.

Example 8.4. A depot is to be located in the unit square $[0,1]^2$. Each night a set of N requests for pickups the following day is made, where N has finite mean. Conditional on $N \geq 1$, the N pickup locations are independent and identically distributed with density $p(\cdot)$ on the unit square. The pickups are completed in a single tour by a van that travels in a straight line from pickup to pickup (Euclidean distance), visiting all pickups before returning to the base. The sequence of pickups is chosen so as to minimize the total travel distance of the van, i.e., the sequence of pickups is the solution to a traveling salesperson problem, starting and finishing at the depot. In this case, the random element ξ consists of the number and locations of pickups, and x gives the Cartesian coordinates of the depot. The goal is to select the depot location to minimize the expected distance traveled by the van.

Here, $Y(x,\xi)$ gives a realization of the distance traveled by the van. We can write

$$Y(x,\xi) = \min_{\pi} Y(x,\xi,\pi), \tag{8.14}$$

where π is a permutation specifying the order in which pickups are visited and $Y(x,\xi,\pi)$ is the resulting distance traveled. (We exclude duplicate permutations that are the reverse of each other in this pointwise minimum.) Each function $Y(\cdot,\xi,\pi)$ is differentiable and in fact has partial derivatives bounded by 2. (To see why, notice that $Y(x,\xi,\pi)$ gives the sum of the distance from x to the first pickup, the distance from the last pickup to x, and the sum of the "internal" distances between the pickups of the permutation. The internal distances do not change as x varies.) Hence, $Y(\cdot,\xi,\pi)$ is Lipschitz with Lipschitz constant 2 for all ξ and π. It then follows from (8.14) that $Y(\cdot,\xi)$ is Lipschitz with Lipschitz constant 2. Furthermore, for fixed x, the set of ξ for which $Y(\cdot,\xi)$ fails to be differentiable at x are such that multiple permutations attain the minimum in (8.14). This set has probability 0 since pickup locations have a density. It follows from our previous discussion that $f(\cdot)$ is differentiable at x and $\nabla f(x) = \mathsf{E}[\nabla Y(x,\xi)]$.

8.4 Known Properties

In this section, we discuss some known properties for well-structured unconstrained optimization problems. Here, "well-structured" means that the sample function enjoys some structural property such as continuity or differentiability. We first investigate under which conditions the optimal solution and value of the SAA problem approach those of the true problem as the sample size n grows. Then, we discuss how quickly this convergence occurs via the Central Limit Theorem (CLT). We also briefly present the convergence of local solutions for both smooth and nonsmooth problems.

8.4.1 Almost Sure Convergence

As we briefly discussed in Sect. 8.2, uniform convergence of the sample average functions is the key condition for establishing convergence of optimal objective values and solutions in SAA. Indeed, one immediate consequence is the consistency of the SAA optimal values, i.e., $v_n^* \to v^*$ a.s. as $n \to \infty$, where v_n^* and v^* are the optimal objective values of the SAA problem (8.2) and the true problem (8.1), respectively. To see why, note that for a fixed sequence $\{\xi_n : n \geq 1\}$, $\{f_n : n \geq 1\}$ can be viewed as a sequence of deterministic functions. Suppose that f_n converges to the true function f uniformly on Θ. Then, for any sequence $\{x_n\} \subset \Theta$ converging to $x \in \Theta$, $f_n(x_n)$ converges to $f(x)$. Many problems, including those with discontinuous sample functions in Sect. 8.2, satisfy this uniform convergence. When the sample function $Y(\cdot,\xi)$ is convex a.s., the pathwise LLN is equivalent to the ULLN on a compact set [55, Corollary 3]. In a problem with non-convex functions, the following result shows that the conditions for the continuity of the true function $f(\cdot)$ discussed in Sect. 8.3 are, in fact, sufficient to ensure the uniform convergence of the approximating functions on a compact set.

Proposition 8.5. *Let Θ be a nonempty compact set. For any fixed $x \in \Theta$, suppose that $Y(\cdot,\xi)$ is continuous at x a.s., and there exists $\delta > 0$ such that the family of random variables $\{Y(y,\xi) : y \in B(x,\delta)\}$ is uniformly integrable. Then $\{f_n(x)\}$ converges to $f(x)$ uniformly on Θ a.s. as $n \to \infty$.*

Proof. The proof can be carried out by adapting the proof of Proposition 7 in [55]. Choose $\tilde{x} \in \Theta$. Let $\{\delta_k \leq \delta(\tilde{x}) : k = 1,2,\ldots\}$ be a sequence of positive numbers decreasing to 0, and define

$$\alpha_k(\xi) = \sup_{x \in B(\tilde{x},\delta_k)} |Y(x,\xi) - Y(\tilde{x},\xi)|.$$

By the continuity of $Y(\cdot,\xi)$ at $\tilde{x}$, $\alpha_k(\xi)$ goes to zero a.s. as k increases. The uniform integrability assumption ensures that $\{\alpha_k(\xi) : k = 1,2\ldots\}$ is uniformly integrable, and hence

$$\lim_{k\to\infty} \mathsf{E}[\alpha_k(\xi)] = \mathsf{E}\left[\lim_{k\to\infty} \alpha_k(\xi)\right] = 0.$$

Note that

$$\sup_{x \in B(\tilde{x},\delta_k)} |f_n(x) - f_n(\tilde{x})| \leq \frac{1}{n}\sum_{i=1}^{n} \alpha_k(\xi_i). \tag{8.15}$$

By the LLN, the right-hand side of (8.15) converges to $\mathsf{E}[\alpha_k(\xi_i)]$ a.s. as $n \to \infty$. Thus, for given $\varepsilon > 0$, there exists a neighborhood V of $\tilde{x}$ such that a.s. for sufficiently large n,

$$\sup_{x \in V \cap \Theta} |f_n(x) - f_n(\tilde{x})| < \varepsilon.$$

Since Θ is compact, there exists a finite number of points $x_1,\ldots,x_m \in \Theta$ and corresponding neighborhoods $V_1,\ldots,V_m$ covering Θ such that a.s. for sufficiently large n,

$$\sup_{x \in V_j \cap \Theta} |f_n(x) - f_n(x_j)| < \varepsilon, j = 1,\ldots,m. \tag{8.16}$$

By Proposition 8.1, $f(\cdot)$ is continuous. Thus, we can choose the neighborhoods $V_1,\ldots,V_m$ in such a way that

$$\sup_{x \in V_j \cap \Theta} |f(x) - f(x_j)| < \varepsilon, j = 1,\ldots,m. \tag{8.17}$$

By the LLN, with probability 1 (w.p.1) for sufficiently large n,

$$|f_n(x_j) - f(x_j)| < \varepsilon, j = 1,\ldots,m. \tag{8.18}$$

Combining (8.16)–(8.18), w.p.1 for sufficiently large n, we have

$$\sup_{x \in \Theta} |f_n(x) - f(x)| < 3\varepsilon.$$

■

When the sample function is continuous, the ULLN implies the continuity of the true function f. However, in general, the continuity of the true function is not a necessary condition for uniform convergence of the approximating function f_n. For example, consider a cumulative distribution function (cdf) $f(x) = P(\xi \leq x)$ and the empirical cdf $f_n(x)$. By the Glivenko–Cantelli Theorem [7, p. 269], f_n converges to f uniformly on $\mathbb{R}$ even if f is discontinuous. Optimizing a discontinuous function is in general a difficult problem, and many practical problems naturally exhibit continuity properties. In this chapter, we therefore focus on problems where f is continuous, unless the domain Θ is a discrete set.

We introduce some notation to proceed to the convergence results below. Let Π_n^* and π^* denote the set of optimal solutions of the SAA and the true problems, respectively. We define the Euclidean distance from a point x to a set B to be

$d(x,B) = \inf_{y \in B} \|x - y\|$, and the distance between two sets $A, B \subset \mathbb{R}^q$ to be $\mathbb{D}(A,B) = \sup\{d(x,B) : x \in A\}$. In the next theorem, we give convergence results based on the continuity of the true function and uniform convergence.

Theorem 8.2 (Theorem 5.3, [57]). *Suppose there exists a compact subset $C \subset \mathbb{R}^d$ such that*

(i) π^* is non-empty and contained in C,
(ii) $\{f_n(x)\}$ converges to $f(x)$ uniformly on C a.s. as $n \to \infty$, and
(iii) for sufficiently large n, Π_n^* is non-empty and contained in C a.s.

Then $v_n^* \to v^*$. Furthermore, if the true function $f(\cdot)$ is continuous on C, then $\mathbb{D}(\Pi_n^*, \pi^*) \to 0$ a.s. as $n \to \infty$.

Proof. Fix $\varepsilon > 0$. Uniform convergence of f_n to f on C ensures that

$$f_n(x) \geq f(x) - \varepsilon$$

for all $x \in C$, for sufficiently large n a.s. The assumption that $\Pi_n^* \subseteq C$ ensures that v_n^* is attained on C for sufficiently large n a.s., so $v_n^* \geq v^* - \varepsilon$ for sufficiently large n a.s. Since ε was arbitrary, $\liminf_n v_n^* \geq v^*$ a.s. Also, since there exists $x^* \in \pi^* \subseteq C$, $v_n^* \leq f_n(x^*) \to v^*$ as $n \to \infty$ a.s. Thus, $v_n^* \to v^*$ as $n \to \infty$ a.s. Turning to convergence of the solution set, suppose that $\mathbb{D}(\Pi_n^*, \pi^*) \nrightarrow 0$. Then there exists $X_n \in \Pi_n^*$ such that for some $\varepsilon > 0$, $d(X_n, \pi^*) \geq \varepsilon$ for all $n \geq 1$. Since C is compact, by passing to a subsequence if necessary, X_n converges to a point $x^* \in C$, and $f(x^*) > v^*$. On the other hand,

$$f_n(X_n^*) - f(x^*) = [f_n(X_n^*) - f(X_n^*)] + [f(X_n^*) - f(x^*)] \tag{8.19}$$

Both the first term and the second term in the right hand side of (8.19) converge to zero by the uniform convergence assumption and continuity of f, respectively. Thus, $v_n^* \to f(x^*) > v^*$, which contradicts the fact that $v_n^* \to v^*$. ∎

Theorem 8.2 ensures that, if X_n^* solves the SAA problem exactly, then $d(X_n^*, \pi^*) \to 0$ a.s. as $n \to \infty$. Moreover, if the true problem has a unique optimal solution x^*, then $X_n^* \to x^*$. When the sample functions are convex, the set of regularity conditions in Theorem 8.2 can be relaxed by using the theory of epi-convergence [57, Theorem 5.4].

Now we consider the case where Θ is a finite set and discuss the convergence of ε-optimal solutions in the SAA method. We first introduce some notation. For $\varepsilon \geq 0$, let

$$\pi^*(\varepsilon) := \{x \in \Theta : f(x) \leq v^* + \varepsilon\}, \quad \Pi_n^*(\varepsilon) := \{x \in \Theta : f_n(x) \leq v_n^* + \varepsilon\} \tag{8.20}$$

denote the ε-optimal solutions for the true and the SAA problems, respectively. Since Θ is finite, the pathwise LLN implies the ULLN. Thus, the a.s. convergence of v_n^* to v^* is guaranteed. Furthermore, asymptotic normality of v_n^* follows under moment conditions if the optimal solution is unique. Also, it can be shown that for any $\varepsilon \geq 0$, $\Pi_n^*(\varepsilon) \subset \pi^*(\varepsilon)$ w.p.1 for n sufficiently large [28]. This means that any ε-optimal solution of the SAA problem is an ε-optimal solution of the true problem for large enough n. In particular, if the true problem has a unique solution x^*, then $\Pi_n^* = \{x^*\}$ w.p.1 for n large enough. But, how quickly does the probability of $\{\Pi_n^*(\varepsilon) \subset \pi^*(\varepsilon)\}$ approach 1 as n increases? Large deviation analysis shows that under a mild regularity condition (essentially finiteness of the moment generating function of $Y(x,\xi)$ at each fixed x), the probability $\mathsf{P}\{\Pi_n^*(\delta) \not\subseteq \pi^*(\varepsilon)\}$ for $0 \leq \delta < \varepsilon$, converges to zero at an exponential rate. We discuss this further in Sect. 8.5.

8.4.2 Convergence Rates for the SAA Method

There exists a well-developed statistical inference for estimators obtained from the SAA approach. From this inference, we can obtain error bounds for obtained solutions and select the sample size n to obtain a desired level of accuracy. The first result below by [32] states that the estimator v_n^* for v^* is negatively biased and the expected value of v_n^* monotonically increases. This monotonicity property of $\mathsf{E}[v_n^*]$ is desirable in the sense that we can expect a tighter lower bound as n increases.

Proposition 8.6. *For all $n \geq 1$, $\mathsf{E}[v_n^*] \leq \mathsf{E}[v_{n+1}^*]$, and $\mathsf{E}[v_n^*] \leq v^*$.*

Proof. Since $\xi_1, \xi_2, \ldots$ are i.i.d.,

$$\begin{aligned}\mathsf{E}[v_{n+1}^*] = \mathsf{E}\left[\min_{x\in\Theta} \frac{1}{n+1}\sum_{i=1}^{n+1} Y(x,\xi_i)\right] &= \mathsf{E}\left[\min_{x\in\Theta} \frac{1}{n+1}\sum_{i=1}^{n+1}\left(\frac{1}{n}\sum_{j\neq i} Y(x,\xi_j)\right)\right] \\ &\geq \frac{1}{n+1}\sum_{i=1}^{n+1} \mathsf{E}\left[\min_{x\in\Theta}\left(\frac{1}{n}\sum_{j\neq i} Y(x,\xi_j)\right)\right] \\ &= \mathsf{E}[v_n^*].\end{aligned}$$

For any $\tilde{x} \in \Theta$, $f_n(\tilde{x}) \geq \min_{x\in\Theta} f_n(x)$. By taking expectation on both sides, we have $v^* = \min_{x\in\Theta} f(x) = \min_{x\in\Theta} \mathsf{E}[f_n(x)] \geq \mathsf{E}[\min_{x\in\Theta} f_n(x)] = \mathsf{E}[v_n^*]$. ■

Next, we discuss the asymptotic behavior of the SAA optimal objective value v_n^*. For a sequence of random variables $\{X_n\}$ and deterministic constants β_n, we say that $X_n = o_p(\beta_n)$, if $X_n/\beta_n \to 0$ in probability. We also say that $X_n = O_p(\beta_n)$, if $\{X_n/\beta_n\}$ is bounded in probability (tight), i.e., for any $\varepsilon > 0$, there exists $M > 0$ such that $\mathsf{P}(|X_n/\beta_n| > M) < \varepsilon$, for all n.

First, assuming that $\mathsf{E}[Y^2(x,\xi)] < \infty$, we have the CLT for any fixed $x \in \Theta$,

$$\sqrt{n}(f_n(x) - f(x)) \xrightarrow{d} Z(x)$$

as $n \to \infty$, where $\xrightarrow{d}$ signifies convergence in distribution, and $Z(x) \sim \mathcal{N}(0, \sigma^2(x))$, $\sigma^2(x) \equiv \mathrm{Var}[Y(x,\xi)]$. The CLT implies that the error $f_n(x) - f(x)$ is of order $O_p(n^{-1/2})$. Under a set of mild regularity conditions, the same canonical convergence rate of v_n^* can be obtained by applying a multidimensional version of the CLT to f.

Theorem 8.3 (Theorem 5.7, [57]). *We suppose that*

(i) Θ *is compact,*
(ii) $\mathsf{E}[Y^2(x,\xi)] < \infty$, *for some* $x \in \Theta$,
(iii) $Y(\cdot,\xi)$ *is Lipschitz on* Θ *with Lipschitz constant* $L(\xi)$ *a.s., and* $\mathsf{E}\left[L^2(\xi)\right] < \infty$.

Then,

$$v_n^* = \inf_{x \in \pi^*} f_n(x) + o_p(n^{-1/2})$$

and

$$\sqrt{n}(v_n^* - v^*) \Rightarrow \inf_{x \in \pi^*} Z(x) \tag{8.21}$$

as $n \to \infty$, *where* Z *is a Gaussian process on* Θ *with* $\mathsf{E}[Z(x)] = 0$ *and* $\mathrm{Cov}(Z(x), Z(y)) = \mathrm{Cov}(Y(x,\xi), Y(y,\xi))$, *for all* $x, y \in \Theta$.

Proof. The essential idea of the proof is to employ a functional CLT for f_n and the Delta method [7] to $V(f)$, where V is the real-valued functional given by $V(g) = \min_{x\in\Theta} g(x)$, for any continuous function g on Θ. ∎

When the true problem has a unique optimal solution x^*, (8.21) implies that v_n^* is asymptotically normally distributed. It again follows from (8.21) that under some uniform integrability conditions, the bias $\mathsf{E}[v_n^*] - v^*$ is $O(n^{-1/2})$. If the true problem has a unique solution, $\mathsf{E}[v_n^*] - v^*$ is $o(n^{-1/2})$, and with additional moments and second order conditions on f, $\mathsf{E}[v_n^*] - v^*$ is, in fact, $O(n^{-1})$.

The SAA optimal solution X_n^* requires a stronger set of conditions to achieve the same asymptotic properties as v_n^*. When f is smooth and has a unique solution, under some regularity conditions, the solution X_n^* of the SAA problem converges to the unique solution x^* of the true problem at the canonical rate $n^{-1/2}$. One of the essential regularity conditions is that the true function f increases quadratically near the unique solution x^*. (It may converge at a faster rate if the function f increases linearly near x^*, as may happen if x^* lies on the boundary of Θ.) We say that the quadratic growth condition is satisfied at $\tilde{x}$ if there exists $\alpha > 0$ and a neighborhood V of $\tilde{x}$ such that for all $x \in \Theta \cap V$,

$$f(x) \geq f(\tilde{x}) + \alpha \|x - \tilde{x}\|^2.$$

If Θ is a convex, full dimensional set and $\tilde{x}$ lies in the interior of Θ, then the quadratic growth condition is equivalent to the second order sufficient optimality condition, i.e., the Hessian matrix $\nabla^2 f(\tilde{x})$ is positive definite. In the convergence result below, we provide conditions that are relatively easy to understand and a sketch of the proof. The readers are referred to [53,57] for the proof under more general regularity conditions.

We say $g : \mathbb{R}^d \to \mathbb{R}$ is Fréchet differentiable at x if there exists a bounded linear operator $D_x g : \mathbb{R}^d \to \mathbb{R}^d$ such that

$$\lim_{\|u\| \downarrow 0} \frac{|g(x+u) - g(x) - D_x g(u)|}{\|u\|} = 0.$$

Theorem 8.4. *Assume that the following hold:*

(i) The true function f has a unique minimizer $x^* \in \Theta$.
(ii) $Y(\cdot,\xi)$ is Lipschitz with Lipschitz constant $L(\xi)$ on Θ a.s., and $\mathsf{E}[L(\xi)] < \infty$.
(iii) $Y(\cdot,\xi)$ is continuously differentiable at any x in a neighborhood of x^* a.s.
(iv) $\mathsf{E}[\|\nabla_x Y(x,\xi)\|^2] < \infty$, for some $x \in \Theta$.
(v) $\nabla_x Y(\cdot,\xi)$ is Lipschitz with Lipschitz constant $K(\xi)$ in a neighborhood of x^* a.s., and $\mathsf{E}\left[K^2(\xi)\right] < \infty$.
(vi) f satisfies the quadratic growth condition at x^*.

Then, $\|X_n^* - x^*\| = O_p(n^{-1/2})$. Furthermore, assume that

(vii) There exists a neighborhood U of x^* and $\alpha > 0$ such that for every u in a neighborhood of zero, the following problem

$$\min_{x \in \Theta} f(x) + u \cdot x \tag{8.22}$$

has an optimal solution $x^*(u) \in U$ and the quadratic growth condition holds at $x^*(u)$.

If $x^*(u)$ is Fréchet differentiable at $u = 0$ and $D_0 x^*(\cdot)$ is continuous,

$$\sqrt{n}(X_n^* - x^*) \Rightarrow D_0 x^*(Z) \tag{8.23}$$

as $n \to \infty$, where Z is a multivariate normal vector with mean 0 and covariance matrix

$$\Sigma = \mathsf{E}\left[(\nabla f(x^*,\xi) - \nabla f(x^*))^{\mathsf{T}} (\nabla f(x^*,\xi) - \nabla f(x^*))\right].$$

Moreover, if $D_0 x^*(\cdot)$ is linear, then $\sqrt{n}(X_n^* - x^*)$ is asymptotically normally distributed.

Proof. By Assumptions (i)–(iii), $X_n^* \to x^*$ a.s. as $n \to \infty$, and $f(\cdot)$ is Lipschitz continuous and continuously differentiable at x^*. Let $\delta_n(x) = f_n(x) - f(x)$.

By the quadratic growth condition (vi) and the generalized mean-value theorem, we have

$$\|X_n^* - x^*\| \le \frac{\sup_{x\in B(x^*,\|X_n^*-x^*\|)} \|\nabla \delta_n(x)\|}{\alpha}.$$

With Assumptions (iv)–(v), by applying the functional CLT and the continuous mapping theorem [7] to $\nabla\delta_n(\cdot)$, we have $\sup_{x\in B(x^*,\|X_n^*-x^*\|)} \|\nabla\delta_n(x)\| = O_p(n^{-1/2})$, and hence $\|X_n^* - x^*\| = O_p(n^{-1/2})$ follows.

By applying the quadratic growth condition (vii) to $x^*(\nabla\delta_n(x^*))$ and using the Lipschitz continuity of $\nabla f(\cdot)$, it can be shown that

$$X_n^* = x^*(\nabla\delta_n(x^*)) + o_p(n^{-1/2}).$$

Since $x^*(u)$ is Fréchet differentiable at $u = 0$,

$$x^*(\nabla\delta_n(x^*)) - x^* = D_0x^*(\nabla\delta_n(x^*)) + o_p(n^{-1/2}).$$

Thus, it follows from the CLT on $\nabla\delta_n(x^*)$ and the continuous mapping theorem that

$$\sqrt{n}(X_n^* - x^*) = D_0x^*(\sqrt{n}\nabla\delta_n(x^*)) + o_p(1) \Rightarrow D_0x^*(Z). \qquad \blacksquare$$

A second-order condition can ensure the quadratic growth condition (vii) for the parameterized objective function $f(x) + u\cdot x$. For example, if Θ is convex, f is twice continuously differentiable, and $\nabla^2 f(x^*)$ is positive definite, Assumption (vii) holds by setting α as the lower bound of the smallest eigenvalue of $\nabla^2 f(x)$ in a neighborhood of x^*. If $x^*(\cdot)$ is Lipschitz, the Fréchet derivative $D_0x^*(\cdot)$ is continuous and linear, and thus the asymptotic normality of X_n^* can be ensured.

8.4.3 *The SAA Method in the Nonconvex Case*

Thus far, the convergence theory for SAA methods that we have presented has been derived under the assumption that we can produce a global minimum of the SAA problem in Θ, and hence the theory can be applied primarily to convex problems. In the nonconvex case, the best that we can hope for from a computational point of view is that we can generate local minimizers of the SAA problems. When the sample function is differentiable almost surely on Θ, the validity of IPA ensures the convergence of the first order points of the SAA problem to those of the true problem. This reaffirms the key principle observed in Sect. 8.3: when IPA is valid, the SAA method is appropriate.

For $x \in \Theta$, $\mathcal{N}(x)$ denotes the normal cone to Θ at x. For x in the interior of Θ, $\mathcal{N}(x) = \{0\}$. For x on the boundary of Θ, $\mathcal{N}(x)$ is the convex cone generated by the outward normals of the faces on which x lies. When Θ is convex,

$$\mathcal{N}(x) = \{y \in \mathbb{R}^d : y^\top(x' - x) \le 0, \text{for all } x' \in \Theta\}.$$

A first-order critical point x of a smooth function f satisfies

$$-\nabla f(x) = z \text{ for some } z \in \mathcal{N}(x),$$

i.e., the direction of most rapid descent lies in the normal cone (of directions we cannot move in without leaving Θ). Let S_n^* and S^* be the set of first-order critical points of the approximating function f_n and the true function f in Θ, respectively. The following theorem states first-order convergence results of the SAA method, and is an immediate result of [55, Proposition 19].

Theorem 8.5. *Suppose there exists a compact subset $C \subset \mathbb{R}^d$ such that*

(i) S^* is non-empty and contained in C,
(ii) the true function $f(\cdot)$ is continuously differentiable on an open set containing C,
(iii) $\{\nabla f_n(x)\}$ converges to $\nabla f(x)$ uniformly on C, a.s. as $n \to \infty$, and
(iv) for sufficiently large n, S_n^* is non-empty and contained in C w.p.1.

Then $\mathbb{D}(S_n^*, S^*) \to 0$ a.s. as $n \to \infty$.

Proof. The proof can be derived from stochastic generalized equations. We do not introduce them here; rather, we present a relatively easier version of the proof with the added assumption that the domain Θ is compact and convex. Suppose that $\mathbb{D}(S_n^*, S^*) \nrightarrow 0$. Since Θ is compact, by passing to a subsequence if necessary, we can assume that there exists a convergent sequence of solutions $\{X_n^* \in S_n^*\}$ such that for some $\varepsilon > 0$, $d(X_n^*, S^*) \geq \varepsilon$ for all $n \geq 1$. Let x^* be a limit point of $\{X_n^*\}$, and then $x^* \notin S^*$. On the other hand, since Θ is convex and each X_n^* satisfies the first order criticality condition, for any $u \in \Theta$

$$\nabla f_n(X_n^*)^\mathsf{T}(u - X_n^*) \geq 0 \text{ w.p.1.}$$

By (ii) and (iii),

$$\nabla f_n(X_n^*)^\mathsf{T}(u - X_n^*) \to \nabla f(x^*)^\mathsf{T}(u - x^*)$$

a.s. as $n \to \infty$. Thus, $\nabla f(x^*)^\mathsf{T}(u - x^*) \geq 0$ for all $x \in \Theta$. But $x^* \notin S^*$ implies that for some $u \in X$ $\nabla f(x^*)^\mathsf{T}(u - x^*) < 0$, which is a contradiction. ■

The assumptions (ii) and (iii) above are satisfied under the sufficient conditions for a valid IPA gradient estimator presented in Sect. 8.3. When the sample path function is continuously differentiable a.s. at any $x \in \Theta$, $\nabla Y(\cdot, \xi)$ is uniformly integrable under the assumptions in Proposition 8.4. By applying Proposition 8.5 to each component of $\nabla Y(\cdot, \xi)$, we can show the continuity of $\nabla f(\cdot)$ and the uniform convergence of $\{\nabla f_n(\cdot)\}$.

Theorem 8.5 implies that the limit point of any solution sequence $\{X_n^* \in S_n^*\}$ must lie in S^*. This does not guarantee that $\{X_n^*\}$ converges almost surely. When there are multiple critical points, the particular critical point chosen from S_n^* depends, among other things, on the optimization algorithm that is used. The existence of a unique first-order critical point can ensure convergence. However, this condition tends to be difficult to verify in practice.

The second order convergence of the SAA method can be obtained by further strengthening the assumptions in Theorem 8.5. Now, we select X_n^* from a set of local minimizers of the SAA problem. By passing to a subsequence if necessary, we assume that $\{X_n^*\}$ converges to some random point $x^* \in \Theta$ a.s. as $n \to \infty$. The additional condition required is there must exist a neighborhood of X_n^* in which X_n^* is a local minimizer and this neighborhood does not shrink to a singleton when $n \to \infty$.

Theorem 8.6 (Theorem 4.1, [5]). *Suppose that the assumptions in Theorem 8.5 hold. Furthermore, assume that for any fixed sample path* $\bar{\xi} = \{\xi_1, \xi_2, \ldots\}$, *there exist* $n_0 > 0$ *and* $\delta > 0$ *such that for all* $n \geq n_0$ *and* $x \in B(X_n^*, \delta) \cap \Theta$, $f_n(X_n^*) \leq f_n(x)$. *Then* x^* *is a local minimum of* $f(\cdot)$ *w.p.1.*

Nonsmooth objective functions arise in a number of interesting stochastic optimization problems such as stochastic programs with recourse and stochastic min-max problems [50]. To close this section, we briefly discuss the local convergence of the SAA method in the nonsmooth setting. When the true and sample functions are continuous and nonsmooth, we can derive convergence results based on the Clarke generalized gradient [10]. For a locally Lipschitz function f, the generalized gradient ∂f can be defined as the convex hull of all the limit points of $\nabla f(x_k)$, where $\{x_k\}$ is any sequence which converges to x while avoiding the points where $\nabla f(x_k)$ does not exist. With some technical definitions, the expectation of the generalized gradient of the sample function can be well-defined [21, 62].

Essentially, the same principle from the smooth case can hold for the nonsmooth problem. When IPA is valid, i.e., $\partial \mathrm{E}[Y(x,\xi)] = \mathrm{E}[\partial_x Y(x,\xi)]$, SAA can be appropriate and the first order convergence can be achieved. A sufficient condition for the validity of IPA is that the sample function is locally Lipschitz continuous with integrable Lipschitz constant. This condition is a fairly general condition in the Lipschitz continuous setting, just as it is in the smooth case.

8.5 SAA Implementation

Implementing the SAA method is conceptually straightforward since only two choices need to be made: the sample size with which to generate the sample-path problem, and the numerical procedure with which to solve the generated sample-path problem. Assuming a numerical procedure is (somehow) chosen using cues laid out in Sects. 8.2 and 8.3, the only remaining task then is choosing an appropriate sample size. Towards making this decision, a reasonable question might be to ask what minimum sample size ensures that the solution resulting from the generated sample-path problem is of a stipulated quality, with a specified probability. In what follows, we present a "minimum sample size" result that answers this question. This is followed by a discussion of certain refined versions of SAA that are aimed at enhancing the implementability of the SAA method.

8.5.1 Sample Size Choice

Recall that for $\varepsilon \geq 0$, $\pi^*(\varepsilon)$ and $\Pi_n^*(\varepsilon)$ denote the ε-optimal solutions for the true and the sample-path problems, respectively. Theorem 8.7 presents an expression for the sample size n that guarantees that $\mathsf{P}\{\Pi_n^*(\delta) \not\subseteq \pi^*(\varepsilon)\} \leq \alpha$, for given $\alpha > 0, \varepsilon > 0$, and a chosen constant $\delta < \varepsilon$. The implication is that when an SAA problem is generated with a sample size exceeding the expression provided, the resulting solution is guaranteed to be ε-optimal with probability exceeding $1-\alpha$. To guide intuition, we present the result only for the setting where Θ is finite. The corresponding expression for the general case follows in a straightforward fashion after making additional assumptions that help to approximate $\{f(x) : x \in \Theta\}$ with $\{f(x) : x \in \tilde{\Theta}\}$, where $\tilde{\Theta}$ is an appropriately chosen finite set that in a certain precise sense "approximates" Θ.

Theorem 8.7 (Theorem 5.18, [51]). *Suppose there exists a constant $\sigma > 0$ such that for any $x \in \Theta \setminus \pi^*(\varepsilon)$, the moment generating function $M_x(t)$ of the random variable $Y(x,\xi) - f(x)$ satisfies $M_x(t) \leq \exp(\sigma^2 t^2/2), \forall t \in \mathbb{R}$. Then, for $\varepsilon > 0, 0 \leq \delta < \varepsilon$, and $\alpha \in (0,1)$, any n satisfying*

$$n \geq \frac{2\sigma^2 \ln(\frac{|\Theta|}{\alpha})}{(\varepsilon - \delta)^2} \tag{8.24}$$

guarantees that $\mathsf{P}\{\Pi_n^*(\delta) \not\subseteq \pi^*(\varepsilon)\} \leq \alpha$.

The proof of Theorem 8.7 proceeds by using the crude bound

$$\begin{aligned} \mathsf{P}\{\Pi_n^*(\delta) \not\subseteq \pi^*(\varepsilon)\} &\leq \sum_{x \in \Theta \setminus \pi^*(\varepsilon)} \mathsf{P}\{f_n(x) \leq v^* + \varepsilon\} \\ &\leq |\Theta| \exp\{-n\eta(\delta,\varepsilon)\} \\ &\leq |\Theta| \exp\{-n(\varepsilon-\delta)^2/2\sigma^2\}, \end{aligned} \tag{8.25}$$

where $\eta(\delta,\varepsilon) = \min_{x \in \Theta \setminus \pi^*(\varepsilon)} I_x(-\delta)$, and $I_x(\cdot)$ is the large deviations rate function of $Y(x,\xi) - f(x)$. The expression in (8.24) then follows upon replacing the left-hand side of (8.25) with α and then solving for the sample size. Note that in choosing a sample size through (8.24), the tolerance δ to which the SAA problem is solved still needs to be chosen by the user. It can also be seen from the expression in (8.24) that the dependence of the minimum sample size on the error probability α is logarithmic, and hence weak.

The sample size directive given by Theorem 8.7, while useful in some SAA settings, can be overly conservative [31, 51], often resulting in a loss in computational efficiency. This is unsurprising considering the crude bound leading to (8.25), and the existence of unknown constants, e.g., σ^2 in the case of finite Θ and several others in the case of continuous Θ, that nevertheless need to be chosen by the user. Such loss in efficiency resulting from the sometimes impractical sample size directives has been one of the primary impediments to SAA's implementability.

8.5.2 *Refined SAA Methods*

With a view towards easier implementation, various refined versions [11, 22, 35, 36, 45] of the SAA method have recently been proposed. In what follows, we discuss one of these paradigms, Retrospective Approximation (RA), in further detail. (The similarly named "Retrospective Optimization" technique was introduced by [20], but to describe the SAA method.)

Recall the efficiency issue associated with the SAA method. SAA dictates that a single sample-path problem be generated with a large enough sample size and solved to adequate tolerance. However, the minimum sample size required to ensure that the resulting solution is of stipulated quality may be so large as to render the procedure not viable. To thwart this difficulty, RA proposes a slight refinement of the SAA paradigm. Instead of solving a single sample-path problem generated with a large enough sample size, RA proposes to generate and solve a sequence of sample-path problems. The sequence of sample-path problems are generated using a nondecreasing sequence of sample sizes $\{m_k\}$, that are then solved to increasing stringency using a sequence of error-tolerances $\{\varepsilon_k\}$ that converge to zero. When the paradigm works as intended, the resulting sequence of solutions approaches the true solution asymptotically. More importantly, the paradigm is constructed to preserve efficiency. The early iterations are efficient because they involve sample-path problems generated with small sample sizes. The later iterations are efficient, at least in principle, due to the use of "warm starts," where solutions from previous iterations are used as initial guesses to the subsequent problems.

Towards further clarification, we now list RA as a nonterminating algorithm.

RA Components:

(i) A procedure for solving a generated sample-path problem to specified tolerance vector ε_k.
(ii) A sequence $\{m_k\}$ of sample sizes tending to infinity.
(iii) A sequence $\{\varepsilon_k\}$ of error-tolerances tending to zero.
(iv) A sequence of weights $\{w_{kj} : j = 1, 2, \dots, k\}$ for each iteration.

RA Logic:

0. Initialize the retrospective iteration number $k = 1$.
1. Generate a sample-path problem with sample size m_k. Use RA component (i) with a "warm start," i.e., with $\overline{X}_{k-1}$ as the initial guess, to solve the generated problem to within error-tolerance ε_k. Obtain a retrospective solution X_k.
2. Use component (iv) to calculate the solution $\overline{X}_k$ as the weighted sum of retrospective solutions $\{X_i\}_{i=1}^k$:

$$\overline{X}_k = \sum_{j=1}^{k} w_{kj} X_j.$$

3. Set $k \leftarrow k + 1$ and go to 1.

Step 1 of the RA listing is deliberately left ambiguous, and is to be made precise depending on the problem context. For example, in the context of using RA within global SO problems, "solving a sample-path problem to within tolerance ε_k" can mean identifying a point X_k whose optimality gap as measured with respect to the objective function $f_{m_k}(x)$ is at most ε_k.

The iterates resulting from the RA paradigm, for the context of global SO, are strongly consistent under conditions similar to those imposed within the SAA method. The proof follows in a rather straightforward fashion from the corresponding theorem [51, Theorem 5.3] in the SAA context in combination with some standard results on M-estimators [52].

Theorem 8.8. *Assume*

A_1. *The feasible region Θ is compact, and the set of global minima $\pi^* \subset \Theta$ of the function f is nonempty.*

A_2. *The sequence of sample functions $\{f_n(x)\}$ is such that the set of global minima Π_n^* of the function f_n is nonempty for large enough n w.p.1.*

A_3. *The functional sequence $\{f_n(x)\} \to f(x)$ uniformly as $n \to \infty$ w.p.1.*

A_4. *The function f is continuous on Θ.*

A_5. *The sequence of sample sizes $\{m_k\}$ and the sequence of error-tolerances $\{\varepsilon_k\}$ in the RA paradigm are chosen to satisfy $\{m_k\} \to \infty$ and $\varepsilon_k \to 0$ as $k \to \infty$.*

A_6. *Given $s > 0$, define the ith sum of the first s weights $w_i(s) = \sum_{j=1}^{s} w_{ij}$ for each $i \geq s$. The weights $\{w_{ij}\}$ are chosen so that $w_i(s) \to 0$ as $i \to \infty$.*

A_7. *The sample-path problems are solved to obtain a retrospective solution X_k satisfying $\|f_{m_k}(X_k) - v_{m_k}^*\| \leq \varepsilon_k$ when $\Pi_{m_k}^* \neq \phi$, with $v_{m_k} = \inf\{f_{m_k}(x) : x \in \Theta\}$.*

Then the sequences $\{f(X_k) - v^*\}$, $\{d(X_k, \pi^*)\} \to 0$ w.p.1. (Assume $d(X_k, \pi^*) = \infty$ if $\Pi_{m_k}^* = \emptyset$.)

The RA method above was presented as a nonterminating algorithm where a sequence of sample sizes $\{m_k\}$ for problem generation and a sequence of error-tolerances $\{\varepsilon_k\}$ relevant during problem solution need to be chosen by the user. This raises the natural question of how these sequences should be chosen to ensure efficiency. Pasupathy [35] partially addresses this question and presents guidelines on choosing these sequences as a function of the convergence rate of the numerical procedure in use. For example, it is shown that for efficiency, it may be best to choose $\varepsilon_k = O(1/\sqrt{m_k})$ when the numerical procedure in use converges at a linear rate. (*Convergence rates* are defined rigorously in Sect. 8.6.2.) Furthermore, when using linearly convergent numerical procedures, efficiency dictates that it is best to choose sample sizes $\{m_k\}$ such that $\limsup m_k/m_{k-1}^p = 0$ for all $p > 1$. Likewise, when using numerical procedures that have superlinear convergence rates, efficiency dictates that it is best to choose $\{m_k\}$ such that $\limsup m_k/m_{k-1}^p < \infty$ for all $p > 1$. We discuss these results in more detail in Sect. 8.6.

More recently, [45] addresses the obvious drawback that the directives provided in [35] are at best asymptotic. In other words, while the results in [35] recommend the rates at which the sample size sequence $\{m_k\}$ and the error-tolerance sequence

$\{\varepsilon_k\}$ should converge to zero, these recommendations still leave a large family of sequences from which to choose. Royset [45] remedies this in the specific context of solving smooth stochastic programs (e.g., when derivatives of the function $Y(x,\xi)$ are observable and $Y(x,\xi)$ is Lipschitz with the Lipschitz constant having finite expectation) with a numerical solver that is linearly convergent (e.g., projected gradient method using Armijo step sizes as detailed in [38]). Using a model that approximates the progress made by the linearly convergent numerical procedure in use, [45] formulates a dynamic program to identify generation-effort/solution-effort trade-off at the beginning of each iteration within RA. The output of the dynamic program includes the sample size that should be used for each generated problem and the computational effort that should be expended toward solving each generated problem.

8.6 Asymptotic Efficiency Calculation

As noted earlier, refined SAA methods like RA, are constructed with a view towards implementation. Does this construction result in any real computational savings? In other words, does RA enjoy provable efficiency gains over the SAA paradigm? In this section, we answer this question in some detail. Towards first providing a benchmark for an asymptotic rate calculation, we present a very concise overview and analysis of stochastic approximation (SA) which, alongside the SAA method, is a standard technique for solving SO problems. This is followed by Sect. 8.6.2 where we discuss the maximum achievable convergence rate by the SAA method. Section 8.6.3 presents the analogous calculation for the RA method.

Towards setting up the problem of identifying asymptotic efficiency, suppose that the optimal solution to the true problem, x^*, is unique, and suppose that we want to obtain a solution that is within a prescribed distance ε from x^*. (All distances are Euclidean unless otherwise noted.) Suppose also that we measure computational effort in terms of the number of simulation replications required, i.e., the number of times $Y(x,\xi)$ is computed, for various x and ξ. Here we take the function $Y(\cdot,\cdot)$ as fixed, rather than allowing it to come from a class of functions as in many varieties of complexity theory; see, e.g., [33]. Moreover, this measure ignores the effort required to compute, e.g., gradients. However, our results will be restricted to *rates* of convergence that ignore proportionality constants, so as long as gradients are obtained through schemes that only proportionally increase the work, e.g., finite differences and infinitesimal perturbation analysis, then our results will be unaffected. Finally, we also ignore the internal computations of an optimization algorithm beyond the simulation effort. Such computations often heavily depend on the dimension of the problem, but since we are fixing f, the dimension is also fixed.

8.6.1 Asymptotic Rates for Stochastic Approximation

A well-understood, general-purpose method for solving stochastic optimization problems, alternative to using the SAA principle, is stochastic approximation [26, 30, 42]. For unconstrained problems in $\mathbb{R}^d$, the classical stochastic approximation algorithm is a simple recursion that produces a sequence of points $\{\tilde{X}_n : n \geq 0\}$, each of which lies in $\mathbb{R}^d$. The recursion requires an initial point $\tilde{X}_0$, a positive gain sequence $\{a_n : n \geq 0\}$, and a sequence of vectors $\{\hat{\nabla} f(\tilde{X}_n) : n \geq 0\}$ in $\mathbb{R}^d$, where $\hat{\nabla} f(\tilde{X}_n)$ is an estimate of $\nabla f(\tilde{X}_n)$. A simple version of a stochastic-approximation recursion for a minimization problem is then

$$\tilde{X}_{n+1} = \tilde{X}_n - a_n \hat{\nabla} f(\tilde{X}_n). \tag{8.26}$$

For the problems we consider here, the gradient estimator can usually be taken to be $\nabla Y(\tilde{X}_n, \xi_n)$, i.e., the gradient of $Y(\cdot, \xi_n)$ evaluated at $\tilde{X}_n$, where $(\xi_n : n \geq 0)$ are i.i.d., since under fairly general conditions (Sect. 8.3), this gradient estimator is unbiased and has other desirable qualities like bounded variance. In that case, if $f(\cdot)$ is smooth, has a unique global minimizer x^*, and $a_n = a/n$ with $a > 0$ sufficiently large, then under additional nonrestrictive conditions,

$$\sqrt{n}(\tilde{X}_n - x^*) \Rightarrow N(0, \Lambda), \tag{8.27}$$

as $n \to \infty$, for a certain $d \times d$ matrix Λ. See [4, Chap. VIII] for an overview of this result and a sketch of how it can be established using the "Ordinary Differential Equation" approach.

The CLT in (8.27) is striking in that the recursion (8.26) is trivial to implement, involves almost no computation beyond the calculation of a sample gradient at each iteration, and is very generally applicable. If the number of iterations of (8.26) is completed in c units of computer time, $n(c)$ grows roughly linearly in c (as would be the case if, e.g., sample gradients are computed in constant time), then a time-changed version of the CLT (8.27) establishes that the resulting SA estimator has an error $\tilde{X}_{n(c)} - x^* = O_p(c^{-1/2})$. Equivalently, the computational effort required to obtain an error of order ε with SA is $O_p(\varepsilon^{-2})$.

It is generally known that the performance of the recursion in (8.26) is highly dependent on the gain sequence $\{a_n\}$. (In fact, even when the gradient estimator $\nabla Y(\tilde{X}_n, \xi_n)$ is directly observable and $a_n = a/n$ $(a > 0)$, convergence to the root fails if the constant a falls below a certain threshold, akin to the parallel-chord method for nonlinear root-finding [34, p. 181].) Accordingly, the last three decades have seen enormous attention given to the question of choosing the gain sequence $\{a_n\}$; see [3, 9, 30, 40] and Chap. 6. While we do not go into any further detail on this question, two key facts stand out. First, within the context of the iteration (8.26), the fastest achievable convergence rate is $O_p(c^{-1/2})$ [40]. Second, a remarkably simple scheme independently developed by [39, 49], and surveyed under the moniker "Polyak–Ruppert averaging" in [4, Chap. VIII], achieves this maximum rate. The scheme involves using the step-size sequence $a_n = a/n^\gamma$ for some $\gamma \in (0,1)$, and then estimating the root x^* via the direct average

$$\overline{X}_n = \frac{1}{n}\sum_{i=1}^{n}\tilde{X}_i.$$

Under mild conditions, the Polyak–Ruppert averaging scheme enjoys a CLT of the same form as (8.27), although with a different covariance matrix Λ. Furthermore, this happens *irrespective* of the value of the constant $a > 0$. (The small-sample performance is, however, seriously affected by the choice of the constant a.) The Polyak–Ruppert averaging scheme also has other optimality properties related to the matrix Λ that appears in the limit; see [4, Chap. VIII].

8.6.2 *Asymptotic Rates for the SAA Method*

As noted in Sect. 8.6.1, the Polyak–Ruppert averaging scheme achieves the maximum possible convergence rate of $O_p(c^{-1/2})$ within the context of stochastic approximation. Loosely speaking, this amounts to requiring $O_p(\varepsilon^{-2})$ computational effort if one wants to obtain ε accuracy. How does the SAA method perform in comparison? Towards setting up this question rigorously, recall the SAA method again—a single sample-path problem is generated with sample size n and solved using a chosen numerical solver. Furthermore, since solving the generated problem to infinite precision is usually impossible in practice, suppose we execute k iterations of the numerical procedure on the generated problem to obtain a solution $X_n(k)$. The total budget expended in the process is then simply $c = n \times k$. It seems clear that, under certain conditions (e.g., numerical procedure cannot solve to infinite precision in a finite number of steps), as the available budget $c \to \infty$, the sample size n and the number of steps k should satisfy $n, k \to \infty$ to ensure that the optimality gap of $X_n(k)$ converges to zero in any reasonable sense. However, what relationship between n and k (for given c) ensures that such convergence happens at the fastest possible rate? Moreover, what is the corresponding maximal rate?

In a recent paper, [46] provide an answer to these questions. Before we summarize their results, let us introduce definitions relating to the convergence rates of the numerical solver in use. These definitions appear in more restrictive form in [46].

Denote the numerical procedure acting on the sample function $f_n(x)$ by the map $A(x) : \Theta \to \Theta$. Let $A^k(x)$ denote the iterate obtained after k successive applications of the map $A(\cdot)$ on the initial iterate x. In all three definitions that follow, we assume that the function $f_n(x)$ attains its infimum $v_n^* := \inf\{f_n(x) : x \in \Theta\}$ and that $f_n(A^k(x)) \to v_n^*$ as $k \to \infty$ for all $x \in \Theta$. Also, to avoid trivialities, assume that $f_n(A^{k+1}(x))$ is different from v_n^* for all k. Denote $Q_t = \limsup_{k\to\infty} |f_n(A^{k+1}(x)) - v_n^*|/|f_n(A^k(x)) - v_n^*|^t$.

Definition 8.1. The numerical procedure $A(x) : \Theta \to \Theta$ is said to exhibit pth-order sublinear convergence if $Q_1 \geq 1$, and there exist constants $p, s > 0$ such that $p = \sup\{r : f_n(A^k(x)) - v_n^* \leq s/k^r \text{ for all } x \in \Theta\}$.

When $f_n(x)$ is convex and Θ is a closed convex set, the subgradient method [33, Sect. 3.2] for nonsmooth convex optimization exhibits sublinear convergence with $p = 1/2$. Similarly, when $f_n(x)$ is strongly convex with $\Theta := \mathbb{R}^d$, the optimal gradient method [33, Sect. 2.2] is sublinear with $p = 2$.

Definition 8.2. The numerical procedure $A(x) : \Theta \to \Theta$ is said to exhibit linear convergence if $Q_1 \in (0,1)$ for all $x \in \Theta$.

The definition of linear convergence implies that there exists a constant θ satisfying $f_n(A(x)) - v_n^* \leq \theta(f_n(x) - v_n^*)$ for all $x \in \Theta$. The projected gradient method with Armijo steps [38] when executed on certain smooth problems exhibits a linear convergence rate.

Definition 8.3. The numerical procedure $A(x) : \Theta \to \Theta$ is said to exhibit superlinear convergence if $Q_1 = 0$ for all $x \in \Theta$. The convergence is said to be pth-order superlinear if $Q_1 = 0$ and $\sup\{t : Q_t = 0\} = p < \infty$ for all $x \in \Theta$.

When $f_n(x)$ is strongly convex and twice Lipschitz continuously differentiable with observable derivatives, Newton's method is second-order superlinear. For settings where the derivative is unobservable, there is a slight degradation in the convergence rate, but Newton's method remains superlinear [6, p. 338].

We are now ready to summarize the main results of [46] through Theorem 8.9, which is in essence a characterization of the maximum achievable convergence rate when using a sublinearly convergent algorithm within the SAA method, and should be juxtaposed with the $O_p(c^{-1/2})$ rate achievable using stochastic approximation as discussed in Sect. 8.6.1.

Theorem 8.9 (Convergence Rate for the SAA Method). *Let the following assumptions hold.*

A_1. *The expectation* $\mathsf{E}[Y^2(x,\xi)] < \infty$ *for all* $x \in \Theta$.
A_2. *The function* $Y(x,\xi)$ *is Lipschitz w.p.1, and has Lipschitz constant* $K(\xi)$ *having finite expectation.*
A_3. *The function* $f_n(x)$ *attains its infimum on* Θ *for each* n *w.p.1.*

Also, let $c = n \times k$ *and* $n/c^{1/(2p+1)} \to a$ *as* $c \to \infty$, *with* $a \in (0,\infty)$. *Then, if the numerical procedure exhibits pth-order sublinear convergence,*

$$c^{p/(2p+1)}(f_n(A^k(x)) - v^*) = O_p(1) \text{ as } c \to \infty.$$

The crucial message given by Theorem 8.9 is that in the context of the SAA method, the maximum achievable convergence rate is $O_p(c^{-p/(2p+1)})$ when the numerical procedure in use exhibits p-th order sublinear convergence. (While Theorem 8.9 does not directly assert that $O_p(c^{-p/(2p+1)})$ is the maximum achievable rate, [46] show this rigorously.) [46] also demonstrate that the corresponding rates when using linearly convergent and pth-order superlinearly convergent procedures are $O_p((c/\log c)^{-1/2})$ and $O_p((c/\log\log c)^{-1/2})$, respectively.

Two observations relating to the assertions in [46] are noteworthy. First, the fastest achievable convergence rate within the SAA method depends on the numerical procedure in use, with faster numerical procedures affording a faster rate. This is not so surprising when one sees that the SAA method splits the available budget ($c = n \times k$) between sampling and solving. Since faster numerical procedures incur a smaller cost to solving, they facilitate attainment of a faster convergence rate. Second, none of the families of numerical procedures considered are capable of attaining the canonical convergence rate $O_p(c^{-1/2})$ that is seen in stochastic approximation. Such degradation from the canonical convergence rate can be explained as the "price" of using a numerical procedure. In other words, unless the numerical procedure used within SAA is capable of infinite precision with only a finite amount of computing effort, there is always a degradation in the convergence rate due to the fact that a non-negligible portion of the budget is expended towards solving the generated problem.

8.6.3 Asymptotic Rates for the RA Method

In this section, we present an analogous analysis for the maximum achievable convergence rates within the RA method. Recall that in the RA method, instead of generating and solving a single sample-path problem as in the SAA method, a sequence of sample-path problems are generated with sample sizes $\{m_k\}$ and solved to corresponding error-tolerances $\{\varepsilon_k\}$. In analyzing the achievable convergence rates within the RA method, we then seek an asymptotic relationship between the error $\|X_k - x^*\|$ incurred at the end of k iterations, and the corresponding total work done C_k. The following result, adapted from [35], captures this relationship as a function of the convergence rate of the numerical procedure in use, but with strict stipulations on the sample-path structure and the ability to observe their derivatives.

Theorem 8.10. *Assume that Assumptions (i)–(vi) of Theorem 8.4 hold. In addition, let the following assumptions hold:*

A_1. *The sample function $f_n(x)$ has a unique minimum X_n^* w.p.1.*

A_2. *When $f_n(x)$ attains a unique minimum X_n^*, $f_n(x)$ is twice differentiable at X_n^*. Furthermore, the matrix of second-order partial derivatives (Hessian) of $f_n(x)$ at X_n^* is positive definite with smallest eigenvalue uniformly bounded away from 0 w.p.1.*

A_3. *The solution X_k obtained from the kth iteration of RA satisfies $\|\nabla f_{m_k}(X_k)\| \le \varepsilon_k$.*

A_4. *The numerical procedure used to solve the sample-path problems in RA exhibits p-th order sublinear convergence or pth-order linear convergence with respect to the observed derivatives.*

A_5. *The sample sizes are increased linearly, i.e., $m_k/m_{k-1} = c > 1$ for all k.*

A_6. *The error-tolerances are chosen so that $\varepsilon_k = O(1/\sqrt{m_k})$.*

Then the sequence of solutions obtained using the RA procedure satisfies $C_k\|X_k - x^*\|^2 = O_p(1)$ *as* $k \to \infty$, *where* C_k *is the total amount of computational work done until the kth iteration and is given by* $C_k = \sum_{i=1}^{k} N_i m_i$. *Here* N_i *is the number of points visited by the numerical procedure during the ith iteration.*

Proof. The proof proceeds along lines very similar to the proof of Theorem 5 in [35]. In what follows, we provide only a proof sketch, and only for the case where the numerical procedure in use exhibits linear convergence. The corresponding proof for the sublinear convergence case follows almost directly after appropriately changing the expression in (8.28).

We first see, since the numerical procedure is assumed to exhibit linear convergence, that

$$N_i = O_p\left(1 + \frac{1}{\log r}\left(\log \frac{\varepsilon_i}{\|\nabla f_{m_i}(X_{i-1})\|}\right)\right), \tag{8.28}$$

for some $r \in (0,1)$. Using the Delta method [7] and Assumptions A_1, A_2, A_3, we write

$$\|\nabla f_{m_i}(X_{i-1})\| = O_p(\|X_i^* - X_{i-1}^*\|) + \varepsilon_{i-1}. \tag{8.29}$$

Since Assumptions (i)-(vi) of Theorem 8.4 hold, $\|X_i^* - x^*\| = O_p(1/\sqrt{m_i})$ and hence $\|X_i^* - X_{i-1}^*\| = O_p(1/\sqrt{m_i} + 1/\sqrt{m_{i-1}})$. Combining this with (8.28) and (8.29) yields

$$N_i = O_p\left(1 + \frac{1}{\log r}\left(\log \frac{\varepsilon_i}{1/\sqrt{m_i} + 1/\sqrt{m_{i-1}} + \varepsilon_{i-1}}\right)\right). \tag{8.30}$$

Now use Assumption A_6 to obtain

$$C_k\|X_k - x^*\|^2 = O_p\left((\sum_{i=1}^{k} m_i)(1/\sqrt{m_k} + \varepsilon_k)^2\right). \tag{8.31}$$

Finally, use (8.31) and Assumption A_5 to conclude that the assertion holds. ■

Theorem 8.10 asserts that, as long as the sample size and error-tolerance sequences are chosen strategically, the error in the obtained solution converges to zero at the canonical rate. This assertion is interesting, since we will recall from Sect. 8.6.2 that the canonical convergence rate is unachievable in the context of the SAA method, barring unlikely contexts where the numerical procedure exhibited exceptionally fast convergence rates. It is also noteworthy that Theorem 8.10 assumes that the derivatives of the sample path are observable. This is to help with terminating the individual iterations of the RA algorithm, and could probably be relaxed further by assuming instead that the derivative is estimated using a consistent estimator appropriately constructed from function observations. The assumption about the numerical procedure exhibiting at least linear convergence is easily satisfied, e.g., projected gradient method [38] for certain smooth problems; and Newton's method [34, Chap. 9] when used on smooth convex programs with observable derivatives.

8.7 Conclusions

We have provided a guide to the principle of SAA for simulation optimization, with a discussion on when SAA might be an appropriate solution method, how the potential for such applicability can be detected, and an appraisal of SAA's implementation and efficiency characteristics. An interesting observation on SAA's applicability is that it can be applied whenever infinitesimal perturbation analysis [17] for gradient estimation can be applied. Loosely speaking, both of these methods become applicable when the sample problems and true problem share characteristics that are important for numerical optimization software, chief among which are continuity, differentiability, and the approximate location of optimal solutions. SAA has a well-developed large-sample theory, both within the global and the local optimality contexts. The latter context seems especially useful within application settings.

On the question of asymptotic efficiency, recent results have established that a straightforward implementation of SAA is inferior to stochastic approximation. This difference in efficiency stems entirely from the fact that SAA, by construction, stipulates that the optimization software being employed uses a fixed sample size irrespective of how close the current solution is to an optimal solution. Towards remedying this, a refinement of SAA called retrospective approximation has been developed. The refinement increases sample sizes at a carefully controlled rate as the numerical optimization proceeds, and in the process recovers the same rate of convergence (up to a multiplicative constant) as stochastic approximation.

Throughout this chapter, we have made the assumption that samples are i.i.d., but that is not essential to the established theory. Indeed, one can apply any of several variance reduction methodologies that induce dependence, and for the most part the theory remains relatively unchanged; see [57]. One can also generate the samples using techniques such as quasi-Monte Carlo and randomized versions thereof [29]. Some of these topics, along with stochastic constraints, are treated in detail in the following chapter.

Acknowledgements This work was partially supported by National Science Foundation grants CMMI-0800688 and CMMI-1200315, and by Singapore MOE Academic Research Fund grant WBS R-266-000-049-133.

References

1. S. Ahmed and A. Shapiro. Solving chance-constrained stochastic programs via sampling and integer programming. In *Tutorials in Operations Research*, pages 261–269. INFORMS, 2008.
2. S. Alexander, T. F. Coleman, and Y. Li. Minimizing CVaR and VaR for a portfolio of derivatives. *Journal of Banknig and Finance*, 30(2):584–605, 2006.
3. S. Andradóttir. A scaled stochastic approximation algorithm. *Management Science*, 42:475–498, 1996.
4. S. Asmussen and P. W. Glynn. *Stochastic Simulation: Algorithms and Analysis*, volume 57 of *Stochastic Modeling and Applied Probability*. Springer, New York, 2007.
5. F. Bastin, C. Cirillo, and P. L. Toint. Convergence theory for nonconvex stochastic programming with an application to mixed logit. *Mathematical Programming B*, 108:207–234, 2006.
6. M. S. Bazaara, H. Sherali, and C. M. Shetty. *Nonlinear Programming: Theory and Algorithms*. John Wiley & Sons, New York, NY., 2006.
7. P. Billingsley. *Probability and Measure*. Wiley, New York, NY., 1995.
8. J. R. Birge and F. Louveaux. *Introduction to Stochastic Programming*. Springer, New York, 1997.
9. M. Broadie, D. M. Cicek, and A. Zeevi. General bounds and finite-time improvement for the Kiefer-Wolfowitz stochastic approximation algorithm. *Operations Research*, 59(5):1211–1224, 2011.
10. F. H. Clarke. *Optimization and Nonsmooth Analysis*. Wiley, New York, 1983.
11. G. Deng and M. C. Ferris. Variable-number sample-path optimization. *Mathematical Programming*, 117:81–109, 2009.
12. J. A. Dieudonné. *Foundations of Modern Analysis*. Academic Press, New York, 1960.
13. M. C. Fu. Gradient estimation. In S. G. Henderson and B. L. Nelson, editors, *Simulation*, Handbooks in Operations Research and Management Science, pages 575–616. Elsevier, Amsterdam, 2006.
14. M. C. Fu and K. Healy. Techniques for optimization via simulation: an experimental study on an (s, S) inventory system. *IIE Transactions*, 29(3):191–199, 1993.
15. M. C. Fu and J. Q. Hu. *Conditional Monte Carlo: Gradient Estimation and Optimization Applications*. Kluwer, Boston, 1997.
16. A. A. Gaivoronski and G. Pflug. Value-at-Risk in portfolio optimization: Properties and computational approach. *Journal of Risk*, 7(2):1–31, 2005.
17. P. Glasserman. *Gradient Estimation Via Perturbation Analysis*. Kluwer, The Netherlands, 1991.
18. G. Gürkan, A. Y. Özge, and S. M. Robinson. Sample-path solution of stochastic variational inequalities. *Mathematical Programming*, 84:313–333, 1999.
19. J. M. Harrison and J. A. Van Mieghem. Multi-resource investment stragies: Operational hedging under demand uncertainty. *European Journal of Operational Research*, 113:17–29, 1999.
20. K. Healy and L. W. Schruben. Retrospective simulation response optimization. In B. L. Nelson, D. W. Kelton, and G. M. Clark, editors, *Proceedings of the 1991 Winter Simulation Conference*, pages 954–957. IEEE, Piscataway, NJ, 1991.
21. T. Homem-de-Mello. Estimation of derivatives of nonsmooth performance measures in regenerative systems. *Mathematics of Operations Research*, 26:741–768, 2001.
22. T. Homem-de-Mello. Variable-sample methods for stochastic optimization. *ACM Transactions on Modeling and Computer Simulation*, 13:108–133, 2003.
23. L. J. Hong, Y. Yang, and L. Zhang. Sequential convex approximations to joint chance constrained programs: A Monte Carlo approach. *Operations Research*, 59(3):617–630, 2011.
24. H. Hu, T. Homem-de-Mello, and S. Mehrotra. Sample average approximation of stochastic dominance constrained programs. *Mathematical Programming*, 133(1–2):171–201, 2012.
25. A. I. Kibzun and Y. S. Kan. *Stochastic Programming Problems with Probability and Quantile Functions*. Wiley, New York, 1996.

26. J. Kiefer and J. Wolfowitz. Stochastic estimation of the maximum of a regression function. *Annals of Mathematical Statistics*, 23:462–466, 1952.
27. S. Kim and S. G. Henderson. The mathematics of continuous-variable simulation optimization. In S. J. Mason, R. R. Hill, L. Moench, and O. Rose, editors, *Proceedings of the 2008 Winter Simulation Conference*, pages 122–132. IEEE, Piscataway, NJ, 2008.
28. A. J. Kleywegt, A. Shapiro, and T. Homem-de-Mello. The sample average approximation method for stochastic discrete optimization. *SIAM Journal on Optimization*, 12:479–502, 2001.
29. M. Koivu. Variance reduction in sample approximations of stochastic programs. *Mathematical Programming*, 103(3):463–485, 2005.
30. H. J. Kushner and G. G. Yin. *Stochastic Approximation and Recursive Algorithms and Applications*. Springer-Verlag, New York, 2nd edition, 2003.
31. J. Luedtke and S. Ahmed. A sample approximation approach for optimization with probabilistic constraints. *SIAM Journal on Optimization*, 19:674–699, 2008.
32. W. K. Mak, D. P. Morton, and R. K. Wood. Monte Carlo bounding techniques for determining solution quality in stochastic programs. *Operations Research Letters*, 24:47–56, 1999.
33. Y. Nesterov. *Introductory Lectures on Convex Optimization*. Kluwer Academic Publishers, Norwell, MA, 2004.
34. J. M. Ortega and W. C. Rheinboldt. *Iterative Solution of Nonlinear Equations in Several Variables*. Academic Press, New York, NY., 1970.
35. R. Pasupathy. On choosing parameters in retrospective-approximation algorithms for stochastic root finding and simulation optimization. *Operations Research*, 58:889–901, 2010.
36. R. Pasupathy and S. Kim. The stochastic root-finding problem: overview, solutions, and open questions. *ACM Transactions on Modeling and Computer Simulation*, 21(3):23, 2011.
37. E. L. Plambeck, B.-R. Fu, S. M. Robinson, and R. Suri. Sample-path optimization of convex stochastic performance functions. *Mathematical Programming*, 75:137–176, 1996.
38. E. Polak. *Optimization: Algorithms and Consistent Approximations*. Springer, New York, NY, 1997.
39. B. T. Polyak. New stochastic approximation type procedures. *Automat. i Telemekh.*, 7:98–107, 1990.
40. B. T. Polyak and A. B. Juditsky. Acceleration of stochastic approximation by averaging. *SIAM Journal on Control and Optimization*, 30:838–855, 1992.
41. A. Ravindran, D. T. Phillips, and J. J. Solberg. *Operations Research: Principles and Practice*. Wiley, New York, NY, 2nd ed. edition, 1987.
42. H. Robbins and S. Monro. A stochastic approximation method. *Annals of Mathematical Statistics*, 22:400–407, 1951.
43. S. M. Robinson. Analysis of sample-path optimization. *Mathematics of Operations Research*, 21:513–528, 1996.
44. S. M. Ross. *Stochastic Processes*. Wiley, New York, 2nd edition, 1996.
45. J. Royset. On sample size control in sample average approximations for solving smooth stochastic programs. *Computational Optimization and Applications*, 55(2):265–309, 2013.
46. J. Royset and R. Szechtman. Optimal budget allocation for sample average approximation. *Operations Research*, 61(3):762–776, 2013.
47. R. Y. Rubinstein and A. Shapiro. Optimization of static simulation models by the score function method. *Mathematics and Computers in Simulation*, 32:373–392, 1990.
48. R. Y. Rubinstein and A. Shapiro. *Discrete Event Systems: Sensitivity Analysis and Stochastic Optimization by the Score Function Method*. Wiley, Chichester, 1993.
49. D. Ruppert. Stochastic approximation. In B. K. Ghosh and P. K. Sen, editors, *Handbook of Sequential Analysis*, pages 503–529. Marcel Dekker, New York, 1991.
50. A. Ruszczyński. A linearization method for nonsmooth stochastic programming problems. *Mathematics of Operations Research*, 12:32–49, 1987.
51. A. Ruszczyński and A. Shapiro, editors. *Stochastic Programming. Handbook in Operations Research and Management Science*. Elsevier, New York, NY, 2003.
52. R. J. Serfling. *Approximation Theorems of Mathematical Statistics*. John Wiley & Sons, Inc., New York, NY, 1980.

53. A. Shapiro. Asymtotic behavior of optimal solutions in stochastic programming. *Mathematics of Operations Research*, 18:829–845, 1993.
54. A. Shapiro. Simulation-based optimization – convergence analysis and statistical inference. *Stochastic Models*, 12(3):425–454, 1996.
55. A. Shapiro. Monte Carlo sampling methods. In A. Ruszczynski and A. Shapiro, editors, *Stochastic Programming*, Handbooks in Operations Research and Management Science. Elsevier, 2003.
56. A. Shapiro. Sample average approximation. In S. I. Gass and M. C. Fu, editors, *Encyclopedia of Operations Research and Management Science*, pages 1350–1355. Springer, 3rd edition, 2013.
57. A. Shapiro, D. Dentcheva, and A. Ruszczyński. *Lectures on Stochastic Programming: Modeling and Theory*. MPS-SIAM Series on Optimization. SIAM-MPS, Philadelphia, PA, 2009.
58. Stochastic Programming Society. Stochastic programming introduction. www.stoprog.org, 2014.
59. J. A. Van Mieghem and N. Rudi. Newsvendor networks: Inventory management and capacity investment with discretionarty activities. *Manufacturing and Service Operations Management*, 4(4):313–335, 2002.
60. S. Vogel. Stability results for stochastic programming problems. *Optimization*, 19(2):269–288, 1988.
61. S. Vogel. A stochastic approach to stability in stochastic programming. *Journal of Computational and Applied Mathematics*, 45:65–96, 1994.
62. R. Wets. Stochastic programming. In G. L. Nemhauser, A. H. G.Rinnooy Kan, and M. J. Todd, editors, *Optimization*, volume 1 of *Handbooks in Operations Research and Management Science*, pages 573–629. Elsevier, 1989.
63. H. Xu and D. Zhang. Smooth sample average approximation of stationary point in nonsmooth stochastic optimization and application. *Mathematical Programming*, 119:371–401, 2009.

Chapter 9
Stochastic Constraints and Variance Reduction Techniques

Tito Homem-de-Mello and Güzin Bayraksan

Abstract We provide an overview of two select topics in Monte Carlo simulation-based methods for stochastic optimization: problems with stochastic constraints and variance reduction techniques. While Monte Carlo simulation-based methods have been successfully used for stochastic optimization problems with deterministic constraints, there is a growing body of work on its use for problems with stochastic constraints. The presence of stochastic constraints brings new challenges in ensuring and testing optimality, allocating sample sizes, etc., especially due to difficulties in determining feasibility. We review results for general stochastic constraints and also discuss special cases such as probabilistic and stochastic dominance constraints. Next, we review the use of variance reduction techniques (VRT) in a stochastic optimization setting. While this is a well-studied topic in statistics and simulation, the use of VRT in stochastic optimization requires a more thorough analysis. We discuss asymptotic properties of the resulting approximations and their use within Monte Carlo simulation-based solution methods.

9.1 Introduction

In this chapter we consider stochastic optimization problems of the form

$$\min_{x\in\Theta} \{f(x) := \mathsf{E}[Y(x,\xi)] \mid g_k(x) := \mathsf{E}[G_k(x,\xi)] \leq 0,\ k \in \mathscr{K}\}, \qquad \text{(SP)}$$

where G_k, $k \in \{0\} \cup \mathscr{K}$ are extended real-valued functions with inputs being the decision vector x and a random vector ξ. For simplicity of exposition, we use $G_0 \equiv Y$ and $g_o \equiv f$. The set of stochastic constraints is determined by the set $\mathscr{K}$. Typically, $\mathscr{K}$ contains a finite number of constraints, i.e., $\mathscr{K} = \{1,2,\ldots,K\}$; however, this

T. Homem-de-Mello (✉)
Universidad Adolfo Ibañez, Santiago, Chile
e-mail: tito.hmello@uai.cl

G. Bayraksan
The Ohio State University, Columbus, OH, USA
e-mail: bayraksan.1@osu.edu

M.C. Fu (ed.), *Handbook of Simulation Optimization*, International Series in Operations Research & Management Science 216, DOI 10.1007/978-1-4939-1384-8_9

does not have to be the case and it can contain uncountably many constraints (see, e.g., Sect. 9.2.3). In this chapter, we use $\mathscr{K} = \emptyset$ to denote (SP) without any stochastic constraints and we set $K = 1$ for the case with a single stochastic constraint. The set of deterministic constraints x must satisfy is denoted by $\Theta \subset \mathbb{R}^{d_x}$ and $\Xi \subset \mathbb{R}^{d_\xi}$ denotes the support of ξ, where d_x and d_ξ are the dimensions of the vectors x and ξ, respectively. We assume that ξ has a known distribution P that is independent of x, and the expectations in (SP), taken with respect to the distribution of ξ, are well-defined and finite for all $x \in \Theta$. We will refer to (SP) as the "true" optimization problem (as opposed to the approximating problems to be discussed in the sequel).

A wide variety of problems can be cast as (SP) depending on $\mathscr{K}$, Θ and G_k, $k \in \{0\} \cup \mathscr{K}$. For example, in a two-stage stochastic linear program with recourse, $\mathscr{K} = \emptyset$, $\Theta = \{Ax = b, x \geq 0\}$, and $G_0(x,\xi) = cx + h(x,\xi)$, where $h(x,\xi)$ is the optimal value of the linear program

$$\begin{aligned} h(x,\xi) = \min_y \;& \tilde{q}y \\ \text{s.t. } & \tilde{W}y = \tilde{r} - \tilde{T}x, \;\; y \geq 0. \end{aligned}$$

Here, ξ is a random vector that is comprised of random elements of $\tilde{q}$, $\tilde{W}$, $\tilde{R}$ and $\tilde{T}$.

In contrast, in a stochastic linear program with a single probabilistic constraint (i.e., $\mathsf{P}(\tilde{A}'x \geq \tilde{b}') \leq \alpha$), we have $K = 1$, $G_0(x,\xi) = cx$ and

$$G_1(x,\xi) = \mathbf{1}\{\tilde{A}'x \geq \tilde{b}'\} - \alpha,$$

where $\mathbf{1}\{E\}$ denotes the indicator function that takes value 1 if the event E happens and 0 otherwise, and $\alpha \in (0,1)$ is a desired probability level. In this case, ξ is comprised of random elements of $\tilde{A}'$ and $\tilde{b}'$. Here, the decision maker requires that the relationship $\tilde{A}'x \geq \tilde{b}'$ be satisfied with probability no more than α.

Monte Carlo simulation-based methods have been successfully used in many different applications of stochastic optimization. The appeal of such methods results from the fact that they often approximate well, with a small number of samples, problems that have very large number of scenarios; see, for instance, [60] for numerical results.

There are multiple ways to use Monte Carlo methods in problem (SP). A generic way of describing them is to construct an approximating problem as follows. Consider a family $\{g_{k,N_k}(\cdot)\}$ of random approximations of the function $g_k(\cdot)$, each $g_{k,N_k}(\cdot)$ being defined as

$$g_{k,N_k}(x) := \frac{1}{N_k} \sum_{j=1}^{N_k} G_k(x,\xi_k^j), \tag{9.1}$$

where $\{\xi_k^1, \ldots, \xi_k^{N_k}\}$ is a sample of size N_k from the distribution[1] of ξ, for $k \in \{0\} \cup \mathscr{K}$. When $\xi_k^1, \ldots, \xi_k^{N_k}$ are mutually independent, the quantity $g_{k,N_k}(x)$ is called a *(standard or crude) Monte Carlo* estimator of $g_k(x)$. Given the family of estimators $\{g_{k,N_k}(\cdot)\}$ defined in (9.1), one can construct the corresponding approximating program

$$\min_{x \in \Theta} \left\{ f_{N_0}(x) \equiv g_{0,N_0}(x) \,|\, g_{k,N_k}(x) \le 0,\ k \in \mathscr{K} \right\}. \tag{9.2}$$

In this chapter we adopt the following notation: N denotes the vector of sample sizes $[N_k]_{k \in \{0\} \cup \mathscr{K}}$; ν^* and ν_N denote the optimal value of (SP) and the optimal value of its approximation (9.2), respectively; similarly, x^* and x_N denote an optimal solution to (SP) and (9.2), respectively. For a particular realization $\{\hat{\xi}_k^1, \ldots, \hat{\xi}_k^{N_k}\}$ of $\{\xi_k^1, \ldots, \xi_k^{N_k}\}$, $k \in \{0\} \cup \mathscr{K}$, we use the notation $\hat{x}_N$ and $\hat{\nu}_N$. Note that for each $x \in \Theta$ the quantity $g_{k,N_k}(x)$ is a random variable, since it depends on the sample $\{\xi_k^1, \ldots, \xi_k^{N_k}\}$. So, the optimal solution(s) x_N and the optimal value of (9.2) ν_N are random as well. Nevertheless, it is possible to study convergence and statistical properties of the approximating problem (9.2). Note that "convergence" here indicates the asymptotic behavior of (9.2) as the sample sizes N_k, $k \in \{0\} \cup \mathscr{K}$, go to infinity. We will sometimes abuse the notation and write $N \to \infty$ as an abbreviated form of the latter condition. The idea of replacing the expectations with sample averages and solving the resulting formulation has a long history, but it has recently become popular under the name of *Sample Average Approximation* (SAA) approach, which is the topic of Chap. 8.

Much of the existing work in the literature focuses on the case where $\mathscr{K} = \emptyset$ in (SP), i.e., the constraints in the problem are deterministic; see, for instance, [38,91,94] and Chap. 8 for reviews of that literature. In this chapter we review some of the existing works that deal with Monte Carlo simulation-based approximations for problems with stochastic constraints. As we shall see, some new issues arise in that context, and the extension of the methods designed for deterministic constraints not only is involved but in some cases is not even known.

We also review some alternative ways to construct the approximating problem (9.2) that do not rely on standard Monte Carlo sampling. The goal of such methods is to obtain estimates that are more accurate than those obtained with standard Monte Carlo sampling, thus allowing the user to obtain a good solution from (9.2) with fewer samples. Given that the computational effort required to solve (9.2) typically grows fast with the sample sizes, the use of such alternative methods becomes crucial for the solution of certain problems.

We close this introduction by acknowledging that this chapter is largely based on [38], which is a more extensive survey of Monte Carlo methods for stochastic optimization. Our treatment here is more detailed with respect to assessment

---

[1] Throughout this chapter, we will use the terminology "sample [of size N] from the distribution of ξ" to indicate a set of N random variables with the same distribution as ξ.

of solution quality and selection of sample sizes for problems with stochastic constraints. We also have an expanded discussion on the asymptotic properties of Monte Carlo sampling-based approximations of stochastic optimization problems using alternative sampling techniques. Finally, we remark that the purpose of this chapter is to give an overview of two topics in Monte Carlo methods for stochastic optimization—namely, problems with stochastic constraints and variance reduction techniques—with the goal of introducing them to students and researchers. Aiming to be accessible to a wider audience, we outline the main results and provide references for a more comprehensive treatment.

The remainder of this chapter is organized as follows. In Sect. 9.2 we study problems with stochastic constraints. We first discuss the general case of expected-value constraints in Sect. 9.2.1. Next, we examine in more detail two special classes of stochastic constraints, namely probabilistic and stochastic dominance constraints, in Sects. 9.2.2 and 9.2.3, respectively. In our review of Monte Carlo methods for stochastic optimization problems with stochastic constraints, we start with formulations, then discuss some properties of the SAA approach, survey solution methods, review choice of sample sizes when appropriate, and finally discuss how to assess the quality of approximate solutions. Monte Carlo methods are often enhanced by the use of variance reduction techniques; the use of such methods in the context of sampling-based stochastic optimization is reviewed in Sect. 9.3. We review antithetic variates, Latin hypercube sampling, quasi-Monte Carlo, importance sampling and the use of control variates. Finally, we end the chapter with some concluding remarks and future research directions in Sect. 9.4.

9.2 Problems with Stochastic Constraints

We start by considering formulation (SP) and the corresponding approximation (9.2) in which $\mathscr{K}$ is a non-empty finite set. Such problems arise naturally in applications where the decision x must satisfy inequalities but the functions defining the inequalities depend on a random parameter such as demand or future prices. Examples include call center staffing problems where the constraints ensure that the expected number of answered calls, which is estimated by sampling, must be at least a certain percentage of the expected total number of received calls [4], and portfolio optimization problems where the constraints are defined by the conditional value-at-risk (CVaR) of the random returns [51]. CVaR constraints can also be used to provide convex approximations to chance constrained problems, as discussed in [71]. There are, however, some classes of problems where the structure of the stochastic constraints plays a fundamental modeling role, as we shall see below in Sects. 9.2.2–9.2.3. Since specialized methods have been developed for these classes of problems, we will study them separately.

9.2.1 Problems with General Expected-Value Constraints

In this section, we discuss some general results that do not exploit any particular structure of the constraints.

The SAA Approach

We first consider the properties of the SAA approach, which as discussed earlier consists of solving (9.2). One of the main concerns is the convergence of optimal values and limit points of optimal solutions to SAA problems in the general case as the sample sizes tend to infinity. This issue was studied two decades ago, for instance, in [48, 90]. While these asymptotic results are very useful, in practice one is naturally concerned with what happens with finite sample sizes; see [100] for some results for the finite sample case.

The main issue that arises when using sampling approximations within the constraints is that of *feasibility*. For example, consider a single constraint of the form $\mathsf{E}[G_1(x,\xi)] \leq 0$, and let $\bar{x}$ be a point on the boundary of the feasibility set so that $\mathsf{E}[G_1(\bar{x},\xi)] = 0$. Moreover, suppose that $G_1(\bar{x},\xi)$ is normally distributed, and consider the sampling approximation $\frac{1}{N_1}\sum_{j=1}^{N_1} G_1(\bar{x},\xi^j)$. Clearly, by the strong law of large numbers, this quantity converges to zero with probability one (w.p.1). However, no matter how large N_1 is, there is always a 50 % chance that $\frac{1}{N_1}\sum_{j=1}^{N_1} G_1(\bar{x},\xi^j) > 0$—in which case $\bar{x}$ is infeasible to the approximating problem. To circumvent the problem, we can replace the constraint $\mathsf{E}[G_1(x,\xi)] \leq 0$ with

$$\mathsf{E}[G_1(x,\xi)] \leq \varepsilon, \tag{9.3}$$

where $\varepsilon \in \mathbb{R}$. A positive value of ε provides a relaxation of the problem, whereas a negative value tightens the problem. Let U^{ε} denote the set defined by the set of $x \in \Theta$ satisfying (9.3), and let $U^{\varepsilon}_{N_1}$ be the corresponding set defined by the sampling approximation. Then, it is possible show that, under proper assumptions—which include compactness of Θ along with other conditions on $G_1(x,\xi)$ such as Lipschitz continuity—one has that, for any $\varepsilon > 0$,

$$\mathsf{P}\left(U^{-\varepsilon} \subseteq U^0_{N_1} \subseteq U^{\varepsilon}\right) \geq 1 - Me^{-\beta\varepsilon^2 N_1} \tag{9.4}$$

for some constants $M > 0$, $\beta > 0$. In other words, the probability that the feasibility set of the approximating problem is "sandwiched" between $U^{-\varepsilon}$ and U^{ε} goes to one exponentially fast. The rate of convergence, of course, depends on the value of ε. We refer to [100] for details.

Solution Methods and Choice of Sample Sizes

Under the SAA approach, given a realization $\{\hat{\xi}^1, \hat{\xi}^2, \ldots,\}$ of $\{\xi^1, \xi^2, \ldots,\}$, one then solves (9.2) for that realization. Any appropriate deterministic algorithm can be used to solve the corresponding problem, depending on the underlying structure (discrete, convex, etc.). In particular, one can use a penalization technique to bring the constraints into the objective function; see, for instance, [61] for a discussion. We are not aware of other approaches designed specifically for general problems with expected-value constraints in mathematical programming.

In contrast to mathematical programming, in the simulation literature often no assumptions regarding the structure of G_k are made. The approximations are obtained through a "black box" response to a simulated system for a given decision x. When $|\Theta|$ is finite and typically of small size, a growing number of recent papers have been studying the stochastic optimization problems with general expected value constraints. Classical *ranking and selection* procedures (see, e.g., [47] and Chap. 2) have been extended to the cases where there are single and multiple stochastic constraints [2, 7]. Ranking and selection procedures aim to find the best solution among a discrete set of decisions with a pre-specified probability of selecting the correct solution. In the presence of stochastic constraints, this also includes determining the feasibility of the decisions. The *optimal computing budget allocation* approach (see [17] and Chap. 3) has also been extended to consider feasibility of the alternatives, e.g., [44, 57], the latter using large deviations theory. Other ongoing work in this area includes looking at relaxation of independent sampling for some procedures [33, 34], cases when $|\Theta|$ is large [85] or when stochastic constraints have some additional structure (e.g., [80]).

Assessing Solution Quality

As mentioned above, optimal solutions of (9.2) converge to their "true" corresponding values under some conditions. More specifically, the distance between x_N—an optimal solution of (9.2)—and the set S^* of optimal solutions of the "true" problem (SP) goes to zero as $N \to \infty$ w.p.1. Thus, we expect x_N to be a good solution for the original problem. It is necessary, however, to assess the quality of such solution in rigorous terms. One of the classic approaches for assessing solution quality in optimization is to bound the candidate solution's optimality gap. If the bound on the optimality gap is sufficiently small, then the candidate solution is of high quality.

In stochastic optimization, Monte Carlo sampling can be used to obtain (statistical) lower bounds. A whole research area has been developed to deal with that issue, starting with [64, 73] (we refer the reader to [8, 38] for reviews). The basic idea is as follows. Consider problem (SP) with $\mathscr{K} = \emptyset$—i.e., there are no stochastic constraints—and let $\hat{x}$ be a "candidate" solution, whose quality we want to assess (for example, we can take $\hat{x}$ as an optimal solution to an independently generated

SAA problem). The *optimality gap* of $\hat{x}$ is given by $f(\hat{x}) - v^*$. Of course, typically, we cannot compute exactly either $f(\hat{x})$ or v^*. We can, however, estimate the gap by calculating

$$\mathscr{G}_N(\hat{x}) \;:=\; f_N(\hat{x}) - v_N, \tag{9.5}$$

using v_N, the optimal value of the approximating problem (9.2) (we write N instead of N_0 since there is no ambiguity). Note that when viewed as an estimator of optimality gap, $\mathscr{G}_N(\hat{x})$ is biased, i.e., $\mathsf{E}[\mathscr{G}_N(\hat{x})] \geq f(\hat{x}) - v^*$. Such bias follows from the fact that $\mathsf{E}[v_N] \leq v^*$. While there are different ways to calculate the above optimality gap estimator, a basic version uses the same independent and identically distributed (i.i.d.) observations $\xi^1, \xi^2, \ldots, \xi^N$ from the distribution of ξ for *both* terms in (9.5). Then, multiple independent estimators $\mathscr{G}_N^k(\hat{x})$ are generated using $N_{\mathscr{G}}$ "batches" of observations $\xi^{k1}, \xi^{k2}, \ldots, \xi^{kN}$, $k = 1, 2, \ldots, N_{\mathscr{G}}$, and these $\mathscr{G}_N^k(\hat{x})$ are averaged to obtain a point estimator of the optimality gap:

$$\bar{\mathscr{G}}(\hat{x}) \;:=\; \frac{1}{N_{\mathscr{G}}} \sum_{k=1}^{N_{\mathscr{G}}} \mathscr{G}_N^k(\hat{x}). \tag{9.6}$$

The sample variance is calculated as usual by $s_{\mathscr{G}}^2 \;:=\; \frac{1}{N_{\mathscr{G}}-1} \sum_{k=1}^{N_{\mathscr{G}}} \left(\mathscr{G}_N^k(\hat{x}) - \bar{\mathscr{G}}(\hat{x})\right)^2$. An approximate $(1-\alpha)$-level confidence interval estimator on the optimality gap of $\hat{x}$ is then obtained by

$$\left[0, \bar{\mathscr{G}}(\hat{x}) + \frac{z_\alpha s_{\mathscr{G}}}{\sqrt{N_{\mathscr{G}}}}\right], \tag{9.7}$$

where z_α denotes a $1-\alpha$ quantile from a standard normal distribution. The resulting point estimator (9.6) and interval estimator (9.7) are called the *Multiple Replications Procedure* (MRP) estimators. A major advantage of MRP is its wide applicability, as it does not require much structure from the original problem (SP). The main assumption is that $Y(x,\xi)$ have finite second moments for all $x \in \Theta$. The framework to implement MRP is relatively simple and step by step instructions can be found, for instance, in [8].

The approach discussed above for the case $\mathscr{K} = \emptyset$ can be extended to the setting of finite $\mathscr{K} \neq \emptyset$ with some adjustments. Suppose there is a single expected-value constraint; i.e., $K = 1$. In [100], the authors suggest the following approach to derive a lower bound. Notice that the optimal value of $\min_{x\in\Theta} \{f(x) \,|\, \mathsf{E}[G_1(x,\xi)] \leq 0\}$ is bounded from below by the optimal value of $\min_{x\in\Theta} \{f(x) + \lambda \mathsf{E}[G_1(x,\xi)]\}$ for any $\lambda \geq 0$. Since the latter problem does not have stochastic constraints, the methods derived for problems with deterministic constraints can be used to obtain a lower bound for its optimal value. For example, by solving M independent approximations of the form

$$v_N(\lambda) := \min_{x\in\Theta}\left\{\frac{1}{N}\sum_{j=1}^{N} Y(x,\xi^j) + \lambda G_1(x,\xi^j)\right\},$$

one can construct confidence intervals for $\mathsf{E}[v_N(\lambda)]$, which in turn is a lower bound for $\min\{\mathsf{E}[Y(x,\xi)] + \lambda\mathsf{E}[G_1(x,\xi)]\}$. Of course, the quality of the overall bound will depend on the value of λ. One possible choice is to take λ as the optimal dual multiplier of the problem

$$\min_{x\in\Theta}\left\{\frac{1}{N}\sum_{j=1}^{N} Y(x,\xi^j)\,\middle|\,\frac{1}{N}\sum_{j=1}^{N} G_1(x,\xi^j) \le 0\right\},$$

where the sample is independent of the samples drawn in each of the M replications. Note that in this case we use the same sample to estimate both the objective function and the constraint. Note also that when the original problem is not convex (e.g., when the feasibility set is finite) such a lower bound can be loose even if N is large.

As in the case of problems with deterministic constraints, an upper bound for the optimal value of (SP) can be obtained simply by evaluating the objective function at any feasible solution. However, as we saw above, in case of stochastic constraints feasibility cannot be guaranteed since the functions defining the constraints cannot be evaluated exactly. One way to ensure feasibility of a given $x \in \Theta$ with a given confidence is to fix $0 < \beta < 1$ and construct a one-sided $100(1-\beta)\%$ confidence interval for $\mathsf{E}[G_1(x,\xi)]$ as [94]

$$\frac{1}{N}\sum_{j=1}^{N} G_1(x,\xi^j) + z_\beta\sqrt{\frac{s_N^2(x)}{N}},$$

where $s_N^2(x)$ is the sample variance of the sample $\{G_1(x,\xi^1),\ldots,G_1(x,\xi^N)\}$. If the above quantity is less than or equal to zero, then we are $100(1-\beta)\%$ confident that x is feasible and consequently the objective value at x yields upper bound for the optimal value of (SP).

Recall that Karush–Kuhn–Tucker (KKT) conditions provide necessary and sufficient conditions for optimality for a class of problems. In the class of problems we consider, the expectations in (SP) and their (sub)gradients can only be approximated by simulation-based methods. Several researchers have looked at determining if a solution is optimal or near optimal by testing KKT conditions, e.g., by developing a series of hypotheses tests for verification [9], or by using a so-called optimality function that has two parts, one that is related to feasibility of the candidate solution with respect to the stochastic constraints and the other used for testing the more general Fritz–John conditions (similar to KKT conditions) for optimality [86].

9.2.2 Problems with Probabilistic Constraints

This class of problems can be written as

$$\min_{x\in\Theta}\{f(x)\,|\,\mathsf{P}(H_k(x,\xi)\le 0)\ge 1-\alpha_k,\ k\in\mathscr{K}\}. \tag{9.8}$$

Clearly, such a problem falls into the framework of (SP), since we can write $\mathsf{P}(H_k(x,\xi)\le 0)\ge 1-\alpha_k$ as $\mathsf{E}[(1-\alpha_k)-\mathbf{1}\{H_k(x,\xi)\le 0\}]\le 0$. While a probabilistic constraint (also called *chance constraint*) is a special case of expected-value constraint, it is possible to exploit its special properties for its analysis and solution. The probabilistic constraints in (9.8) are used in situations where violation of the constraint inside the probability has a qualitative instead of a quantitative nature, i.e., it matters whether the constraints are violated or not; the amount of violation is less important. Such constraints are often used to model service level or reliability restrictions, such as "demand must be satisfied in at least 95 % of the cases."

It is important to distinguish between the case of separate chance constraints as in (9.8) and that of joint constraints of the form $\mathsf{P}(H_1(x,\xi)\le 0,\ldots,H_K(x,\xi)\le 0)\ge 1-\alpha$. Note that the latter can be represented as $\mathsf{P}(H(x,\xi)\le 0)\ge 1-\alpha$ by defining $H(x,\xi):=\max\{H_1(x,\xi),\ldots,H_K(x,\xi)\}$, although some properties such as differentiability may be lost in such representation. Still, using this representation, we can assume that $K=1$, which is a particular case of (9.8).

Problems with probabilistic constraints have been studied for decades, starting with the work of Charnes and Cooper [16]. As pointed out by Ahmed and Shapiro [1], the two major difficulties with such problems are that (a) evaluating $\mathsf{P}(H_k(x,\xi)\le 0)$ can be difficult (e.g., it may involve multidimensional integrals), and (b) the sets $\{x:\mathsf{P}(H_k(x,\xi)\le 0)\ge 1-\alpha_k\}$ may be nonconvex. This subject is very rich; we refer to [84] for a thorough discussion, and to [1] for a more recent view. For the purposes of this survey, we will review work that uses Monte Carlo methods to approximate the chance constraints.

The SAA Approach

We start by discussing a direct SAA approach for this class for problems. To simplify the discussion, we shall assume that there is only one chance constraint, i.e., $K=1$, and we drop the index k from the notation. Specifically, we consider the problem

$$\min_{x\in\Theta}\left\{\frac{1}{N}\sum_{j=1}^{N}Y(x,\xi^j)\,\middle|\,\frac{1}{N}\sum_{j=1}^{N}\mathbf{1}\{H(x,\xi^j)\le 0\}\ge 1-\gamma\right\}, \tag{9.9}$$

and as before consider the behavior of the optimal value and optimal solutions of (9.9) as function of N. Note that we have replaced the term $1-\alpha$ on the right-hand side with $1-\gamma$, where γ is a parameter. By allowing γ to be different from α we obtain a problem that is either more relaxed or more tight than the original one, which is relevant when considering feasibility issues.

It is important to point out that the analysis discussed in Sect. 9.2.1 cannot be used here since the function $\mathbf{1}\{H(\cdot,\xi^j)\le 0\}$ is not continuous. Nevertheless, some convergence results can be derived in this setting as well. For example, in [78] it is shown that if (a) both Y and H are continuous in x, (b) the set Θ is compact and (c) there exists an optimal solution x^* such that, given any neighborhood of x^*, there exists some x in that neighborhood such that $\mathsf{P}(H(x,\xi)\le 0) > 1-\alpha$, then we have that

- $\nu_N \to \nu^*$ w.p.1 and
- $\mathrm{dist}(x_N, S^*) \to 0$ w.p.1,

where $\mathrm{dist}(x,A)$ is the distance from a point x to a set A, defined as $\inf_{a\in A}\|x-a\|$. Under certain assumptions—such as compactness of Θ and Lipschitz continuity of H with respect to x, or finiteness of Θ—an exponential convergence of the type (9.4) holds in this setting as well [62].

Solution Methods and Choice of Sample Sizes

A natural question that arises is, how to solve the approximating problem (9.9)? Let $\{\hat{\xi}^1,\hat{\xi}^2,\ldots\}$ be a realization of $\{\xi^1,\xi^2,\ldots\}$, and consider problem (9.9) defined for that specific realization. Here we need to distinguish between the two cases $\gamma=0$ and $\gamma>0$. When $\gamma=0$, problem (9.9) can be equivalently written as

$$\min_{x\in\Theta}\left\{\frac{1}{N}\sum_{j=1}^{N} Y(x,\hat{\xi}^j)\,:\, H(x,\hat{\xi}^j)\le 0,\ j=1,\ldots,N\right\}. \tag{9.10}$$

Depending on the structure of H—for example, when H is linear or convex in x—this can be a very tractable problem. The downside, of course, is that the replacement of $\alpha>0$ with $\gamma=0$ yields a more conservative model. Still, [15] improves upon an original result in [13], showing that, given $0<\delta\le 1$, by choosing

$$N \ge \frac{2}{\alpha}\left(\log\frac{1}{\delta}+d_x\right)$$

(recall that d_x is the dimension of x in (SP)), the optimal solution to (9.10) is feasible to the original problem (9.8) with probability at least $1-\delta$, regardless of the distribution of ξ when $H(\cdot,\xi)$ is convex. In [70], it is shown that a better value for N, which grows as $\log(1/\alpha)$ instead of $1/\alpha$, can be obtained under further assumptions on H and the distribution of ξ. Campi and Garatti [14] and Pagnoncelli et al. [79] discuss approaches to remove some of the sampled constraints in order to obtain a less conservative problem.

The situation is rather different when $\gamma > 0$ in (9.9). It is easy to see that convexity is lost because of the presence of the indicator functions even if H has nice properties. Note however that (9.9) is still a chance-constrained problem, where the underlying distribution is the empirical distribution defined by $\{\hat{\xi}^1,\ldots,\hat{\xi}^N\}$. Thus, any method proposed for chance-constrained problems with finite number of scenarios can be used to solve (9.9), e.g., by finding the so-called *p-efficient points* corresponding to the constraints [23], or by strengthening the integer programming formulation for the problem when $H(x,\xi)$ is of the form $\max_i\{\xi_i - h_i(x)\}$ [63], i.e., separating the random component from the function.

A different sampling-based approach for chance constrained problems is proposed in [42] for the situation where, given $x \in \Theta$, the function $H(x,\xi)$ is differentiable at x with probability one. Note that when H represents joint chance constraints such an assumption typically will not hold when ξ has discrete distribution, because of the "kinks" of the max function. When this assumption (and some others) do hold, the constraint $\mathsf{P}(H(x,\xi) \le 0) \ge 1-\alpha$ can be written as a difference of convex (DC) functions. As a result, the original problem can be approximated (by successive linearization of one of the functions in the DC formulation) by a sequence of convex problems, and in the limit one obtains a KKT point for the original problem. Because the functions in the DC formulation cannot be evaluated exactly, a sampling-based approach is proposed to solve each of these convex problems.

Recently, problems with probabilistic constraints have also been studied from the perspective of ranking and selection procedures; see [41].

Assessing Solution Quality

It is possible to derive statistical lower and upper bounds for the optimal value of (9.8), v^*. For a given $x \in \Theta$ consider the estimator

$$\hat{p}(x) := \frac{1}{N}\sum_{j=1}^{N} \mathbf{1}\{H(x,\xi^j) > 0\}$$

of $p(x) := \mathsf{P}(H(x,\xi) > 0)$. As discussed before, any feasible x yields an upper bound for the optimal value of the problem. A given $x \in \Theta$ is feasible for (9.8) if $p(x) \le \alpha$. In our context, computing $p(x)$ is impractical so we would like to use the estimator $\hat{p}(x)$. One way of doing this is to construct a one-sided $100(1-\beta)\%$ confidence interval for $p(x)$ similarly to the idea described for general expected-value constrained problems, i.e., given a realization $\{\hat{\xi}^1,\hat{\xi}^2,\ldots\}$ of $\{\xi^1,\xi^2,\ldots\}$, compute the value of $\hat{p}(x)$ corresponding to that sample and check whether

$$\hat{p}(x) + z_\beta \sqrt{\frac{\hat{p}(x)(1-\hat{p}(x))}{N}} \leq \alpha, \tag{9.11}$$

where we used the fact that $\sum_{j=1}^N \mathbf{1}\{H(x,\xi^j) > 0\}$ has a binomial distribution with parameters N and $p(x)$ and assumed that N is sufficiently large to ensure that the binomial distribution can be well approximated by a normal distribution with mean $Np(x)$ and variance $Np(x)(1-p(x))$. If (9.11) is satisfied, we are $100(1-\beta)\%$ confident that x is feasible.

More efficient techniques have been developed by exploiting the structure of the probabilistic constraints. Following [71], let $B(k;p,n)$ denote the cumulative distribution function of the binomial distribution with parameters n and p. Given $0 < \beta < 1$, define the quantity $U(x) := \sup\{\rho \in [0,1] \,|\, B(N\hat{p}(x);\rho,N) \geq \beta\}$ (note that $U(x)$ is random). Then, it is possible to show that $\mathsf{P}(p(x) < U(x)) \geq 1-\beta$. This suggests the following procedure: given a realization $\{\hat{\xi}^1,\hat{\xi}^2,\ldots\}$ of $\{\xi^1,\xi^2,\ldots\}$, compute the values of $\hat{p}(x)$ and $U(x)$ corresponding to that sample; if $U(x) \leq \alpha$, then we are $100(1-\beta)\%$ confident that x is feasible.

The calculation of lower bounds for the optimal value of (9.8) can in principle follow the method described in Sect. 9.2.1, but again the lack of convexity implies that Lagrangian-based bounds are not useful due to the existence of an optimality gap. The following alternative approach is suggested in [71, 78]. Consider problem (9.8), and suppose the objective function is deterministic (call it f). Consider M independent estimators of the optimal value ν^* of (9.8), defined by the optimal value of (9.9) with M independent samples of size N each. Let ν_N^j be the optimal value of the jth replication, and let $\nu_N^{(1)} \leq \ldots \leq \nu_N^{(M)}$ denote the corresponding order statistics. The procedure calls for using $\nu_N^{(L)}$ as a statistical lower bound for the optimal value of (9.8), as it can be shown that

$$\mathsf{P}\left(\nu_N^{(L)} \leq \nu^*\right) \geq 1 - B(L-1;\theta_N,M),$$

where θ_N is defined as $B(\lceil \gamma N \rceil;\alpha,N)$ and as before $B(k;p,n)$ denotes the cumulative distribution function of the binomial distribution with parameters n and p. This suggests the following procedure: given M independent realizations of $\{\xi^1,\xi^2,\ldots,\xi^N\}$, let $\nu_N^{(j)}$ be the jth smallest value among the corresponding optimal values of the approximating problems. Choose L, M and N in a such a way that $B(L-1;\theta_N,M) \leq \beta$; then, we obtain that $\nu_N^{(L)}$ is a lower bound for ν^* with confidence of at least $100(1-\beta)\%$. Of course, there are many such possible choices for L, M and N, and an appropriate choice must take into account computational considerations; we refer to [78] for a comprehensive discussion.

9.2.3 Problems with Stochastic Dominance Constraints

We turn now to the class of optimization problems with *stochastic dominance* constraints. Stochastic dominance is used to compare the distributions of two random variables (e.g., see [66]), thus providing a way to measure risk. Dentcheva and Ruszczyński [21, 22] first introduced optimization problems with stochastic dominance constraints as an attractive approach for managing risks in an optimization setting. While pursuing expected profits, one avoids high risks by choosing options that are preferable to a random benchmark. Recently, optimization models using stochastic dominance have increasingly been the subject of theoretical considerations and practical applications in areas such as finance, energy, and transportation.

Review of Stochastic Dominance

For completeness, we briefly review the main concepts of stochastic dominance. Given a real-valued random variable Z, we write the cumulative distribution function of Z as $F_1(Z;\eta) := \mathsf{P}(Z \le \eta)$. Furthermore, for $n \ge 2$, define recursively the functions

$$F_n(Z;\eta) := \int_{-\infty}^{\eta} F_{n-1}(Z;t)\,dt,$$

assuming that the first $n-1$ moments of Z are finite. We then say that Z stochastically dominates another random variable W in nth order (denoted $Z \succeq_{(n)} W$) if

$$F_n(Z;\eta) \;\le\; F_n(W;\eta) \quad \text{for all } \eta \in \mathbb{R}.$$

Let $(a)^+ = \max\{a,0\}$. It is useful to note the equivalence given in [74] for $n \ge 2$

$$F_j(Z;\eta) \;=\; \frac{1}{(j-1)!}\mathsf{E}\left[\left((\eta - Z)^+\right)^{j-1}\right], \quad j = 2,\dots,n, \tag{9.12}$$

which implies that the condition that the first $n-1$ moments of Z are finite suffices to ensure that $F_n(Z;\eta) < \infty$ for all η.

The concept of stochastic dominance is also related to utility theory [99], which hypothesizes that for each rational decision maker there exists a utility function u such that the (random) outcome Z is preferred to the (random) outcome W if $\mathsf{E}[u(Z)] \ge \mathsf{E}[u(W)]$. Often the decision maker's exact utility function is not known; in such cases one would say that Z is preferred to W if $\mathsf{E}[u(Z)] \ge \mathsf{E}[u(W)]$ for all u belonging to a certain set of functions. This set of functions is determined by the risk attitude—for example, a *risk-averse* decision maker's utility function is nondecreasing and concave. To see the connection with the notions of stochastic dominance defined above (for $n = 1,2$), let $\mathscr{U}_1$ be the set of all nondecreasing functions $u : \mathbb{R} \mapsto \mathbb{R}$ and let $\mathscr{U}_2$ be the set of all nondecreasing concave functions $u : \mathbb{R} \mapsto \mathbb{R}$. Then, it is well known that

$$Z \succeq_{(n)} W \iff \mathsf{E}[u(Z)] \ge \mathsf{E}[u(W)], \quad \forall u \in \mathscr{U}_n, \; n = 1,2, \tag{9.13}$$

whenever the expectations exist. Stochastic dominance is also closely related to concepts of stochastic ordering; for example, the condition $\mathsf{E}[u(Z)] \geq \mathsf{E}[u(Y)]$ for all $u \in \mathscr{U}_2$ is called stochastic increasing concave order (see, e.g., [89]).

Basic Properties and Reformulations

Using the above concepts, an optimization model with stochastic dominance constraints can then be formulated as follows [21, 22]:

$$\begin{aligned} & \min f(x) \\ & \text{s.t. } H(x,\xi) \succeq_{(n)} W \\ & \quad\; x \in \Theta, \end{aligned} \tag{9.14}$$

where the random variable W represents a benchmark. The cases that have received most attention in the literature are $n = 1$ and $n = 2$. The difficulties with the $n = 1$ case are similar to those arising with probabilistic constraints, notably nonconvexity. The case $n = 2$, on the other hand, is a convex problem so long as $f(\cdot)$ is convex, $H(\cdot,\xi)$ is concave and the set Θ is convex—indeed, the equivalence (9.12) allows us to write the problem with expected value constraints, yielding

$$\begin{aligned} & \min f(x) \\ & \text{s.t. } \mathsf{E}\left[(\eta - H(x,\xi))^+\right] \leq \mathsf{E}\left[(\eta - W)^+\right] \quad \forall \eta \in \mathbb{R} \\ & \quad\; x \in \Theta, \end{aligned} \tag{9.15}$$

which is a convex program.

In principle, problem (9.15) falls into the general framework of Sect. 9.2.1, as it contains expected-value constraints. A major difference, however, is the fact that (9.15) has one constraint for each $\eta \in \mathbb{R}$, i.e., it has uncountably many constraints. This issue is circumvented when the random variable W has finitely many outcomes $w_1,\ldots,w_r$, in which case it suffices to write the constraints in (9.15) only for $\eta = w_j$, $j = 1,\ldots,r$, thus yielding a problem with finitely many expected-value constraints [21] . When the distribution of ξ also has finite support, the expectations in (9.15) can be written as sums, so the problem becomes deterministic. When W has infinitely many outcomes, it is natural to resort to sampling methods; note however that the analysis is more delicate than that described in Sect. 9.2.1 since it involves not only approximating the expectations but also sampling over the set of (uncountably many) constraints. We will discuss this issue further shortly.

It is also useful to consider the case when the function H in (9.14) is vector-valued, which we write as $(H_1(x,\xi),\ldots,H_m(x,\xi))$. This situation occurs in many practical settings—for example, when $H(x,\xi)$ is a linear function of the form

$A(\xi)x$, where $A(\xi)$ indicates a random matrix. Of course, in this case, W is also an m-dimensional random vector. We have then two alternatives. One is to write the problem with m one-dimensional stochastic dominance constraints; i.e., $H_j(x,\xi) \succeq_{(n)} W_j$, $j = 1,\dots,m$. Even though such a formulation provides a direct extension of the unidimensional case seen above, it disregards the dependence among the components H_j of H. Alternatively, we can use concepts of multivariate stochastic dominance. One such concept is that of *convex dominance* introduced by Hu, Homem-de-Mello, and Mehrotra [43]. Given m-dimensional random vectors Z and W and a convex set $\mathscr{C} \subset \mathbb{R}^m$, we say that Z dominates W in nth order linearly with respect to $\mathscr{C}$ if

$$v^\mathsf{T} Z \succeq_{(n)} v^\mathsf{T} W \quad \text{for all } v \in \mathscr{C}. \tag{9.16}$$

Note that the notion of convex dominance includes as a particular case the concept of positive linear dominance (see, e.g., [66]), which corresponds to $\mathscr{C} = \mathbb{R}^m_+$. Under convex dominance, problem (9.15) is then written as

$$\begin{aligned} &\min f(x) \\ &\text{s.t. } \mathsf{E}\left[\left(\eta - v^\mathsf{T} H(x,\xi)\right)^+\right] \le \mathsf{E}\left[\left(\eta - v^\mathsf{T} W\right)^+\right] \quad \forall \eta \in \mathbb{R},\ \forall v \in \mathscr{C} \\ &\qquad x \in \Theta. \end{aligned} \tag{9.17}$$

When W has finitely many outcomes $w_1,\dots,w_r$ and the set $\mathscr{C}$ is polyhedral, the dominance relationship (9.16) for $n = 2$ can be written as [39]

$$\mathsf{E}\left[\left(v_k^\mathsf{T} w_j - v_k^\mathsf{T} H(x,\xi)\right)^+\right] \le \mathsf{E}\left[\left(v_k^\mathsf{T} w_j - v_k^\mathsf{T} W\right)^+\right] \quad j = 1,\dots,r,\ k = 1,\dots,K,$$

where $v_1,\dots,v_K$ are certain vectors in the set $\mathscr{C}$. Thus, in that case the problem still has finitely many expected-value constraints.

The SAA Approach and Solution Methods

An analysis of sampling approximations to problem (9.17) (which includes (9.15) as a particular case) is provided in [43]. The corresponding sample average approximation is written as

$$\begin{aligned} &\min f(x) \\ &\text{s.t. } \frac{1}{N}\sum_{j=1}^{N}\left((v_k^i)^\mathsf{T} W^i - (v_k^i)^\mathsf{T} H(x,\xi^j)\right)^+ \le \frac{1}{N}\sum_{j=1}^{N}\left((v_k^i)^\mathsf{T} W^i - (v_k^i)^\mathsf{T} W^j\right)^+ \\ &\qquad\qquad i = 1,\dots,N,\ k = 1,\dots,K \\ &\qquad x \in \Theta. \end{aligned} \tag{9.18}$$

In the above formulation, $\{(\xi^j,\ W^j)\},\ j=1,\dots,N$ is a sample of size N from $(\xi,\ W)$, and the $\{v_k^i\}$, $i=1,\dots,N$ are certain vectors in $\mathscr{C}$. Typically, the v_k^i vectors are unknown in advance; to remedy the problem, a cutting-surface algorithm based on the ideas in [39] is proposed, which converges in finitely many iterations to an optimal solution of (9.18) [43]. Moreover, the feasibility set U_N of (9.18) satisfies

$$\mathsf{P}\left(U^{-\varepsilon} \subseteq U_N \subseteq U^{\varepsilon}\right) \geq 1 - Me^{-\beta\varepsilon^2 N},$$

where U^{ε} is the feasibility set corresponding to the dominance constraint in (9.15) perturbed by ε on the right-hand side.

Assessing Solution Quality

Statistical lower and upper bounds (e.g., to assess solution quality) can also be derived for the approximation (9.18). Procedures that are based on similar ideas to those discussed in Sect. 9.2.1—more specifically, a Lagrangian-based relaxation for the lower bound, and the objective value of a feasible solution for the upper bound—but with the necessary adaptation to the setting of (9.17) are proposed in [43]. An alternative procedure for the case where H is real-valued (rather than vector-valued) but the set Θ is nonconvex (for example, discrete) is discussed in [101]. The basic idea is to formulate a hypothesis test to check feasibility of a given solution, and then use a multiple-replication procedure similar to that described in Sect. 9.2.1—but modified to discard certain replications—to calculate an optimality gap.

9.3 Variance Reduction Techniques

Monte Carlo sampling-based approximations and algorithms can be significantly improved by reducing the variability of the estimates they generate. Variance reduction techniques have a long history in the simulation and statistics literature. The main goal of such methods is to provide estimators of values associated with a random variable that have better properties than the standard Monte Carlo estimators. Consider for example the quantity $f(x)$ defined in (SP) for a fixed $x \in \Theta$ and its sample average estimator defined in (9.1) (recall that we set $f \equiv g_0$ and $Y \equiv G_0$). When the sample $\{\xi^1,\dots,\xi^N\}$ is independent and identically distributed, its variance is given by

$$\mathrm{Var}[f_N(x)] = \frac{\mathrm{Var}[Y(x,\xi)]}{N}.$$

Although $\mathrm{Var}[Y(x,\xi)]$ is typically unknown, it can be estimated by a sample variance as follows:

$$S_N^2(x) := \frac{\sum_{i=1}^N [Y(x,\xi^i) - f_N(x)]^2}{N-1}.$$

The above estimator is unbiased, i.e., $\mathsf{E}[S_N^2(x)] = \mathrm{Var}[Y(x,\xi)]$.

Of course, it is desirable to have estimators with as small variance as possible. While this is the case in the context of pointwise estimation, it is even more so in the case of optimization, since poor estimates of the objective (or of its derivatives) may lead to slow convergence of an algorithm. Clearly, if $\mathrm{Var}[Y(x,\xi)]$ is large then $\mathrm{Var}[f_N(x)]$ will be large as well, unless the sample size can be chosen to counterbalance that effect. In many cases, however, choosing a large sample size is not practical, as the evaluation of $Y(x,\hat{\xi})$ for a given $\hat{\xi}$ can be costly.

The goal of variance reduction techniques is to derive estimators $f_N(x)$, v_N, etc. with smaller variance than those obtained with standard Monte Carlo. While in some cases this is accomplished by exploiting the structure of the problem, some general techniques do exist. We discuss next some variance reduction methods, mostly in the context of stochastic optimization problems though we do provide some background for the case of pointwise estimators (i.e., for a fixed $x \in \Theta$).

In our presentation below, we revert to the case where there are no stochastic constraints ($\mathscr{K} = \emptyset$ in (SP)). Also, for some of the methods discussed below we assume that the vector ξ has independent components. When such an assumption does not hold one can often write the components of ξ as functions of some independent uniform random variables; see, for instance, [10].

9.3.1 Antithetic Variates

Antithetic Variates (AV) aim to reduce variance by inducing correlations. Suppose N is even and components of ξ are independent. Instead of using N i.i.d. random variates, the AV estimator aims to use $N/2$ negatively correlated pairs $(\underline{\xi}^j, \overline{\xi}^j)$, $j = 1,\dots,N/2$. This is typically achieved by generating $N/2$ i.i.d. random vectors $U^1,\dots,U^{N/2}$ of dimension d_ξ distributed uniformly over $[0,1]^{d_\xi}$, and their corresponding antithetic variates from the opposite end of the distribution (taken component-wise), $1-U^1,\dots,1-U^{N/2}$, which are also i.i.d. distributed uniformly over $[0,1]^{d_\xi}$; note that $(U^j, 1-U^j)$, $j = 1,\dots,N/2$ are negatively correlated. Then, regular variate generation techniques are utilized to generate the pairs $(\underline{\xi}^j, \overline{\xi}^j)$ using $(U^j, 1-U^j)$. For the case with dependent components of ξ, we refer to [87].

In contrast to the standard Monte Carlo estimator of $\mathsf{E}[Y(x,\xi)]$ with variance $N^{-1}\sigma^2(x) := N^{-1}\mathrm{Var}[Y(x,\xi)]$, the antithetic variates estimator

$$f_{N,\mathrm{AV}}(x) = \frac{1}{N/2}\sum_{j=1}^{N/2} \frac{Y(x,\underline{\xi}^j) + Y(x,\overline{\xi}^j)}{2}$$

has variance $N^{-1}\sigma^2(x) + N^{-1}\mathrm{Cov}(Y(x,\underline{\xi}^j), Y(x,\overline{\xi}^j))$. Therefore, as long as

$$\mathrm{Cov}(Y(x,\underline{\xi}^j), Y(x,\overline{\xi}^j)) < 0,$$

the antithetic estimator has a smaller variance than its crude Monte Carlo counterpart and they both produce unbiased estimators of the expectation.

The degree of variance reduction depends on the extent to which the negative correlation between the pair $(\underline{\xi}^j, \overline{\xi}^j)$ is preserved after $Y(x,\cdot)$ is applied to this pair. In [36], it is argued that the negative correlation can be preserved for a class of two-stage stochastic linear programs with stochasticity only on the right-hand side, and computational results are presented. This is because $Y(x,\cdot)$ is a monotone function of the right-hand side of the second stage problem for this class of problems. In general, if $Y(x,\cdot)$ is a bounded and monotone function in each of its arguments that is not constant in the interior of its domain, variance reduction can be achieved using AV [58]. The work in [36] has been expanded by also considering v_N, the optimized sample means, where [28] analytically shows the extent of variance reduction using AV on a newsvendor problem and present computational results for two-stage stochastic linear programs. The computations in [49] indicate that when the monotonicity in the objective function is lost, AV can increase (e.g., double) the variance. However, when AV is effective, combination of AV with other variance reduction techniques such as randomized quasi-Monte Carlo is found to be very effective. The use of AV in the context of assessing solution quality for two-stage stochastic linear programs is investigated in [97]. These papers indicate that antithetic variates can result in modest variance reduction with minimal computational effort for a class of stochastic optimization problems.

9.3.2 Latin Hypercube Sampling

A fairly general way of obtaining estimators with smaller variance is based on the concept of stratified sampling (see, for instance, [27] and references therein). Generally speaking, the idea is to partition the sample space and fix the number of samples on each component of the partition, which should be proportional to the probability of that component. This way we ensure that the number of sampled points on each region will be approximately equal to the *expected* number of points to fall in that region. It is intuitive that such a procedure yields smaller variance than crude Monte Carlo; for proofs see [27]. Notice, however, that although theoretically appealing, implementing such a procedure is far from trivial, since the difficulty is to determine the partition as well as to compute the corresponding probabilities.

There are many variants of this basic method; a classical one is the so-called *Latin Hypercube Sampling* (LHS) approach, introduced in [65]. The LHS method operates as follows. Suppose we want to draw N samples from a random vector ξ with d_ξ independent components, each of which has a Uniform(0,1) distribution. The algorithm consists repeating the two steps below for each dimension $j = 1,\dots,d_\xi$:

1. Generate

$$X^1 \sim U\left(0,\frac{1}{N}\right),\ X^2 \sim U\left(\frac{1}{N},\frac{2}{N}\right),\dots,X^N \sim U\left(\frac{N-1}{N},1\right);$$

2. Let $\xi_j^i := X^{\pi(i)}$, where π is a random permutation of $1,\ldots,N$.

In [65], it is shown that each sample ξ_j^i (viewed as a random variable) has *the same distribution* as ξ_j, which in turn implies the estimators generated by the LHS method are unbiased. In case of arbitrary distributions, the above procedure is easily modified by drawing the sample as before and applying an inversion method.

Under some conditions, the LHS method does indeed reduce the variance compared to standard Monte Carlo [65]. Asymptotically (i.e., as the sample size N goes to infinity), LHS is never worse than standard Monte Carlo [76, 96], even without the assumptions of [65]. More specifically, $V_{LHS} \le N/(N-1)V_{MC}$, where V_{LHS} and V_{MC} are respectively the variances under LHS and standard Monte Carlo. Thanks to such properties (and to the simplicity of the method), the LHS technique has been widely used in simulation, showing up even in off-the-shelf spreadsheet-based software.

One drawback of using the LHS method is that, by construction, the generated samples are *not* independent—indeed, variance is reduced precisely because of the correlation introduced by the method. Lack of independence implies that classical statistical results such as the Central Limit Theorem (CLT) do not apply directly to the resulting estimator; consequently, confidence intervals cannot be built in the standard way. It is worthwhile mentioning that [75] proves a version of the CLT for LHS estimators, which is useful from the perspective of rates of convergence but not necessarily for the construction of confidence intervals as the result involves some quantities that are difficult to estimate. In practice, in order to derive confidence intervals one typically performs multiple independent replications (for example, m replications, each with a sample of size N) and applies the classical theory to the data set consisting of the LHS average estimator from each replication.

The use of LHS in stochastic optimization, while not as widespread as in simulation, has demonstrated the benefits of that approach, as reported in papers such as [5, 40, 60, 93, 101]. The use of LHS for assessing solution quality in stochastic programming was examined theoretically and empirically in [25, 97]. The effect of using LHS in the context of the newsvendor model was studied in detail in [28], where some general conclusions are drawn about the effectiveness of that approach for stochastic optimization.

It is possible to study convergence of estimators of optimal values and optimal solutions of (9.2) with sampling that is not necessarily i.i.d. under a general theoretical framework . Throughout the discussion below, assume that Θ is compact. Suppose that for each $x \in \Theta$ we have that

$$f_N(x) \to f(x) \quad \text{w.p.1.} \tag{9.19}$$

Suppose also that the function $Y(\cdot,\xi)$ is Lipschitz. Then, we have that

(i) $f_N(x) \to f(x)$ uniformly on Θ w.p.1;
(ii) $v_N \to v^*$ w.p.1;
(iii) $\text{dist}(x_N, S^*) \to 0$ w.p.1.

Such a result is well known; see, e.g., [88, pp. 67–70]. Under further assumptions on $Y(\cdot,\xi)$ and Θ (for example, $Y(\cdot,\xi)$ is piecewise linear and the support of the distribution of ξ is finite, or the feasible set Θ is finite), the conclusion (iii) can be replaced with $x_N \in S^*$ w.p.1 for N large enough. For reference, we shall call that condition the "piecewise-linear/finite-set assumption" for short.

The study of rates of convergence of estimators of optimal values and optimal solutions in this context is introduced by Homem-de-Mello [37]. In what follows we summarize the results from that paper, to where we refer the reader for details. Suppose that for each $x \in \Theta$, there exist a number $C_x > 0$ and a function $\gamma_x(\cdot)$ such that $\gamma_x(0) = 0$, $\gamma_x(z) > 0$ if $z > 0$, and

$$\mathsf{P}\left(|f_N(x) - f(x)| \geq \delta\right) \leq C_x\, e^{-N\gamma_x(\delta)} \qquad \text{for all } N \geq 1 \text{ and all } \delta > 0, \tag{9.20}$$

i.e., the probability that the deviation between $f_N(x)$ and $f(x)$ is bigger than δ goes to zero exponentially fast with N (notice that (9.20) implies that $f_N(x)$ converges in probability to $f(x)$). Suppose also that the function $Y(\cdot,\xi)$ is Lipschitz. Then, it can be shown that given $\varepsilon > 0$, there exist constants $K > 0$ and $\alpha > 0$ such that

$$\mathsf{P}\left(\mathrm{dist}(x_N, S^*) \geq \varepsilon\right) \leq Ke^{-\alpha N} \qquad \text{for all } N \geq 1. \tag{9.21}$$

In other words, if an exponential rate of convergence holds for the pointwise estimators $f_N(x)$, then it will hold for the estimators of optimal solutions. The result in (9.21) can be strengthened in case under the piecewise-linear/finite-set assumption, as it can be shown that

$$\mathsf{P}\left(x_N \notin S^*\right) \leq Ke^{-\alpha N} \qquad \text{for all } N \geq 1.$$

Results for convergence of estimators of optimal value can be derived under the piecewise-linear/finite-set assumption. A simplified description of the results from [37] goes as follows. Suppose that the "true" problem (SP) has a unique optimal solution x^*, and that

$$\frac{f_N(x^*) - f(x^*)}{\sigma_N(x^*)} \xrightarrow{d} \mathcal{N}(0,1). \tag{9.22}$$

where $\sigma_N^2(x) := \mathrm{Var}[f_N(x)]$ and $\xrightarrow{d}$ denotes convergence in distribution. Then,

$$\frac{\nu_N - \nu^*}{\sigma_N(x^*)} \xrightarrow{d} \mathcal{N}(0,1), \tag{9.23}$$

where $\mathcal{N}(\mu,\sigma^2)$ denotes the normal distribution with mean μ and variance σ^2. Similarly to the result for estimators of optimal solution, the above results states that if a CLT holds for the pointwise estimators $f_N(x)$, then it will hold for the estimators of optimal values.

The above framework allows us to study convergence of estimators of optimal values and optimal solutions of (9.2) under LHS. Indeed, it can be shown that condition (9.19) holds under finiteness of second moment of $Y(x,\xi)$; condition (9.20) holds under certain monotonicity assumptions on $Y(x,\cdot)$; and condition (9.22) holds if the random variable $Y(x^*,\xi)$ is bounded, and the function $Y(x^*,\cdot)$ is not additive.

9.3.3 Quasi-Monte Carlo

Quasi-Monte Carlo (QMC) methods have a long history as tools to approximate integrals, and as such have been widely used in many areas. Describing all the nuances and the properties of such methods would fall out of the scope of this chapter; thus, we only provide a brief discussion. We refer to [24, 58, 72] for comprehensive treatments of QMC concepts. To set the stage, consider again the function $Y(x,\xi)$ and assume that ξ is a random vector with independent components, each with uniform distribution on $[0,1]^{d_\xi}$. Consider the problem of estimating $f(x) := \mathsf{E}[Y(x,\xi)]$ for a fixed x.

The basic idea of QMC is to calculate a sample average estimate as in the standard Monte Carlo, but instead of drawing a random sample from the uniform distribution on $[0,1]^{d_\xi}$, a certain set of points $\hat{\xi}^1,\ldots,\hat{\xi}^N$ on space $[0,1]^{d_\xi}$ is carefully chosen. The deterministic estimate

$$f_{N,\mathrm{QMC}}(x) := \frac{1}{N}\sum_{i=1}^{N} Y(x,\hat{\xi}^i) \tag{9.24}$$

is constructed. A key result is the so-called Koksma–Hlawka inequality, which, roughly speaking, states that the quality of the approximation given by $f_{N,\mathrm{QMC}}(x)$ depends on the quality of the chosen points (measured by the difference between the corresponding empirical measure and the uniform distribution, which is quantified by the so-called *star-discrepancy*) as well as on the nature of the function $Y(x,\cdot)$ (measured by its total variation). A great deal of the research on QMC methods aims at determining ways to construct *low-discrepancy sequences*, i.e., sequences of points $\hat{\xi}^1,\hat{\xi}^2,\ldots$ for which the star-discrepancy is small for all N. Particular types of sequences that have proven valuable are the so-called *digital nets* and also *lattice rules*. We briefly describe them next.

Let $b \geq 2$ be an arbitrary integer, called the base. An *elementary interval in base* b (in dimension s) is a subinterval E of $[0,1]^s$ of the form

$$E = \prod_{j=1}^{s}\left[\frac{a_j}{b^{d_j}}, \frac{a_j+1}{b^{d_j}}\right]$$

for nonnegative integers $\{a_j\}$ and $\{d_j\}$ such that $a_j < b^{d_j}$ for all j. The volume of E is $b^{-\sum_j d_j}$. Next, let t and m be nonnegative integers such that $t \leq m$. A finite sequence of b^m points is a (t,m,s)-*net* in base b if every elementary interval in base b of volume b^{t-m} contains exactly b^t points of the sequence. A sequence of points $u^1, u^2, \ldots$ is a (t,s)-*sequence* in base b if, for all integers $k \geq 0$ and $m > t$, the set of points consisting of the u^n such that $kb^m \leq n < (k+1)b^m$ is a (t,m,s)-net in base b. Such (t,m,s)-nets can be shown to yield approximations such that the integration error is of order $(\log N)^s/N$.

Many of the lattice rules in $[0,1]^s$ used in practice are generated as follows: $\xi_i = (i/N)v_0 \mod 1$ for $i = 0, \ldots, N-1$ where v_0 is the generating vector, whose components are integers between 0 and $N-1$. There are some general principles to choose the generating vector. A new approach introduced in [56], whereby the lattice rules are tailored to the function being integrated, leads to convergence rates that sometimes approach the theoretical optimal rate of $1/N^2$, provided the function is sufficiently smooth.

Despite the theoretical attractiveness of QMC methods with respect to error rates, an issue that arises when using such techniques in practice is the fact that the bounds provided by the Koksma–Hlawka inequality involve difficult-to-compute quantities such as the total variation of $Y(x,\cdot)$. In other words, they yield qualitative rather than quantitative results; hence, obtaining a good estimate of the error may be difficult. A common way to overcome this issue is to incorporate some randomness into the choice of QMC points. By doing so, errors can be estimated using standard methods, e.g., via multiple independent replications. Some choices for randomizing the points of the QMC sequence include the so-called Cranley–Patterson procedure—where every number in the sequence is perturbed (modulo 1) by a single number generated from a Uniform(0,1) distribution—and scrambling the digits of each number in the sequence in a particular way; we refer to [54] for details.

Even with the randomization, one should be cautious when using QMC methods since oftentimes these techniques "backfire" in problems with moderate or large dimensionality if not used properly. This can be explained by the fact that the error rates depend on the dimensionality—for example, the $(\log N)^{d_\xi}/N$ rate seen above. In such cases, one may try to determine the *effective dimension* of the problem, i.e., the number of variables that account for most of the variability, and then apply a QMC strategy only for those variables. Such notion can be made precise, see for instance [76, 77]. Moreover, the theoretical rates derived for QMC often rely on smoothness of the integrand, which may not always be present. Still, when used properly such techniques can be highly valuable, resulting in estimators that are orders of magnitude better than standard Monte Carlo.

Similarly to what happens with standard random numbers, generating a good QMC sequence may not be simple. Some sequences are easy to generate but the more powerful ones require sophisticated methods. Fortunately, public software is available—here we mention [29,53,55,59], where libraries can be found.

A few papers study the application of QMC methods to stochastic optimization problems. In [46], empirical results are provided for the use of Hammersley sequences (one form of QMC). In [26], the aforementioned concept of effective dimension is used to develop an algorithm for two-stage stochastic programs that attempt to determine the "important variables" in the problem based on dual information. The remaining variables are "padded" with either Monte Carlo or Latin Hypercube sampling; a rigorous analysis of such strategy can be found in [25]. As in the case of LHS discussed in Sect. 9.3.2, theoretical results on convergence are harder to obtain than in the Monte Carlo case due to the loss of the i.i.d. property. In [81, 82], the authors show that, under mild assumptions, the estimator function f_N constructed with QMC points *epiconverges* to the true function f, which guarantees convergence of optimal values and optimal solutions under appropriate further conditions. In [49], those results are applied to the case where the QMC sequence is randomized with the Cranley–Patterson procedure. The numerical results in those papers also suggest considerable gains in terms of rates of convergence when using QMC methods. Homem-de-Mello [37] studies the rates of convergence of estimators of optimal values under randomized QMC, using the general framework for non-i.i.d. sampling described in Sect. 9.3.2. Results are provided for a specific QMC method for which a CLT exists, so that (9.22) holds. As it turns out, in that case it is possible to show that the estimators of optimal values converge at the rate $[(\log N)^{d_\xi - 1}/N^3]^{1/2}$. One particular difficulty that arises when using QMC methods for stochastic programming problems lies in the fact that such problems do not have smooth integrands. Recent work sheds new light on that issue by showing that, under certain assumptions, all terms of the so-called ANOVA decomposition of such functions except the residual term are actually infinitely differentiable [35].

9.3.4 Importance Sampling

Importance Sampling (IS) aims to reduce variance by "shifting" the samples to the most important regions. Suppose again the aim is to estimate $\mathsf{E}[Y(x,\xi)]$ for a given $x \in \Theta$ and suppose ξ has density p.[2] Then, $f(x) = \mathsf{E}[Y(x,\xi)] = \int_\Xi Y(x,\xi)p(\xi)d\xi$. Now consider another density q over Ξ with the property that $p(E) = 0$ for every set E for which $q(E) = 0$, and rewrite $\mathsf{E}[Y(x,\xi)] = \int_\Xi Y(x,\xi)\mathscr{L}(\xi)q(\xi)d\xi$. Here, $\mathscr{L}(\xi) = \frac{p(\xi)}{q(\xi)}$ is the *likelihood ratio*, which we assume is well-defined (for this, we may set $\mathscr{L}$ to zero whenever both p and q are zero). Instead of the usual Monte Carlo estimator $f_N(x) = \frac{1}{N}\sum_{j=1}^N Y(x,\xi^j)$ that uses an i.i.d. sample $\xi^1, \xi^2, \ldots, \xi^N$ from the density p, the importance sampling estimator

$$f_{N,\mathrm{IS}}(x) = \frac{1}{N}\sum_{j=1}^{N} Y(x,\tilde{\xi}^j)\mathscr{L}(\tilde{\xi}^j)$$

[2]Such an assumption is made just for simplicity of notation; the general case can be dealt with via Radon–Nikodym derivatives, see for instance [3].

uses an i.i.d. sample $\tilde{\xi}^1, \tilde{\xi}^2, \ldots, \tilde{\xi}^N$ from the new density q. Note that both estimators are unbiased. However, not all choices of density q will lead to a reduction in variance. In fact, if q is not chosen appropriately, it might lead to an increase in variance. Therefore, finding an appropriate density q is critical to IS and much of the research in this area is directed to this issue.

To understand how to find q, consider the following facts. First, $\mathsf{E}[Y(x,\tilde{\xi})\mathscr{L}(\tilde{\xi})] = \mathsf{E}[Y(x,\xi)]$ and $\mathsf{E}[Y^2(x,\tilde{\xi})\mathscr{L}^2(\tilde{\xi})] = \mathsf{E}[Y^2(x,\xi)\mathscr{L}(\xi)]$. Therefore, the variance of the IS estimator is given by

$$\mathrm{Var}[f_{N,\mathrm{IS}}(x)] = \frac{1}{N}\left[\mathsf{E}[Y^2(x,\xi)\mathscr{L}(\xi)] - (\mathsf{E}[Y(x,\xi)])^2\right], \tag{9.25}$$

and the variance is reduced if and only if $\mathsf{E}[Y^2(x,\xi)\mathscr{L}(\xi)] < \mathsf{E}[Y^2(x,\xi)]$. If we want to minimize the variance—i.e., have zero variance—in (9.25), we need to have q such that $\mathsf{E}[Y^2(x,\xi)\mathscr{L}(\xi)] = (\mathsf{E}[Y(x,\xi)])^2$. When $Y(x,\cdot)$ is nonnegative, this results in the optimal zero-variance density

$$q^*(\xi) = \frac{p(\xi)Y(x,\xi)}{\mathsf{E}[Y(x,\xi)]}.$$

This optimal density, however, requires the knowledge of an unknown quantity, $\mathsf{E}[Y(x,\xi)]$, and is therefore unattainable. Nevertheless, this gives the intuition that q should be approximately proportional to $p(\xi)Y(x,\xi)$ in order to achieve variance reduction even though the proportionality constant may not be known. Here we see the difficulty of using IS in the optimization context: it is clear that the above expression for $q^*(\xi)$ cannot hold for all $x \in \Theta$, so the same IS density may be good for some x but bad for others.

In the context of pointwise estimation, there is significant work on how to determine the IS distribution in the literature as this is critical to the success of IS. We briefly mention some of that work without getting into much detail. In exponential titling, the IS distribution is restricted to belong to an exponential family of distributions [31,95]. In [88], a method is discussed which parameterizes the IS distribution and solves a stochastic optimization problem to determine the parameters of this distribution in order to minimize variance. This optimization can be done via a sampling-based method and perturbation analysis; see, e.g., [30]. Other approaches in the literature to obtain the IS distribution include nonparametric methods [67,102] and minimization of the Kullback–Leibler distance to the optimal distribution [20]. Importance sampling techniques have been very successful in the estimation of rare event probabilities, especially due to the connection with large deviations theory; see, for instance, [32] and also [12] for a recent survey of that area.

Importance sampling is one of the earliest variance reduction methods applied to two- and multi-stage stochastic linear programs [18, 19, 45]. In [19], the authors suggest using an additive approximation of $Y(x,\xi)$ given by

$Y(x,\overline{\xi}) + \sum_{i=1}^{d_\xi} \Delta_i Y(x,\xi)$, where $\overline{\xi}$ is a base case and each $\Delta_i Y(x,\xi)$ gives the marginal effect of the ith element of ξ. They suggest finding the marginal effects by $\Delta_i Y(x,\xi) = Y(x,\overline{\xi}_i) - Y(x,\overline{\xi})$, where $\overline{\xi}_i$ agrees with the base case $\overline{\xi}$ in all elements except for the ith element, which agrees with ξ. This results in solving $d_\xi + 1$ linear programs to determine q, one for each marginal and one for the base case. The authors argue that in the context of power generation, IS can capture rare events such as power supply down and high demands better than crude Monte Carlo, which are supported by the computations in [45]. Higle [36] applies this method to a wider range of problems, and the computations indicate that the method can be effective in reducing variance of $\mathsf{E}[Y(x,\xi)]$ estimates for some problems, but it can also actually increase variance for some other problems.

As mentioned earlier, a major difficulty that arises when employing IS methods in the context of iterative algorithms for more general stochastic optimization problems is that changes in x throughout an optimization routine can result in different IS distributions for different x. The idea of using *trust regions* on which the same IS distribution can be used, but the results are inconclusive [92]. In [6], IS techniques are employed for a chance-constrained problem in which the violation probabilities α_k in (9.8) are very small, so the methods for choice of sample sizes discussed in Sect. 9.2.2 are not practical. They show that, for the application they study, it is possible to construct IS distributions that are good for all x in an *outer approximation* of the feasibility set of (9.8)—as discussed in that paper, trying to find an IS distribution that is good for all $x \in \Theta$ is not only unnecessary but also counterproductive. Recent work applying IS within the stochastic dual dynamic programming algorithm for multi-stage stochastic programs with nested mean-CVaR objectives shows promising results [50]. As only relatively few scenarios under random sampling contribute to estimating CVaR, an IS scheme provides better estimators by concentrating the sampling to the important regions. A thorough study of sequential IS methods in the context of optimization remains an open research area [11].

9.3.5 *Control Variates*

Like antithetic variates, Control Variates (CV) aim to reduce variance by inducing correlations. In the case of control variates, though, this is achieved by introducing a *control variable* that can either be negatively or positively correlated with $Y(x,\xi)$. Let C denote the control variable and λ be a scalar. Suppose $\mathsf{E}[C] = 0$. Note that if the mean of the control variable is known, which is often the case, then it can be subtracted from it to obtain a variable with zero mean. The control variate estimator of $\mathsf{E}[Y(x,\xi)]$ is given by

$$f_{N,\mathrm{CV}}(x) = \frac{1}{N} \sum_{j=1}^{N} \left(Y(x,\xi^j) + \lambda C^j \right).$$

For any given λ, $f_{N,\text{CV}}(x)$ is an unbiased estimator with variance

$$\frac{1}{N}\left(\sigma^2(x)+\lambda^2\text{Var}[C]+2\lambda\text{Cov}(Y(x,\xi),C)\right). \tag{9.26}$$

We can minimize this variance by setting λ to $\lambda^* = \frac{-\text{Cov}(Y(x,\xi),C)}{\text{Var}[C]}$. By substituting λ^* back in (9.26), we see that as long as C and $Y(x,\xi)$ are correlated, the variance of the CV estimator

$$\text{Var}[f_{N,\text{CV}}(x),\lambda^*] = \frac{1}{N}\left(\sigma^2(x) - \frac{\text{Cov}^2(Y(x,\xi),C)}{\text{Var}[C]}\right)$$

is less than the variance of the crude MC estimator, $N^{-1}\sigma^2(x)$. Notice that even though $\text{Var}[C]$ may be known, $\text{Cov}(Y(x,\xi),\mathscr{C})$ is unknown but can be estimated, resulting in an estimator of λ^*. Unfortunately, when an estimator of λ^* is used, $f_{N,\text{CV}}(x)$ is no longer unbiased. However, this can still yield significant variance reduction and the resulting CV estimator obeys a CLT of the form

$$\sqrt{N}\left(f_{N,\text{CV}}(x) - \mathsf{E}[Y(x,\xi)]\right) \xrightarrow{d} \mathscr{N}(0,\text{Var}[f_{N,\text{CV}}(x),\lambda^*]),$$

due to a result of [69].

A number of control variates to estimate $\mathsf{E}[Y(x,\xi)]$ that are cheap to compute and are quite effective across a number of test problems are presented in [36]. In [92], linear control variates are used to obtain more accurate estimators of the gradient, Hessian, and the value of $\mathsf{E}[Y(x,\xi)]$ at a current solution point x in each iteration of a Monte Carlo sampling-based method to solve two-stage stochastic linear programs. Similarly, in [83], a subgradient-inequality-based linear control variate within stratified sampling is used to estimate $\mathsf{E}[Y(x,\xi)]$ at each iteration's solution x of an algorithm for a class of two-stage stochastic convex programs. Both papers show that control variates significantly reduce variance (up to more than 1,000 times in some cases) and allow these Monte Carlo sampling-based solution procedures to be numerically more viable. Control variates have also been successfully used in ranking and selection procedures [68, 98].

9.4 Conclusions

In this chapter, we have surveyed the current landscape of Monte Carlo simulation-based methods for stochastic optimization problems with stochastic constraints and the use of variance reduction techniques in Monte Carlo approximations of stochastic optimization problems. For stochastically constrained problems, we reviewed asymptotic and finite sample size properties, and discussed solution methods and assessment of solution quality. Variance reduction techniques can significantly improve the quality of the Monte Carlo simulation-based estimators. We reviewed their use in the stochastic optimization setting, discussing asymptotic properties and their use within Monte Carlo simulation-based solution methods. In particular, we discussed the use of antithetic variates, Latin hypercube sampling, quasi-Monte Carlo, importance sampling, and control variates. One topic we skipped in VRT is the use of common random numbers (CRN), which aims to reduce variance by inducing positive correlations via correlated observations and typically applies when we are comparing two or more alternative systems (see, e.g., [52]). In the context of stochastic optimization, CRN techniques can be efficiently used, for instance, in assessing solution quality (see, e.g., [64]) or when ranking and selecting a system among a discrete set of alternatives [47].

We note that both topics covered in this chapter are active areas of research and new work is frequently appearing to address unresolved issues. Being able to efficiently solve problems with stochastic constraints brings us closer to solving many real-world problems with multiple performance measures that involve randomness. Variance reduction techniques help make Monte Carlo sampling-based methods considerably more reliable and efficient; therefore, they are very important for the practical success of Monte Carlo methods in stochastic optimization. These two important topics are being tackled by the mathematical programming and simulation communities. Broadly speaking, in mathematical programming one typically exploits certain structures found in the problem, whereas the simulation literature typically works with problems that do not have much structure and can only be evaluated with expensive simulations. Both approaches have their merits and can be preferable in certain situations. That said, the line between the two has been becoming blurry, as this book illustrates. One area of future research is to further explore and exploit the connections between the simulation and the more structured mathematical programming approaches. For the increased success and applicability of these methods, data-driven methods need to be further developed along with software linking real-world data to stochastic optimization models to Monte Carlo simulation-based methods.

Acknowledgements This work is supported in part by the National Science Foundation under Grant CMMI-1345626, and by Conicyt-Chile under grant Fondecyt 1120244.

References

1. S. Ahmed and A. Shapiro. Solving chance-constrained stochastic programs via sampling and integer programming. In *Tutorials in Operations Research*, pages 261–269. INFORMS, 2008.
2. S. Andradóttir and S.-H. Kim. Fully sequential procedures for comparing constrained systems via simulation. *Naval Research Logistics*, 57(5):403–421, 2010.
3. S. Asmussen and P. Glynn. *Stochastic Simulation*. Springer, New York, 2007.
4. J. Atlason, M. Epelman, and G. Henderson. Call center staffing with simulation and cutting plane methods. *Annals of Operations Research*, 127:333–358, 2004.
5. T. G. Bailey, P. Jensen, and D. Morton. Response surface analysis of two-stage stochastic linear programming with recourse. *Naval Research Logistics*, 46:753–778, 1999.

6. J. Barrera, T. Homem-de-Mello, E. Moreno, B. K. Pagnoncelli, and G. Canessa. Chance-constrained problems and rare events: an importance sampling approach. Preprint available at Optimization Online (http://www.optimization-online.org), 2014.
7. D. Batur and S.-H. Kim. Finding feasible systems in the presence of constraints on multiple performance measures. *ACM Transactions on Modeling and Computer Simulation*, 20, 2010. Article No. 13.
8. G. Bayraksan and D. P. Morton. Assessing solution quality in stochastic programs via sampling. In *Tutorials in Operations Research*, pages 102–122. INFORMS, 2009.
9. B. Bettonvil, E. del Castillo, and J. P. Kleijnen. Statistical testing of optimality conditions in multiresponse simulation-based optimization. *European Journal of Operational Research*, 199(2):448–458, 2009.
10. B. Biller and S. Ghosh. Multivariate input processes. In S. G. Henderson and B. L. Nelson, editors, *Simulation*, volume 13 of *Handbooks in Operations Research and Management Science*, chapter 5. Elsevier Science Publishers B.V., Amsterdam, Netherlands, 2006.
11. J. R. Birge. *Particle Methods for Data-Driven Simulation and Optimization*, volume 8 of *INFORMS TutORials in Operations Research*, chapter 5, pages 92–102. INFORMS, Hannover, MD, 2012.
12. J. Blanchet and H. Lam. State-dependent importance sampling for rare-event simulation: An overview and recent advances. *Surveys in Operations Research and Management Science*, 17 (1):38–59, 2012.
13. G. Calafiore and M. C. Campi. Uncertain convex programs: randomized solutions and confidence levels. *Mathematical Programming*, 102(1, Ser. A):25–46, 2005.
14. M. C. Campi and S. Garatti. A sampling-and-discarding approach to chance-constrained optimization: feasibility and optimality. *Journal of Optimization Theory and Applications*, 148(2):257–280, 2011.
15. M. C. Campi, S. Garatti, and M. Prandini. The scenario approach for systems and control design. *Annual Reviews in Control*, 33:149–157, 2009.
16. A. Charnes and W. W. Cooper. Chance-constrained programming. *Management Science*, 5: 73–79, 1959.
17. C. Chen and L. Lee. *Stochastic Simulation Optimization: An Optimal Computing Budget Allocation*. World Scientific Publishing Co. Pte. Ltd., 2011.
18. G. Dantzig and G. Infanger. Multi-stage stochastic linear programs for portfolio optimization. *Annals of Operations Research*, 45:59–76, 1993.
19. G. B. Dantzig and P. W. Glynn. Parallel processors for planning under uncertainty. *Annals of Operations Research*, 22:1–21, 1990.
20. P. T. de Boer, D. P. Kroese, S. Mannor, and R. Y. Rubinstein. A tutorial on the cross-entropy method. *Annals of Operations Research*, 134:19–67, 2005.
21. D. Dentcheva and A. Ruszczyński. Optimization with stochastic dominance constraints. *SIAM Journal on Optimization*, 14(2):548–566, 2003.
22. D. Dentcheva and A. Ruszczyński. Optimality and duality theory for stochastic optimization problems with nonlinear dominance constraints. *Mathematical Programming*, 99:329–350, 2004.
23. D. Dentcheva, A. Ruszczyński, and A. Prékopa. Concavity and efficient points of discrete distributions in probabilistic programming. *Mathematical Programming*, 89:55–77, 2000.
24. J. Dick and F. Pillichshammer. *Digital Nets and Sequences*. Cambridge University Press, 2010.
25. S. S. Drew. *Quasi-Monte Carlo Methods for Stochastic Programming*. PhD thesis, Northwestern University, 2007.
26. S. S. Drew and T. Homem-de-Mello. Quasi-Monte Carlo strategies for stochastic optimization. In L. F. Perrone, F. P. Wieland, J. Liu, B. G. Lawson, D. M. Nicol, and R. M. Fujimoto, editors, *Proceedings of the 2006 Winter Simulation Conference*, pages 774–782, IEEE, Piscataway, NJ, 2006.
27. G. Fishman. *Monte Carlo: Concepts, Algorithms and Applications*. Springer-Verlag, New York, NY, 1997.

28. M. B. Freimer, J. T. Linderoth, and D. J. Thomas. The impact of sampling methods on bias and variance in stochastic linear programs. *Computational Optimization and Applications*, 51 (1):51–75, 2012.
29. I. Friedel and A. Keller. Fast generation of randomized low-discrepancy point sets. In *Monte Carlo and quasi-Monte Carlo methods, 2000 (Hong Kong)*, pages 257–273. Springer, Berlin, 2002. Software available at http://www.multires.caltech.edu/software/libseq/.
30. M. Fu. Optimization via simulation: A review. *Annals of Operations Research*, 53:199–247, 1994.
31. P. Glasserman. *Monte Carlo Metods in Financial Engineering*. Springer, 2003.
32. P. Glasserman, P. Heidelberger, and P. Shahabuddin. Asymptotically optimal importance sampling and stratification for pricing path dependent options. *Journal of Mathematical Finance*, 9(2):117–152, 1999.
33. C. Healey, S. Andradóttir, and S.-H. Kim. Minimal switching procedures for constrained ranking and selection under independent and common random numbers. Technical report, Georgia Institute of Technology, 2014.
34. C. Healey, S. Andradóttir, and S.-H. Kim. Selection procedures for simulations with multiple constraints under independent and correlated sampling. *ACM Transactions on Modeling and Computer Simulation*, 24(3), 2014.
35. H. Heitsch, H. Leovey, and W. Römisch. Are Quasi-Monte Carlo algorithms efficient for two-stage stochastic programs? Available at the *Stochastic Programming e-Print Series*, www.speps.org, 2013.
36. J. Higle. Variance reduction and objective function evaluation in stochastic linear programs. *INFORMS Journal on Computing*, 10:236–247, 1998.
37. T. Homem-de-Mello. On rates of convergence for stochastic optimization problems under non-i.i.d. sampling. *SIAM Journal on Optimization*, 19(2):524–551, 2008.
38. T. Homem-de-Mello and G. Bayraksan. Monte Carlo sampling-based methods for stochastic optimization. *Surveys in Operations Research and Management Science*, 19:56–85, 2014.
39. T. Homem-de-Mello and S. Mehrotra. A cutting surface method for linear programs with polyhedral stochastic dominance constraints. *SIAM Journal on Optimization*, 20(3):1250–1273, 2009.
40. T. Homem-de-Mello, V. L. de Matos, and E. C. Finardi. Sampling strategies and stopping criteria for stochastic dual dynamic programming: a case study in long-term hydrothermal scheduling. *Energy Systems*, 2:1–31, 2011.
41. L. J. Hong and B. L. Nelson. Chance constrained selection of the best. Working paper, Northwestern University, 2014.
42. L. J. Hong, Y. Yang, and L. Zhang. Sequential convex approximations to joint chance constrained programs: A Monte Carlo approach. *Operations Research*, 59(3):617–630, 2011.
43. J. Hu, T. Homem-de-Mello, and S. Mehrotra. Sample average approximation of stochastic dominance constrained programs. *Mathematical Programming, Ser. A*, 133:171–201, 2011.
44. S. Hunter and R. Pasupathy. Optimal sampling laws for stochastically constrained simulation optimization. *INFORMS Journal on Computing*, 25(3):527–542, 2013.
45. G. Infanger. Monte Carlo (importance) sampling within a Benders decomposition algorithm for stochastic linear programs. *Annals of Operations Research*, 39:69–95, 1992.
46. J. Kalagnanam and U. Diwekar. An efficient sampling technique for off-line quality control. *Technometrics*, 39(3):308–319, 1997.
47. S.-H. Kim and B. Nelson. Selecting the best system. In S. Handerson and B. Nelson, editors, *Handbooks in Operations Research and Management Science, Volume 13: Simulation*. Elsevier, 2006.
48. A. J. King and R. T. Rockafellar. Asymptotic theory for solutions in statistical estimation and stochastic programming. *Mathematics of Operations Research*, 18:148–162, 1993.
49. M. Koivu. Variance reduction in sample approximations of stochastic programs. *Mathematical Programming*, 103(3):463–485, 2005.

50. V. Kozmík and D. Morton. Evaluating policies in risk-averse multi-stage stochastic programming, *Mathematical Programming,* published online, 2014.
51. P. Krokhmal, J. Palmquist, and S. Uryasev. Portfolio optimization with conditional value-at-risk objective and constraints. *Journal of Risk*, 4:43–68, 2002.
52. A. M. Law. *Simulation Modeling and Analysis*. McGraw-Hill, 5th edition, 2014.
53. P. L'Ecuyer. *SSJ Users Guide (Package hups: Tools for Quasi-Monte Carlo).* University of Montreal, 2009. Software users guide, available at http://www.iro.umontreal.ca/~simardr/ssj/indexe.html.
54. P. L'Ecuyer and C. Lemieux. Recent advances in randomized quasi-Monte Carlo methods. In M. Dror, P. L'Ecuyer, and F. Szidarovszky, editors, *Modeling Uncertainty: An Examination of Stochastic Theory, Methods, and Applications*, pages 419–474. Kluwer Academic Publishers, Boston, 2002.
55. P. L'Ecuyer and D. Munger. LatticeBuilder: A general software tool for constructing rank-1 lattice rules. *ACM Transactions on Mathematical Software*, 2012. To appear. Software available at http://www.iro.umontreal.ca/~simardr/latbuilder/latbuilder.html.
56. Monte Carlo and Quasi Monte Carlo Methods 2010, H. Wozniakowski and L. Plaskota, Eds., Springer-Verlag, Berlin, 133–159, 2012.
57. L. H. Lee, N. A. Pujowidianto, L.-W. Li, C.-H. Chen, and C. M. Yap. Approximate simulation budget allocation for selecting the best design in the presence of stochastic constraints. *IEEE Transactions on Automatic Control*, 57:2940–2945, 2012.
58. C. Lemieux. *Monte Carlo and Quasi-Monte Carlo Sampling*. Springer Series in Statistics. Springer, New York, 2009.
59. C. Lemieux, M. Cieslak, and K. Luttmer. *RandQMC Users Guide: A Package for Randomized Quasi-Monte Carlo Methods in C*. University of Waterloo, 2004. Software users guide, available at http://www.math.uwaterloo.ca/~clemieux/randqmc.html.
60. J. T. Linderoth, A. Shapiro, and S. J. Wright. The empirical behavior of sampling methods for stochastic programming. *Annals of Operations Research*, 142(1):215–241, 2006.
61. Y. Liu, H. Xu, and J. J. Ye. Penalized sample average approximation methods for stochastic mathematical programs with complementarity constraints. *Mathematics of Operations Research*, 36(4):670–694, 2011.
62. J. Luedtke and S. Ahmed. A sample approximation approach for optimization with probabilistic constraints. *SIAM Journal on Optimization*, 19(2):674–699, 2008.
63. J. Luedtke, S. Ahmed, and G. Nemhauser. An integer programming approach for linear programs with probabilistic constraints. *Mathematical Programming*, 122(2):247–272, 2010.
64. W. K. Mak, D. P. Morton, and R. K. Wood. Monte Carlo bounding techniques for determining solution quality in stochastic programs. *Operations Research Letters*, 24:47–56, 1999.
65. M. D. McKay, R. J. Beckman, and W. J. Conover. A comparison of three methods for selecting values of input variables in the analysis of output from a computer code. *Technometrics*, 21: 239–245, 1979.
66. A. Müller and D. Stoyan. *Comparison Methods for Stochastic Models and Risks*. John Wiley & Sons, Chichester, 2002.
67. J. C. Neddermeyer. Computationally efficient nonparametric importance sampling. *Journal of the American Statistical Association*, 104(486):788–802, 2009.
68. B. Nelson and J. Staum. Control variates for screening, selection, and estimation of the best. *ACM Transactions on Modeling and Computer Simulation*, 16:52–75, 2006.
69. B. L. Nelson. Control variate remedies. *Operations Research*, 38:359–375, 1990.
70. A. Nemirovski and A. Shapiro. Scenario approximation of chance constraints. In G. Calafiore and F. Dabbene, editors, *Probabilistic and Randomized Methods for Design Under Uncertainty*, pages 3–48. Springer, London, 2005.
71. A. Nemirovski and A. Shapiro. Convex approximations of chance constrained programs. *SIAM Journal on Optimization*, 17(4):969–996, 2006.
72. H. Niederreiter. *Random Number Generation and Quasi-Monte Carlo Methods*. SIAM, Philadelphia, PA, 1992.

73. V. Norkin, G. Pflug, and A. Ruszczyński. A branch and bound method for stochastic global optimization. *Mathematical Programming*, 83:425–450, 1998.
74. W. Ogryczak and A. Ruszczyński. On consistency of stochastic dominance and mean-semideviation models. *Mathematical Programming Series B*, 89(2):217–232, 2001.
75. A. B. Owen. A central limit theorem for Latin hypercube sampling. *Journal of the Royal Statistical Society, Ser. B*, 54:541–551, 1992.
76. A. B. Owen. Latin supercube sampling for very high-dimensional simulations. *ACM Transactions on Modeling and Computer Simulation*, 8:71–102, 1998.
77. A. B. Owen. The dimension distribution and quadrature test functions. *Statistica Sinica*, 13 (1):1–17, 2003.
78. B. K. Pagnoncelli, S. Ahmed, and A. Shapiro. Computational study of a chance constrained portfolio selection problem. *Journal of Optimization Theory and Applications*, 142(2): 399–416, 2009.
79. B. K. Pagnoncelli, D. Reich, and M. C. Campi. Risk-return trade-off with the scenario approach in practice: a case study in portfolio selection. *Journal of Optimization Theory and Applications*, 155(2):707–722, 2012.
80. C. Park and S.-H. Kim. Penalty function with memory for discrete optimization via simulation with stochastic constraints. Technical report, Georgia Institute of Technology, 2014.
81. T. Pennanen. Epi-convergent discretizations of multistage stochastic programs. *Mathematics of Operations Research*, 30:245–256, 2005.
82. T. Pennanen and M. Koivu. Epi-convergent discretizations of stochastic programs via integration quadratures. *Numerische Mathematik*, 100:141–163, 2005.
83. P. Pierre-Louis, D. Morton, and G. Bayraksan. A combined deterministic and sampling-based sequential bounding method for stochastic programming. In *Proceedings of the 2011 Winter Simulation Conference*, pages 4172–4183, IEEE, Piscataway, NJ, 2011.
84. A. Prékopa. Probabilistic programming. In A. Ruszczynski and A. Shapiro, editors, *Handbook of Stochastic Optimization*. Elsevier Science Publishers B.V., Amsterdam, Netherlands, 2003.
85. N. A. Pujowidianto, S. Hunter, R. Pasupathy, L. Lee, and C.-H. Chen. On approximating optimal sampling laws for stochastically constrained simulation optimization on large finite sets. Technical report, Virginia Polytechnic Institute and State University, 2013.
86. J. Royset. Optimality functions in stochastic programming. *Mathematical Programming*, 135 (1–2):293–321, 2012.
87. R. Rubinstein, G. Samordnitsky, and M. Shaked. Antithetic variates, mutlivariate dependence and simulation of stochastic systems. *Management Science*, 311(1):66–77, 1985.
88. R. Y. Rubinstein and A. Shapiro. *Discrete Event Systems: Sensitivity Analysis and Stochastic Optimization by the Score Function Method*. John Wiley & Sons, Chichester, England, 1993.
89. M. Shaked and J. G. Shanthikumar. *Stochastic Orders and their Applications*. Academic Press, Boston, 1994.
90. A. Shapiro. Asymptotic behavior of optimal solutions in stochastic programming. *Mathematics of Operations Research*, 18:829–845, 1993.
91. A. Shapiro. Monte Carlo sampling methods. In A. Ruszczynski and A. Shapiro., editors, *Stochastic Programming*, volume 10 of *Handbooks in Operations Research and Management Science*. Elsevier Science Publishers B.V., Amsterdam, Netherlands, 2003.
92. A. Shapiro and T. Homem-de-Mello. A simulation-based approach to two-stage stochastic programming with recourse. *Mathematical Programming*, 81:301–325, 1998.
93. A. Shapiro, T. Homem-de-Mello, and J. C. Kim. Conditioning of convex piecewise linear stochastic programs. *Mathematical Programming*, 94:1–19, 2002.
94. A. Shapiro, D. Dentcheva, and A. Ruszczyński. *Stochastic Programming: Modeling and Theory*. SIAM Series on Optimization, 2009.
95. D. Siegmund. Importance sampling in the Monte Carlo study of sequential tests. *The Annals of Statistics*, 4(4):673–684, 1976.
96. M. L. Stein. Large sample properties of simulations using Latin hypercube sampling. *Technometrics*, 29:143–151, 1987.

97. R. Stockbridge and G. Bayraksan. Variance reduction in Monte Carlo sampling-based optimality gap estimators for two-stage stochastic linear programming. Technical report, the Ohio State University, 2014.
98. S. Tsai and B. Nelson. Fully sequential selection procedures with control variates. *IIE Transactions*, 42:71–82, 2010.
99. J. von Neumann and O. Morgenstern. *Theory of Games and Economic Behavior*. Princeton University Press, Princeton, NJ, 2nd edition, 1947.
100. W. Wang and S. Ahmed. Sample average approximation of expected value constrained stochastic programs. *Operations Research Letters*, 36:515–519, 2008.
101. L. Zhang and T. Homem-de-Mello. An optimal path model for the risk-averse traveler, 2014. Preprint available at Optimization Online (http://www.optimization-online.org).
102. P. Zhang. Nonparametric importance sampling. *Journal of the American Statistical Association*, 91:1245–1253, 1996.

Chapter 10
A Review of Random Search Methods

Sigrún Andradóttir

Abstract This chapter provides a brief review of random search methods for simulation optimization. We start by describing the structure of random search when system performance is estimated via simulation. Next, we discuss methods for solving simulation optimization problems with discrete decision variables and one (stochastic) performance measure, with emphasis on simulated annealing. Finally, we expand our scope to address simulation optimization problems with continuous decision variables and/or multiple (stochastic) performance measures.

10.1 Introduction

This chapter describes the use of random search to optimize complex stochastic systems whose expected performance under any particular system design is unavailable in closed form, and instead must be estimated via computer simulation. Thus, if Θ denotes the set of all possible system designs and $f(x) = \mathsf{E}[Y(x,\xi)]$ denotes the expected system performance under each design $x \in \Theta$, then we aim to solve the optimization problem

$$\min_{x\in\Theta} f(x) \tag{10.1}$$

under the assumption that the values of the objective function f will be estimated using simulation. We will outline the generic structure of random search methods and provide a review of a representative set of specific random search approaches.

Random search methods involve repeatedly sampling and evaluating system designs based on the observed history (i.e., the designs that have been sampled so far and their estimated performance) in search of the best feasible design. They are well suited for solving simulation optimization problems where the objective function often has little known structure (and hence derivatives are unavailable) and the optimization procedure must identify improved solutions with guidance from the

S. Andradóttir (✉)
Georgia Institute of Technology, Atlanta, GA, USA
e-mail: sa@gatech.edu

M.C. Fu (ed.), *Handbook of Simulation Optimization*, International Series in Operations Research & Management Science 216, DOI 10.1007/978-1-4939-1384-8_10

estimated performance of the solutions considered so far. The scope of applicability of random search is broad, and includes deterministic and stochastic optimization problems with discrete and/or continuous decision parameters.

The outline of this chapter is as follows. In Sect. 10.2, we describe the structure of random search methods. In Sect. 10.3, we review a representative set of random search methods for solving discrete simulation optimization problems. In Sect. 10.4, we discuss example procedures for solving simulation optimization problems with either continuous decision variables or multiple performance measures. An earlier overview of random search for simulation optimization emphasizing desirable features such as convergence and efficiency can be found in Andradóttir [14].

10.2 Structure of Random Search Methods for Simulation Optimization

In this section, we describe the structure of random search methods applied to solve the simulation optimization problem (10.1). We will let $\mathscr{S}_n$ denote the sampling strategy used to select candidate system designs in iteration $n \geq 1$ of the algorithm, with M_n specifying the number of such designs. The sampling strategy can be updated adaptively as the algorithm learns from the results so far, and does not need to be chosen in advance of execution. Similarly, M_n is a parameter of the sampling strategy $\mathscr{S}_n$, and consequently does not need to be chosen in advance. The sampling strategy can be used both to identify new, promising system designs, and also to obtain improved objective function estimates for previously sampled designs (by re-sampling them). The following generic random search algorithm can also be found in Andradóttir [14].

Generic Random Search Algorithm for Simulation Optimization:

Step 0 (initialize): Choose the initial sampling strategy $\mathscr{S}_1$ and let $n = 1$.

Step 1 (sample): Select $x_n^{(1)}, \ldots, x_n^{(M_n)} \in \Theta$ according to the sampling strategy $\mathscr{S}_n$.

Step 2 (simulate): Estimate $f(x_n^{(i)})$, for $i = 1, \ldots, M_n$, using simulation.

Step 3 (update): Use the simulation results obtained in Step 2 to compute an estimate of the optimal solution x_n^* and to choose an updated sampling strategy $\mathscr{S}_{n+1}$. Let $n = n + 1$ and go to Step 1.

The primary difference among random search methods involves the choice of the sampling strategies $\{\mathscr{S}_n\}$. Also, the generic algorithm does not include a stopping criterion. This is consistent with the fact that convergence results for random search are typically asymptotic in nature. In practice, it is of course necessary to augment the algorithm with a suitable stopping criterion.

In the remainder of this chapter, we will describe specific random search procedures for simulation optimization. As much of the random search literature focuses on problems with discrete decision variables, we start by considering such problems in Sect. 10.3. Then we consider extensions to continuous feasible regions and multiple objectives in Sect. 10.4.

10.3 Discrete Simulation Optimization

In this section, we review example procedures for discrete simulation optimization problems with one (stochastic) performance measure. (In addition, there may be deterministic performance measures whose values do not need to be estimated via simulation. They can be incorporated into the constraint set Θ.) We start by discussing research on how the well-known simulated annealing algorithm can be applied to solve discrete simulation optimization problems in Sect. 10.3.1. Then we briefly review certain other random search procedures for solving the optimization problem (10.1) in Sect. 10.3.2.

10.3.1 Simulated Annealing

The simulated annealing algorithm dates back to the pioneering work by Metropolis et al. [61]. Since then, a large body of literature has appeared on simulated annealing, including important works by Kirkpatrick et al. [55], Mitra et al. [62], Hajek [42], and others. A basic version of the simulated annealing algorithm for solving a deterministic optimization problem of the form (10.1) is provided below. We will need the following notation:

- for all $x \in \Theta$, $N(x) \subset \Theta$ is a set of "neighbors" of x (alternatively, $N(x)$ is a "neighborhood" of x);
- for all $x \in \Theta$, $R(x, \cdot)$ is a probability distribution on $N(x)$;
- $\{T_n\}$ is a sequence of strictly positive numbers;
- w.p. = with probability;
- $(y)^+ = \max\{0, y\}$ for all $y \in \mathbb{R}$.

Basic Simulated Annealing Algorithm for Deterministic Optimization:

Step 0 (initialize): Choose an initial system design $x_0 \in \Theta$ and let $n = 1$.
Step 1 (sample): Select a candidate solution $x'_n \in N(x_{n-1})$ according to the probability distribution $R(x_{n-1}, \cdot)$.
Step 2 (update): Let

$$x_n = \begin{cases} x'_n & \text{w.p. } \exp\left(-\frac{(f(x'_n)-f(x_{n-1}))^+}{T_n}\right), \\ x_{n-1} & \text{otherwise,} \end{cases}$$

$n = n + 1$, and go to Step 1.

It is clear that the Basic Simulated Annealing Algorithm described above fits within the framework outlined in Sect. 10.2 with $M_n = 1$ and $x_n^* = x_n$ for all $n \in \mathbb{N}$, except that a "simulate" step is of course not needed here, as the exact objective function values are available. Simulated annealing is designed to solve global optimization problems in the presence of (possibly multiple) local optimal solutions. In each iteration n, the algorithm generates a candidate solution x_n', and then decides whether to stay at the current solution x_{n-1} or move to the candidate solution. More specifically, if the candidate solution x_n' satisfies $f(x_n') \le f(x_{n-1})$, so that x_n' is better than x_{n-1}, then x_n' becomes the new estimated optimal solution. On the other hand, if $f(x_n') > f(x_{n-1})$, so that the candidate solution is worse than x_{n-1}, then there is nevertheless a chance that x_n' will be chosen as a new estimate of the optimal solution. This "hill-climbing" feature is designed to allow the algorithm to escape from locally optimal solutions that are not globally optimal (i.e., solutions x satisfying $f(x) < f(x')$ for all $x' \in N(x)$ but $f(x) > \inf_{x' \in \Theta} f(x')$). The probability $\exp(-(f(x_n') - f(x_{n-1}))^+/T_n)$ of making such a hill-climbing move depends both on how inferior the candidate solution is relative to the current solution (i.e., on the magnitude of $(f(x_n') - f(x_{n-1}))^+$) and on the current "temperature" T_n, with hill-climbing moves being less likely for worse candidate solutions and for smaller temperatures. Hill-climbing decisions can be made by generating a uniform random number U_n on the interval $[0,1]$, accepting the candidate solution x_n' if $U_n \le \exp(-(f(x_n') - f(x_{n-1}))^+/T_n)$, and rejecting it otherwise (in which case the current solution x_{n-1} remains the estimate of the optimal solution). In addition to requiring that the temperature T_n be strictly positive for all n, most of the simulated annealing literature assumes that $T_n \to 0$ as $n \to \infty$ at a logarithmic rate, see for example Hajek [42].

The previous discussion addressed the use of simulated annealing to solve deterministic optimization problems. In the remainder of this section, we will review simulated annealing algorithms designed to solve optimization problems with noisy objective function values, as is the case in simulation optimization.

Bulgak and Sanders [22] present a heuristic simulated annealing approach and use it to solve a buffer allocation problem. They deal with the noise in the objective function evaluations by using confidence intervals to ensure that the difference in performance is statistically significant when the candidate state has a better objective function estimate than the current state. Haddock and Mittenthal [43] also present a heuristic simulated annealing method together with numerical results. Their method employs a different "update" step than the Basic Simulated Annealing Algorithm described above (motivated by the work of Glauber [36]), and also uses a rapidly decreasing temperature sequence $\{T_n\}$ (their sequence decreases at an exponential rate, rather than at a logarithmic rate).

We now turn to simulated annealing approaches that are provably convergent when applied to solve discrete simulation optimization problems. For all a and $b \neq 0$, let $\mathcal{N}(a, b^2)$ denote a normal random variable with mean a and variance b^2. Gelfand and Mitter [34] show that if the transition matrix $R(\cdot,\cdot)$ is irreducible, the temperature sequence $\{T_n\}$ converges to zero, and the noise in the estimate of the difference $f(x'_n) - f(x_{n-1})$ in performance between the candidate and current states in iteration n has a $\mathcal{N}(0, \sigma_n^2)$ distribution where $\sigma_n = o(T_n)$ as $n \to \infty$, then their simulated annealing procedure with noisy objective function estimates converges in probability to the set Θ^* of global optimal solutions provided that the same algorithm using exact objective function values converges to Θ^* in probability.

Gutjahr and Pflug [41] also present convergence results for the simulated annealing algorithm when the noise in the estimated objective function values needed in iteration n has the normal $\mathcal{N}(0, \sigma_n^2)$ distribution. They show that when $\sigma_n = O(n^{-\gamma})$, where $\gamma > 1$, then the simulated annealing algorithm with noisy objective function evaluations has the same asymptotic performance as when exact objective functions are available (and hence converges in probability to Θ^* if the temperature sequence $\{T_n\}$ is chosen properly). They also generalize their convergence result to noise distributions that are symmetric around zero and more peaked around zero than the $\mathcal{N}(0, \sigma_n^2)$ distribution satisfying $\sigma_n = O(n^{-\gamma})$, where $\gamma > 1$ (i.e., for all $\varepsilon > 0$, the noise is more likely to take values in $(-\varepsilon, \varepsilon)$ than the specified $\mathcal{N}(0, \sigma_n^2)$ random variable).

Gelfand and Mitter [34] and Gutjahr and Pflug [41] assume that the variance σ_n^2 in the objective function evaluations required in iteration n of the simulated annealing algorithm converges to zero as n grows. Thus, more precise estimates are required for larger values of n, which typically results in more computation time per iteration. Fox and Heine [33] provide convergence guarantees for simulated annealing applied to solve discrete simulation optimization problems that do not require a restrictive variance assumption. However, they assume that the objective function values are restricted to a finite set (they also consider relaxing this assumption). Each time an estimate of an objective function value $f(x)$ is needed, they generate a few more observations of $f(x)$ and average them with observations of $f(x)$ obtained in earlier iterations to obtain the desired estimate of $f(x)$. They show that this variant of simulated annealing converges in probability to Θ^* if the same algorithm with the exact objective function values converges in probability to Θ^*.

Alrefaei and Andradóttir [6] present two simulated annealing algorithms for discrete simulation optimization. Beyond using noisy objective function estimates, rather than the exact objective function values, these algorithms differ from the Basic Simulated Annealing Algorithm presented above in two important ways, namely the temperature sequence is constant (i.e., $T_n = T > 0$ for all n) and the choice of the estimate of the optimal solution x_n^* is decoupled from the sequence $\{x_n\}$ used to search the state space Θ for the optimal solution (see also the Generic Random Search Algorithm described in Sect. 10.2). Alrefaei and Andradóttir [6] prove that their algorithms converge almost surely to Θ^* and provide numerical results comparing their algorithms with each other and with the methods analyzed by Gelfand and Mitter [34], Gutjahr and Pflug [41], and Fox and Heine [33].

Alkhamis et al. [5] study a simulated annealing algorithm that employs confidence intervals to determine whether the difference between the estimated objective function values at the current and candidate solutions is statistically significant.

They prove convergence in probability to Θ^* when the noise the objective function evaluations in iteration n converges to zero sufficiently fast relative to the temperature sequence $\{T_n\}$, and also provide numerical results. Ahmed and Alkhamis [1] analyze a simulated annealing approach with a constant temperature and with decoupled sequences $\{x_n\}$ and $\{x_n^*\}$ (see Alrefaei and Andradóttir [6]) that uses the two-stage ranking and selection procedure by Dudewicz and Dalal [30] to decide how many objective function observations are collected from the current and candidate solutions in iteration n. Alkhamis and Ahmed [4] continue this work by combining the approach with constant temperature and decoupled $\{x_n\}$ and $\{x_n^*\}$ sequences of Ahmed and Alkhamis [1] with the confidence interval approach of Alkhamis et al. [5]. Wang and Zhang [76] study a simulated annealing approach where a hypothesis test is used to determine whether to stay at the current state or move to the candidate state.

Prudius and Andradóttir [70] study two simulated annealing algorithms for discrete simulation optimization with decreasing temperatures $\{T_n\}$ and with decoupled $\{x_n\}$ and $\{x_n^*\}$ sequences. The two algorithms differ in that one uses only the data collected on the objective function values at the current and candidate solutions x_{n-1} and x_n' in the current iteration n to decide on the next current point x_n (no averaging), whereas the other one uses data collected on the values of $f(x_{n-1})$ and $f(x_n')$ in iterations 1 through n to decide on x_n (averaging). Both algorithms are shown to converge almost surely, and numerical results show that using all available data on the objective function values (as in averaging) does not necessarily improve performance (because the associated reduction in noise is not necessarily beneficial).

As was mentioned at the beginning of this section, the literature on simulated annealing for deterministic optimization is vast, and several researchers have studied the application of simulated annealing to solve simulation optimization problems. In addition to the contributions reviewed so far in this section, other works on simulated annealing for noisy response functions include Painton and Diwekar [66], who incorporate a penalty function to account for noise in the objective function estimates, Rosen and Harmonosky [71], who combine simulated annealing with response surface methodology, and Branke et al. [21], who consider a different "update" step than the Basic Simulated Annealing Algorithm above under a known variance assumption.

10.3.2 Other Developments

In this section, we briefly review certain other random search methods that have been developed for solving discrete simulation optimization problems. Several of these techniques are reviewed in more detail in other chapters in this volume, viz. Chaps. 2, 11, and 12. Additional material on random search for simulation optimization can, for example, be found in the reviews by Jin and Branke [52] and Bianchi et al. [20].

Stochastic ruler methods constitute a class of random search methods for discrete simulation optimization. Like simulated annealing, they involve sampling a single candidate point x_n' in each iteration n and deciding whether to accept this point (so that $x_n = x_n'$) or reject it (so that $x_n = x_{n-1}$). However, unlike simulated annealing, this decision is not made by comparing estimated objective function values at the current and candidate points. Instead, estimated objective function values at the candidate point x_n' are compared with observations of the "stochastic ruler," which is a uniform random variable whose range covers (approximately) the range of the estimated objective function values. The original stochastic ruler method was proposed by Yan and Mukai [83] and proven to converge to Θ^* in probability. Alrefaei and Andradóttir [7, 8] have studied modified versions of the stochastic ruler method that involve less work per iteration (i.e., the maximum number of comparisons with the stochastic ruler in iteration n is constant, rather than diverging to infinity with n) and decoupling the $\{x_n\}$ and $\{x_n^*\}$ sequences (see Sect. 10.2). They prove that their approaches converge almost surely to Θ^* and provide numerical results comparing the approaches.

Most random search methods for simulation optimization compare system designs based on performance estimates, and can thus be regarded as *ascent/descent methods*. For example, Gong et al. [37] present and analyze a "stochastic comparison" method that resembles the stochastic ruler method of Yan and Mukai [83] except that each iteration involves comparisons between objective function estimates at the current and candidate points, rather than comparisons with a stochastic ruler. For other related work, see Andradóttir [11–13] and Homem-de-Mello [44].

Deterministic optimization features various methods that involve partitioning the feasible region, including branch-and-bound (see, e.g., Nemhauser and Wosley [63]) and random search (see, e.g., Pintér [69]). *Partitioning methods* have also been developed for simulation optimization. For example, Norkin et al. [64] present a *branch-and-bound* method that involves partitioning the feasible region Θ into subsets, estimating upper and lower bounds on the optimal objective function value within each subset, choosing the estimated optimal solution from the subset with the smallest upper bound, and further partitioning the "record subset" that has the smallest lower bound. The estimates of the upper and lower bounds improve when the subset remains in the partition, and eventually converge to the actual values of the upper and lower bounds. Moreover, the upper and lower bounds are tight for singletons. Norkin et al. [64] prove that their method converges almost surely to Θ^*, discuss the choice of upper and lower bounds in various settings, and provide illustrative examples. This work is continued by Norkin et al. [65], who provide additional analysis, discussion about bound estimation, and examples.

Shi and Ólafsson [73] present a *nested partitions* method for simulation optimization. Like branch-and-bound, their approach involves partitioning one subset ("the most promising region") in most iterations. The other subsets are then combined into one "surrounding region." Then sample designs are collected from each region and their performance estimated via simulation. The promising index is estimated for each region as the best estimated performance of the designs sampled from the region, and the region with the best promising index will be the new most promising region. If the most promising region is a singleton, then it can clearly not be partitioned further. If it is the surrounding region, then the method backtracks to either its super-region or the entire feasible region. Otherwise, the most promising region is partitioned. Shi and Ólafsson [73] prove that their method converges almost surely to Θ^*. Pichitlamken and Nelson [68] present a nested partitions method that differs from the original method of Shi and Ólafsson [73] in that a ranking and selection method is used to select the best sampled solution from each subset, hill-climbing and restart steps are added, and the sequence $\{x_n^*\}$ is chosen differently. Moreover, Xu and Nelson [80] present and analyze a method that combines the branch-and-bound and nested partition approaches.

Hong and Nelson [46] present a random search algorithm for local simulation optimization named *COMPASS* (for "convergent optimization via most-promising-area stochastic search"). The feasible region is composed of vectors with integer elements, and the approach is local in that the aim is to identify a design with better performance than any point with Euclidian distance of one away from that design. COMPASS keeps track of all designs sampled so far and performance estimates at these designs. In each iteration, new designs are sampled from the portion of the feasible region that is closer to the design with the best estimated performance than to any other sampled point (again in Euclidian distance, with adjustments to ensure that the sampling region is bounded) and simulation results are obtained for the newly sampled points (and possibly also for previously sampled points). Hong and Nelson [46] prove convergence w.p.1 to a local optimal solution and provide numerical examples. This work is continued by Hong [45], Hong et al. [47], and Xu et al. [78, 79] who improve the efficiency of the original COMPASS approach (see also the discussion of Xu [77] in Sect. 10.4.1 below).

Andradóttir and Prudius [15] discuss the need for balancing exploration (global search), exploitation (local search), and estimation (of objective function values at promising points) within simulation optimization, and then present two versions of *BEESE* (for "balanced explorative and exploitative search with estimation"), a random search approach designed to achieve such a balance. The two approaches are called R-BEESE (for Random-BEESE) and A-BEESE (Adaptive-BEESE). Both methods switch between global search and local search for improved system designs, with R-BEESE doing so at random and A-BEESE doing so adaptively based on recent progress made via global and local search. Both methods also add an estimation component via resampling of the design with the best estimated performance and by ensuring that sufficient data has been collected at the estimated optimal solution (moreover, A-BEESE involves more local search than its deterministic variant A-BEES, which also adds an estimation component). Andradóttir and Prudius [15] prove that their methods converge almost surely to Θ^* and provide numerical results.

Model-based methods form another class of random search techniques. These methods maintain a probabilistic model on the solutions space Θ that is used to generate candidate solutions, whose estimated performance is in turn used to update the probabilistic model. The cross-entropy method of Rubinstein and Kroese [72] involves the use of a parametric family of distributions and relies on the relationship between optimization and rare event simulation. In each iteration, solutions are sampled from the current model in the parametric family. The performance of these solutions is estimated, a sample quantile is computed, and the solutions whose estimated performance exceeds the sample quantile are used to obtain an updated model in the parametric family. The model update involves approximating an optimal importance sampling distribution on the set of solutions whose performance exceeds the sample quantile via the use of the Kullback–Leibler (cross-entropy) distance. Hu et al. [48] provide a model-based algorithm for simulation optimization that differs from the cross-entropy method primarily in the updating of the model. They provide convergence analysis that is applicable both when Θ is discrete and continuous.

10.4 Extensions

In this section, we review example procedures for solving simulation optimization problems with either continuous decision variables or multiple (stochastic) performance measures. We start by discussing research on continuous simulation optimization problems in Sect. 10.4.1. Then we review certain procedures for solving simulation optimization problems with multiple performance measures in Sect. 10.4.2.

10.4.1 Continuous Simulation Optimization

Several researchers have studied the use of random search methods to solve continuous simulation optimization problems. The simplest form of random search is *pure (non-adaptive) random search*, where solutions are sampled repeatedly from a fixed distribution on the feasible region Θ (e.g., the search does not utilize information gathered in previous iterations to guide the search for improved solutions). Baumert and Smith [19] present a pure random search approach that estimates the objective function value at each sampled solution x by averaging all observations that are within a certain distance from x. The sampled point with the best estimated objective function value is chosen as the estimated optimal solution. Baumert and Smith [19] discuss at what rate the distance should decrease in order for the method to converge in probability. Their work was continued by Andradóttir and Prudius [16] who provide further analysis of the (deterministic) shrinking ball method of Baumert and Smith [19], develop and analyze the stochastic shrinking ball method, and provide numerical results.

Chia and Glynn [24] study the rate of convergence of pure random search as a function of the number m of sampled points and number n of observations at each point, with the estimated optimal solution being the point with the best estimated objective function value. They identify at what rates m and n should grow to achieve the best rate of convergence. Similarly, Ensor and Glynn [31] study the choice of m and n in grid search. Cheng [23] also studies the asymptotic behavior of pure random search, addresses implementation issues, and provides numerical examples. In related research, Yakowitz et al. [81] study how the number of points vs. number of observations per point should be selected in search approaches that use low-dispersion sequences to select points. They also discuss a sequential version of their approach and the use of different numbers of observations to estimate the objective function value at different sampled points.

Alexander et al. [3] develop a pure random search procedure that iteratively samples solutions from Θ and then compares the incumbent and sampled solutions using increasingly precise (as the number of iterations grows) estimates of the objective function values at these solutions; the point with the better estimate becomes the new estimate of the optimal solution. They show that their procedure is globally convergent w.p.1. Ghate and Smith [35] study a generalized simulated annealing procedure that also involves comparing estimated objective function values at the incumbent and sampled solutions in each iteration, with the estimate being more precise for larger numbers of iterations. They prove convergence in probability and provide numerical results. Various other authors also study methods that move between current and sampled solutions based on estimated objective function values at those points, see for example Gurin [38], Gurin and Rastrigin [39], Devroye [28], and Marti [59] (and Devroye [29] for related work with finite Θ).

Yakowitz and Lugosi [82] develop a method that in certain iterations samples new solutions from a fixed global distribution (as in pure random search) and ensures that every sampled point has a sufficient number of observations, and in other iterations it adaptively *resamples* previously sampled points. The estimate of the optimal solution is the most recently sampled point. They prove that their method is globally convergent in probability. Andradóttir and Prudius [16] present the Adaptive Search with Resampling (ASR) method and prove that it is globally convergent w.p.1. Their method includes both sampling and resampling steps (similar to the approach of Yakowitz and Lugosi [82]), but the search is adaptive, only promising sampled points are "accepted" for further consideration (and hence additional observations are not collected at points that are not promising), and the estimated optimal solution is the best point sampled so far. Numerical results suggest that the ASR method performs better in practice than the earlier approach. Hu and Andradóttir [49] improve further on the ASR method by allowing previously accepted points to be discarded once better points have been found. They prove that their Adaptive Search with Resampling and Discarding (ASRD) method is convergent w.p.1 and provide numerical results indicating that the addition of discarding leads to substantial improvements in performance.

Huang et al. [51], Sun et al. [74], and Xu [77] all use Kriging meta-models and random search to solve simulation optimization problems. More specifically, Huang et al. [51] propose the SKO (Sequential Kriging Optimization) approach, where each iteration starts with a kriging meta-model of the objective function, identifies a solution that maximizes an Expected Improvement (EI) function (described in

Sect. 2.7 of [51]), and then either terminates the search (if the EI is small) or updates the kriging model using the new data point. Sun et al. [74] propose and analyze the GPS (Gaussian Process-based Search) algorithm whose sampling strategy takes into account how likely feasible solutions are to improve on the current best estimate of the optimal solution based on the current kriging model. The GPS approach can be used for both continuous and discrete simulation optimization. In related work, Xu [77] presents the SKOPE (Stochastic Kriging for OPtimization Efficiency) sampling approach and integrates this approach with the AHA discrete simulation optimization method of Xu et al. [79].

We conclude this section by briefly mentioning other methods that can be used to solve continuous simulation optimization problems. Methods that involve partitioning the feasible region have been developed by Deng and Ferris [27] and Kabirian and Ólafsson [54]. More specifically, Deng and Ferris [27] adapt the DIRECT (DIviding RECTangles) algorithm of Jones et al. [53] to simulation optimization, and Kabirian and Olafsson [54] present and analyze a golden region search algorithm for continuous simulation optimization. Model-based methods (see Sect. 10.3.2, Rubinstein and Kroese [72], and Hu et al. [48]) can be used for both discrete and continuous simulation optimization. Finally, Ferris et al. [32] and Deng and Ferris [26] discuss continuous simulation optimization algorithms that involve successive quadratic approximations of the objective function.

10.4.2 Simulation Optimization with Multiple Objectives

In this section, we review certain random search approaches designed for solving simulation optimization problems with multiple (stochastic) performance measures. Ahmed et al. [2] consider a problem with a deterministic objective function (cost) and stochastic constraints (on system performance). They present a simulated annealing approach for solving such problems, where a hypothesis testing step is added after candidate generation to determine if the candidate solution is feasible with the desired confidence. Baeslar and Sepúlveda [17] present a *goal programming* framework to handle multiple stochastic performance measures. A goal value is specified for each performance measure and the original (multi-objective) optimization problem is translated into a (single-objective) optimization problem where a weighted sum of the deviations from the specified goals is minimized (possibly after normalization to address discrepancies between measurement units for the different performance measures). This optimization problem is then solved using a genetic algorithm. Baeslar and Sepúlveda [18] continue this research by applying their methodology to optimize a cancer treatment center.

Multiple authors have proposed random search methods aiming to identify *Pareto optimal* solutions to simulation optimization problems with multiple performance measure. More specifically, a solution x dominates another solution x' if no objective performs worse at x than at x', and at least one objective performs strictly better at x than at x'. A solution is Pareto optimal if it is not dominated by any other solution.

Gutjahr [40] presents a Stochastic Pareto Simulated Annealing (SPSA) based on the Pareto Simulated Annealing (PSA) approach of Czyzak and Jaszkiewicz [25], which maintains a search set and a solution set, and in each iteration a candidate solution is generated in the neighborhood of each element of the search set, the estimated performance of the candidate and current points are compared, and the search and solution sets are updated (the update of the search set involves hill-climbing with weights computed by the algorithm). Gutjahr [40] also specifies a Stochastic Pareto Ant Colony Approach (SP-ACO) and compares the two approaches with a brute-force approach. Other approaches for multi-objective simulation optimization based on simulated annealing include Alrefaei et al. [9, 10] and Mattila et al. [60]. Lee et al. [56] present an approach based on multiobjective evolutionary algorithms (MOEA) (see, e.g., Zhou et al. [84]) and use it to solve an aircraft spare parts allocation problem, and Lee et al. [57] study a multi-objective COMPASS approach (see Sect. 10.3.2 for discussion of the original COMPASS approach).

Another approach to handling multiple performance measures is to designate one as the objective and the others as constraints. Li et al. [58] combine COMPASS with a penalty-function approach for handling *constraints*, and prove almost sure convergence of the resulting approach. Vieira et al. [75] also present and analyze an adaptation of COMPASS designed to handle one constraint. Park and Kim [67] present the penalty function with memory (PFM) approach for handling stochastic constraints. This approach can be combined with a random search approach designed for solving unconstrained problems (the authors combine it with the nested partitions approach, see Sect. 10.3.2) and differs from the approach of Li et al. [58] primarily in that the penalty function does not diverge at feasible solutions. Hu and Andradóttir [50] combine the ASRD framework discussed in Sect. 10.4.1 with a penalty function approach and prove that their framework guarantees almost sure convergence when applied to solve simulation optimization problems with stochastic constraints.

Acknowledgements The author gratefully acknowledges support from the National Science Foundation under Grant CMMI–0856600. The author also thanks Liujia Hu for helpful comments.

References

1. M. A. Ahmed and T. M. Alkhamis. Simulation-based optimization using simulated annealing with ranking and selection. *Computers and Operations Research*, 29:387–402, 2002.
2. M. A. Ahmed, T. M. Alkhamis, and M.Hasan. Optimizing discrete stochastic systems using simulated annealing and simulation. *Computers and Industrial Engineering*, 32:823–836, 1997.
3. D. L. J. Alexander, D. W. Bulger, J. M. Calvin, H. E. Romejin, and R. L. Sherriff. Approximate implementations of pure random search in the presence of noise. *Journal of Global Optimization*, 31:601–612, 2005.
4. T. M. Alkhamis and M. A. Ahmed. Simulation-based optimization using simulated annealing with confidence interval. In R. G. Ingalls, M. D. Rossetti, J. S. Smith, and B. A. Peters, editors, *Proceedings of the 2004 Winter Simulation Conference*, 514–519, 2004.

5. T. M. Alkhamis, M. A. Ahmed, and V. K. Tuan. Simulated annealing for discrete optimization with estimation. *European Journal of Operational Research*, 116:530–544, 1999.
6. M. H. Alrefaei and S. Andradóttir. A simulated annealing algorithm with constant temperature for discrete stochastic optimization. *Management Science*, 45:748–764, 1999.
7. M. H. Alrefaei and S. Andradóttir. A modification of the stochastic ruler method for discrete stochastic optimization. *European Journal of Operational Research*, 133:160–182, 2001.
8. M. H. Alrefaei and S. Andradóttir. Discrete stochastic optimization using variants of the stochastic ruler method. *Naval Research Logistics*, 52:344–360, 2005.
9. M. H. Alrefaei and A. H. Diabat. A simulated annealing technique for multi-objective simulation optimization. *Applied Mathematics and Computation*, 215:3029–3035, 2009.
10. M. Alrefaei, A. Diabat, A. Alawneh, R. Al-Aomar, and M. N. Faisal. Simulated annealing for multi objective stochastic optimization. *International Journal of Science and Applied Information Technology*, 2:18–21, 2013.
11. S. Andradóttir. A method for discrete stochastic optimization. *Management Science*, 41:1946–1961, 1995.
12. S. Andradóttir. A global search method for discrete stochastic optimization. *SIAM Journal on Optimization*, 6:513–530, 1996.
13. S. Andradóttir. Accelerating the convergence of random search methods for discrete stochastic optimization. *ACM Transactions on Modeling and Computer Simulation*, 9:349–380, 1999.
14. S. Andradóttir. An overview of simulation optimization via random search. Chapter 20 in S. G. Henderson and B. L. Nelson, editors, *Handbooks in Operations Research and Management Science: Simulation*. North-Holland, 2006.
15. S. Andradóttir and A. A. Prudius. Balanced explorative and exploitative search with estimation for simulation optimization. *INFORMS Journal on Computing*, 21:193–208, 2009.
16. S. Andradóttir and A. A. Prudius. Adaptive random search for continuous simulation optimization. *Naval Research Logistics*, 57:583–604, 2010.
17. F. F. Baeslar and J. A. Sepúlveda. Multi-response simulation optimization using stochastic genetic search within a goal programming framework. In J. A. Joines, R. R. Barton, K. Kang, and P. A. Fishwick, editors, *Proceedings of the 2000 Winter Simulation Conference*, 788–794, 2000.
18. F. F. Baeslar and J. A. Sepúlveda. Multi-objective simulation optimization for a cancer treatment center. In B. A. Peters, J. S. Smith, D. J. Medeiros, and M. W. Rohrer, editors, *Proceedings of the 2001 Winter Simulation Conference*, 1405–1411, 2001.
19. S. Baumert and R. L. Smith. Pure random search for noisy objective functions. Technical Report, University of Michigan, 2002.
20. L. Bianchi, M. Dorigo, L. M. Gambardella, and W. J. Gutjahr. A survey on metaheuristics for stochastic combinatorial optimization. *Natural Computing*, 8:239–287, 2009.
21. J. Branke, S. Meisel, and C. Schmidt. Simulated annaeling in the presence of noise. *Journal of Heuristics*, 14:627–654, 2008.
22. A. A. Bulgak and J. L. Sanders. Integrating a modified simulated annealing algorithm with the simulation of a manufacturing system to optimize buffer sizes in automatic assembly systems. In M. Abrams, P. Haigh, and J. Comfort, editors, *Proceedings of the 1988 Winter Simulation Conference*, 684–690, 1988.
23. R. C. H. Cheng. Fitting statistical models of random search in simulation studies. *ACM Transactions on Modeling and Computer Simulation* 23:article 15, 2013.
24. Y. L. Chia and P. W. Glynn. Limit theorems for simulation-based optimization via random search. *ACM Transactions on Modeling and Computer Simulation*, 23:article 16, 2013.
25. P. Czyzak and A. J. Jaszkiewicz. Pareto simulated annealing – A metaheuristic technique for multiple-objective combinatorial optimization. *Journal of Multi-Criteria Decision Analysis*, 7:34–47, 1998.
26. G. Deng and M. C. Ferris. Adaptation of the UOBYQA algorithm for noisy functions. In L. F. Perrone, F. P. Wieland, J. Liu, B. G. Lawson, D. M. Nicol, and R. M. Fujimoto, editors, *Proceedings of the 2006 Winter Simulation Conference*, 312–319, 2006.

27. G. Deng and M. C. Ferris. Extension of the DIRECT optimization algorithm for noisy functions. In S. G. Henderson, B. Biller, M. H. Hsieh, J. Shortle, J. D. Tew, and R. R. Barton, editors, *Proceedings of the 2007 Winter Simulation Conference*, 497–504, 2007.
28. L. Devroye. On the convergence of statistical search. *IEEE Transactions on Systems, Man and Cybernetics*, 6:46–56, 1976.
29. L. Devroye. Probabilistic search as a strategy selection procedure. *IEEE Transactions on Systems, Man and Cybernetics*, 6:315–321, 1976.
30. E. J. Dudewicz and S. R. Dalal. Allocation of observations in ranking and selection with unequal variances. *Sankhya*, B37:28–78, 1975.
31. K. B. Ensor and P. W. Glynn. Stochastic optimization via grid search. *Lectures in Applied Mathematics*, 33:89–100, 1997.
32. M. C. Ferris, T. S. Munson, and K. Sinapiromsaran. A practical approach to sample-path simulation optimization. In J. A. Joines, R. R. Barton, K. Kang, and P. A. Fishwick, editors, *Proceedings of the 2000 Winter Simulation Conference*, 795–804, 2000.
33. B. L. Fox and G. W. Heine. Probabilistic search with overrides. *The Annals of Applied Probability*, 5:1087–1094, 1995.
34. S. B. Gelfand and S. K. Mitter. Simulated annealing with noisy or imprecise energy measurements. *Journal of Optimization Theory and Applications*, 62:49–62, 1989.
35. A. Ghate and R. L. Smith. Adaptive search with stochastic acceptance probabilities for global optimization. *Operations Research Letters*, 36:285–290, 2008.
36. R. J. Glauber. Time-dependent statistics of the Ising model. *Journal of Mathematical Physics*, 4:294–307, 1963.
37. W.-B. Gong, Y.-C. Ho, and W. Zhai. Stochastic comparison algorithm for discrete optimization with estimation. *SIAM Journal on Optimization*, 10:384–404, 1999.
38. L. S. Gurin. Random search in the presence of noise. *Engineering Cybernetics*, 4:252–260, 1966.
39. L. S. Gurin and L. A. Rastrigin. Convergence of the random search method in the presence of noise. *Automation and Remote Control*, 26:1505–1511, 1965.
40. W. J. Gutjahr. Two metaheuristics for multiobjective stochastic combinatorial optimization. *Stochastic Algorithms: Foundations and Applications*, Lecture Notes in Computer Science, 3777:116–125, 2005.
41. W. J. Gutjahr and G. Ch. Pflug. Simulated annealing for noisy cost functions. *Journal of Global Optimization*, 8:1–13, 1996.
42. B. Hajek. Cooling schedules for optimal annealing. *Mathematics of Operations Research*, 13:311–329, 1988.
43. J. Haddock and J. Mittenthal. Simulation optimization using simulated annealing. *Computers and Industrial Engineering*, 22:387–395, 1992.
44. T. Homem-de-Mello. Variable-sample methods for stochastic optimization. *ACM Transactions on Modeling and Computer Simulation*, 13:108–133, 2003.
45. L. J. Hong. Discrete optimization via simulation using coordinate search. In M. E. Kuhl, N. M. Steiger, F. B. Armstrong, and J. A. Joines, editors, *Proceedings of the 2005 Winter Simulation Conference*, 803–810, 2005.
46. L. J. Hong and B. L. Nelson. Discrete optimization via simulation using COMPASS. *Operations Research*, 54:115–129, 2006.
47. L. J. Hong, B. L. Nelson, J. Xu. Speeding up COMPASS for high-dimensional discrete optimization via simulation. *Operations Research Letters*, 38:550–555, 2010.
48. J. Hu, M. C. Fu, and S. I. Marcus. A model reference adaptive search method for stochastic global optimization. *Communications in Information and Systems*, 8:245–276, 2008.
49. L. Hu and S. Andradóttir. Adaptive random search with discarding for continuous simulation-based optimization. Submitted for publication.
50. L. Hu and S. Andradóttir. Simulation-based continuous optimization with stochastic constraints. Working paper.
51. D. Huang, T. T. Allen, W. I. Notz, and N. Zeng. Global optimization of stochastic black-box systems via sequential kriging meta-models. *Journal of Global Optimization*, 34:441–466, 2006.

52. Y. Jin and J. Branke. Evolutionary optimization in uncertain environments. *IEEE Transactions on Evolutionary Computation*, 9:303–317, 2005.
53. D. R. Jones, C. D. Perttunen, and B. E. Stuckman. Lipschitzian optimization without the Lipschitz constant. *Journal of Optimization Theory and Application*, 79:157–181, 1993.
54. A. Kabirian and S. Ólafsson. Continuous optimization via simulation using Golden Region search. *European Journal of Operational Research*, 208:19–27, 2011.
55. S. Kirkpatrick, C. D. Gelatt, and M. P. Vecchi. Optimization by simulated annealing. *Science*, 220:671–680, 1983.
56. L. H. Lee, E. P. Chew, S. Teng, and Y. Chen. Multi-objective simulation-based evolutionary algorithm for an aircraft spare parts allocation problem. *European Journal of Operational Research*, 189:476–491, 2008.
57. L. H. Lee, E. P. Chew, and H. Li. Multi-objective COMPASS for discrete optimization via simulation. In S. Jain, R. R. Creasey, J. Himmelspach, K. P. White, and M. Fu, editors, *Proceedings of the 2011 Winter Simulation Conference*, 4070–4079, 2011.
58. J. Li, A. Sava, and X. Xie. Simulation-based discrete optimization of stochastic discrete event systems subject to non closed-form constraints. *IEEE Transactions on Automatic Control*, 54:2900–2904, 2009.
59. K. Marti. Minimizing noisy objective functions by random search methods. *Zeitschrift fur Angewandte Mathematik and Mechanik*, 62:T377–T380, 1982.
60. V. Mattila, K. Virtanen, and R. P. Hämäläinen. A simulated annealing algorithm for noisy multiobjective optimization. *Journal of Multi-criteria Decision Analysis*, 20:255–276, 2013.
61. N. Metropolis, A. Rosenbluth, M. Rosenbluth, A. Teller, and E. Teller. Equation of state calculations by fast computing machines. *Journal of Chemical Physics*, 21:1087–1092, 1953.
62. D. Mitra, F. Romeo, and A. Sangiovanni-Vincentelli. Convergence and finite-time behavior of simulated annealing. *Advances in Applied Probability*, 18:747–771, 1986.
63. G. L. Nemhauser and L. A. Wosley. *Integer and Combinatorial Optimzation*. Wiley, 1999.
64. V. I. Norkin, Y. M. Ermoliev, and A. Ruszczyński. On optimal allocation of indivisibles under uncertainty. *Operations Research*, 46:381–395, 1998.
65. V. I. Norkin, G. Ch. Pflug, and A. Ruszczyński. A branch and bound method for stochastic global optimization. *Mathematical Programming*, A83:425–450, 1998.
66. L. Painton and U. Diwekar. Stochastic annealing for synthesis under uncertainty. *European Journal of Operational Research*, 83:489–502, 1995.
67. C. Park and S.-H. Kim. Handling stochastic constraints in discrete optimization via simulation. In S. Jain, R. R. Creasey, J. Himmelspach, K. P. White, and M. Fu, editors, *Proceedings of the 2011 Winter Simulation Conference*, 4217–4226, 2011.
68. J. Pichitlamken and B. L. Nelson. A combined procedure for optimization via simulation. *ACM Transactions on Modeling and Computer Simulation*, 13:155–179, 2003.
69. J. Pintér. Convergence qualification of adaptive partition algorithms in global optimization. *Mathematical Programming*, 56:343–360, 1992.
70. A. A. Prudius and S. Andradóttir. Averaging frameworks for simulation optimization with applications to simulated annealing. *Naval Research Logistics*, 59:411–429, 2012.
71. S. L. Rosen and C. M. Harmonosky. An improved simulated annealing simulation optimization method for discrete parameter stochastic systems. *Computers and Operations Research*, 32:343–358, 2005.
72. R. Y. Rubinstein and D. P. Kroese. *The Cross-Entropy Method: A Unified Approach to Combinatorial Optimization, Monte-Carlo Simulation, and Machine Learning*. Springer, 2004.
73. L. Shi and S. Ólafsson. Nested partitions method for stochastic optimization. *Methodology and Computing in Applied Probability*, 2:271–291, 2000.
74. L. Sun, L. J. Hong, and Z. Hu. Optimization via simulation using Gaussian process-based search. In S. Jain, R. R. Creasey, J. Himmelspach, K. P. White, and M. Fu, editors, *Proceedings of the 2011 Winter Simulation Conference*, 4139–4150, 2011.
75. H. Vieira Jr., K. H. Kienitz, M. C. N. Belderrain. Discrete-valued, stochastic-constrained simulation optimization with COMPASS. In S. Jain, R. R. Creasey, J. Himmelspach, K. P. White, and M. Fu, editors, *Proceedings of the 2011 Winter Simulation Conference*, 4196–4205, 2011.

76. L. Wang and L. Zhang. Stochastic optimization using simulated annealing with hypothesis test. *Applied Mathematics and Computation*, 174:1329–1342, 2006.
77. J. Xu. Efficient discrete optimization via simulation using stochastic kriging. In C. Laroque, J. Himmelspach, R. Pasupathy, O. Rose, and A. M. Uhrmacher, editors, *Proceedings of the 2012 Winter Simulation Conference*, 466–477, 2012.
78. J. Xu, L. J. Hong, and B. L. Nelson. Industrial strength COMPASS: A comprehensive algorithm and software for optimization via simulation. *ACM Transactions on Modeling and Computer Simulation* 20:article 3, 2010.
79. J. Xu, B. L. Nelson, and L. J. Hong. An adapative hyperbox algorithm for high-dimensional discrete optimization via simulation problems. *INFORMS Journal on Computing*, 25:133–146, 2013.
80. W. L. Xu and B. L. Nelson. Empirical stochastic branch-and-bound for optimization via simulation. In B. Johansson, S. Jain, J. Montoya-Torres, J. Hugan, and E. Yücesan, editors, *Proceedings of the 2010 Winter Simulation Conference*, 983–994, 2010.
81. S. Yakowitz, P. L'Ecuyer, and F. Vázquez-Abad. Global stochastic optimization with low-dispersion point sets. *Operations Research*, 48:939–950, 2000.
82. S. Yakowitz and E. Lugosi. Random search in the presence of noise, with application to machine learning. *SIAM Journal on Scientific and Statistical Computing*, 11:702–712, 1990.
83. D. Yan and H. Mukai. Stochastic discrete optimization. *SIAM Journal on Control and Optimization*, 30:594–612, 1992.
84. A. Zhou, B.-Y. Qu, H. Li, S.-Z. Zhao, P. N. Suganthan, and Q. Zhang. Multiobjective evolutionary algorithms: A survey of the state of the art. *Swarm and Evolutionary Computation*, 1:32–49, 2011.

Chapter 11
Stochastic Adaptive Search Methods: Theory and Implementation

Zelda B. Zabinsky

Abstract Random search algorithms are very useful for simulation optimization, because they are relatively easy to implement and typically find a "good" solution quickly. One drawback is that strong convergence results to a global optimum require strong assumptions on the structure of the problem.

This chapter begins by discussing optimization formulations for simulation optimization that combines *expected* performance with a measure of *variability*, or risk. It then summarizes theoretical results for several adaptive random search algorithms (including pure adaptive search, hesitant adaptive search, backtracking adaptive search and annealing adaptive search) that converge *in probability* to a global optimum on ill-structured problems. More importantly, the complexity of these adaptive random search algorithms is *linear* in dimension, on average.

While it is not possible to *exactly* implement stochastic adaptive search with the ideal linear performance, this chapter describes several algorithms utilizing a Markov chain Monte Carlo sampler known as hit-and-run that *approximate* stochastic adaptive search. The first optimization algorithm discussed that uses hit-and-run is called improving hit-and-run, and it has polynomial complexity, on average, for a class of convex problems. Then a simulated annealing algorithm and a population based algorithm, both using hit-and-run as the candidate point generator, are described. A variation to hit-and-run that can handle mixed continuous/integer feasible regions, called pattern hit-and-run, is described. Pattern hit-and-run retains the same convergence results to a target distribution as hit-and-run on continuous domains. This broadly extends the class of the optimization problems for these algorithms to mixed continuous/integer feasible regions.

Z.B. Zabinsky (✉)
University of Washington, Seattle, WA, USA
e-mail: zelda@u.washington.edu

M.C. Fu (ed.), *Handbook of Simulation Optimization*, International Series in Operations Research & Management Science 216, DOI 10.1007/978-1-4939-1384-8_11

11.1 Introduction

Computer simulations are often used to model complex systems, and to identify design decisions that optimize performance. The complex systems often involve uncertainty, and so stochastic discrete-event simulations are useful to explore the design space without conducting physical experiments. Optimization of the design decisions is complicated for several reasons. First, the performance of the system typically has multiple objectives, so the optimization formulation is challenging. Second, once a performance function is specified, it is not known explicitly, but can only be estimated through replications of the computer simulation. And third, the performance function lacks structure that is often used in optimization algorithms, such as linearity or convexity. As such, the performance function may have many "local" optima. Also, the design decisions may include both continuous and integer variables, which influences the type of optimization algorithm that is appropriate.

This chapter first discusses the modeling issues inherent in optimizing a complex system with a high degree of uncertainty. If the performance function, or objective function, has a lot of noise, then optimizing the *expected* performance may need to be tempered with a measure of variability or risk. When the constraints on the system also involve a high degree of uncertainty, the formulation of the optimization problem may not be a straightforward extension of a deterministic optimization model. Also, complex systems often involve multiple objectives, which impacts the optimization model as well as appropriate algorithms.

Once an optimization formulation has been specified, then an algorithm is chosen to numerically explore the decision space. Random search algorithms, also known as stochastic search methods, are popular for ill-structured black-box global optimization problems, because they are straightforward to implement and usually find a relatively good solution quickly. These algorithms have been inspired by physics, such as simulated annealing and interacting particle algorithms, as well as by biology, including genetic algorithms, evolutionary programming, particle swarm, and ant colony optimization. Random search algorithms can be used for mixed continuous and discrete variables. Typically random search algorithms sacrifice a guarantee of optimality for finding a good solution quickly with convergence in probability. However, random search methods have been shown to have a potential to solve large-scale problems efficiently in a way that is not possible for deterministic algorithms [18]. Whereas it is known that a deterministic method for global optimization is NP-hard [48], there is evidence that a stochastic algorithm can be executed in polynomial time on average, with a high probability of getting a solution close to the optimum.

The second section of this chapter focuses on theoretical results regarding the computational complexity of stochastic adaptive search algorithms for ill-structured, "black-box" deterministic global optimization problems. The main idea of this section is that an ideal complexity of a *linear* number of function evaluations in terms of dimension, on average, is theoretically possible with pure adaptive search. This result is for a deterministic objective function (no noise), so only one function evaluation is needed at any solution. Generalizations of pure adaptive search are also presented (including hesitant adaptive search, backtracking adaptive search, and annealing adaptive search) with useful analyses that provide insight into the behavior of random search algorithms.

Section 11.4 of this chapter describes several random search algorithms for global optimization (see also [57]). A major challenge for random search algorithms is the difficulty in efficiently sampling the feasible region; even sampling a uniformly distributed point in a convex bounded feasible region is difficult. Hit-and-run [45] is a well-known Markov chain Monte Carlo (MCMC) sampler with fast convergence properties for sampling from bounded sets. The algorithms described in this section make use of hit-and-run to sample feasible points, and include a parameter that effectively increases the probability of sampling near the global optimum. These algorithms strive to approximate the linear performance of pure adaptive search and annealing adaptive search, and make use of the asymptotic convergence properties of hit-and-run to a target sampling distribution. Section 11.4.1 describes improving hit-and-run, the simplest optimization algorithm using hit-and-run. Section 11.4.2 generalizes improving hit-and-run by adding an acceptance/rejection step; casting it in the form of simulated annealing with hit-and-run as the candidate point generator. Section 11.4.3 generalizes the algorithm further by allowing a population of points. The population of candidate points interact through a modified acceptance/rejection step. Again, hit-and-run is used as the candidate point generator. Whereas hit-and-run operates on continuous sets, a variation called pattern hit-and-run operates on mixed continuous/integer sets. Pattern hit-and-run is described in Sect. 11.4.4. As with hit-and-run, pattern hit-and-run converges to a target probability distribution on a mixed continuous/integer set. Thus, pattern hit-and-run may also be used in the simulated annealing framework and the interacting particle algorithm for a broad class of mixed continuous/integer optimization problems.

Finally, Sect. 11.5 addresses preliminary work incorporating the estimation of a noisy function with repeated replications of a simulation into an adaptive random search algorithm, and Sect. 11.6 concludes the chapter.

11.2 Optimization Models for Complex Systems with Noise

Before going into a detailed discussion of random search algorithms, I would like to comment on the modeling aspect of an optimization formulation when the complex system of interest involves high degrees of uncertainty.

When there is a large degree of uncertainty, then some lessons from the financial optimization literature may be applied, and the performance of the system would include not only the expected value, but some measure of variability or risk. Classic models of portfolio investment optimization consider two objectives: maximize expected return and minimize risk of the investments. In simulation optimization, where there is a high degree of uncertainty, it is important to consider a measure of risk in addition to the *expected* performance of the system in the optimization formulation. For example, consider comparing two designs; one has an expected cost of 100 with a standard deviation of 20, whereas the second has a higher expected cost of 105 with a smaller standard deviation of 1. (Note, the distribution of cost may not be symmetric.) For practical purposes, the second design may be preferred because a small degradation in cost is offset by the benefit of more consistency, i.e., less risk.

For notational purposes, consider a general optimization problem (P), defined as,

$$(P) \qquad \min_{x \in \Theta} f(x)$$

where x is a vector of n decision variables, Θ is an n-dimensional feasible region and assumed to be nonempty, and f is a real-valued function defined over Θ.

It is common to use the expected performance of the system as an objective function in simulation optimization, i.e., $f(x) = \mathsf{E}[Y(x,\xi)]$, where $Y(x,\xi)$ is returned by a simulation and ξ is the noise term. Typically, a sample mean over a number of replications R, such as

$$\hat{f}(x) = \frac{\sum_{r=1}^{R} Y(x,\xi_r)}{R},$$

is used to estimate the expected performance.

An alternative to using expected performance is to use quantiles in the objective function to incorporate uncertainty [20], e.g., minimize a cost threshold such that the probability the actual cost is below the threshold exceeds some probability:

$$\begin{aligned} &\min && z \\ &\text{subject to} && \mathsf{P}(Y(x,\xi) \le z) \ge \alpha. \end{aligned} \tag{11.1}$$

If $\alpha = 0.5$, then the quantile z is just the median, and for symmetric distributions, the median equals the mean. However, if $\alpha = 0.9$, then the interpretation is that 90 % of the time the actual cost is below the cost threshold z. The quantile constraint in (11.1) is a particular example of chance constraints, which are discussed in detail in Chap. 9. In finance, quantiles correspond to value-at-risk (VaR) [40], i.e., in terms of a cumulative distribution function F_Y of a random variable Y and a probability level $\alpha \in (0,1)$, the value-at-risk $\mathrm{VaR}_\alpha(Y)$ is given by the α-quantile $q_\alpha(Y)$:

$$\mathrm{VaR}_\alpha(Y) = q_\alpha(Y) = \min\{z | F_Y(z) \ge \alpha\}.$$

The conditional value-at-risk, abbreviated CVaR, is defined as the expectation of Y in the conditional distribution of its upper α-tail, given by

$$\mathrm{CVaR}_\alpha(Y) = \frac{1}{1-\alpha} \int_0^1 \mathrm{VaR}_\beta(Y) d\beta.$$

In [40], CVaR is shown to have desirable properties that make it an effective objective function that incorporates uncertainty. It has been used extensively as a risk function in financial optimization, and could be useful in simulation optimization.

When there is randomness in the constraints, then the notion of feasibility comes into play. Suppose a noisy constraint is formulated using an expected value approach, i.e., $\mathsf{E}[Y(x,\xi)] \leq 0$. Then for a symmetric function, approximately 50 % of the time the constraint would be violated. A conservative approach is to consider a worst-case analysis, as with robust optimization. Then the model would include constraints of the form, $\max_\xi Y(x,\xi) \leq 0$. A less conservative approach is to move the noisy constraints into the objective with penalty coefficients, as with goal programming.

Goal programming is another way to include multiple objectives into the optimization formulation. For example, suppose there are three objectives, f_1, f_2, and f_3. The first step in goal programming is to create a goal for each objective, such as $f_1 = 100$. The next step is to create an objective function that penalizes deviations from the goals. In this example, the objective function would include a term such as $|f_1 - 100|$, or a quadratic penalty term $(f_1 - 100)^2$. There are many versions of goal programming [19, 38], including one-sided goals and hierarchical goals, that can be adapted to a simulation optimization model.

Another common multi-objective formulation is to combine the multiple objectives into a single objective with the use of weights, e.g., $f(x) = w_1 f_1 + w_2 f_2 + w_3 f_3$. When the multiple objectives are convex in the decision variables, then various combinations of the weights can be used to explore the efficient frontier, i.e., to obtain a set of Pareto-optimal solutions. However, when the objectives are not convex, then a straightforward combination of weights may miss solutions of interest on the efficient frontier. Then a method of moving objectives into constraints and manipulating the right-hand sides of the constraints can be used to explore the efficient frontier. This method has been called the ε-method [12], which is effective for both continuous and integer-valued variables in the decision space.

For the rest of this chapter, we focus on the general optimization problem (P), recognizing that the objective function $f(x)$ and constraint set Θ could include a variety of formulations. For convenience, we denote the global optimal solution to (P) by (x_*, y_*) where

$$x_* = \arg\min_{x\in\Theta} f(x) \quad \text{and} \quad y_* = f(x_*) = \min_{x\in\Theta} f(x).$$

11.3 Performance Analyses of Stochastic Adaptive Search Methods

A strong motivation for random search algorithms is that they can find a "good" solution very quickly. This section provides analyses of performance of several stochastic adaptive search methods, including pure random search, pure adaptive search, hesitant adaptive search, backtracking adaptive search and annealing adaptive search. The analysis is theoretical in nature, but provides insights into how quickly a random search algorithm might perform, and how the quality of the solution is related to iteration. The insights are used in algorithm development using MCMC samplers in Sect. 11.4.

11.3.1 Pure Random Search and Pure Adaptive Search

We start by contrasting pure random search (PRS) with pure adaptive search (PAS), and showing that the average computation of PRS is *exponential* in dimension whereas the average computation of PAS is *linear* in dimension. This motivates the development of algorithms to approximate the ideal performance of PAS.

Pure random search, or "blind search," is the simplest and most obvious random search method for the global optimization problem (P), as discussed in [47]. Pure random search was first defined in 1958 by Brooks [8] and discussed in [16, 17] for problems with continuous variables. PRS is also easily defined for problems with discrete variables, as in [47, 55]. Pure random search samples repeatedly from the feasible region Θ, typically according to a uniform sampling distribution (continuous or discrete). More generally, PRS samples independently, from a distribution δ on the domain Θ. As in [52, 55], it is convenient to use the range distribution ρ, where $\rho(A) = \delta(f^{-1}[A])$ for each measurable subset A of Θ. We also use the cumulative distribution of the range associated with ρ and denoted by p, so

$$p(y) = \rho((-\infty, y)) = \delta(\{x \in \Theta : f(x) \leq y\})$$

for $y \in \mathbb{R}$. When δ is the uniform distribution, then $p(y)$ reduces to the ratio of the n-dimensional volume of the set with points less than y to the volume of Θ.

It can be shown that PRS converges to within an ε distance of the global optimum with probability one [46, 55]. Note that, for a finite problem (e.g., Θ is discrete and bounded), ε may equal zero, whereas for a continuous problem, ε is taken as a small positive value. Pure random search is often used to determine starting points for local search methods, referred to as multi-start algorithms.

Even though pure random search converges almost surely [47], it will take a long time. The expected number of sample points needed to sample within ε of the optimum grows exponentially in dimension. To understand this exponential computation, consider the expected number of iterations until a point is within ε distance of the global minimum, denoted $\mathsf{E}[N(y_* + \varepsilon)]$, as a measure of computational complexity. Each iteration of PRS generates one sample point in Θ and performs exactly one function evaluation, so this performance measure captures the majority of the computational effort. (This assumes that it is relatively easy to generate sample points from δ on Θ.) The random variable $N(y)$, the number of iterations until a point is first sampled with an objective function value of y or less, has a geometric distribution [15], where the probability of a "success" is $p(y)$, the probability of generating a point in the level set $\Theta(y)$, where $\Theta(y) = \{x \in \Theta : f(x) \leq y\}$. Then the expected number of iterations until a point is first sampled within ε distance of the optimum is

$$\mathsf{E}[N(y_* + \varepsilon)] = \frac{1}{p(y_* + \varepsilon)}. \quad (11.2)$$

This relationship supports the natural intuition that, as the target region close to the global optimum gets small (i.e., $p(y_* + \varepsilon)$ gets small), the expected number of iterations gets large (inversely proportional). The probability of failure to achieve $y_* + \varepsilon$ after k iterations is $(1 - p(y_* + \varepsilon))^k$. The variance of the number of iterations until a point first lands in $\Theta(y)$ is

$$\text{Var}[N(y)] = \frac{1 - p(y)}{p(y)^2}.$$

This supports numerical observations that PRS (and perhaps other random search methods) experiences large variation; as the probability $p(y)$ decreases, the variance of $N(y)$ increases.

Now we can see that the performance of PRS depends on how $p(y_* + \varepsilon)$ is impacted by dimension. Consider a continuous global optimization problem where the domain Θ is an n-dimensional ball of radius 1, and the area within ε distance of the global optimum is a ball of radius ε, for $0 < \varepsilon \leq 1$. Using a uniform sampling distribution on this problem, $p(y_* + \varepsilon) = \varepsilon^n$ for $0 < \varepsilon \leq 1$, and the expected number of iterations until a sample point falls within the ε-ball is $(1/\varepsilon)^n$, an exponential function of dimension.

To expand this example, suppose the objective function on the n-dimensional unit ball satisfies a Lipschitz condition with constant K. Then the probability of sampling a point within ε of the global optimum is $p(y_* + \varepsilon) = (\varepsilon/K)^n$ [55] and the expected number of PRS iterations is also exponential in the dimension of the problem,

$$\mathsf{E}[N(y_* + \varepsilon)] = (K/\varepsilon)^n. \quad (11.3)$$

Now consider the performance of PRS on a discrete problem, specifically the traveling salesperson problem (TSP) with N cities and subsequently $(N-1)!$ possible points in the domain Θ. If there is a unique minimum, then $p(y_*) = 1/(N-1)!$ and the expected number of PRS iterations to first sample the minimum is $(N-1)!$, which explodes in N.

All random search algorithms are meant to improve upon PRS. However, the "No Free Lunch" theorems in Wolpert and Macready [50] show that, in essence, no single algorithm can improve on any other when averaged across all possible problems. As a consequence of these theorems, in order to improve upon PRS, we must either restrict the class of problems, have some prior information, or *adapt* the algorithm as properties of a specific problem are observed.

In contrast to pure random search, where each sample is independent and identically distributed, we next consider pure adaptive search, where each sample depends on the one immediately before it. Pure adaptive search (PAS) was introduced in [36] for convex programs and later analyzed for global optimization problems with Lipschitz continuous functions in [60] and for finite domains in [65]. Pure adaptive search, by definition, generates candidate points that are strictly improving. More formally, we have

Pure Adaptive Search (PAS)

Step 0. Set $k = 0$. Generate X_0 according to probability measure δ on Θ. Set $Y_0 = f(X_0)$.

Step 1. Generate X_{k+1} according to the normalized restriction of δ to $\Theta^-_{k+1} = \{x \in \Theta : f(x) < Y_k\}$. Set $Y_{k+1} = f(X_{k+1})$.

Step 2. If the stopping criterion is met, stop. Otherwise, increment k and return to Step 1.

PAS demonstrates the potential performance for a random search algorithm. Direct implementation of PAS is difficult because sampling over an irregular shape, such as an improving level set $\{x : f(x) < y\}$, is a challenge, whereas sampling over regular shapes, such as hyperrectangles, is easy. However, the analysis shows the value of being able to find points in the improving level set; the number of iterations of PAS is *linear* in dimension, which is an exponential improvement over the number of iterations of PRS [55].

The expected number of iterations of PAS to get within ε of the optimum is

$$\mathsf{E}[N(y_* + \varepsilon)] = \ln\left(1/p(y_* + \varepsilon)\right), \tag{11.4}$$

which illustrates the logarithmic improvement over PRS, as in (11.2). And for the Lipschitz-continuous example on the n-dimensional unit ball, the natural logarithm provides linearity in dimension:

$$\mathsf{E}[N(y_* + \varepsilon)] = n \ln\left(K/\varepsilon\right),$$

as compared to the expression for PRS in (11.3).

A more complete analysis of PAS for Lipschitz continuous objective functions, showing it is a non-homogeneous Poisson process, can be found in Theorem 2.9, [55], with an analogous analysis for finite domains (see Corollary 2.9, [55]). The *linearity* result for pure adaptive search implies that adapting the search to sample improving points is very powerful.

To gain insight into the power of steady improvement, consider an algorithm that is a combination of PRS and PAS. We refer to the combined PRS-PAS algorithm as p-PAS because it has probability p of sampling according to PAS (making an improvement) and probability $1 - p$ of sampling according to PRS where $0 \leq p \leq 1$. When $p = 1$, the p-PAS algorithm reduces to PAS, and when $p = 0$ it is simply

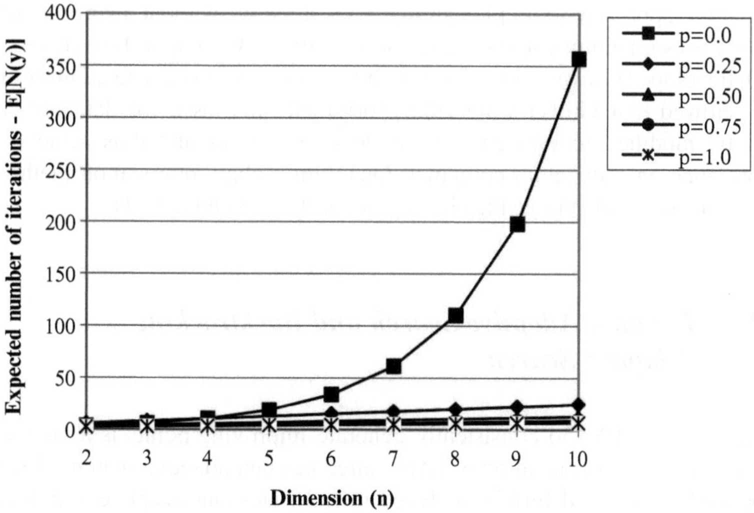

Fig. 11.1 The expected number of iterations for the p-PAS algorithm for dimensions from 2 to 10, and with several values of p, where $p = 0$ is PRS and $p = 1$ is PAS

PRS. The p-PAS algorithm also is a closer representation of practical algorithms that cannot ensure consistent improvement.

An analysis of the p-PAS algorithm in terms of $0 < p \leq 1$ provides a way to quantify the value of steady improvement. The expected number of iterations to sample within ε of the optimum for p-PAS in the continuous case is

$$\mathsf{E}[N(y_* + \varepsilon)] = 1 + \frac{1}{p} \ln \left(1 - p + \frac{p}{p(y_* + \varepsilon)} \right). \tag{11.5}$$

This expression quantifies the impact of steady improvement on performance. This can also be derived using hesitant adaptive search (discussed in the following Sect. 11.3.2) using a bettering probability of $b(y) = (1 - p)p(y) + p$.

The expected number of iterations to convergence as a function of dimension is plotted for a specific problem in Fig. 11.1 [55, Chap. 4.1], which illustrates the exponential complexity of PRS when $p = 0$, as contrasted with the linear growth of p-PAS for $p > 0$; thus, only a small probability of sampling in the improving region is needed to dramatically improve performance, and there is a diminishing return as p exceeds 0.5 and gets closer to 1.0.

The idealistic linear property of PAS has inspired the design of several algorithms with the goal of approximating PAS. Several algorithms have used MCMC samplers embedded in an optimization context to approximate the performance of PAS [22, 39], and even more algorithms have used the MCMC sampler hit-and-run, in continuous and discrete versions, for optimization [29, 32, 33, 41, 44, 62]. These are discussed in more detail in Sect. 11.4.

Another approach to implementing PAS directly, without using an MCMC sampler, uses quantum computing for optimization. Baritompa, Bulger, and Wood [2, 10] developed Grover's adaptive search for optimization on a quantum computer. They showed that Grover's algorithm (originally proposed for database search) could be modified to efficiently generate PAS iterates and thus achieve linear performance on a quantum computer. Quantum global optimization methods are an exciting area and enlarge the class of stochastic search methods.

11.3.2 Hesitant Adaptive Search and Backtracking Adaptive Search

The property of PAS to consistently generate improving points is a challenge to implement, so a generalization of PAS, called hesitant adaptive search (HAS), was introduced in [11] and further analyzed in [53]. Hesitant adaptive search relaxes strict improvement by including a bettering probability $b(y)$ that may depend on objective function value. This allows $b(y)$ to reflect the difficulty of sampling an improving point when y is close to the global minimum. The interpretation of $1 - b(y)$ is the hesitation probability. PAS is a special case of HAS when $b(y) \equiv 1$.

Hesitant Adaptive Search (HAS)

Step 0. Set $k = 0$. Generate $X_0 \in \Theta$ according to probability measure δ on Θ. Set $Y_0 = f(X_0)$.

Step 1. Generate X_{k+1} according to the normalized restriction of δ on the improving set $\Theta_k^- = \{x : x \in \Theta \text{ and } f(x) < Y_k\}$ with probability $b(Y_k)$, and otherwise set $X_{k+1} = X_k$. Set $Y_{k+1} = f(X_{k+1})$.

Step 2. If a stopping criterion is met, stop. Otherwise, increment k and return to Step 1.

The behavior of HAS is a marked Poisson process, where the stochastic process of Y_k, the objective function on the kth iteration, improves, hesitates, then improves again. The expected number of iterations of HAS to get within ε of the optimum for a continuous global optimization problem [11] is nicely expressed as

$$\mathsf{E}[N(y_* + \varepsilon)] = 1 + \int_{(y_*+\varepsilon,\infty)} \frac{d\rho(t)}{b(t)p(t)}.$$

When the bettering probability equals one and the sampling distribution δ is uniform, then the expression for HAS gives back the linearity result of PAS. The probability distribution of the number of iterations to termination in a general mixed continuous-discrete case is derived in [53], and a summary is provided in [55].

The analysis of HAS has been used in [58] to develop a stopping and restarting strategy of a multi-start method that considers tradeoffs between computational effort and the probability of obtaining the global optimum. Many algorithms use multi-start, with independently sampled initial starting points, combined with a local search algorithm, or even multi-start with simulated annealing at a low temperature. A common question is how many multi-starts to run, and when to stop a single run of the algorithm. The idea in [58] is to use the HAS analysis to model the behavior of the algorithm with specific parameters that are updated from observations of the algorithm. Then the analysis can evaluate the tradeoffs between computational effort and probability of obtaining the global optimum, to determine whether to stop a run and restart with a new run, or whether to terminate completely. A framework called Dynamic Multistart Sequential Search was developed and numerical tests demonstrate its viability for global optimization with black-box functions. The HAS theory provides information regarding the benefit of performing more runs, based on the observed performance of the algorithm.

A further generalization of HAS, named backtracking adaptive search (BAS) was developed to more closely capture the behavior of algorithms that accept non-improving points with a certain probability, in order to escape local barriers and more freely move around the space. The backtracking adaptive search algorithm was first analyzed for discrete problems in [26], and subsequent analyses in [3, 9, 51] provide a complete characterization of the distribution of the number of iterations to convergence for a general mixed continuous-discrete problem. The BAS algorithm uses both a bettering probability $b(y)$, as in HAS, and a worsening probability $w(y)$, while the hesitation probability is $h(y) = 1 - b(y) - w(y)$.

Backtracking Adaptive Search (BAS)

Step 0. Set $k = 0$. Generate $X_0 \in \Theta$ according to probability measure δ on Θ. Set $Y_0 = f(X_0)$.

Step 1. Generate X_{k+1} according to the normalized restriction of δ to

$$\begin{aligned}&\{x \in \Theta : f(x) > Y_k\} \text{ with probability } w(Y_k),\\&\{x \in \Theta : f(x) = Y_k\} \text{ with probability } h(Y_k),\\&\{x \in \Theta : f(x) < Y_k\} \text{ with probability } b(Y_k),\end{aligned}$$

where $w + h + b = 1$. Set $Y_{k+1} = f(X_{k+1})$.

Step 2. If a stopping criterion is met, stop. Otherwise, increment k and return to Step 1.

The analyses of BAS, HAS and PAS provide insights into the impact of consistent improvement and the sampling distribution on the performance. It has already been used to inspire many algorithms and stopping criterion, and may be further used to more accurately model tradeoffs between exploration and exploitation of algorithms. At the time of writing this, preliminary work is ongoing to incorporate randomness due to noise in the objective function into the analysis of HAS and BAS. The goal is to include tradeoffs in the estimation error, as well as the exploration and exploitation, to design better algorithms.

11.3.3 Annealing Adaptive Search

Another generalization of PAS is annealing adaptive search (AAS), which is a theoretical abstraction of simulated annealing. In contrast to PAS and HAS, which consider points sampled from improving regions of Θ, AAS generates points from the entire feasible set Θ according to a Boltzmann distribution parameterized by temperature.

In our context with objective function $f(x)$ and feasible region Θ, the density for the Boltzmann distribution with parameter $T > 0$ is

$$\frac{e^{-f(x)/T}}{\int_\Theta e^{-f(z)/T}dz}.$$

The algorithm was introduced as adaptive search in [42], and it was proven that the record values (best values observed) stochastically dominate those of PAS and hence inherit the ideal linear complexity of PAS. The analysis was extended to include the total number of sample points (including non-improving points) in [44], and the name AAS was coined in [52]. The expected number of sample points of AAS is also *linear* on average under certain conditions on the cooling schedule for the temperature parameter. The cooling schedule that maintains a *linear* complexity is analytically derived for a broad class of problems, including both continuous and discrete domains, and uses the HAS analysis to provide bounds on performance (see [44], and summarized in [52, 55, 56]).

Annealing Adaptive Search (AAS)

Step 0. Set $k = 0$. Generate X_0 according to the uniform distribution (i.e., Boltzmann distribution with temperature parameter ∞) on Θ. Set $Y_0 = f(X_0)$ and set $T_0 = \tau(Y_0)$, where τ is the cooling schedule.

Step 1. Generate X_{k+1} according to a Boltzmann distribution with temperature parameter T_k on Θ. Set

$$Y_{k+1} = \begin{cases} f(X_{k+1}) & \text{if } f(X_{k+1}) < Y_k, \\ Y_k & \text{otherwise.} \end{cases}$$

Set temperature parameter $T_{k+1} = \tau(Y_{k+1})$.

Step 2. If the stopping criterion is met, stop. Otherwise, increment k and return to Step 1.

The concepts of PAS, HAS, BAS and AAS provide motivation for developing algorithms that have certain sampling distributions, whether it is sampling from a uniform distribution, a Boltzmann distribution, or a general δ distribution on Θ. The challenge is how to efficiently sample from these distributions. There are known methods for sampling from regular shapes, such as hyperrectangles, but sampling over irregular shapes, such as improving level sets, is difficult. Even sampling from a hyperrectangle embedded within a hyperrectangle, where the dimensions of the inner box are unknown, is difficult. A Markov chain Monte Carlo sampler is a standard method used to generate points that approximate a desired distribution over an arbitrary set. The algorithms in the next section use MCMC samplers to approximate sampling from the desired distributions.

11.4 Optimization Algorithms using Hit-and-Run

Motivated by the theoretical analyses of PAS, HAS, BAS and AAS in Sect. 11.3, we use an MCMC sampling approach to approximate a sampling distribution (e.g., uniform, or Boltzmann) that provides good performance.

Hit-and-run [45] is an MCMC sampling technique that iteratively generates a sequence of points in a set by taking steps of random length in randomly generated directions. Hit-and-run can generate a sequence of points that asymptotically approach a uniform distribution on open sets, and thus could be used to approximate PAS sampling uniformly on improving level sets. The time for hit-and-run to converge to a uniform distribution on convex sets is polynomially bounded, and it is considered the most efficient algorithm known to date for generating an asymptotically uniform distribution in a convex set [27, 28]. This suggests it may be an efficient approximation of PAS for optimization. The improving hit-and-run algorithm, motivated by PAS, is discussed in Sect. 11.4.1.

With the use of a filter, specifically acceptance–rejection of candidate points, hit-and-run can be used to approximate arbitrary multivariate distributions, including the Boltzmann distribution [6]. Thus, by adding the Metropolis criterion commonly used in simulated annealing [31], hit-and-run can be used to approximate AAS. The algorithm using hit-and-run within a simulated annealing framework is referred to as hide-and-seek [41], and is discussed in Sect. 11.4.2. A cooling schedule for hide-and-seek is provided in [44] that uses the analysis of AAS and HAS. Hit-and-run in a simulated annealing framework was applied to convex sets and shown to be polynomially bounded [7, 22].

Hit-and-run was also used in a population-based simulated annealing context, called the interacting particle algorithm [32, 33], discussed in Sect. 11.4.3. The computational advantage of a population of points is demonstrated in [32]. The interacting particle algorithm, as in simulated annealing, has a temperature parameter. The temperature is allowed to heat and cool using a meta-control approach [33, 34], and is summarized in Sect. 11.4.3.

Hit-and-run was originally defined for continuous sets, but later expanded to mixed continuous/integer sets, with variations called discrete hit-and-run [4] and pattern hit-and-run [29, 30], which retain the same convergence properties of asymptotically converging to a target probability distribution. A description of pattern hit-and-run is provided in Sect. 11.4.4. Some preliminary analyses on the mixing time of pattern hit-and-run and some numerical results using pattern hit-and-run for mixed continuous/integer global optimization [29] are also mentioned. Pattern hit-and-run may also be used in the simulated annealing framework and the interacting particle algorithm.

11.4.1 Improving Hit-and-Run (IHR)

The initial application of hit-and-run to optimization was called improving hit-and-run (IHR) [62], and was successfully applied to black-box optimization engineering design problems [54, 59, 63]. The main idea behind improving hit-and-run is to embed hit-and-run within an optimization procedure, to attempt to approximate the linear performance of pure adaptive search. The term "improving" was coined to indicate that the sequence of points were improving with regard to their objective function values as in pure adaptive search.

If hit-and-run was allowed to run long enough on each improving level set to generate a nearly uniform point, then one would expect it would be a good approximation to PAS, and only a linear number of such improving iterations would be needed, on average. Instead of generating long sequences of hit-and-run on each improving region, IHR takes the other extreme, and reduces the hit-and-run sequence to a length of *one*. It repeats until an improving point is found, and then updates the best objective function value to indicate the next improving level set. Although the sequence of points generated per iteration is too short to closely approximate uniformity, the hope is that the algorithm may still inherit a polynomial complexity. In fact, for the class of positive definite quadratic programs, the expected number of *function evaluations* for IHR is polynomial in dimension, in particular, $O(n^{5/2})$ [62].

The following gives a more detailed description of IHR.

Improving Hit-and-Run (IHR)

Step 0. Set $k = 0$. Generate $X_0 \in \Theta$, and set $Y_0 = f(X_0)$.

Step 1. Generate a random direction vector D_k from the multivariate normal distribution with mean 0 and covariance matrix H^{-1}. If $H = I$, this is equivalent to generating a direction uniformly distributed on the boundary of a unit hypersphere.

Step 2. Generate a candidate point $W_{k+1} = X_k + \lambda D_k$ by sampling uniformly over the line set

$$L_k = \{x : x \in \Theta \text{ and } x = X_k + \lambda D_k, \lambda \text{ a real scalar}\}.$$

If $L_k = \emptyset$, go to Step 1.

Step 3. Update the current point X_{k+1} with the candidate point if it is improving, i.e., set

$$X_{k+1} = \begin{cases} W_{k+1} & \text{if } f(X_k + \lambda D_k) < Y_k \\ X_k & \text{otherwise} \end{cases}$$

and set $Y_{k+1} = f(X_{k+1})$.

Step 4. If the stopping criterion is met, stop. Otherwise, increment k and return to Step 1.

Steps 1 and 2 describe how hit-and-run generates a candidate point. A random direction is generated in Step 1, and a random point is generated on the line set created by intersecting the direction with the feasible region in Step 2. There are several variations of the algorithm, which are discussed in [55, 61, 62]. Different direction distributions are discussed in [22–24]. Instead of choosing a point uniformly on the line set in Step 2, Wang et al. [49] explore other distributions in "adaptive parametric improving hit-and-run."

Step 3 is the acceptance/rejection step that moves only to improving points. The acceptance/rejection step in IHR may be considered a special case of simulated annealing with temperature equal to zero, because only improving points are accepted. In the next section, this acceptance/rejection step is modified to accept non-improving points with some probability, as in the Metropolis criterion for simulated annealing.

The analysis of PAS was used to show that the expected number of function evaluations of IHR is $O(n^{5/2})$ for a certain class of convex programs [62]. The insight from the PAS analysis is that IHR is relatively insensitive to small perturbations in the objective function, so if the overall trend of the objective is convex, then the performance of IHR is not hindered by many small local optima. The intuition is that there is no need to closely approximate an improving direction (such as a gradient search), because as long as the direction chosen in IHR has a high probability of intersecting the improving region, we would expect polynomial time performance, on average. This suggests that IHR may perform well for simulation optimization, where the randomness in the objective function due to the simulation will not have a big impact, as long as the objective function itself is nearly convex.

11.4.2 Simulated Annealing with Hit-and-Run

Whereas IHR samples points according to hit-and-run and only accepts improving points, Romeijn and Smith [41, 42] embed hit-and-run into a simulated annealing algorithm. Their algorithm, called hide-and-seek, uses hit-and-run to generate a candidate point that is then accepted or rejected based on the Metropolis criterion with a temperature parameter. Even though the acceptance probability for simulated annealing is interpreted as aiding the algorithm to escape local optima, simulated annealing has also been successfully applied to convex programs. Bertsimas and Vempala [7] and Kalai and Vempala [22] used hit-and-run as a candidate generator in a simulated annealing-type algorithm for solving convex programs. In [22], simulated annealing is shown to converge quickly, and under certain conditions, the Boltzmann distribution is proven to be optimal for annealing.

One of the advantages of using hit-and-run as the generator within simulated annealing is that it involves no parameters that need fine tuning; all it needs is a random direction and a random step length. In this sense it is a robust generator. While most simulated annealing generators are local in the sense that only immediate neighbors or a small subset of the domain has a possibility of being generated, hit-and-run has a positive probability of sampling anywhere in the domain on a single iteration.

From a theoretical perspective, an advantage of hit-and-run is that it can converge to any arbitrary distribution, and in particular a Boltzmann distribution for a fixed temperature [6]. Thus, if hit-and-run was run long enough in each iteration of hide-and-seek, it would be close to a Boltzmann distribution, and the linear complexity of AAS would be well-approximated. Hit-and-run embedded as a candidate generator in simulated annealing has both analytical and numerical success.

The following gives a more detailed description of hide-and-seek.

Hide-and-Seek

Step 0. Set $k = 0$. Generate $X_0 \in \Theta$, and set $Y_0 = f(X_0)$ and $T_0 \in (0, \infty)$.

Step 1. Generate a random direction vector D_k from the multivariate normal distribution with mean 0 and covariance matrix H^{-1}. If $H = I$, this is equivalent to generating a direction uniformly distributed on the boundary of a unit hypersphere.

Step 2. Generate a candidate point $W_{k+1} = X_k + \lambda D_k$ by sampling uniformly over the line set

$$L_k = \{x : x \in \Theta \text{ and } x = X_k + \lambda D_k, \lambda \text{ a real scalar}\}.$$

If $L_k = \emptyset$, go to Step 1.

Step 3. Update the current point X_{k+1} by accepting or rejecting the candidate point as follows,

$$X_{k+1} = \begin{cases} W_{k+1} & \text{with probability } \mathsf{P}_{T_k}(X_k, W_{k+1}), \\ X_k & \text{otherwise,} \end{cases}$$

where $\mathsf{P}_{T_k}(X_k, W_{k+1}) = \min\{1, \exp(f(X_k) - f(W_{k+1}))/T_k\}$.

Update the temperature according to a cooling schedule, i.e., set $T_{k+1} = \tau(T_k)$, and set $Y_{k+1} = f(X_{k+1})$. Also update the best point so far, $Y_{Best} = \min\{Y_{Best}, Y_{k+1}\}$.

Step 4. If the stopping criterion is met, stop. Otherwise, increment k and return to Step 1.

The same variations to direction distributions and sampling points on the line set discussed in Sect. 11.4.1 can be applied to hide-and-seek. In [22], an estimate of the covariance matrix is used in Step 1 for the direction distribution.

There are also many variations of the cooling schedule, denoted $\tau(T_k)$, in Step 3. Simulated annealing algorithms that generate candidate points in a local neighborhood of the current point rely on accepting non-improving points to escape a basin surrounding a local minimum. For these algorithms, the cooling schedule is typically slow to provide convergence to a global minimum. However, hit-and-run has a positive probability of sampling anywhere in the space, so convergence to a global minimum can be proven without strict assumptions on the cooling schedule.

Bélisle [5] proves that the sequence of objective function values $\{f(X_k), k = 0,1,\ldots\}$ converges in probability to the global minimum y_*, as long as the cooling schedule converges to 0 in probability i.e., for all $\varepsilon > 0$ and $x_0 \in S$,

$$\mathsf{P}(f(X_k) > y_* + \varepsilon | X_0 = x_0) \to 0 \text{ as } k \to \infty.$$

A slightly weaker form of convergence is to investigate the best point sampled, in contrast to the last point sampled in the sequence. For instance, if hide-and-seek uses a constant temperature value that is very large, it behaves like pure random search, which converges to the global optimum when keeping track of the best point sampled. The best objective function value sampled by hide-and-seek converges to the global minimum [41], i.e.,

$$Y_{Best} \to y_* \text{ almost surely, as } k \to \infty$$

for a broad class of problems, cooling schedules and direction distributions.

These convergence results do not put strong restrictions on how quickly the cooling schedule converges to zero. This implies that the cooling schedule can be fast, and may even be non-monotonic (i.e., heat and cool), as long as it eventually cools to zero. It is important to realize that the cooling schedule has an effect on the *rate* of convergence of hide-and-seek. The convergence rate also depends on, of course, the specific objective function and constraints in the problem.

The cooling schedule in [44] chooses the temperature for the next iteration so that the probability of generating an improving point under the Boltzmann distribution is at least a constant, such as $1-\alpha$. Consequently the expected number of function evaluations is linear in the dimension of the problem, assuming hit-and-run has converged to a Boltzmann distribution. Since approximating a Boltzmann distribution is computationally more difficult when the temperature is small, the cooling schedule strategy in [44] keeps the temperature as high as possible while maintaining the $1-\alpha$ probability of improvement. The derivation of the temperature uses a worst-case analysis, where a worst-case problem is defined for both continuous and integer domains. The worst-case problem needs several pieces of information: dimension n, the current record value Y_{Best}, the Lipschitz constant K for a continuous problem, or an analogous constant $\tilde{K}$ for a discrete problem, the n-dimensional volume of the feasible set, and the optimal value y_*. For a specific problem, we certainly know the dimension, and at the kth iteration, we know the record value Y_{Best}. An upper bound on K or $\tilde{K}$ can be estimated (see [37]). Similarly, an upper bound on the volume of the feasible region may be used while maintaining the linearity properties of the resulting cooling schedule. The main difficulty is that the optimal function value y_* is unknown; however, y_* can be estimated using order statistics as in [41]. The computational results in [44] illustrate the benefit of using this cooling schedule strategy as compared to other cooling schedules in the literature.

11.4.3 Population-Based Simulated Annealing (Interacting Particle Algorithm) with Hit-and-Run

Another use of hit-and-run in a global optimization context is in a population-based method called the interacting particle algorithm. The interacting particle algorithm in [32–34] is based on Feynman–Kac systems in statistical physics [13, 14] and combines the ideas of simulated annealing with population-based algorithms (such as genetic algorithms).

In order to explore the benefits of a population-based method, a numerical study was performed in [32] comparing the interacting particle algorithm to simulated annealing with multi-start. Making multiple restarts can also be viewed as having multiple particles instead of a single particle; however when the restarts are independent there is no interaction between the particles. The numerical results in [32] indicate that there is a significant computational benefit to the interaction. However, there is a need for more research to determine the best population size.

A detailed description of the interacting particle algorithm with hit-and-run follows.

Interacting Particle Algorithm with Hit-and-Run

Step 0. Initialization: Set $k = 0$. Generate N points, $X_0^i \in \Theta$, for $i = 1,\ldots,N$, independently. Set $T_0 \in (0,\infty)$.

Step 1. N-Particle Exploration: From each point X_k^i, perform a single hit-and-run step to generate candidate point W_{k+1}^i, for $i = 1,\ldots,N$.

Step 2. Temperature Parameter Update: Update the temperature $T_{k+1} = \tau(T_k)$, where $\tau(T_k)$ is a function that may depend on all of the information generated by the interacting-particle algorithm up to iteration k, including historical particle locations, objective function values and temperature values.

Step 3. N-Particle Selection: Select the locations for the next population of points by setting $X_{k+1}^i \leftarrow W_{k+1}^j$ with probability $s_j(T_{k+1})$, where

$$s_j(T_{k+1}) = \left(\frac{G_{T_{k+1}}(W_{k+1}^j)}{\sum_{i=1}^N G_{T_{k+1}}(W_{k+1}^i)} \right)$$

and $G_{T_{k+1}}(w) = \exp(-f(w)/T_{k+1})$. Set $Y_{k+1}^i = f(X_{k+1}^i)$, and also update the best point so far, $Y_{Best} \leftarrow \min\{Y_{Best}, Y_{k+1}^i\}$.

Step 4. If the stopping criterion is met, stop. Otherwise, increment k and return to Step 1.

As in simulated annealing, the interacting particle algorithm has a temperature parameter that must approach zero for the population to converge. However, it does not need to approach zero monotonically. In [33, 34], a meta-control approach (based on [25]) was developed to control the temperature parameter in order to speed convergence with a high probability of accuracy. Instead of choosing a cooling schedule a priori, the meta-control methodology dynamically determines the changes in the temperature parameter (heating and cooling) by adapting it to the state of the sampling process. The methodology controls the evolution of the probability density function of the particles with the temperature parameter to make the algorithm's search process satisfy user-defined performance criteria. The criteria in [33] include the expected objective function value of the particles, the spread of the particles, and the algorithm running time. Numerical results indicate improved performance by allowing heating and cooling of the temperature parameter through the meta-control methodology [33].

The combination of the interacting particle algorithm with the meta-control methodology reduces the necessity to have a cooling schedule that works for all problems. The meta-control methodology can adapt the cooling schedule to the specific problem based on observed function evaluations. More research is needed to adaptively control the parameters of the generation mechanism as well as the selection mechanism.

11.4.4 Pattern Hit-and-Run (PHR)

The theoretical and computational success of hit-and-run on continuous sets has led to many extensions and variations. Andersen and Diaconis [1] proposed a generalization of hit-and-run algorithms for MCMC samplers. They describe choosing the candidate point according to the target density π restricted to the line set determined by the direction vector, as in [5]. The choice of the random direction is allowed to have a very general distribution, and the concept of a one-dimensional Euclidean line determined by the direction vector is generalized to include subsets of Θ.

A form of a generalized line on an integer set is developed in [4] using a bidirectional random walk (biwalk), and the algorithm is called discrete hit-and-run. While discrete hit-and-run has the desired asymptotic convergence properties to a target distribution on a bounded integer set, the construction of a biwalk requires a lot of computation. This led to pattern hit-and-run, which uses "patterns" to generate a biwalk and can be implemented very efficiently [29, 30]. Pattern hit-and-run can also be extended to a mixed continuous/integer set and retains the asymptotic convergence properties to a target distribution. Because the analysis of

PAS, HAS, BAS and AAS holds for continuous and discrete variables, pattern hit-and-run provides an efficient MCMC sampler that can be used in an optimization algorithm to approximate a sampling distribution on a mixed continuous/integer domain.

It is easy to embed pattern hit-and-run (PHR) into an optimization algorithm by replacing Steps 1 and 2 in IHR, hide-and-seek, or the interacting particle algorithm with the following steps. The acceptance/rejection step in Step 3 remains the same.

Given a starting point $x \in \Theta$, PHR has two steps to generate a candidate point w.

Pattern Hit-and-Run (PHR)

Step 1. Generate a pattern that determines a biwalk from x. A pattern consists of a step length vector, denoted by $V = (v_1, \ldots, v_n)$, and a permutation of n coordinate dimensions, denoted by $I = \{I_1, \ldots, I_n\}$ with $I_i \in \{1, \ldots, n\}$ and $I_i \neq I_j$ for $i \neq j$. Two methods for generating V are:

1. Sphere Biwalk generates V by choosing a uniform direction on an n-dimensional hypersphere of radius R, then generating a uniform point on the diameter defined by this direction, and rounding the random point to the nearest mixed continuous/integer lattice point.
2. Box Biwalk generates V uniformly on an n-dimensional mixed continuous/integer box $C = [-c_1, c_1] \times \cdots \times [-c_n, c_n]$.

Step 2. Generate a candidate point w, uniformly distributed on the biwalk intersected with Θ.

Step 1 of PHR determines a biwalk using a pattern consisting of a step length for each dimension, and an order of dimensions. For example, in a three dimensional problem where dimensions 1 and 2 are constrained to be integer valued, and dimension 3 is real-valued, the random step lengths may be $v_1 = 5, v_2 = 4$, and $v_3 = 1.7$. The biwalk may first move along dimension 1, then 3, then 2, denoted $I = \{1, 3, 2\}$. Generating a pattern biwalk in Step 1 is analogous to generating a random direction in hit-and-run. In [29], it is proven that, as the mesh of a lattice in the feasible region gets finer and approaches a continuum, the stochastic process of points generated by PHR converge to the stochastic process of points generated by hit-and-run.

Step 2 of PHR generates a candidate point along the intersection of the biwalk with Θ. As in continuous hit-and-run, if Θ is ill-structured (e.g., nonconvex or disconnected), then usually Θ is enclosed in a hyperrectangle, and an acceptance/rejection on the line set in the hyperrectangle is used to determine a uniform point on the line set that is also in Θ. For hit-and-run, the additional computation due to the one-dimensional acceptance–rejection is analyzed in [24], and it is shown that the size of the enclosing box is not a critical factor to the overall computation. Another variation to reduce the number of rejected sample points on the line is to incorporate a slice sampler into hit-and-run; for details, see [24, 61]. These modifications can also be applied to the one-dimensional sampling on a biwalk of PHR.

If Θ is a polytope, then the end points of the line set can easily be determined using a modified minimum ratio test, and a uniform point can be easily generated by mapping to the line segment. This mapping for continuous hit-and-run was extended to PHR to map a point onto the biwalk on a polytope, for details see [30]. Thus the biwalk is never actually generated, and the pattern is used in the mapping to efficiently generate a candidate point.

Since the rate of convergence of hit-and-run to a uniform distribution on a convex set is polynomial in dimension, and since PHR converges to hit-and-run as the mesh of a lattice approaches a continuous set, we have hope that PHR also exhibits polynomial time convergence for some types of discrete sets. To date, a few special sets have been identified. In [29], PHR is proven to converge to a uniform distribution in polynomial time, $O(n^3)$, on an integer lattice of a hyperrectangle in n dimensions. It is well known that generating a point uniformly distributed on an integer lattice of a hyperrectangle $H = \{x : l_i \leq x_i \leq u_i, \text{ for } i = 1,\ldots,n\}$ is very easy when the upper and lower bounds (u_i and l_i) are known. However, suppose the bounds are *not* known, and it is possible to enclose the target hyperrectangle in a larger one whose sides are twice the length. Then an n-dimensional acceptance/rejection sampling on the larger hyperrectangle would need an exponential number of sample points to obtain one in the smaller set. In contrast, PHR could provide an approximately uniform point in polynomial time. PHR also has a polynomial rate of convergence to a uniform distribution on an integer lattice intersected with a truncated cube. The proof in [30] relies on analysis in [35].

It is straightforward to implement PHR in an optimization algorithm (such as IHR, simulated annealing, or the interacting particle algorithm) when the feasible region is mixed continuous/integer constrained in a polytope, or enclosed in a larger region. A speculation is that, if the improving level sets of an integer program are nested hyperrectangles, then the algorithm would have nearly polynomial performance. Recognizing that PHR has polynomial-time performance on a truncated cube suggests that there may be an appropriate definition of a "well-rounded" integer program for which there is polynomial complexity, on average. This is an open research issue.

Numerical tests have been performed using IHR with PHR on 18 test problems with dimensions varying between 5 and 20 [30]. IHR with PHR and sphere or box biwalk outperformed discrete hit-and-run with a random biwalk [4], separating continuous and integer variables, and a step function approach [43]; for details, see [29]. The use of pattern hit-and-run in an optimization framework is application to simulation optimization problems, allowing continuous and integer variables.

11.5 Estimation and Optimization

The algorithms described thus far may be used in simulation optimization, but there has been no mention of how to estimate the objective function at each sample point. One approach is to use a fixed number of replications of the simulation to perform throughout the optimization algorithm, but there is reason to vary the number of replications depending on the observed function values. As with classical statistics, the more replications, the tighter the confidence interval on the estimate of the function, but, when integrated into an optimization algorithm, there is no need to have the same tight confidence intervals on regions that are of little interest (with large function values). An algorithm named Probabilistic Branch and Bound (PBnB) was introduced in [64] and the analysis expanded in [21] to derive confidence intervals based on a random sample in the domain, and dynamically updated replications.

In [21], the PBnB algorithm is designed to estimate a user-defined quantile, and approximate the corresponding level set of solutions, at a user-specified confidence level. This algorithm can be applied to mixed continuous/integer domains. Consider the example used in (11.1), where the user wants to identify a cost threshold z such that the probability the actual cost is below z exceeds 90 %, i.e., identify the best 10 % solutions. A simulation is used to estimate cost, and PBnB iteratively determines the sample points to execute the simulation, and the number of replications. The algorithm provides confidence intervals on the value of z, which tighten as more iterations are performed. In addition, the set of best 10 % solutions is estimated by hyperrectangles. During the execution of the algorithm, subregions are pruned when there is a high probability that they do not belong to the target level set, subregions are maintained when there is a high probability that they *do* belong to the target level set, and subregions are partitioned for more sampling and replications when their status is uncertain.

An illustration of approximating the set of the best 10 % solutions on a two-dimensional sinusoidal function is illustrated in Fig. 11.2 (taken from [21]). Two dimensions are chosen to visualize the partitioning and level set approximation easily. A nonconvex, multi-modal sinusoidal function is used as a test function with two cases; one case has no noise, and the second case has random noise according to a standard normal distribution.

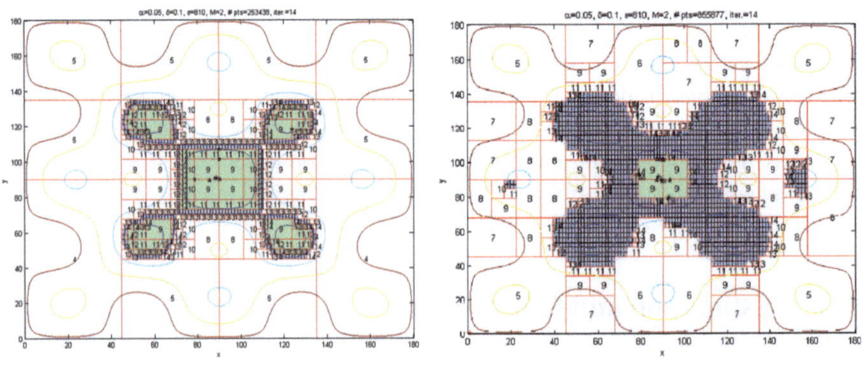

Fig. 11.2 Sinusoidal function with no noise on the *left*, and $\mathcal{N}(0,1)$ noise on the *right*

In Fig. 11.2, each rectangle represents a subregion in the algorithm. The number in a subregion represents the iteration that the subregion is pruned or maintained. The white rectangles are pruned, the light gray (green in color) rectangles are maintained, and the dark gray (blue in color) rectangles are the last current subregions. In Fig. 11.2, the best 10 % level set with no noise is approximated almost perfectly, but the noisy version shows a larger dark gray region where we are not confident whether the solutions belong to the top 10 % or not. In this way, the algorithm provides information to the user about the uncertainty associated with the objective function. Research in extending the versions of PBnB in [21, 64] is ongoing.

11.6 Conclusions

Random search methods are useful for black-box global optimization problems, and combined with estimation techniques, can be readily applied to noisy functions. Theoretical analyses in PAS, HAS and AAS can be used to guide the algorithm development, and allocate computation appropriately. The use of MCMC samplers, such as hit-and-run, pattern hit-and-run, and variations, can be embedded in optimization algorithms to approximate the linear performance in theory. The analyses can be used to develop stopping criterion, as in [58], and to derive analytical cooling schedules, as in [44]. The interacting particle algorithm with meta-control benefits from having a population of points, and dynamically heats and cools the temperature parameter as function values are observed. Adapting these algorithms to simulation optimization, as in PBnB, involves integrating an estimation scheme into the optimization algorithm to efficiently search and estimate the objective function.

Acknowledgement This work was supported in part by the National Science Foundation under Grant CMMI-1235484.

References

1. H. C. Andersen and P. Diaconis. Hit and run as a unifying device. *Journal de la societe francaise de statistique & revue de statistique appliquee*, 148(4):5–28, 2007.
2. W. P. Baritompa, D. W. Bulger, and G. R. Wood. Grover's quantum algorithm applied to global optimization. *SIAM Journal of Optimization*, 15(4):1170–1184, 2005.
3. W. P. Baritompa, D. W. Bulger, and G. R. Wood. Generating functions and the performance of backtracking adaptive search. *Journal of Global Optimization* 37:159–175, 2007.
4. S. Baumert, A. Ghate, S. Kiatsupaibul, Y. Shen, R. L. Smith, and Z. B. Zabinsky. Discrete hit-and-run for generating multivariate distributions over arbitrary finite subsets of a lattice. *Operations Research*, 57(3):727–739, 2009.
5. C. J. P. Bélisle. Convergence theorems for a class of simulated annealing algorithms on R^d. *J. Applied Probability*, 29:885–895, 1992.

6. C. J. P. Bélisle, H. E. Romeijn, and R. L. Smith. Hit-and-run algorithms for generating multivariate distributions. *Mathematics of Operations Research*, 18:255–266, 1993.
7. D. Bertsimas and S. Vempala. Solving convex programs by random walks. *Journal of the ACM*, 51(4):540–556, 2004.
8. S. H. Brooks. A discussion of random methods for seeking maxima. *Operations Research*, 6:244–251, 1958.
9. D. W. Bulger, R. D. Alexander, W. P. Baritompa, G. R. Wood, and Z. B. Zabinsky. Expected hitting times for backtracking adaptive search. *Optimization*, 53(2):189–202, 2004.
10. D. W. Bulger, W. P. Baritompa, and G. R. Wood. Implementing pure adaptive search with Grover's quantum algorithm. *Journal of Optimization Theory and Applications*, 116:517–529, 2003.
11. D. W. Bulger and G. R. Wood. Hesitant adaptive search for global optimization. *Mathematical Programming*, 81:89–102, 1998.
12. K. Deb. *Multi-Objective Optimization Using Evolutionary Algorithms*. Wiley, New York, 2001.
13. P. Del Moral. *Feynman-Kac Formulae: Genealogical and Interacting Particle Systems with Applications*. Springer, New York, 2004.
14. P. Del Moral and L. Miclo. On the convergence and applications of generalized simulated annealing. *SIAM Journal of Control and Optimization*, 37(4):1222–1250, 1999.
15. J. L. Devore. *Probability and Statistics for Engineering and the Sciences*. 4th Edition, Wadsworth, Inc. Belmont, CA, 1995.
16. L. C. W. Dixon and G. P. Szegö. *Towards Global Optimization*. North-Holland, Amsterdam, 1975.
17. L. C. W. Dixon and G. P. Szegö. *Towards Global Optimization 2*. North-Holland, Amsterdam, 1978.
18. M. E. Dyer and A. M. Frieze. Computing the volume of convex bodies: a case where randomness provably helps. *Proceedings of Symposia in Applied Mathematics*, 44:123–169, 1991.
19. F. S. Hillier and G. J. Lieberman. *Introduction to Operations Research*. 9th edition, McGraw-Hill, 2010.
20. Y. C. Ho, Q. C. Zhao, and Q. S. Jia. *Ordinal Optimization: Soft Optimization for Hard Problems*. Springer, 2007.
21. H. Huang and Z. B. Zabinsky. Adaptive probabilistic branch and bound with confidence intervals for level set approximation. In R. Pasupathy, S.-H. Kim, A. Tolk, R. Hill, and M. E. Kuhl, editors, *Proceedings of the 2013 Winter Simulation Conference*, pages 980–991, 2013.
22. A. Kalai and S. Vempala. Convex optimization by simulated annealing. *Mathematics of Operations Research*, 31(2):253–266, 2006.
23. D. E. Kaufman and R. L. Smith. Direction choice for accelerated convergence in hit-and-run sampling. *Operations Research*, 46(1):84–95, 1998.
24. S. Kiatsupaibul, R. L. Smith, and Z. B. Zabinsky, An analysis of a variation of hit-and-run for uniform sampling from general regions. *ACM Transactions on Modeling and Computer Simulation (ACM TOMACS)*, 21:3, 16:1–16:11, 2011.
25. W. Kohn, Z. B. Zabinsky, and V. Brayman. Meta-control of an optimization algorithm. *Journal of Global Optimization*, 34(2):293–316, 2006.
26. B. P. Kristinsdottir, Z. B. Zabinsky, and G. R. Wood. Discrete backtracking adaptive search for global optimization. In *Stochastic and Global Optimization, dedicated to the 70th anniversary of Professor J. Mockus,* edited by G. Dzemyda, V. Saltenis, and A. Zilinskas, Kluwer Academic Publishers, 147–174, 2002.
27. L. Lovász. Hit-and-run mixes fast. *Mathematical Programming*, 86:443–461, 1999.
28. L. Lovász and S. Vempala. Hit-and-run from a corner. *SIAM Journal of Computing*, 35(4):985–1005, 2006.
29. H. O. Mete, Y. Shen, Z. B. Zabinsky, S. Kiatsupaibul, and R. L. Smith. Pattern discrete and mixed hit-and-run for global optimization. *Journal of Global Optimization*, 50(4):597–627, 2011.

30. H. O. Mete and Z. B. Zabinsky. Pattern hit-and-run for sampling efficiently on polytopes. *Operations Research Letters*, 40(1):6–11, 2012.
31. N. Metropolis, A. Rosenbluth, M. Rosenbluth, A. Teller, and E. Teller. Equation of state calculations by fast computing machines. *Journal of Chemical Physics*, 21: 1087–1090, 1953.
32. O. Molvalioglu, Z. B. Zabinsky, and W. Kohn. Multi-particle simulated annealing. In *Models and Algorithms for Global Optimization*, edited by A. Törn, J. Zilinskas, and A. Zilinskas, Springer, New York, 215–222, 2007.
33. O. Molvalioglu, Z. B. Zabinsky, and W. Kohn. The interacting-particle algorithm with dynamic heating and cooling. *Journal of Global Optimization*, 43:329–356, 2009.
34. O. Molvalioglu, Z. B. Zabinsky, and W. Kohn. Meta-control of an interacting-particle algorithm. *Nonlinear Analysis: Hybrid Systems*, 4(4):659–671, 2010.
35. B. Morris and A. Sinclair, Random walks on truncated cubes and sampling 0–1 knapsack solutions. *SIAM Journal on Computing*, 34(1):195–226, 2004.
36. N.R. Patel, R. L. Smith, and Z. B. Zabinsky. Pure adaptive search in Monte Carlo optimization. *Mathematical Programming*, 4:317–328, 1988.
37. J. D. Pintér. *Global Optimization in Action (Continuous and Lipschitz Optimization: Algorithms, Implementations and Applications)*. Kluwer Academic Publishers, Dordrecht / Boston / London, 1996.
38. R. L. Rardin. *Optimization in Operations Research*. Prentice Hall, New Jersey, 1998.
39. D. Reaume, H. E. Romeijn, and R. L. Smith. Implementing pure adaptive search for global optimization. *Journal of Global Optimization*, 20(1):33–47, 2001.
40. R. T. Rockafellar. Coherent approaches to risk in optimization under uncertainty. *Tutorials in Operations Research*, INFORMS, 38–61, 2007.
41. H. E. Romeijn and R. L. Smith. Simulated annealing for constrained global optimization. *Journal of Global Optimization*, 5:101–126, 1994.
42. H. E. Romeijn and R. L. Smith. Simulated annealing and adaptive search in global optimization. *Probability in the Engineering and Informational Sciences*, 8:571–590, 1994.
43. H. E. Romeijn, Z. B. Zabinsky, D. L. Graesser, and S. Neogi. New reflection generator for simulated annealing in mixed-integer/continuous global optimization. *Journal of Optimization Theory and Applications*, 101(2):403–427, 1999.
44. Y. Shen, S. Kiatsupaibul, and R. L. Smith. An analytically derived cooling schedule for simulated annealing. *Journal of Global Optimization*, 38:333–365, 2007.
45. R. L. Smith. Efficient Monte Carlo procedures for generating points uniformly distributed over bounded regions. *Operations Research*, 32:1296–1308, 1984.
46. F. J. Solis and R. J. B. Wets. Minimization by random search techniques. *Mathematics of Operations Research*, 6:19–30, 1981.
47. J. C. Spall. *Introduction to Stochastic Search and Optimization: Estimation, Simulation and Control*. Wiley, Hoboken, New Jersey, 2003.
48. S. A. Vavasis. Complexity issues in global optimization: a survey. *Handbook of Global Optimization*, edited by R. Horst, and P. M. Pardalos, Kluwer Academic Publishers, Netherlands, 27–41, 1995.
49. W. Wang, A. Ghate, and Z. B. Zabinsky. Adaptive parameterized improving hit-and-run for global optimization. *Optimization Methods and Software (OMS)*, 24:4–5, 569–594, 2009.
50. D. H. Wolpert and W. G. Macready. No free lunch theorems for optimization. *IEEE Transactions on Evolutionary Computation*, 1:67–82, 1997.
51. G. R. Wood, D. W. Bulger, W. P. Baritompa, and D. L. Alexander. Backtracking adaptive search: the distribution of the number of iterations to convergence. *Journal of Optimization Theory and Applications*, 128(3):547–562, 2006.
52. G. R. Wood and Z. B. Zabinsky. Stochastic adaptive search. In *Handbook of Global Optimization Volume 2*, edited by P. M. Pardalos and H. E. Romeijn, Kluwer Academic Publishers, Dordrecht, Netherlands, 231–249, 2002.
53. G. R. Wood, Z. B. Zabinsky, and B. P. Kristinsdottir. Hesitant adaptive search: the distribution of the number of iterations to convergence. *Mathematical Programming*, 89(3):479–486, 2001.

54. Z. B. Zabinsky. Stochastic methods for practical global optimization. *Journal of Global Optimization*, 13:433–444, 1998.
55. Z. B. Zabinsky. *Stochastic Adaptive Search for Global Optimization*. Kluwer Academic Publishers, Boston, 2003.
56. Z. B. Zabinsky. Random search algorithms. In *Wiley Encyclopedia of Operations Research and Management Science*, edited by J. J. Cochran, L. A. Cox, Jr., P. Keskinocak, J. P. Kharoufeh, and J. C. Smith, John Wiley & Sons, 2011.
57. Z. B. Zabinsky, Stochastic search methods for global optimization. In *Wiley Encyclopedia of Operations Research and Management Science*, edited by J. J. Cochran, L. A. Cox, Jr., P. Keskinocak, J. P. Kharoufeh and J. C. Smith, John Wiley & Sons, 2011.
58. Z. B. Zabinsky, D. Bulger, and C. Khompatraporn. Stopping and restarting strategy for stochastic sequential search in global optimization. *Journal of Global Optimization*, 46(2):273–286, 2010.
59. Z. B. Zabinsky, D. L. Graesser, M. E. Tuttle, and G. I. Kim. Global optimization of composite laminate using improving hit and run. In *Recent Advances in Global Optimization*, edited by C. A. Floudas, P. M. Pardalos, Princeton University Press, Princeton, NJ, 343–365, 1992.
60. Z. B. Zabinsky and R. L. Smith. Pure adaptive search in global optimization. *Mathematical Programming*, 53:323–338, 1992.
61. Z. B. Zabinsky and R. L. Smith. Hit-and-run methods. In S. I. Gass and M. C. Fu, editors, *Encyclopedia of Operations Research & Management Science*, pages 721–729. Springer, 3rd edition, 2013.
62. Z. B. Zabinsky, R. L. Smith, J. F. McDonald, H. E. Romeijn, and D. E. Kaufman. Improving hit and run for global optimization. *Journal of Global Optimization*, 3:171–192, 1993.
63. Z. B. Zabinsky, M. E. Tuttle, and Khompatraporn, C., A case study: composite structure design optimization. In *Global Optimization: Scientific and Engineering Case Studies,* edited by Janos Pintér, Springer Science and Business Media, LLC, 507–528, 2006.
64. Z. B. Zabinsky, W. Wang, Y. M. Prasetio, A. Ghate, and J. W. Yen. Adaptive probabilistic branch and bound for level set approximation. In S. Jain, R. R. Creasey, J. Himmelspach, K. P. White, and M. Fu, editors, *Proceedings of the 2011 Winter Simulation Conference*, pages 4146–4157, 2011.
65. Z. B. Zabinsky, G. R. Wood, M. A. Steel, and W. P. Baritompa. Pure adaptive search for finite global optimization. *Mathematical Programming*, 69:443–448, 1995.

Chapter 12
Model-Based Stochastic Search Methods

Jiaqiao Hu

Abstract Model-based algorithms are a class of stochastic search methods that have successfully addressed some hard deterministic optimization problems. However, their application to simulation optimization is relatively undeveloped. This chapter reviews the basic structure of model-based algorithms, describes some recently developed frameworks and approaches to the design and analysis of a class of model-based algorithms, and discusses their extensions to simulation optimization.

12.1 Introduction

In this chapter, we address the problem of finding the values of a set of design parameters that attain the optimum of an objective function, written in the following general form:

$$x^* \in \arg\max_{x \in \Theta} h(x), \tag{12.1}$$

where Θ is the feasible region, which is often a non-empty compact subset of $\mathbb{R}^d$, and $h : \Theta \to \mathbb{R}$ is a bounded, deterministic objective function. Stochastic optimization often refers to the case where the objective function h itself takes the form of an expectation

$$h(x) = \mathsf{E}[H(x, \xi)],$$

where ξ is a random variable representing the stochastic uncertainty of the system, which for example could be a sample path, and only estimates of the "noisy" sample performance H are available. In this chapter, simulation optimization refers to the special case of stochastic optimization when the sample performance H is assessed in a path-wise manner through computer simulation. In contrast with

J. Hu (✉)
Stony Brook University, Stony Brook, NY, USA
e-mail: jqhu@ams.sunysb.edu

M.C. Fu (ed.), *Handbook of Simulation Optimization*, International Series in Operations Research & Management Science 216, DOI 10.1007/978-1-4939-1384-8_12

deterministic optimization, simulation optimization problems are characterized by random uncertainties in their performance measures, and the objective functions are often highly nonlinear with respect to the underlying decision variables.

When the objective function is differentiable, a well-known class of methods for solving simulation optimization problems is stochastic approximation [4, 5, 28, 29], the topic of Chaps. 6 and 7. These methods mimic the classical gradient algorithms in deterministic optimization and rely on the estimation of the gradient of the objective function. Because they are gradient-based, these methods generally find local optimal solutions. Sample average approximation [14, 27] is another approach that often exploits structural information such as differentiability, linearity or convexity. The main idea of this approach is to transform a simulation optimization problem into a deterministic one by expending a large amount of simulation effort on each visited solution to obtain a precise estimate of the objective function value. The resulting deterministic counterpart of the original stochastic problem is then solved by a deterministic optimization algorithm.

Due to the limited structural knowledge for general simulation optimization problems, it is natural to adapt random search methods from deterministic optimization to these types of problems. A random search method is usually recursive and approximates the optimal solution by a sequence of iterates (e.g., candidate solutions, promising subsets, probability models) generated according to a specified random mechanism. These methods differ primarily in the type of iterates an algorithm produces and in the choices of the random strategy used to generate the iterates. Because random search methods typically only rely on the objective function values rather than structural information such as convexity and differentiability, they are robust, easy to implement, and can be applied to a broad range of optimization problems with very different characteristics.

From an algorithmic point of view, a random search algorithm can be broadly classified as being either *instance-based* or *model-based* [44]. In instanced-based algorithms, an iterate corresponds to a single or a subset of candidate solution(s), and the construction of new iterates depends explicitly on iterates generated in previous iterations. These include both population-based algorithms such as genetic algorithms [10], which produce a collection of candidate solutions at each iteration, and methods like nested partitions [36] that are based on repeatedly identifying a promising subset of the feasible region as the search proceeds. Currently, random search-based simulation optimization is primarily dominated by instance-based methods, with numerous algorithms proposed in the literature and their behaviors relatively well studied and understood. In addition to the two aforementioned methods, some typical examples include response surface methods [3], simulated annealing [26], tabu search [9], stochastic ruler methods [38], stochastic comparison [11], and the COMPASS algorithm [15]; see Chaps. 2, 10, and 11, as well as [2] for a review of this class of methods.

While the field of simulation optimization has significantly evolved in terms of instance-based algorithms, less attention has been devoted to the study of model-based methods. Unlike instance-based algorithms, the model-based methods are based on sampling candidate solutions from an intermediate (usually parameterized) probability distribution over the feasible region. The idea is to iteratively modify the distribution model based on the sampled solutions to bias the search towards regions containing high quality solutions. Therefore, each iterate in a model-based algorithm corresponds to a distribution function, which can be abstractly viewed as a "model" characterizing the promising regions of the solution space. Examples of this type of method include ant colony optimization (ACO) [8], estimation of distribution algorithms (EDAs) [30], annealing adaptive search (AAS) [32, 39], probability collectives (PCs) [37], and the cross-entropy (CE) method [35]. These algorithms have successfully solved some hard nonlinear, non-differentiable problems and are becoming increasingly prominent in deterministic optimization.

Unfortunately, because model-based algorithms are generally heuristic in nature, very few of them have found their way into the simulation optimization literature. In this chapter, we aim to stimulate new research ideas in this area by presenting some recently developed frameworks and approaches for the design and analysis of a class of model-based algorithms. As a starting point, we introduce these developments in the context of deterministic optimization with the objective function h being viewed as a "black box" that returns the exact function value for a specified solution. We then proceed by providing specific examples and algorithms to illustrate the key modifications needed, as well as issues and challenges involved in extending model-based algorithms to general simulation optimization settings.

12.2 A Brief Review of Model-Based Methods

The basic idea of model-based algorithms is to use a sequence of probability distribution functions to successively characterize the promising regions of the solution space. So in a model-based algorithm, it is the probability distribution rather than candidate solutions (as in an instance-based algorithm) that is propagated from one iteration to another. Most algorithms that fall into this category are iterative methods involving the basic steps of the framework below.

Basic Model-Based Optimization Framework

Step 1. Randomly generate a population of candidate solutions $X^{(k)}$ from g_k, where g_k is the probability distribution on Θ at the kth iteration.
Step 2. Evaluate/estimate the performance of generated candidate solutions.
Step 3. Update g_k based on the performance of the sampled solutions in $X^{(k)}$ to construct a new distribution g_{k+1}; increase k by 1 and reiterate from Step 1.

Note that since a population of candidate solutions is generated at each step, such algorithms retain the primary strengths of population-based approaches such as genetic algorithms, while providing more flexibility and robustness in exploring the entire feasible region (i.e., via sampling from g_k). Clearly, a major algorithmic question in model-based algorithms is how the update in Step 3 is carried out. Important practical implementation issues are the efficient construction/representation of the probability distribution g_k and efficient sampling from g_k over the feasible region Θ. How these issues are addressed is what differentiates particular approaches.

Pure random search [39] (see also Chap. 11) can be viewed as one of the simplest model-based methods, in the sense that g_k is taken to be a fixed uniform distribution over Θ (not updated at all). The algorithm proceeds by generating a sequence of uniformly distributed random points over the feasible region and using the best candidate solution obtained thus far as an estimate of the optimal solution. Although the idea behind the algorithm is simple, the complexity of the algorithm increases exponentially with the dimension of the solution space.

A nontrivial improvement of pure random search is the annealing adaptive search (AAS) algorithm [32, 39], described in detail in Chap. 11, which replaces the fixed uniform distribution in a pure random search method by a sequence of Boltzmann distributions parameterized by iteration-varying temperatures T_k. These Boltzmann distributions are constructed in such a way that as the temperature decreases to zero, the sequence of distributions will become more concentrated on the set of optimal solutions. So solutions sampled from Boltzmann distributions with small values of T_k will be close to the optimum with high probability. For the class of Lipschitz optimization problems, it has been shown that the expected number of iterations required by AAS to achieve a given level of precision increases at most linearly in the problem dimension [32, 39]. However, the idealized AAS is not readily implementable in practice for solving optimization problems, because the problem of sampling exactly from a given Boltzmann distribution is known to be very difficult. This implementation issue has been addressed in a number of papers (see e.g., [39, 40] and the references therein), and the basic idea is to use Markov chain Monte Carlo techniques to sample asymptotically from the Boltzmann distribution.

The cross-entropy (CE) method [35] was originally motivated by the problem of estimating probabilities of rare events in simulation [33], before it was discovered that it could be modified to solving deterministic optimization problems. The key idea of CE is to use a family of parameterized distributions to successively approximate an optimal (importance sampling) distribution concentrated only on the set of (near) optimal solutions, which is carried out by iteratively estimating the optimal parameter that minimizes the Kullback–Leibler (KL) divergence between the parameterized distribution and the target optimal distribution. Since its introduction, there have been extensive developments regarding implementation and practical applications of CE (see [35]). Those that are particularly relevant to our discussion include the adaptation of CE to handle stochastic network combinatorial optimization problems [34], the application of the method to solving buffer allocation problems in a simulation-based environment [1], and the work of [31], which uses CE as a direct policy search approach to solving stochastic dynamic programming problems. The literature analyzing the convergence properties of the CE method is relatively sparse, with most of the existing results limited to specific settings; see, e.g., [13] for a convergence proof of a variational version of CE in the context of estimation of rare event probabilities, and [7] for probability one convergence proofs of CE for discrete optimization problems. General convergence and asymptotic rate results for CE were recently obtained in [20, 22] by relating the algorithm to recursions of stochastic approximation type (see Sect. 12.4).

Two other well-established model-based methods are the ant colony optimization (ACO) [8] and the estimation of distribution algorithms (EDAs) [30]. ACO was inspired by the behavior of a colony of biological ants, which are capable of solving shortest path problems by exchanging their local information indirectly through a certain chemical substance called pheromone. ACO is frequently applied to solving combinatorial problems, e.g., the traveling salesman problem. In such problems, the generation of candidate solutions (tours) is based on the series of random moves performed by a collection of artificial ants called agents, which are controlled by an empirical distribution constructed based on each agent's local experience. ACO has been formally extended to stochastic settings for solving stochastic combinatorial optimization problems. One such extension is called the stochastic ant colony optimization (S-ACO) [12], which uses Monte-Carlo sampling to estimate the expectation involved in evaluating the objective function. The probability one convergence of S-ACO to the global optimal solution has been established in [12].

EDAs inherit the spirit of genetic algorithms (GAs), but eliminate the crossover and mutation operators to avoid the disruption of partial solutions. In EDAs, a new population of candidate solutions are generated according to the probability distribution induced or estimated from the promising solutions selected from the previous generation. Unlike CE, EDAs often take into account the interrelations between the underlying decision variables needed to represent the individual candidate solutions. At each iteration of the algorithm, a high-dimensional probabilistic model that better represents the interdependencies between the decision variables is induced; this step constitutes the most crucial and difficult part of the method. We refer the reader to [30] for a review of the way in which different probabilistic models are used as EDA instantiations. A proof of convergence of a class of EDAs, under the idealized infinite population assumption, can be found in [41].

There are many other model-based algorithms proposed for optimization. Some notable examples include probability collectives (PCs) [37], particle swarm optimization [25], the particle filtering approach [42], and gradient-guided stochastic search [43]. A complete description of all of them is outside the scope of this chapter. Instead, we will focus our discussions on some recently developed approaches that allow us to arrive at a systematic framework to study a class of model-based algorithms in a uniform manner. We review specific algorithms under the framework, present their asymptotic convergence properties, and discuss their adaptations to simulation optimization.

12.3 A Model Reference Optimization Framework

We begin with a description of the model reference adaptive search (MRAS) method introduced by Hu et al. [18], which will serve as a basis for subsequent discussions.

Model-based methods construct a sequence of distributions $\{g_k\}$ with some desired convergence properties (ideally including $g_k \to g^*$ as $k \to \infty$, with g^*

being a limiting distribution assigning all of its probability mass to the set of optimal solutions). Some common approaches for constructing such a sequence include: (a) various proportional selection schemes—used in ACOs, EDAs, and PCs; (b) Boltzmann selection schemes—used in AAS and the continuous EDA algorithm in [6]; and (c) optimal importance sampling measure—primarily used in the CE method.

However, in all cases, the obvious difficulties are that the sequence $\{g_k\}$ often depends on h, which may not be available in any explicit form, and that the problem of sampling exactly from even a known (but arbitrary) distribution g_k is in general intractable. In [18], a general approach called model reference adaptive search (MRAS) is proposed, where these difficulties are circumvented by sampling from a *surrogate distribution* that approximates g_k. The idea of MRAS is to specify a family of parameterized distributions $\{f_\varphi(\cdot), \varphi \in \Phi\}$ (with Φ being the parameter space) and then project g_k onto the family to obtain a sequence of sampling distributions $\{f_{\varphi_k}\}$ with desired convergence properties. The projection is carried out by finding an optimal parameter φ_k that minimizes the Kullback–Leibler (KL) divergence between g_k and the parameterized family $\{f_\varphi(\cdot), \varphi \in \Phi\}$ (cf. also [35]), i.e.,

$$\varphi_k = \arg\min_{\varphi \in \Phi} \mathscr{D}(g_k, f_\varphi),$$

where

$$\mathscr{D}(g_k, f_\varphi) := \int_\Theta \ln \frac{g_k(x)}{f_\varphi(x)} g_k(x) \nu(dx) = \mathsf{E}_{g_k}\left[\ln \frac{g_k(X)}{f_\varphi(X)}\right], \tag{12.2}$$

ν is the Lebesgue/discrete measure on Θ, and $\mathsf{E}_g[\cdot]$ is the expectation taken with respect to the density/mass functions g. The hope is that the parameterized family is specified with some structure so that once the parameter is determined, sampling from each of these distributions should be a relatively easy task. Moreover, the task of updating the entire distribution can be simplified to the task of updating its associated parameters, and the sequence $\{g_k\}$, henceforth referred to as *reference distributions*, is only used implicitly to guide the parameter updating procedure.

Note that there are other ways of constructing surrogate distributions. For example, in AAS, Markov chain Monte Carlo (MCMC) techniques are frequently used to approximate the target Boltzmann distribution at each iteration [32], and in traditional EDAs, empirical distribution models are directly constructed to generate new candidate solutions [30, 41]. However, these algorithms can also be accommodated within the above model reference framework by projecting the target distributions onto a family of parameterized distributions. Specifically, under such a framework, the three example sequences of reference distributions $\{g_k\}$ can be expressed in the following recursive forms:

(a) proportional selection scheme: $g_k(x) = \dfrac{S(h(x))g_{k-1}(x)}{\mathsf{E}_{g_{k-1}}[S(h(X))]}$;

(b) Boltzmann selection:

$$g_k(x) = \frac{e^{h(x)/T_k}}{\int_{x\in\Theta} e^{h(x)/T_k}\nu(dx)} = \frac{e^{h(x)(\frac{1}{T_k}-\frac{1}{T_{k-1}})} g_{k-1}(x)}{\mathsf{E}_{g_{k-1}}\left[e^{h(X)(\frac{1}{T_k}-\frac{1}{T_{k-1}})}\right]} := \frac{S(h(x))g_{k-1}(x)}{\mathsf{E}_{g_{k-1}}[S(h(X))]},$$

with $\{T_k\}$ being a sequence of parameters determined by an annealing schedule;

(c) importance sampling measure: $g_k(x) = \dfrac{S(h(x))f_{\varphi_k}(x)}{\mathsf{E}_{\varphi_k}[S(h(X))]}$,

where $S(\cdot)$ is a non-negative increasing (possibly iteration-varying) function, and $\mathsf{E}_{\varphi}[\cdot]$ is the expectation taken with respect to f_{φ}. The algorithm instantiation considered in [18] uses the recursive procedure corresponding to case (a) to construct the g_k sequence. This form of reference distributions weights g_k by the value of the performance measure by giving more mass to solutions with good performance. The resulting g_{k+1} has the property that it improves the expected performance of g_k, since

$$\mathsf{E}_{g_{k+1}}[S(h(X))] = \frac{\mathsf{E}_{g_k}[S^2(h(X))]}{\mathsf{E}_{g_k}[S(h(X))]} \geq \mathsf{E}_{g_k}[S(h(X))].$$

This property ensure the convergence of the sequence $\{g_k\}$ to a degenerate distribution concentrated on the set of optimal solutions. The general structure of the MRAS method is outlined below.

Basic Model Reference Adaptive Search (MRAS) Optimization Framework

Step 0. Select a parameterized family $\{f_{\varphi}\}$ and the $\{g_k\}$ sequence with desired convergence properties.

Step 1. Given φ_k, generate N candidate solutions $X_k^1,\ldots,X_k^N$ by sampling from f_{φ_k}.

Step 2. Update the parameter φ_{k+1} based on the sampled solutions by minimizing the KL-divergence

$$\varphi_{k+1} = \arg\min_{\varphi\in\Phi} \mathscr{D}(g_{k+1}, f_{\varphi});$$

set $k \leftarrow k+1$ and reiterate from Step 1 until a stopping criterion is satisfied.

In some model-based algorithms such as CE and EDAs, there is often an additional selection step embedded in the above procedure. The idea is to concentrate the computational effort on the set of promising solutions by using only a portion of the samples—the set of "elite" samples—to update the probability model at Step 2. Also note that in general, the expectation involved in the KL-divergence (cf. (12.2)) cannot be evaluated exactly. So in practice, the objective function in the minimization problem at Step 2 is often replaced by its sample average approximation. This leads to an estimator of φ_{k+1} that is biased for any finite sample size N. This bias issue will be discussed later in Sect. 12.5.

12.3.1 Convergence Result

For distributions in the natural exponential family (NEF), the minimization in Step 2 of the basic MRAS framework can be carried out in analytical form, which makes the method easy to implement efficiently.

Definition 12.1. A parameterized family $\{f_\varphi(\cdot), \varphi \in \Phi \subseteq \mathbb{R}^m\}$ on Θ is called a natural exponential family if there exist mappings $\Gamma : \mathbb{R}^d \to \mathbb{R}^m$ and $K : \mathbb{R}^m \to \mathbb{R}$ such that $f_\varphi(x) = \exp\left(\varphi^T \Gamma(x) - K(\varphi)\right)$, where $K(\varphi) = \ln \int_\Theta \exp(\varphi^T \Gamma(x)) \nu(dx)$ is a normalization constant.

The NEF has many interesting properties. For the purpose of our discussion, we recall that $K(\varphi)$ is strictly convex in the interior of Φ and the mean parameter function defined by

$$m(\varphi) := \nabla_\varphi K(\varphi) = \mathsf{E}_\varphi[\Gamma(X)] \tag{12.3}$$

is a one-to-one invertible mapping of φ. Intuitively, $m(\varphi)$ is essentially a transformed version of the sufficient statistic $\Gamma(x)$, whose value contains all information necessary in estimating the parameter φ. For example, in the univariate normal distribution $\mathcal{N}(\mu, \sigma^2)$ with mean μ and variance σ^2, it can be seen that $\Gamma(x) = (x, x^2)^T$ and $\varphi = (\mu/\sigma^2, -1/(2\sigma^2))^T$. Thus, given the value of $m(\varphi)$, the equation $m(\varphi) = \mathsf{E}_\varphi[\Gamma(X)] = (\mu, \sigma^2 + \mu^2)^T$ can be uniquely solved for μ and σ^2.

When NEFs are used in the framework with the sample size N adaptively increasing, the convergence result for the instantiation of MRAS considered in [18] takes the form

$$\lim_{k \to \infty} m(\varphi_k) = \Gamma(x^*) \ \ w.p.1. \text{ as } k \to \infty.$$

Since $m(\varphi)$ is one-to-one, this shows that the sequence of sampling distributions $\{f_{\varphi_k}\}$ will converge to a limiting distribution f_{φ^*} for g_k sequences of proportional selection scheme type. In addition, it has been argued in [18] that in many special cases of interest, the limiting distribution turns out to be a degenerate distribution on the set of optimal solutions. For example, when multivariate normal distributions with mean vector μ_k and covariance matrix Σ_k are used as parameterized distributions, the convergence result translates to $\lim_{k\to\infty} \mu_k = x^*$ and $\lim_{k\to\infty} \Sigma_k = \mathbf{0}_{n\times n}$ w.p.1, where $\mathbf{0}_{n\times n}$ represents an n-by-n zero matrix.

12.3.2 Simulation Optimization

In [19], the MRAS method has been generalized to stochastic settings where the objective function $h(x)$ in (12.1) can only be estimated, e.g., via a simulation model or real-time observations. We begin by providing a high-level description of the method.

Stochastic MRAS Framework for Simulation Optimization

Step 0. Select a parameterized family $\{f_\varphi\}$ and an idealized $\{g_k\}$ sequence with desired convergence properties.

Step 1. Given φ_k, generate N candidate solutions $\Lambda_k := \{X_k^1, \ldots, X_k^N\}$ by sampling from f_{φ_k}.

Step 2. Take M_k simulation observations for each $x \in \Lambda_k$ and estimate $h(x)$ by the sample average $\overline{H}(x) := \frac{1}{M_k}\sum_{j=1}^{M_k} H(x, \xi_j)$, where ξ_j is the simulation noise in the jth replication run.

Step 3. Update the parameter φ_{k+1} based on the sampled solutions by minimizing the KL-divergence

$$\varphi_{k+1} = \arg\min_{\varphi \in \Phi} \mathscr{D}(\tilde{g}_{k+1}, f_\varphi);$$

set $k \leftarrow k+1$ and reiterate from Step 1 until a stopping criterion is satisfied.

The basic structure of the stochastic MRAS method (SMRAS) is similar to that of MRAS for deterministic optimization, the main addition being the requirement of an additional performance estimation step (i.e., Step 2 above). So in addition to the sample size N used in MRAS, we also need to specify the number of simulation replications to be allocated to each sampled candidate solution. At each iteration, the sample mean based on M_k observations is used to estimate the true performance $h(x)$. Another modification from the original MRAS method occurs at Step 3, where a distribution $\tilde{g}_k$ is used as an approximation of the idealized g_k distribution in minimizing the KL-divergence. The sequence $\{\tilde{g}_k\}$ is obtained by replacing the true objective function $h(x)$ in the construction of g_k with its sample average approximation $\overline{H}(x)$.

There are two general types of approaches for handling the simulation noise involved in evaluating the objective function. One type of approaches (e.g., sample average approximation) relies on highly precise estimates of the objective function values by allocating a significant amount of simulation replications to each visited solution. The other type of approaches (e.g., stochastic approximation) does not require precise performance estimates, but generally involves some forms of averaging, so that the estimation error due to simulation noise will automatically cancel out over the course of a large number of iterations. The SMRAS method falls in between these two types of approaches and requires increasingly precise estimates of the objective function values as the search progresses. In particular, for the proportional selection scheme type $\{g_k\}$ sequence, conditions on $\{M_k\}$ that ensure the convergence of the method are provided in [19]. It has been shown that when the simulation noises (not necessarily i.i.d.) satisfy the large deviation principle, M_k is required to grow linearly with the number of iterations k.

As compared with the MRAS method for deterministic optimization, the updating formula for φ_{k+1} (or equivalently $m(\varphi_{k+1})$, due to the invertibility of the mapping m) in SMRAS has an additional simulation error term. For general reference distribution sequences, a large deviation approach similar to that of [19] can be applied to determine the conditions on $\{M_k\}$ under which the error term converges to zero with probability one. Again, when NEFs are used in the SMRAS framework and $\{g_k\}$ is constructed using the proportional selection scheme, essentially the same convergence result as stated in Sect. 12.3.1, i.e.,

$$\lim_{k\to\infty} m(\varphi_k) = \Gamma(x^*) \;\; w.p.1. \;\text{ as } k \to \infty,$$

has been established under some appropriate conditions on the algorithm input parameters, the sample size N, and the number of simulation replications M_k; see [19].

12.4 A Stochastic Approximation Framework

In this section, we present a stochastic approximation (cf. Chap. 6) framework to analyze a general class of model-based algorithms. We show that a slight modification of the MRAS method introduced in Sect. 12.3 will lead to an interesting connection between model-based algorithms and the well-known stochastic approximation method. This connection implies that for a general non-differentiable (deterministic) optimization problem, a model-based algorithm implicitly transforms the underlying problem into an equivalent stochastic optimization problem with smooth differentiable structures, and the algorithm itself can be viewed as a gradient-based recursion over a transformed parameter space for solving the equivalent smoothed problem. This interpretation of model-based algorithms not only explains why these algorithms work well for hard optimization problems with little structure, but also allows us to investigate their theoretical properties such as convergence and rate of convergence by using theory and tools from stochastic approximation.

The main idea of the stochastic approximation framework is to replace the reference distribution sequence $\{g_k\}$ in the original MRAS method by a sequence $\{\hat{g}_k\}$ of the form:

$$\hat{g}_{k+1}(x) = \alpha_k g_{k+1}(x) + (1-\alpha_k) f_{\varphi_k}(x), \;\; \alpha_k \in [0,1] \; \forall k, \tag{12.4}$$

where f_{φ_k} is the sampling distribution obtained at the kth iteration of an algorithm and α_k is a mixture coefficient. Note that by taking $\hat{g}_{k+1}$ as the weighted average of g_{k+1} and f_{φ_k}, it is "forced" to stay close to both distributions. Thus, if $\hat{g}_{k+1}$ is used in place of g_{k+1} in minimizing the KL-divergence $\mathscr{D}(\hat{g}_{k+1}, f_\varphi)$ at Step 2 of the MRAS method, the new sampling distribution $f_{\varphi_{k+1}}$ obtained will not deviate significantly from the current sampling distribution f_{φ_k}. Since an NEF distribution f_φ is uniquely characterized by its associated parameter φ, the above intuition can be formally stated in terms of the difference between the two successive mean parameter functions (cf. (12.3)) of the projected probability distributions [20, 22]:

$$m(\varphi_{k+1}) - m(\varphi_k) = -\alpha_k \nabla_\varphi \mathscr{D}(g_{k+1}, f_\varphi)|_{\varphi=\varphi_k}. \tag{12.5}$$

Equation (12.5) explicitly brings out the updating direction of the parameter functions at each step, which is in the direction of the negative gradient of the iteration-varying objective function for the minimization problem $\min_{\varphi\in\Phi} \mathscr{D}(g_{k+1}, f_\varphi)\ \forall k$. This suggests that regardless of the type of decision variables involved in the original problem (12.1), algorithms conforming to the framework are essentially gradient-based recursions for solving a sequence of optimization problems on the parameter space Φ with smooth differentiable structures. In the special case of the CE method, i.e., when g_{k+1} in the right-hand-side of recursion (12.4) is replaced with $S(h(x))f_{\varphi_k}(x)/\mathsf{E}_{\varphi_k}[S(h(X))]$, it can be seen that (12.5) becomes

$$m(\varphi_{k+1}) - m(\varphi_k) = \alpha_k \frac{\mathsf{E}_{\varphi_k}[S(h(X))(\Gamma(X) - m(\varphi_k))]}{\mathsf{E}_{\varphi_k}[S(h(X))]} = \alpha_k \nabla_\varphi \ln \mathsf{E}_\varphi[S(h(X))]\big|_{\varphi=\varphi_k}. \tag{12.6}$$

So the updating direction is in the gradient of the objective function for the maximization problem $\max_{\varphi\in\Phi} \ln \mathsf{E}_\varphi[S(h(X))]$. The optimal solution of this optimization problem is a parameter φ^* whose associated sampling distribution f_{φ^*} assigns maximum probability to the set of optimal solutions of (12.1).

Letting $\eta_k := m(\varphi_k)$ and using the invertibility of the mapping m, we can write (12.5) in the abstract form

$$\eta_{k+1} = \eta_k - \alpha_k L(\eta_k), \tag{12.7}$$

where $L(\eta_k)$ represents the gradient of the underlying (possibly iteration-varying) objective function at η_k, which, for the three example $\{g_k\}$ sequences discussed in Sect. 12.3, takes the general form

$$L(\eta_k) = \frac{\mathsf{E}_{\varphi_k}[S(h(X))G(X, \eta_k)]}{\mathsf{E}_{\varphi_k}[S(h(X))]} \tag{12.8}$$

for some appropriate function $G(x, \eta_k)$. In actual implementation, expectations are replaced by sample averages based on Monte Carlo sampling, (12.7) becomes stochastic approximation with direct gradient estimation:

$$\tilde{\eta}_{k+1} = \tilde{\eta}_k - \alpha_k \tilde{L}(\tilde{\eta}_k), \tag{12.9}$$

where $\tilde{L}$ is an estimator for L based on sampled candidate solutions. The most straightforward estimator is

$$\tilde{L}(\tilde{\eta}_k) = \frac{N^{-1}\sum_{i=1}^{N} S(h(X_k^i))G(X_k^i, \tilde{\eta}_k)}{N^{-1}\sum_{i=1}^{N} S(h(X_k^i))}.$$

Thus, it is clear that the rich body of tools and results from stochastic approximation [28,29] can be incorporated into the framework to analyze model-based algorithms.

12.4.1 Convergence Results

Convergence of the CE Method

The convergence of the CE algorithm has been studied in [20, 22] by writing (12.9) in the form of a generalized Robbins–Monro algorithm in terms of the true gradient, a bias term, and a noise term caused by Monte Carlo random sampling, and then following a standard ordinary differential equation (ODE) argument (cf. e.g., [4,5,28]). Basically, it has been shown in [22] that the asymptotic behavior of CE is governed by the properties of a limit set of an underlying ODE. In addition, if the limit set consists purely of isolated equilibrium points of the ODE, then the sequence of $\{\tilde{\eta}_k\}$ generated by CE will converge to a unique limiting point η^* w.p.1. Under such a condition, the following asymptotic convergence rate result has also been established in [22]:

$$k^{\frac{\tau}{2}}(\tilde{\eta}_k - \eta^*) \xrightarrow{d} \mathcal{N}(0, \Sigma) \text{ as } k \to \infty,$$

where $\tau \in (0,1)$ is an appropriate constant and Σ is a positive definite covariance matrix.

Model-Based Annealing Random Search (MARS)

The stochastic approximation framework also allows for a lot of flexibility in developing provably convergent algorithms that perform well in practice. For example, [21] has investigated the use of Boltzmann distributions as reference models in the framework to address the implementation difficulty of annealing adaptive search, leading to a new globally convergent algorithm called model-based annealing random search (MARS). The algorithm complements existing research based on MCMC sampling techniques in the sense that it samples candidate solutions from a sequence of NEF distributions that approximates the target Boltzmann distributions, whereas MCMC techniques are sequential sampling procedures that sample directly from the Boltzmann distributions. The major steps of the MARS algorithm are given next.

Basic MARS Algorithm

Step 0. Select an NEF distribution family $\{f_\varphi, \varphi \in \Phi\}$, a sequence of temperature parameters $\{T_k\}$, and a gain sequence $\{\alpha_k\}$.
Step 1. Given φ_k, generate N candidate solutions $X_k^1, \ldots, X_k^N$ by sampling from f_{φ_k}.
Step 2. Update the parameter

$$\varphi_{k+1} = \arg\min_{\varphi \in \Phi} \mathscr{D}(\hat{g}_{k+1}, f_\varphi);$$

set $k \leftarrow k+1$ and reiterate from Step 1.

At Step 2 of MARS, the reference distribution $\hat{g}_{k+1}$ is given by $\hat{g}_{k+1}(x) = \alpha_k \bar{g}_{k+1}(x) + (1-\alpha_k) f_{\varphi_k}(x)$, which is the mixture of the current sampling distribution f_{φ_k} with $\bar{g}_{k+1}$, an empirical estimate of the true Boltzmann distribution $g_{k+1}(x) := \frac{e^{h(x)}/T_k}{\int_{\mathbf{X}} e^{h(x)/T_k} dx}$ based on the sampled solutions $X_k^1, \ldots, X_k^N$.

In light of Eq. (12.5), the mean parameter function $m(\varphi_{k+1})$ corresponding to the new parameter φ_{k+1} obtained at Step 2 of MARS can be viewed as an iterate generated by the gradient recursion

$$m(\varphi_{k+1}) = m(\varphi_k) - \alpha_k \nabla_\varphi \mathscr{D}(\bar{g}_{k+1}, f_\varphi)|_{\varphi=\varphi_k}. \tag{12.10}$$

By the properties of NEFs, the gradient in (12.10) can be expressed in terms of a true gradient term involving the Boltzmann distribution g_{k+1} and an error term due to random sampling, leading to a Robbins–Monro type stochastic approximation algorithm

$$\begin{aligned} m(\varphi_{k+1}) &= m(\varphi_k) - \alpha_k \Big(m(\varphi_k) - \mathsf{E}_{g_{k+1}}[\Gamma(X)] + \mathsf{E}_{g_{k+1}}[\Gamma(X)] - \mathsf{E}_{\bar{g}_{k+1}}[\Gamma(X)] \Big) \\ &= m(\varphi_k) - \alpha_k \nabla_\varphi \mathscr{D}(g_{k+1}, f_\varphi)|_{\varphi=\varphi_k} - \alpha_k \big(\mathsf{E}_{g_{k+1}}[\Gamma(X)] - \mathsf{E}_{\bar{g}_{k+1}}[\Gamma(X)] \big). \end{aligned} \tag{12.11}$$

Note that (12.11) generalizes a typical stochastic approximation recursion in that the function $\mathscr{D}(g_{k+1}, f_\varphi)$ may change shape with k. This time-varying feature of MARS actually turns out to be a desirable property, because the idealized sequence of Boltzmann distributions $\{g_k\}$ converges to a limiting distribution g^* as k goes to infinity. This will in turn imply the convergence of the sequence of the optimal solutions $\{\varphi_k\}$ to a global optimizer φ^*. Under some appropriate conditions on the algorithm input parameters, the following convergence result has been obtained in [21]:

$$\lim_{k \to \infty} m(\varphi_k) = \Gamma(x^*) \;\; w.p.1.$$

In addition, for a polynomially increasing sample size $N = ak^{\beta}$ and a gain sequence of the form $\alpha_k = c/k^{\alpha}$ for constants $a > 0$, $c > 0$, $\alpha \in (\frac{1}{2}, 1)$, and $\beta > \alpha$, an asymptotic normality result of the following form is also given in [21]:

$$k^{\frac{\alpha+\beta}{2}} \left(m(\varphi_k) - \Gamma(x^*) \right) \xrightarrow{d} \mathcal{N}(0, \Sigma) \text{ as } k \to \infty,$$

where Σ is a positive definite covariance matrix.

12.4.2 *Simulation Optimization*

The simulation optimization setting, where $H(x, \xi)$ is obtained in a simulation replication, requires an additional simulation allocation rule $\{M_k\}$, which allocates M_k simulation observations to each of the N candidate solutions generated at the kth iteration. Thus, in constructing gradient estimators, if the true performance at a sampled solution X_k^i is replaced by the sample average

$$\overline{H}_k(X_k^i) = \frac{1}{M_k} \sum_{j=1}^{M_k} H(X_k^i, \xi_j),$$

then an estimator of the true gradient $L(\tilde{\eta}_k)$ in (12.8) will take the form

$$\overline{L}_N(\tilde{\eta}_k) = \frac{N^{-1} \sum_{i=1}^{N} S(\overline{H}_k(X_k^i)) G(X_k^i, \tilde{\eta}_k)}{N^{-1} \sum_{i=1}^{N} S(\overline{H}_k(X_k^i))}.$$

Consequently, the gradient iteration (12.9) can be carried out at each step by replacing the true performance at a sampled solution X_k^i by its sample average approximation $\overline{H}_k(X_k^i)$, leading to a recursion of the form

$$\tilde{\eta}_{k+1} = \tilde{\eta}_k - \alpha_k \tilde{L}_N(\tilde{\eta}_k) + \alpha_k \left(\tilde{L}_N(\tilde{\eta}_k) - \overline{L}_N(\tilde{\eta}_k) \right). \tag{12.12}$$

Under some appropriate conditions on the allocation rule $\{M_k\}$, the expectation (conditional on the current sampled solutions) of the error term $\tilde{L}_N(\tilde{\eta}_k) - \overline{L}_N(\tilde{\eta}_k)$ goes to 0 as $k \to \infty$. Thus by treating the simulation noise $\tilde{L}_N(\tilde{\eta}_k) - \overline{L}_N(\tilde{\eta}_k)$ as a vanishing bias term, the (possibly local) convergence and convergence rate analysis of recursion (12.12) can be studied along the same line as in the deterministic optimization case.

Application of MARS to Finite-Horizon Markov Decision Processes

To illustrate the adaptation of the stochastic approximation framework to simulation setting, we modify and extend the MARS algorithm to a stochastic setting and present a simulation-based algorithm called approximate stochastic annealing (ASA) for solving finite-horizon Markov decision processes (MDPs) [16]. The idea is to interpret an MDP as a stochastic optimization problem on the (randomized) policy space (where candidate solutions are policies) and then use ASA as a specific optimization strategy to directly search the policy space to find good policies.

Consider a discrete-time finite $\mathscr{H}$-horizon stochastic system $x_{t+1} = f(x_t, a_t, w_t)$ for $t = 0, 1, \ldots, \mathscr{H} - 1$, where x_t represents the system state at time t taking values from a finite state space $\mathscr{X}$, a_t is the control applied at time t chosen from a finite action set $\mathscr{A}$, $\{w_t\}$ is a sequence of random vectors representing the stochastic uncertainty of the system, and f is the next-state transition function. Let $R_t(x_t, a_t, w_t)$ be the one-stage reward for action a_t taken in state x_t at time t. Define Π as the set of non-stationary deterministic Markovian policies $\pi = \{\pi_t,\ t = 0, \ldots, \mathscr{H} - 1\}$, where each $\pi_t : \mathscr{X} \to \mathscr{A}$ is a function that specifies the action to be applied at time t for each $x \in \mathscr{X}$. For an initial state $x_0 = x$, the expected total reward (value function) associated with a policy π is given by

$$V^{\pi}(x) := \mathsf{E}\Big[\sum_{t=0}^{\mathscr{H}-1} R_t(x_t, \pi_t(x_t), w_t)\big|x_0 = x\Big]. \tag{12.13}$$

The objective is to find an optimal policy $\pi^* \in \Pi$ that maximizes the expected total reward for a given state x, i.e.,

$$V^{\pi^*}(x) = \max_{\pi \in \Pi} V^{\pi}(x). \tag{12.14}$$

At each iteration, ASA searches for improved policies by sampling from a probability distribution function $\phi(\pi, q)$ over the policy space Π, where q is a parameter vector taking values from some parameter space. The distribution is then modified using a Boltzmann selection scheme based on the simulated/estimated value functions of the sampled policies. One simple way to specify the parameterized distribution $\phi(\pi, q)$ is to use an $|\mathscr{X}|$-by-$|\mathscr{A}|$-by-$\mathscr{H}$ stochastic matrix q_k, whose (i, j, t)th entry $q_k(i, j, t)$ specifies the probability that the action a_j is applied to state x_i at time t. Such a stochastic matrix q_k gives rise to a probability mass function over Π:

$$\phi(\pi, q_k) := \prod_{t=0}^{\mathscr{H}-1} \prod_{i=1}^{|\mathscr{X}|} \prod_{j=1}^{|\mathscr{A}|} \big[q_k(i, j, t)\big]^{I\{\pi \in \Pi_{i,j}(t)\}} \quad \forall \pi \in \Pi, \tag{12.15}$$

where $I\{\cdot\}$ is the indicator function and $\Pi_{i,j}(t) := \{\pi : \pi_t(x_i) = a_j\}$ denotes the set of deterministic policies that assign action a_j to state x_i at time t. Thus, as in the MARS algorithm, the goal is to iteratively update the entries $q_k(i,j,t)$ so that $\phi(\pi,q_k)$ will be a close approximation to the desired Boltzmann distribution

$$g_{k+1}(\pi) = \frac{e^{V^\pi/T_k}}{\sum_{\pi\in\Pi} e^{V^\pi/T_k}}.$$

This results in the following adaptive policy search procedure.

Approximate Stochastic Annealing (ASA) Algorithm for Solving MDPs

Step 0. Sclect a sequence of temperature parameters $\{T_k\}$ and a gain sequence $\{\alpha_k\}$.
Step 1. Given q_k, sample N policies $\Lambda_k := \{\pi_k^1,\ldots,\pi_k^N\}$ from $\phi(\pi,q_k)$.
Step 2. For each $\pi \in \Lambda_k$, perform M_k independent simulation replication runs and let $\bar{V}_k^\pi = \frac{1}{M_k}\sum_{l=1}^{M_k} V_l^\pi$, where V_l^π is an estimate of the value function V^π obtained in the lth replication run.
Step 3. Update the q matrix

$$q_{k+1} = \arg\min_q \mathscr{D}\big(\hat{g}_{k+1},\phi(\cdot,q)\big).$$

Set $k \leftarrow k+1$ and reiterate from Step 1.

Note that at Step 3, the KL-divergence is with respect to $\hat{g}_{k+1}(\pi) = \alpha_k\tilde{g}_{k+1}(\pi) + (1-\alpha_k)\phi(\pi,q_k)$, where $\tilde{g}_{k+1}$ is an empirical estimate of the true Boltzmann distribution g_{k+1} based on the sampled policies in Λ_k and value function estimates V_l^π.

Since the parameterized distribution $\phi(\pi,q_k)$ given in (12.15) belongs to NEFs, it is not difficult to show that the entries of the q_k matrix updated at Step 3 satisfy the following recursion:

$$\begin{aligned} q_{k+1}(i,j,t) - q_k(i,j,t) &= \alpha_k\Big[\mathsf{E}_{\tilde{g}_{k+1}}[I\{\pi\in\Pi_{i,j}(t)\}] - q_k(i,j,t)\Big] \\ &= \alpha_k\Big[\mathsf{E}_{g_{k+1}}[I\{\pi\in\Pi_{i,j}(t)\}] - q_k(i,j,t)\Big] \\ &\quad - \alpha_k\Big[\mathsf{E}_{g_{k+1}}[I\{\pi\in\Pi_{i,j}(t)\}] - \mathsf{E}_{\tilde{g}_{k+1}}[I\{\pi\in\Pi_{i,j}(t)\}]\Big]. \end{aligned} \quad (12.16)$$

Since g_{k+1} assigns more weight to policies with better performance, the first term on the right-hand-side of the second equality implies that the entries of q_k are updated in a direction that "pursues" the optimal policy π^*, whereas the second term can be viewed as a noise term caused by the approximation error between $\tilde{g}_{k+1}$ and g_{k+1}. Thus the convergence analysis of ASA essentially boils down to the issue of inspecting whether the Boltzmann distribution g_{k+1} can be closely approximated by its empirical estimate $\tilde{g}_{k+1}$. In particular, under some mild conditions on the algorithm input parameters, the following result is obtained in [16]:

$$q_k(i,j,t) \to I\{\pi^* \in \Pi_{i,j}(t)\} \ \forall i,j,t \ \text{ as } k\to\infty \text{ w.p.1},$$

which indicates that the sequence of stochastic matrices q_k generated by the algorithm will converge to a limiting matrix assigning unit probability mass to the optimal policy π^*.

12.5 A Stochastic Averaging Approach

As we have seen in previous sections, convergence analysis of model-based algorithms typically requires a sample size N that increases polynomially with the number of algorithm iterations. In practice, this translates to using a per-iteration computational effort that grows without bound as the number of iterations increases, which may have a negative impact on the algorithm's practical performance, especially in the setting where simulation/function evaluations are expensive.

This efficiency issue has been tackled in [17], where the basic idea is to maintain a population of probability distribution models (rather than just a single model as in a typical model-based algorithm) at each iteration and then adaptively allocate a given computing budget among different models in order to maximize the expected performance of the algorithm. In this section, we present an approach that aims to improve the sampling efficiency of model-based algorithms from a different perspective, focusing primarily on reducing the number of candidate solutions generated per iteration. This is carried out through embedding a stochastic averaging procedure within model-based algorithms to make more efficient use of the past sampling information. The material in this section is based on [23, 24].

For simplicity, we consider the general reference distribution introduced in Sect. 12.3:

$$g_k(x) = \frac{S(h(x))g_{k-1}(x)}{\mathsf{E}_{g_{k-1}}[S(h(X))]}.$$

By expanding the above recursion, we can write g_k in terms of the initial distribution g_0 as

$$g_k(x) = \frac{S_k(h(x))g_0(x)}{\mathsf{E}_{g_0}[S_k(h(X))]}, \tag{12.17}$$

where S_k is some appropriate iteration-varying function that depends on S. Substituting (12.17) into the minimization problem $\min_\varphi \mathscr{D}(g_{k+1}, f_\varphi)$ and dropping terms that are constants with respect to φ, it can be seen that the parameter φ_{k+1} obtained at Step 2 of the MRAS method of Sect. 12.3 can be equivalently obtained by solving the following optimization problem

$$\varphi_{k+1} = \arg\max_{\varphi\in\Theta} \left(Q_{k+1}(\varphi) := \int_{x\in\Theta} S_{k+1}(h(x)) \ln f_\varphi(x) dx \right). \tag{12.18}$$

As discussed previously, in model-based algorithms, the integral involved in the Q-function $Q_{k+1}(\varphi)$ is estimated by generating N i.i.d. candidate solutions $X_k^1,\dots,X_k^N$ from f_{φ_k}, and then replacing $Q_{k+1}(\varphi)$ by its sample average approximation

$$\bar{Q}_{k+1}(\varphi) := \frac{1}{N}\sum_{i=1}^{N}\frac{S_{k+1}(h(X_k^i))}{f_{\varphi_k}(X_k^i)}\ln f_\varphi(X_k^i).$$

Although $\bar{Q}_{k+1}(\varphi)$ is an unbiased estimator of $Q_{k+1}(\varphi)$, the corresponding optimization step will lead to an estimator of φ_{k+1} that is biased for any finite sample size N, because the optimal solution to (12.18) involves a ratio of integrals/expectations. Consequently, common implementations of these algorithms either require hundreds or even thousands of candidate solutions to be generated per iteration [6, 35], or require the use of a sample size N that increases at least polynomially with k in order to reduce the ratio bias effect [13, 18, 21, 22].

This bias issue has been addressed in [23, 24] by replacing the sample average approximation $\bar{Q}_k(\varphi)$ with the stochastic averaging procedure

$$\hat{Q}_{k+1}(\varphi) = (1-\beta_k)\hat{Q}_k(\varphi) + \beta_k\frac{1}{N}\sum_{i=1}^{N}\frac{S_{k+1}(h(X_k^i))}{f_{\varphi_k}(X_k^i)}\ln f_\varphi(X_k^i), \tag{12.19}$$

with $\hat{Q}_1(\varphi) := \frac{1}{N}\sum_{i=1}^{N}\left(S_1(h(X_0^i))/f_{\varphi_0}(X_0^i)\right)\ln f_\varphi(X_0^i)$, where β_k is a step size constant satisfying $\beta_k \in (0,1]\ \forall k$. Note that this procedure incrementally updates the current estimate of the Q-function as new sampling information becomes available at each iteration. In addition, due to the recursive nature of (12.19), all candidate solutions generated in the previous iterations contribute to the estimation of the Q-function $Q_{k+1}(\varphi)$. Consequently, it is reasonable to expect that the number of samples per iteration N can be significantly reduced or even held at a small constant value.

It is interesting to note that when NEF is used, $\hat{Q}_{k+1}(\varphi)$ can be expressed as a linear combination of the parameter vector φ and the function $K(\varphi)$:

$$\hat{Q}_{k+1}(\varphi) = \varphi^T\mathscr{S}_{k+1} - K(\varphi)\mathscr{R}_{k+1},$$

where the quantities $\mathscr{S}_k$ and $\mathscr{R}_k$ can be computed via the respective recursions

$$\mathscr{S}_{k+1} = \mathscr{S}_k + \beta_k\left(\frac{1}{N}\sum_{i=1}^{N}\frac{S_{k+1}(h(X_k^i))}{f_{\varphi_k}(X_k^i)}\Gamma(X_k^i) - \mathscr{S}_k\right),$$

$$\mathscr{R}_{k+1} = \mathscr{R}_k + \beta_k\left(\frac{1}{N}\sum_{i=1}^{N}\frac{S_{k+1}(h(X_k^i))}{f_{\varphi_k}(X_k^i)} - \mathscr{R}_k\right),$$

with $\mathscr{S}_1 := \frac{1}{N}\sum_{i=1}^{N}S_1(h(X_0^i))/f_{\varphi_0}(X_0^i)\Gamma(X_0^i)$ and $\mathscr{R}_1 := \frac{1}{N}\sum_{i=1}^{N}S_1(h(X_0^i))/f_{\varphi_0}(X_0^i)$. Thus, by substituting $\hat{Q}_{k+1}(\varphi)$ for $Q_{k+1}(\varphi)$ in (12.18), we have the following optimization problem:

$$\varphi_{k+1} = \arg\max_{\varphi \in \Theta} \left(\varphi^T \mathscr{S}_{k+1} - K(\varphi)\mathscr{R}_{k+1} \right),$$

whose unique closed-form solution is given by

$$m(\varphi_{k+1}) = \frac{\mathscr{S}_{k+1}}{\mathscr{R}_{k+1}} \text{ or equivalently } \varphi_{k+1} = m^{-1}\left(\frac{\mathscr{S}_{k+1}}{\mathscr{R}_{k+1}}\right).$$

The above stochastic averaging idea has been combined with the MARS algorithm of Sect. 12.4, leading to a two-time-scale stochastic approximation type of algorithm called MARS with stochastic averaging (MARS-SA).

MARS with Stochastic Averaging (MARS-SA)

Step 0. Select temperature parameters $\{T_k\}$ and gain sequences $\{\alpha_k\}$ and $\{\beta_k\}$.
Step 1. Generate N i.i.d candidate solutions $\Lambda_k := \{X_k^1, \dots, X_k^N\}$ from $f_{\varphi_k}(x)$.
Step 2. Update $\mathscr{S}_{k+1}$ and $\mathscr{R}_{k+1}$ according to the recursions:

$$\mathscr{S}_{k+1} = \mathscr{S}_k + \beta_k \left(\frac{1}{N} \sum_{x \in \Lambda_k} \frac{e^{h(x)/T_{k+1}}}{f_{\varphi_k}(x)} \Gamma(x) - \mathscr{S}_k \right),$$

$$\mathscr{R}_{k+1} = \mathscr{R}_k + \beta_k \left(\frac{1}{N} \sum_{x \in \Lambda_k} \frac{e^{h(x)/T_{k+1}}}{f_{\varphi_k}(x)} - \mathscr{R}_k \right).$$

Step 3. Compute a new parameter φ_{k+1} as

$$m(\varphi_{k+1}) = \alpha_k \frac{\mathscr{S}_{k+1}}{\mathscr{R}_{k+1}} + (1 - \alpha_k) m(\varphi_k).$$

Set $k \leftarrow k+1$ and reiterate from Step 1.

By exploiting the connection of MARS-SA to stochastic approximation method, it is shown in [23, 24] that the algorithm converges globally even when the per iteration sample size N is held at a small constant value. In addition, preliminary empirical results reported in [24] indicate that the new algorithm can be more efficient (in terms of the number of performance evaluations) than the original MARS algorithm.

12.6 Conclusions and Open Research Questions

In this chapter, we have provided an overview of three model-based optimization methods and discussed their extensions to simulation optimization. In particular, the MRAS method discussed in Sect. 12.3 offers a general framework to design and implement model-based algorithms, whereas the stochastic approximation method presented in Sect. 12.4 provides a systematic approach to analyze the convergence properties of these algorithms. In addition, we have also outlined in Sect. 12.5 a stochastic averaging procedure that addresses the estimator bias issue in model-based algorithms and aims to make these algorithms computationally more efficient. These methods are illustrated through a number of exemplary algorithms that are both provably convergent and exhibit promising empirical performance.

There are many challenging research issues that remain to be addressed. For example, the theoretical convergence and empirical performance of model-based algorithms are greatly influenced by the choices of reference distributions. So a natural research question is how the reference distributions should be chosen for specific problems. Moreover, since existing convergence results are all asymptotic in nature, a theoretical issue is to study whether finite-time performance bounds (e.g., similar to those in stochastic adaptive search [39]) can also be developed for these algorithms.

Model-based algorithms generally do not make use of problem structure, whereas in a continuous-variable simulation optimization setting, there may be additional information available (e.g., stochastic gradient estimates, Lipschitz continuity) from problem knowledge. It is well-known that effective use of structure may dramatically improve the solution efficiency. Therefore, another interesting research direction is to investigate how to incorporate problem structure information into model-based algorithms, as well as to identify classes of problems for which this can be done in a systematic manner.

The extension of model-based algorithms to simulation optimization is carried out in a relatively straightforward manner by introducing an additional simulation allocation sequence $\{M_k\}$ to obtain increasingly precise performance estimates as the search proceeds. This motivates the design of new model-based algorithms that do not rely on expending additional simulation effort on performance estimation ($M_k = 1$ for all k), e.g., in a way that the simulation noise will act like martingale difference noise in stochastic approximation and automatically average out over a large number of iterations. This is yet another avenue of research that merits further investigation.

Acknowledgements This work was supported in part by the National Science Foundation (NSF) under Grant CMMI-1130761 and by the Air Force Office of Scientific Research (AFOSR) under Grant FA95501010340.

References

1. G. Allon, D. P. Kroese, T. Raviv, and R. Y. Rubinstein. Application of the cross-entropy method to the buffer allocation problem in a simulation-based environment. *Annals of Operations Research*, 134:137–151, 2005.
2. S. Andradóttir. An overview of simulation optimization with random search. In S. G. Henderson and B. L. Nelson, editors, *Handbooks in Operations Research and Management Science: Simulation*. Elsevier, 2006.

3. R. R. Barton and M. Meckesheimer. Metamodel-based simulation optimization. In S. G. Henderson and B. L. Nelson, editors, *Handbooks in Operations Research and Management Science: Simulation*. Elsevier, 2006.
4. M. Benaim. A dynamical system approach to stochastic approximations. *SIAM Journal on Control and Optimization*, 34:437–472, 1996.
5. V. S. Borkar. *Stochastic approximation: a dynamical systems viewpoint*. Cambridge University Press; New Delhi: Hindustan Book Agency, 2008.
6. Y. Cai, X. Sun, and P. Jia. Probabilistic modeling for continuous eda with boltzmann selection and kullback-leibeler divergence. In *Proceedings of the 8th Annual Conference on Genetic and Evolutionary Computation*, pages 389–396, 2006.
7. A. Costa, O. D. Jones, and D. Kroese. Convergence properties of the cross-entropy method for discrete optimization. *Operations Research Letters*, 35(5):573–580, 2007.
8. M. Dorigo and L. M. Gambardella. Ant colony system: A cooperative learning approach to the traveling salesman problem. *IEEE Transactions on Evolutionary Computation*, 1:53–66, 1997.
9. F. W. Glover. Tabu search: A tutorial. *Interfaces*, 20:74–94, 1990.
10. D. E. Goldberg. *Genetic Algorithms in Search, Optimization, and Machine Learning*. Kluwer Academic Publishers, Boston, MA, 1989.
11. W. B. Gong, Y. C. Ho, and W. Zhai. Stochastic comparison algorithm for discrete optimization with estimation. *SIAM Journal on Optimization*, 10:384–404, 1999.
12. W. J. Gutjahr. A converging aco algorithm for stochastic combinatorial optimization. In *Proceedings of SAGA 2003 Stochastic Algorithms: Foundations and Applications*, pages 10–25, 2003.
13. T. Homem-de-Mello. A study on the cross-entropy method for rare-event probability estimation. *INFORMS Journal on Computing*, 19:381–394, 2007.
14. T. Homem-de-Mello. On rates of convergence for stochastic optimization problems under non-independent and identically distributed sampling. *SIAM Journal on Optimization*, 19:524–551, 2008.
15. L. J. Hong and B. L. Nelson. Discrete optimization via simulation using compass. *Operations Research*, 54:115–129, 2006.
16. J. Hu and H. S. Chang. An approximate stochastic annealing algorithm for finite horizon markov decision processes. In *Proceedings of the 49th IEEE Conference on Decision and Control*, pages 5338–5343, 2010.
17. J. Hu, H. S. Chang, M. C. Fu, and S. I. Marcus. Dynamic sample budget allocation in model-based optimization. *Journal of Global Optimization*, 50:575–596, 2011.
18. J. Hu, M. C. Fu, and S. I. Marcus. A model reference adaptive search method for global optimization. *Operations Research*, 55:549–568, 2007.
19. J. Hu, M. C. Fu, and S. I. Marcus. A model reference adaptive search method for stochastic global optimization. *Communications in Information and Systems*, 8:245–276, 2008.
20. J. Hu and P. Hu. On the performance of the cross-entropy method. In *Proceedings of the 2009 Winter Simulation Conference*, pages 459–468. IEEE, Piscataway, NJ, 2009.
21. J. Hu and P. Hu. Annealing adaptive search, cross-entropy, and stochastic approximation in global optimization. *Naval Research Logistics*, 58:457–477, 2011.
22. J. Hu, P. Hu, and H. S. Chang. A stochastic approximation framework for a class of randomized optimization algorithms. *IEEE Transactions on Automatic Control*, 57:165–178, 2012.
23. J. Hu and C. Wang. Discrete optimization via approximate annealing adaptive search with stochastic averaging. In *Proceedings of the 2011 Winter Simulation Conference*, pages 4206–4216. IEEE, Piscataway, NJ, 2011.
24. J. Hu, E. Zhou, and Q. Fan. Model-based annealing random search with stochastic averaging. *ACM Transactions on Modeling and Computer Simulation*, forthcoming, 2014.
25. J. Kennedy and R. Eberhart. Particle swarm optimization. In *Proceedings of IEEE International Conference on Neural Networks*, pages 1942–1948, 1995.
26. S. Kirkpatrick, C. D. Gelatt, and M. P. Vecchi. Optimization by simulated annealing. *Science*, 220:671–680, 1983.

27. A. Kleywegt, A. Shapiro, and T. Homem-de-Mello. The sample average approximation method for stochastic discrete optimization. *SIAM Journal on Optimization*, 12:479–502, 2001.
28. H. J. Kushner and D. S. Clark. *Stochastic Approximation Methods for Constrained and Unconstrained Systems and Applications*. Springer-Verlag, New York, 1978.
29. H. J. Kushner and G. G. Yin. *Stochastic Approximation Algorithms and Applications*. Springer-Verlag, New York, 1997.
30. P. Larrañaga and J. A. Lozano. *Estimation of Distribution Algorithms: A New Tool for Evolutionary Computation*. Kluwer Academic Publisher, Boston, MA, 2002.
31. S. Mannor, R. Y. Rubinstein, and Y. Gat. The cross-entropy method for fast policy search. In *Proceedings of the 20th International Conference on Machine Learning*, pages 512–519, 2003.
32. H. E. Romeijn and R. L. Smith. Simulated annealing and adaptive search in global optimization. *Probability in the Engineering and Informational Sciences*, 8:571–590, 1994.
33. R. Y. Rubinstein. Optimization of computer simulation models with rare events. *European Journal of Operational Research*, 99:89–112, 1997.
34. R. Y. Rubinstein. Combinatorial optimization, ants and rare events. In S. Uryasev and P. M. Pardalos, editors, *Stochastic Optimization: Algorithms and Applications*, pages 304–358, 2001.
35. R. Y. Rubinstein and D. P. Kroese. *The Cross-Entropy Method: A Unified Approach to Combinatorial Optimization, Monte-Carlo Simulation, and Machine Learning*. Springer, New York, 2004.
36. L. Shi and S. Ólafsson. Nested partitions method for global optimization. *Operations Research*, 48:390–400, 2000.
37. D. H. Wolpert. Finding bounded rational equilibria part i: Iterative focusing. In T. Vincent, editor, *Proceedings of the Eleventh International Symposium on Dynamic Games and Applications*, 2004.
38. D. Yan and H. Mukai. Stochastic discrete optimization. *SIAM Journal on Control and Optimization*, 30:594–612, 1992.
39. Z. B. Zabinsky. *Stochastic Adaptive Search for Global Optimization*. Kluwer Academic Publishers, 2003.
40. Z. B. Zabinsky, R. L. Smith, J. F. McDonald, H. E. Romeijn, and D. E. Kaufman. Improving hit-and-run for global optimization. *Journal of Global Optimization*, 3:171–192, 1993.
41. Q. Zhang and H. Mühlenbein. On the convergence of a class of estimation of distribution algorithm. *IEEE Transactions on Evolutionary Computation*, 8:127–136, 2004.
42. E. Zhou, M. C. Fu, and S. I. Marcus. A particle filtering framework for randomized optimization algorithms. In *Proceedings of the 2008 Winter Simulation Conference*, pages 647–654. IEEE, Piscataway, NJ, 2008.
43. E. Zhou and J. Hu. Gradient guided adaptive stochastic search. *IEEE Transactions on Automatic Control*, 59:1818–1832, 2014.
44. M. Zlochin, M. Birattari, N. Meuleau, and M. Dorigo. Model-based search for combinatorial optimization: A critical survey. *Annals of Operations Research*, 131:373–395, 2004.

Chapter 13
Solving Markov Decision Processes via Simulation

Abhijit Gosavi

Abstract This chapter presents an overview of simulation-based techniques useful for solving Markov decision processes (MDPs). MDPs model problems of sequential decision-making under uncertainty, in which decisions made in each state collectively affect the trajectory of the states visited by the system over a time horizon of interest. Traditionally, MDPs have been solved via dynamic programming (DP), which requires the transition probability model that is difficult to derive in many realistic settings. The use of simulation for solving MDPs allows us to bypass the transition probability model and solve large-scale MDPs considered intractable to solve by traditional DP. The simulation-based methodology for solving MDPs, which like DP is also rooted in the Bellman equations, goes by names such as reinforcement learning, neuro-DP, and approximate or adaptive DP. We begin with a description of algorithms for infinite-horizon discounted reward MDPs, followed by the same for infinite-horizon average reward MDPs. Then we present a discussion on finite-horizon MDPs. For each problem considered, we present a step-by-step description of a selected group of algorithms. In making this selection, we have attempted to blend the old and the classical with more recent developments. Finally, after touching upon extensions and convergence theory, we conclude with a brief summary of some applications and directions for future research.

13.1 Introduction

Reinforcement learning (RL) and approximate dynamic programming (ADP), also called adaptive DP by some authors, are closely related research fields that have been successfully applied to many practical problems addressing sequential decision-making under uncertainty. The central ideas in these fields are closely tied to solving control problems in discrete-event dynamic systems where the underlying problem revolves around finding the optimal control (action) in each state visited by the system. These problems were initially studied by Richard Bellman [10], who also formulated what is now known as the Bellman equation. Much of the

A. Gosavi (✉)
Missouri University of Science and Technology, Rolla, MO, USA
e-mail: gosavia@mst.edu

M.C. Fu (ed.), *Handbook of Simulation Optimization*, International Series in Operations Research & Management Science 216, DOI 10.1007/978-1-4939-1384-8_13

methodology in RL and ADP is tied to solving some version of this equation. These problems are often called Markov decision processes/problems (MDPs). The solution methods invented by Bellman [11] and Howard [45] are called value iteration and policy iteration, respectively; collectively, they are called dynamic programming (DP) methods.

When the number of states and/or the number of actions is very large, DP suffers from the curse of dimensionality, i.e., it becomes difficult to apply DP in an exact sense. This is because DP requires the transition probability matrices, which can become huge for large-scale problems, too large to be stored or manipulated. It is on these problems that simulation can play a major role in solution methods. The key role that simulation plays is in avoiding the transition probabilities. It is well-known that for complex systems, producing simulators is significantly easier than developing exact mathematical models, i.e., the transition probabilities.

RL algorithms can run in simulators and have the potential to break the curse of dimensionality to produce optimal or near-optimal solutions. In particular, we will focus on methods in which the number of actions is finite and relatively small, e.g., a dozen. The algorithms will be based on the Bellman equation but will *not* need the transition probability model.

A significant body of literature in the area of RL and ADP appears to be divided into two branches. The first seeks to use algorithms on a real-time basis within the system, i.e., while the system is running (on-line). This branch is more closely associated with the name RL and is popular within the computer science and artificial intelligence (robotics) community. The other branch works primarily in an off-line sense and seeks to solve large-scale MDPs where the transition probabilities can be estimated but a naïve application of DP does not work. This branch is more closely associated with the name ADP and finds applications in electrical, industrial, and mechanical engineering. Another somewhat synthetic approach to study the differences between these two branches is to consider the function used: RL algorithms for the most part work with the Q-function defined by Eq. (13.1) or (13.2), whereas most ADP algorithms work with the value function of DP (e.g., Definition 13.1).

In the simulation community, one is usually interested in the algorithms belonging to the RL variety, because the Q-function can help avoid the transition probabilities. Note, however, that there are important exceptions to this, e.g., evolutionary policy iteration algorithm [25] and many model-building algorithms (see e.g., [84]). Regardless of whether the Q-function or the value function is used, in the simulation community, the interest lies in problems where the transition probability model is not easy to generate. As such, in this chapter, we limit ourselves to discussing algorithms that can bypass the transition probability model. As stated above, the algorithms we discuss will be limited to finite action spaces. It is also important to note that we will employ the RL algorithm in an off-line sense within the simulator. Hence, one assumes that the distributions of the random variables that form inputs to the system are available.

Some of the earlier books that cover the topics of RL and simulation-based algorithms include [12, 24, 30, 79, 82]. Numerous books have also appeared that primarily focus on the ADP aspects, and some of these include [66, 74]. The latest edition of the second volume of [15] has an entire chapter dedicated to ADP (and also to RL). See also [21] for policy search via simulation optimization. RL has also been surveyed in journal articles [35, 49, 52].

In writing this chapter, we have had to make some conscious choices regarding the selection of algorithms from the vast array now available in the literature. We have sought to blend our discussion of classical algorithms based on Q-functions and value iteration with that of the more recent developments in the area of policy iteration. We study the three main classes of objective functions commonly studied in this area: the infinite-horizon discounted reward, the infinite-horizon average reward, and the finite-horizon total reward. A highlight of our presentation is a step-by-step description of combining function approximators with algorithms of the Q-Learning type—an important topic from the perspective of attacking large-scale problems within simulators. Another highlight is the discussion on the finite horizon and average reward problems that are of significant interest in the operations research community.

The rest of this chapter is organized as follows. In Sect. 13.2, we present some background material including notation. Algorithms related to discounted reward, average reward, and total reward (finite horizon) are presented in Sects. 13.3, 13.4, and 13.5, respectively. Some simple numerical results are presented in Sect. 13.6. Sects. 13.7 and 13.8 present short discussions on extensions and convergence theory, respectively. Sect. 13.9 concludes this chapter with a discussion of applications and topics for future research and open problems.

13.2 Background

In this section, we present some background for simulation-based optimization of MDPs. In an MDP, the system transitions from one state to another in a dynamic fashion. The decision-maker is required to select an action from a set of actions (where the set contains at least two actions) in a subset of states. Those states in which actions have to be chosen are called decision-making states. Henceforth, by states, we will mean decision-making states, since for analyzing MDPs, it is sufficient to observe the transitions that occur from one decision-making state to another. Thus, as a result of selecting an action, the system transitions to another state (which can be the same state)—usually in a probabilistic manner. (In this chapter, we will confine our discussion to those MDPs in which the transitions are probabilistic, since they are more interesting in a simulation-based context.) The probability of transitioning (moving/jumping) from one state to another under the influence of an action is called the one-step transition probability, or simply the transition probability. The transition probabilities are collectively referred to as the transition probability model.

During a transition from one state to another, the system receives a (one-step) *immediate reward*, which is either zero, positive, or negative. Negative rewards can be viewed as costs. The decision-maker's goal is to choose actions in each state in a fashion that optimizes the value of some performance metric of interest. The actions chosen in the different states visited by the system are stored in a *policy*, and the decision-maker is interested in the policy that optimizes the performance metric of interest, so the solution to an MDP is an optimal policy. As stated above, our interest in this chapter lies in simulation-based methods that can help determine the optimal policy without generating the transition probabilities, and thereby solve complex large-scale MDPs with finite action sets, considered intractable via traditional DP methods.

The performance metric's value generally depends on the actions chosen in each state, the immediate rewards, the time horizon of interest and whether the time value of money is considered. When the time value of money is taken into account in the calculations, we have a discounted performance metric, while when it is ignored, we have an undiscounted performance metric. The time value of money will be discussed later in more detail. We now present some of the fundamental notation needed in this chapter.

13.2.1 Notation and Assumptions

A deterministic policy is one in which the decision-maker selects a fixed (deterministic) action in each decision-making state, in contrast to a *stochastic* policy in which in each decision-making state, each action is chosen with some fixed probability (such that the probability of selecting all actions sums to one for every state). A *stationary* policy is one in which the action selected in a state does not change with time. In general, we will be interested in finding stationary deterministic policies, so henceforth when we refer to a policy, we mean a stationary deterministic policy unless explicitly specified otherwise.

Let $\mathscr{S}$ denote the finite set of states visited by the system, $\mathscr{A}(i)$ the finite set of actions permitted in state i, and $\mu(i)$ the action chosen in state i when policy μ is pursued. We define $\mathscr{A} \equiv \cup_{i \in \mathscr{S}} \mathscr{A}(i)$. Further let $r(.,.,.) : \mathscr{S} \times \mathscr{A} \times \mathscr{S} \to \mathbb{R}$ denote the immediate reward and $p(.,.,.) : \mathscr{S} \times \mathscr{A} \times \mathscr{S} \to [0,1]$ denote the associated transition probability. Then the *expected* immediate reward earned in state i when action a is chosen in it can be expressed as:

$$\bar{r}(i,a) = \sum_{j=1}^{|\mathscr{S}|} p(i,a,j) r(i,a,j).$$

We will make the following two assumptions about the problems considered in this chapter.

- Both the state-space, $\mathscr{S}$, and the action-space, $\mathscr{A}$, are finite.
- The Markov chain of every policy is regular, i.e., the transition probability of the Markov chain can be raised to some finite power such that all elements in the resulting matrix become strictly positive [42].

13.2.2 Performance Metrics

We now define the three metrics of interest to us. The first metric is the expected total *discounted* reward over an infinitely long time horizon. The discount factor is a well-known mechanism used in MDP theory to capture the time value of money. It is to be interpreted as follows. If one earns z dollars at a time τ time periods after the current time, then the current value of those z dollars will be

$$z\left(\frac{1}{1+\kappa}\right)^{\tau},$$

where κ is the rate of interest. We denote $1/(1+\kappa)$ by γ. For the MDP, the assumption is that transition from one state to another requires one time period, i.e., $\tau = 1$. Hence, the discount factor that will be used after one state transition will be γ.

Definition 13.1. The expected total discounted reward of a policy μ starting at state i in an MDP over an infinitely long time horizon is:

$$V_{\mu}(i) \equiv \liminf_{k\to\infty} \mathsf{E}_{\mu}\left[\sum_{s=1}^{k} \gamma^{s-1} r(x_s,\mu(x_s),x_{s+1}) \middle| x_1 = i\right],$$

where γ is the (one-step) discount factor, E_{μ} denotes the expectation operator over the trajectory induced by policy μ, and x_s denotes the state occupied by the system before the sth transition (jump) in the trajectory occurs.

In a discounted reward MDP, the goal is to maximize this performance metric for all values of $i \in \mathscr{S}$, i.e., it appears that there are multiple objective functions, but fortunately, it can be shown that when all policies have regular Markov chains, there exists a stationary, deterministic optimal policy that maximizes the value of the above metric for all starting states simultaneously [15].

We now define the other popular performance metric for infinite time horizons: the expected reward per transition over an infinitely long time horizon, commonly known as the "average reward." In this metric, the time value of money is ignored.

Definition 13.2. The average (expected) reward of a policy μ per transition in an MDP, starting at state i over an infinitely long time horizon, is defined as:

$$\rho_\mu(i) \equiv \liminf_{k\to\infty} \frac{1}{k} \mathsf{E}_\mu \left[\sum_{s=1}^{k} r(x_s,\mu(x_s),x_{s+1}) | x_1 = i \right].$$

If the Markov chain of the policy is regular, the average reward does not depend on the starting state, i.e., $\rho_\mu(i) = \rho_\mu$ for all $i \in \mathscr{S}$. The goal thus becomes to maximize the average reward.

The third performance metric of interest here is that of the expected *total* reward over a *finite* time (i.e., number of stages) horizon, which is sought to be maximized. It depends on the starting state in the problem, which is assumed to be known, and is defined as follows:

Definition 13.3. The expected total reward over a finite horizon of T time periods for a policy μ when the starting state is i is defined as:

$$\phi_\mu(i) \equiv \mathsf{E}_\mu \left[\sum_{s=1}^{T} r(x_s,\mu(x_s),x_{s+1}) \middle| x_1 = i \right],$$

where s is generally known as the stage, and the starting state i is fixed for the problem.

13.2.3 Bellman Equations

The theory of DP is rooted in the famous Bellman equations, which were originally presented in the form of the value functions (see [11]). For the simulation-based context, it is the Q-function that is more useful, and hence we present the Bellman equations in the Q-format. We present the first equation for the discounted reward MDP.

Theorem 13.1. *For a discounted reward MDP, there exists a function $Q : \mathscr{S} \times \mathscr{A} \to \mathbb{R}$ such that the following set of equations have a unique solution*

$$Q(i,a) = \sum_{j=1}^{|\mathscr{S}|} p(i,a,j) \left[r(i,a,j) + \gamma \max_{b\in\mathscr{A}(j)} Q(j,b) \right] \quad \forall (i,a) \tag{13.1}$$

and the policy μ, defined by $\mu(i) \in \arg\max_{a\in\mathscr{A}(i)} Q(i,a)$ for all $i \in \mathscr{S}$, is an optimal policy for the MDP.

The associated result for the average reward MDP is:

Theorem 13.2. *For an average reward MDP, there exists a function $Q : \mathscr{S} \times \mathscr{A} \to \mathbb{R}$ and a scalar $\rho^* \in \mathbb{R}$ such that a solution exists for the following equations:*

$$Q(i,a) = \sum_{j \in \mathscr{S}} p(i,a,j) \left[r(i,a,j) + \max_{b \in \mathscr{A}(j)} Q(j,b) - \rho^* \right] \quad \forall (i,a). \tag{13.2}$$

Further ρ^* *equals the optimal average reward of the MDP, and the policy* μ*, defined by* $\mu(i) \in \arg\max_{a \in \mathscr{A}(i)} Q(i,a)$ *for all* $i \in \mathscr{S}$*, is an optimal policy for the MDP.*

We note that $Q(\cdot,\cdot)$ is commonly called the Q-factor, Q-value, or the state-action value in the literature. Finding the optimal values of these quantities holds the key to solving the MDP.

For the finite-horizon problem, we need a somewhat enhanced style of notation to account for the stage. In a finite-horizon problem, the state-action pair, (i,a), will be replaced by the state-stage-action triple, (i,s,a), where s is the stage index and takes values in the set $\mathscr{T} = \{1,2,\ldots,T\}$. The notation for the immediate reward, the transition probability, and the Q-function will need to account for this triple. In such problems, we will assume that there is no decision-making to be performed in stage $T+1$ and that the Q-value in that stage, regardless of the state or action, will be zero. *The starting state, i.e., when* $s=1$*, will be assumed to be known with certainty in finite-horizon problems.* The following is the Bellman equation for a finite-horizon undiscounted problem.

Theorem 13.3. *There exists a function* $Q : \mathscr{S} \times \mathscr{T} \times \mathscr{A} \to \mathbb{R}$ *such that a solution exists for the following equations: For all* $i \in \mathscr{S}$*, all* $s \in \mathscr{T}$*, and all* $a \in \mathscr{A}(i,s)$*:*

$$Q(i,s,a) = \sum_{j \in \mathscr{S}} p(i,s,a,j,s+1) \left[r(i,s,a,j,s+1) + \max_{b \in \mathscr{A}(j,s+1)} Q(j,s+1,b) \right], \tag{13.3}$$

where $Q(j,T+1,b) = 0$ *for all* $j \in \mathscr{S}$ *and* $b \in \mathscr{A}(j,T+1)$*. The policy* μ*, defined by* $\mu(i,s) \in \arg\max_{a \in \mathscr{A}(i,s)} Q(i,s,a)$ *for all* $i \in \mathscr{S}$ *and all* $s \in \mathscr{T}$*, is an optimal policy for the MDP.*

13.3 Discounted Reward MDPs

In this section, we present some of the key (simulation-based) RL algorithms for solving discounted reward MDPs. We begin with the classical Q-Learning algorithm, which is based on value iteration, along with a discussion on how it can be combined with function approximation. This is followed by a popular, but heuristic, algorithm called SARSA(λ), based on the notion of the temporal difference learning algorithm TD(λ). Thereafter, we present two algorithms based on policy iteration: one is based on the classical modified policy iteration approach and the other based on the actor-critic algorithm. We conclude this section with a relatively new algorithm that combines ideas of genetic algorithms within policy iteration.

13.3.1 Q-Learning

The idea of Q-Learning [88] is based on iteratively solving the Bellman equation presented in Eq. (13.1) while avoiding the transition probabilities. We will first present the intuition underlying the derivation of this algorithm and then present the steps more formally.

Clearly, Eq. (13.1) contains the transition probabilities that we seek to avoid in simulators. The main idea underlying Q-Learning is to use a Robbins–Monro stochastic approximation algorithm [70] to estimate the mean via samples without *directly* summing the samples. If x^k denotes the kth sample and $\hat{x}^k$ denotes the estimate after k samples, then the update is as follows:

$$\hat{x}^{k+1} \leftarrow (1-\alpha^k)\hat{x}^k + \alpha^k x^k,$$

where α^k denotes the step size in the kth iteration, which is a small positive scalar less than 1 that must satisfy the following conditions:

$$\sum_{k=1}^{\infty} \alpha^k = \infty; \quad \sum_{k=1}^{\infty} \left(\alpha^k\right)^2 < \infty.$$

Examples of step-size rules that satisfy the conditions above include

$$\alpha^k = \frac{A}{B+k}(B \geq A \geq 0), \quad \alpha^k = \frac{\log k}{k}(k \geq 2).$$

In practice, finding the right step-size often requires some experimentation; see Chaps. 6 and 7 for further discussion.

To apply the Robbins–Monro update for estimating the Q-factors in Eq. (13.1), it is necessary to express the right hand side of the Bellman equation as an expectation as follows:

$$Q(i,a) = \mathsf{E}_{i,a}\left[r(i,a,j) + \gamma \max_{b \in \mathscr{A}(j)} Q(j,b)\right] \quad \forall (i,a),$$

where the expectation operator $\mathsf{E}_{i,a}[\cdot]$ is over the random state transitions that can occur from state i under the influence of action a. Then the Q-Learning algorithm is as follows:

$$Q^{k+1}(i,a) = (1-\alpha^k)Q^k(i,a) + \alpha^k\left[r(i,a,j) + \gamma \max_{b \in \mathscr{A}(j)} Q^k(j,b)\right], \tag{13.4}$$

where the terms in the square brackets represent the sample. Note that (13.4) does not contain the transition probabilities. In a simulator, where every action is selected in each state with the same probability, it can be shown that as k tends to infinity, the algorithm converges to the unique solution of Eq. (13.1), i.e., the Bellman equation, thereby solving the problem.

The above presentation ignores subtleties that we now note. First, there are multiple Q-factors being estimated simultaneously here. Furthermore, in any given update, the terms within the square brackets potentially contain Q-factors of *other* state-action pairs. In general, the Q-factors of the other state-action pairs will not have been updated with the same frequency, which is true of synchronous updating used in DP. Updating of this nature is called asynchronous updating, and the differences in the frequencies arise from the fact that the trajectory pursued in the simulator is random. In the simulation-based setting, the haphazard order of updating can rarely be avoided, but in practice, it is necessary to maintain rough equality in the frequencies with which state-action pairs are visited. The convergence analysis of the algorithm must take all of this into account.

Not surprisingly, all convergence proofs require that all state-action pairs be visited infinitely often in the limit. Also, since we are dealing with simulation noise, all convergence proofs ensure convergence only with probability (w.p.) 1. Fortunately, a number of proofs for convergence have been worked out under some rather mild conditions [12, 16] that can be ensured within simulators for the case in which the Q-factor for each state-action pair is stored separately—a scenario generally referred to as the "look-up table" case. When the state-action space is very large, e.g., of the order of hundreds of thousands or millions, it is impossible to store each Q-factor separately, and one must then use a function-approximation scheme. When function approximation schemes are used, showing convergence becomes significantly more challenging, although some progress has been made even in this direction recently (see the chapter on ADP in [15]). The general Q-Learning algorithm is presented below.

Basic Q-Learning Algorithm

Step 0. Input $k_{\max}$= total number of iterations. Initialize iteration count $k = 0$ and all Q-values to 0, i.e., for all (l,u), where $l \in \mathscr{S}$ and $u \in \mathscr{A}(l)$, set $Q^k(l,u) = 0$. Start system simulation at any arbitrary state.

Step 1. For current state i, select action a w.p. $1/|\mathscr{A}(i)|$, and simulate to reach next state j, receiving reward $r(i,a,j)$.

Step 2. Update the Q-value of (i,a) via Eq. (13.4). Increment k by 1.
If $k < k_{max}$, then set $i \leftarrow j$ and return to Step 1; otherwise, go to Step 3.

Step 3. For each $l \in \mathscr{S}$, select $d(l) \in \arg\max_{b \in \mathscr{A}(l)} Q^k(l,b)$.
The policy (solution) generated by the algorithm is d.
Stop.

Alternative action–selection strategies for Step 1 will be discussed later.

13.3.2 Q-Learning with Function Approximation

As stated above, function approximation is necessary when the state-action space is large. The function approximation approach hinges on using what is known as a basis-function representation of the Q-function, along with the steepest-descent rule over the Bellman error, which is essentially the sum of squared errors in regression theory applied to the Bellman equation. We now present the key ideas. To simply exposition, we drop the superscript k in this subsection.

Basis Functions

The Q-factor can be represented via basis-functions and their weights using a *linear architecture* as follows:

$$Q_w(i,a) = \sum_{m=1}^{n} w(m,a)\phi(m,a), \tag{13.5}$$

where $\{\phi(\cdot,\cdot)\}$ denote the basis functions and $\{w(\cdot,\cdot)\}$ the weight functions, and n should be much smaller than the size of the state space. The actual set of basis functions is problem dependent. We now provide a simple example where the state has a single dimension.

Example with a Linear Architecture

Consider an MDP with a single-dimensional state, i, and two actions. Let $n = 2$, where for both values of a, the analyst chooses to use the following architecture:

$$\phi(1,a) = 1; \quad \phi(2,a) = i.$$

Thus, $Q_w(i,1) = w(1,1) + w(2,1)i$; $Q_w(i,2) = w(1,2) + w(2,2)i$.

Bellman Error

The following model is used to represent the Q-factor:

$$Q(i,a) = \sum_{j=1}^{|\mathscr{S}|} p(i,a,j)\left[r(i,a,j) + \gamma \max_{b\in\mathscr{A}(j)} Q_w(j,b)\right] \quad \forall (i,a),$$

where $Q_w(\cdot,\cdot)$ denotes an estimate of the true Q-factor. The Bellman error, BE, is then defined as follows:

$$BE \equiv \frac{1}{2} \sum_{i \in \mathscr{S}, a \in \mathscr{A}(i)} [Q(i,a) - Q_w(i,a)]^2.$$

Thus, the Bellman error denotes one-half times the sum of the squared difference between the hypothesized model of the Q-factor and that given by the function approximator. This is the classical sum of squared errors of regression and the coefficient of $1/2$ is popular in the machine learning community because it simplifies the resulting algorithm. The following expression, the detailed derivation of which can be found in [12], is commonly used to minimize the Bellman error (the first use was seen in [90] and now extensively used in the RL literature):

$$\frac{\partial BE}{\partial w(m,a)} = -\frac{\partial Q_w(i,a)}{\partial w(m,a)} \left[r(i,a,j) + \gamma \max_{b \in \mathscr{A}(j)} Q_w(j,b) - Q_w(i,a) \right] \text{ for all } (m,a). \tag{13.6}$$

The above is generally combined with the following steepest-descent algorithm:

$$w(m,a) \leftarrow w(m,a) - \alpha \frac{\partial BE}{\partial w(m,a)} \text{ for all } (m,a), \tag{13.7}$$

where α is the step size. For using the above, we must determine the expressions for $\frac{\partial Q_w(i,a)}{\partial w(m,a)}$, which can be done easily from the architecture defined in Eq. (13.5). In general, it is easy to see from Eq. (13.5) that for linear architectures:

$$\frac{\partial Q_w(i,a)}{\partial w(m,a)} = \phi(m,a).$$

Then, for the example MDP considered above, we obtain the following definitions for the partial derivatives:

$$\frac{\partial Q_w(i,a)}{\partial w(1,a)} = 1; \quad \frac{\partial Q_w(i,a)}{\partial w(2,a)} = i. \tag{13.8}$$

We now describe the algorithm using the above example as a specific case. We note the basis-function representation is conceptual; we do not store $Q_w(i,1)$ or $Q_w(i,2)$ in the computer's memory. Rather, only the following four scalars are stored: $w(1,1)$, $w(2,1)$, $w(1,2)$, and $w(2,2)$.

Note that in the above, α uses the step-size rules discussed previously. Furthermore, note that the rule to update weights (i.e., Eqs. (13.9) and (13.10)) is derived from Eqs. (13.6) to (13.8). The above is a popular algorithm, and can be extended easily (using the above equations) to more complex basis functions that can express a non-linear architecture, e.g.,

$$Q_w(i,1) = w(1,1) + w(2,1)i + w(3,1)(i)^2; Q_w(i,2) = w(1,2) + w(2,2)i + w(3,2)(i)^2.$$

Algorithm for the 2-Action Linear Architecture Example

Step 0. Input $k_{\max}$= total number of iterations. Initialize iteration count $k = 0$ and the weights for action 1, i.e., $w(1,1)$ and $w(2,1)$, to small random numbers, and set the corresponding weights for action 2 to the same values. Start system simulation at any arbitrary state.

Step 1. For current state i, select action a w.p. $1/|\mathscr{A}(i)|$, and simulate to reach next state j, receiving reward $r(i,a,j)$.

Step 2. Compute the following:

$$Q_{\text{old}} \leftarrow w(1,a) + w(2,a)i.$$

Then, set $$Q_{\text{next}} \leftarrow \max\{Q_{\text{next}}(1), Q_{\text{next}}(2)\}, \text{ where}$$

$$Q_{\text{next}}(1) = w(1,1) + w(2,1)j; \quad Q_{\text{next}}(2) = w(1,2) + w(2,2)j.$$

Then, update the two weights as follows:

$$w(1,a) \leftarrow w(1,a) + \alpha\left(r(i,a,j) + \gamma Q_{\text{next}} - Q_{\text{old}}\right) 1; \tag{13.9}$$

$$w(2,a) \leftarrow w(2,a) + \alpha\left(r(i,a,j) + \gamma Q_{\text{next}} - Q_{\text{old}}\right) i. \tag{13.10}$$

Increment k by 1. If $k < k_{\max}$, then set $i \leftarrow j$, go to Step 1; otherwise, go to Step 3.

Step 3. The policy learned, μ, is virtually stored within the weights. To determine the action prescribed in a state i for any $i \in \mathscr{S}$, compute the following:

$$\mu(i) \in \arg\max_{a \in \mathscr{A}(i)} \left[w(1,a) + w(2,a)i\right].$$

Before concluding this subsection, we note that the steepest-descent technique, coupled with Bellman error (discussed above), closely resembles the approach adopted in the RL community that uses neurons (also called linear neural networks), via the Widrow–Hoff (adenaline) rule [58, 93], for approximating the Q-function.

13.3.3 SARSA(λ)

We now present a heuristic algorithm which is known to have strong empirical performance. It is based on the concept of TD(λ), popular in RL [80], and the SARSA algorithm [71]. We first explain the concept of TD(λ) and follow that by SARSA.

The notion of TD(λ) is based on the idea that the immediate reward (also called "feedback" in the machine learning community), computed within the simulator after a state transition occurs, can be used to update *all* state-action pairs in the system. Note that in Q-Learning, only the Q-factor of the most recently visited state-action pair is updated when the immediate reward is obtained from a state transition. This is called a single-step update. In contrast, in TD(λ), the impact of the update for any state-action pair is proportional to how recently it was visited, something that is measured with the "recency traces."

We present this idea somewhat more formally now. If $W^k(i,a)$ denotes the iterate in the kth iteration for the state-action pair, (i,a), the Robbins–Monro update is:

$$W^{k+1}(i,a) \leftarrow W^k(i,a) + \alpha^k \left[\text{feedback}\right], \tag{13.11}$$

where the feedback really depends on the objective function at hand. In the TD(λ) update, where $\lambda \in [0,1]$, one simulates an infinitely long trajectory, and the feedback takes on the following form:

$$\text{feedback } = R_k + \lambda R_{k+1} + \lambda^2 R_{k+2} + \cdots, \tag{13.12}$$

where R_k is a term that depends on the algorithm and the iteration index k. Note that in Q-Learning, $\lambda = 0$, and $R_k = r(i,a,j) + \gamma \max_{b \in \mathscr{A}(j)} Q(j,b) - Q(i,a)$. When $\lambda > 0$ (but less than 1), we have a multi-step update, also called a TD(λ) update. In such an update, *all* states in the system are updated after every state transition in the simulator, and the feedback from multiple state transitions is used in updating all (or as we will see later, all but one) states. In the algorithm that follows, we will use such an updating mechanism.

Within the simulator, the SARSA algorithm uses a policy which is initially fully stochastic but gradually becomes greedy with respect to the Q-factors. This is also called an ε-greedy form of action selection (or policy). Furthermore, in SARSA, the feedback contains the Q-factor of the next state-action pair visited in the simulator; this is different than in Q-Learning, where the feedback contains the *maximum* Q-factor of the next state.

Formally, an ε-greedy form of action selection can be described as follows. In a state i, one selects the greedy action

$$\arg\max_{u \in \mathscr{A}(i)} Q(i,u)$$

w.p. P^k and any one of the remaining actions w.p.

$$\frac{1-P^k}{|\mathscr{A}(i)|-1}.$$

Furthermore, the probability of selecting non-greedy actions is gradually diminished to zero. A potential rule for the probability that can achieve this is: $P^k = 1 - B/k$, where for instance $B = 0.5$.

SARSA (λ) Algorithm

Step 0. Input $k_{\max}$= total number of iterations. Initialize iteration count $k = 0$ and all the Q-values, $Q(l,u)$ for all $l \in \mathscr{S}$ and $u \in \mathscr{A}(l)$, to arbitrary values. Set the recency trace, $e(l,u)$, to 0 for all $l \in \mathscr{S}$ and $u \in \mathscr{A}(l)$. Start system simulation at any arbitrary state.

Step 1. For current state i, select action $a \in \mathscr{A}(i)$ via an ε-greedy form of action selection. Simulate action a, and let the next state be j, with $r(i,a,j)$ being the immediate reward. Select an action $b \in \mathscr{A}(j)$ via the ε-greedy form of action selection. Then, compute the feedback, δ, and update $e(i,a)$, the recency trace for (i,a), as follows:

$$\delta \leftarrow r(i,a,j) + \gamma Q(j,b) - Q(i,a);$$

$$e(i,a) \leftarrow e(i,a) + 1. \tag{13.13}$$

Step 2. Update $Q(l,u)$ all $l \in \mathscr{S}$ and $u \in \mathscr{A}(l)$ using

$$Q(l,u) \leftarrow Q(l,u) + \alpha\delta e(l,u). \tag{13.14}$$

Then, update the recency traces for all $l \in \mathscr{S}$ and $u \in \mathscr{A}(l)$ using

$$e(l,u) \leftarrow \lambda\gamma e(l,u).$$

Increment k by 1. If $k < k_{\max}$, then set $i \leftarrow j$, go to Step 1; otherwise, go to Step 3.

Step 3. For each $l \in \mathscr{S}$, select $d(l) \in \arg\max_{b \in \mathscr{A}(l)} Q(l,b)$. The policy (solution) generated by the algorithm is d. Stop.

Original Version of SARSA

SARSA(0), which can be interpreted as a special case of SARSA(λ) with $\lambda = 0$, was in fact the original version of SARSA [71]. It uses one-step updates, and hence no eligibility traces are needed in it. Furthermore, it can be shown to converge to the optimal policy under some mild conditions on how the ε-greedy action selection is performed (see [75] for details).

Steps in SARSA are similar to those for SARSA(λ) above with the following differences: No eligibility traces are required, and in Step 2, only the Q-factor of the current state-action pair, (i,a), is updated as shown below:

$$Q(i,a) \leftarrow Q(i,a) + \alpha\delta.$$

The multi-step updating of TD(λ), used in SARSA(λ), can speed up the updating in the sense that fewer state-transitions may be necessary in the simulator for convergence. But it also has some drawbacks: First, after each transition, all state-action pairs have to be updated, which is computationally intensive (unlike single-step updating in which only one state-action pair is updated after one state-transition). Secondly, and more importantly, SARSA (λ) requires the storage of the recency traces, which can be challenging via function approximation. On the other hand, SARSA, which does not require these traces, can be handily combined with function approximators, via the Bellman error approach (discussed in the previous section). Thus, overall, the usefulness of employing multi-step updating in simulators has not been resolved comprehensively in the literature and remains an open issue for future study.

Finite Trajectories

A version of SARSA (λ) that often performs better in practice uses finite trajectories. In such a version, the trajectory for any state is allowed to end (or is truncated) when that state is revisited. This concept has been discussed extensively in [76] in the context of the "replacing traces." In such a finite trajectory algorithm, the updating of the traces in the steps above is performed as follows: Eq. (13.13) is replaced by

$$e(i,a) = 1 \text{ and for all } u \in \mathscr{A}(i) \setminus \{a\},\ e(i,u) = 0;$$

in all other respects, the algorithm is identical to that described above.

The above mechanism for updating traces ensures that when a state is revisited, the feedback prior to that visit is discarded when that particular state is to be updated in the future. Essentially, the intuition underlying this is that the effect of an action in a state should only be measured by the impact the action produces in terms of the cumulative rewards generated by it, and this impact should terminated when that state is revisited, because a new (different) action will be selected when the state is revisited. Updating of this nature has been used widely in artificial intelligence; see e.g., [91]. This sort of updating can be tied to the original idea of TD(λ), defined in Eq. (13.12), by noting that the trajectory is now a finite one that ends when the state concerned is revisited.

The notion of TD(λ) has been discussed in the context of the value function, but not the Q-factors, in much of the literature [12, 79]. In the simulation-based setting, however, one is interested in the Q-factors, and hence SARSA(λ) holds more appeal to the simulation community. In the next subsection, we will discuss another application of TD(0)—also in the context of Q-factors.

13.3.4 Approximate Policy Iteration

We now present an algorithm based on ideas underlying policy iteration—a well-known DP technique that is based on selecting a policy, evaluating its value function (also called policy evaluation), and then moving on to a better policy. An algorithm called modified policy iteration [86] uses an iterative approach, rooted in value iteration, to evaluate the value function of a given policy. The classical form of policy iteration on the other hand performs the policy evaluation in one step by solving the Bellman equation for the given policy, which is also called the Poisson equation. This step requires the transition probabilities that we seek to avoid here. Modified policy iteration, however, is more relevant in the simulation-based context, since one can invoke the Q-factor version of the Poisson equation, and then use a Q-Learning-like approach to solve the Poisson equation, thereby avoiding the transition probabilities.

In this section, we present a Q-factor version of the modified policy iteration algorithm that can be used within a simulator. Algorithms belonging to this family have also been called Q-P-Learning [30], originally in the average reward context [31]. The algorithm we present is closely related to approximate policy iteration (API), which is based on the value function of DP, rather than Q-factors. API has been discussed extensively in [12]. Unfortunately, API based on the value function cannot be used directly when the transition probability model is not available, which is the case of interest here. When the transition probability model is available, it is unclear why one needs a simulator, since efficient methods for DP are available in the literature. Hence, we restrict our discussion to the Q-factor version here which can be implemented within simulators, bypassing the transition probability model.

The central idea is to start with any randomly selected policy and evaluate its Q-factors via the Poisson equation. In the algorithm, the given policy, whose Q-factors that are being evaluated, will be stored in the form of the P-factors (a name used to distinguish them from the Q-factors that are simultaneously being estimated within the simulator). A clear separation of the two types of Q-factors allows one to perform function approximation. The Q-factor version of the Poisson equation (or the Bellman equation for a given policy) for policy μ is:

$$Q(i,a) = \sum_{j=1}^{|\mathscr{S}|} p(i,a,j)\left[r(i,a,j) + \gamma Q(j,\mu(j))\right] \quad \forall (i,a).$$

Thus, when a given policy, μ, is available, the algorithm will seek to solve the above equation via a form of Q-Learning. As discussed in the context of Q-Learning, one can use the Robbins–Monro algorithm in a simulator to solve this equation without any need for the transition probabilities. When the Q-factors are evaluated, i.e., the above equation is solved, a new policy is generated via the policy improvement step [15]. We now present a step-by-step description of the Q-factor version of API.

Approximate Policy Iteration (API) Algorithm Using Q-Factors

Step 0. Input $k_{\max}$= number of iterations per policy evaluation and $E_{\max}$ = total number of policy evaluations, $\alpha \in (0,1)$. Initialize policy evaluations count $E = 0$ and all the P-values, $P(l,u)$ for all $l \in \mathscr{S}$ and $u \in \mathscr{A}(l)$, to arbitrary values.

Step 1 (Policy Evaluation). Start fresh simulation at any initial state. Initialize all the Q-values, $Q(l,u)$, to 0, and k, the number of iterations within a policy evaluation, to 0.

Step 1a. For current state i, select action a w.p. $1/|\mathscr{A}(i)|$, and simulate to reach next state j, receiving reward $r(i,a,j)$.

Step 1b. Update $Q(i,a)$ via

$$Q(i,a) \leftarrow (1-\alpha)Q(i,a) + \alpha\left[r(i,a,j) + \gamma Q\left(j, \arg\max_{b \in \mathscr{A}(j)} P(j,b)\right)\right]. \quad (13.15)$$

Step 1c. Set $k \leftarrow k+1$. If $k < k_{\max}$, set $i \leftarrow j$ and go to Step 1a; else go to Step 2.

Step 2 (Policy Improvement). Set for all $l \in \mathscr{S}$ and all $u \in \mathscr{A}(l)$,

$$P(l,u) \leftarrow Q(l,u); \quad E \leftarrow E+1.$$

If E equals $E_{\max}$, then go to Step 3; otherwise, go to Step 1.

Step 3. For each $l \in \mathscr{S}$, select $d(l) \in \arg\max_{b \in \mathscr{A}(l)} Q(l,b)$. The policy (solution) generated by the algorithm is d. Stop.

In practice, the algorithm exhibits robust behavior but may be time-consuming, since a number of policies are generally evaluated before the algorithm converges, and each policy evaluation requires numerous iterations. The convergence of this algorithm, which is a special case of TD(0) adapted to Q-factors, can be shown along the lines of the convergence of Q-Learning [39]. Furthermore, as in Q-Learning, one can use function approximation; separate approximators would be needed for the Q- and the P-function.

13.3.5 Actor-Critic Algorithm

We now present an API algorithm that is based on policy iteration but is much faster because it performs the policy evaluation via only one iteration. The algorithm has been called the actor-critic or the adaptive critic in the literature. This algorithm has evolved over time with ideas from [8, 53, 89, 95]. We present the most modern version, which has some proven convergence properties.

The algorithm stores the value function and a substitute (proxy) for the action–selection probability. Here $J(i)$ will denote the value function for state i, and $H(i,a)$

for all $i \in \mathscr{S}$ and all $a \in \mathscr{A}(i)$ will denote the quantities (proxies) used to select actions in the states visited. We present the step-by-step algorithm.

Approximate Policy Iteration (API) via Actor-Critic

Step 0. Input $k_{\max}$= total number of iterations, $\alpha \in (0,1)$. Initialize iteration count $k=0$ and all J-values and H-values to 0, i.e., for all l, where $l \in \mathscr{S}$, and $u \in \mathscr{A}(l)$, set $J(l) \leftarrow 0$ and $H(l,u) \leftarrow 0$. Initialize a scalar, $\bar{H}$, to the largest possible value such that $\exp(\bar{H})$ can be stored in the computer's memory without overflow. Start system simulation at any arbitrary state.

Step 1. For current state i, select action a w.p.

$$\frac{\exp(H(i,a))}{\sum_{b\in\mathscr{A}(i)}\exp(H(i,b))},$$

and simulate to reach next state j, receiving reward $r(i,a,j)$. The above style of action selection is called the Gibbs-softmax method of action selection.

Step 2. (Critic update) Increment k by 1. Update $J(i)$ via

$$J(i) \leftarrow (1-\alpha)J(i) + \alpha\left[r(i,a,j) + \gamma J(j)\right].$$

Step 3. (Actor update) Update $H(i,a)$ using a step size, β, that shares a special relationship with α (discussed below):

$$H(i,a) \leftarrow H(i,a) + \beta\left[r(i,a,j) + \gamma J(j) - J(i)\right].$$

If $H(i,a) < -\bar{H}$, set $H(i,a) = -\bar{H}$; if $H(i,a) > \bar{H}$, set $H(i,a) = \bar{H}$.

Step 4. If $k < k_{max}$, then set $i \leftarrow j$, go to Step 1; otherwise, go to Step 5.

Step 5. For each $l \in \mathscr{S}$, select $d(l) \in \arg\max_{b\in\mathscr{A}(l)} H(l,b)$. The policy (solution) generated by the algorithm is d. Stop.

The algorithm's ε-convergence to optimality can be shown when the step-sizes share the following relationship in addition to the usual conditions of stochastic approximation and some other conditions [53]:

$$\lim_{k\to\infty}\frac{\beta^k}{\alpha^k} = 0.$$

The above requires that β converge to 0 faster than α. It is not difficult to find step sizes that satisfy these conditions. One example that satisfies all of these conditions required in [53] is: $\alpha^k = log(k)/k$ and $\beta^k = A/(B+k)$. Although the initial policy is a stochastic policy, under ideal conditions of convergence, the algorithm converges to a deterministic stationary optimal policy in the limit as $k \to \infty$.

Note that the update in Step 4 ensures that the $H(\cdot,\cdot)$ values are projected onto the interval $[-\bar{H},\bar{H}]$. This is necessary because in the simulator the values of one or more elements of the matrix $H(\cdot,\cdot)$ become unbounded.

The algorithm can be combined with function approximation, using ideas related to Bellman error, by using separate approximators for $J(\cdot)$ and $H(\cdot,\cdot)$. One difficulty with this algorithm is that convergence to optimality depends on finding a sufficiently large value for $\bar{H}$. Note that the algorithm requires computation of $\exp(H(i,a))$. If the theoretical value for $\bar{H}$ that leads to an optimal solution is so large that the computer overflows in trying to compute $\exp(\bar{H})$, one obtains sub-optimal policies.

13.3.6 Evolutionary Policy Iteration

We conclude this section with a relatively recent development that combines ideas from genetic algorithms (GAs) and policy iteration. The GA is a widely used meta-heuristic [44] based on the principles of genetic evolution. It is typically used in discrete optimization when the number of solutions is large, but a mechanism is available to estimate the objective function at each solution. The GA is a very popular algorithm known to have the ability of generating good solutions when the number of solutions is very large. The other attractive feature of GA is that it is simple to code.

The algorithm of interest here is Evolutionary Policy Iteration (EPI) due to [25]. It uses simulation to evaluate the value function of each state for a given policy. The overall scheme is similar to that used in a typical GA, and in what follows, we present an informal explanation.

One starts with an arbitrarily selected population (collection) of n policies. Thereafter, an *elite* policy is generated from the population. This is the best policy in the current population; the elite policy and how it is generated is a special feature of this algorithm. Via mutation, $(n-1)$ policies are generated from the existing population; this is called offspring generation in GAs. The elite policy and the $(n-1)$ newly generated policies are considered to be the new population, and the algorithm repeats the steps described above performing another iteration. The algorithm terminates when the last K (a large number, e.g., 20) iterations have not produced any change in the elite policy.

Estimating the value function of a given policy, μ, for a given state, i, can be performed via one simulation trajectory as shown below (it is also possible to use a one long trajectory to update all states, but we restrict our discussion to the case of a given state). The output generated by the rth trajectory will be denoted by $\hat{V}^r_\mu(i)$. In practice, one needs to simulate numerous trajectories, and obtain the average of the values generated by each trajectory. Thus, if R trajectories are simulated, the average will be:

$$\tilde{V}_\mu(i) = \frac{\sum_{r=1}^{R} \hat{V}^r_\mu(i)}{R}.$$

The routine described below must be applied for each state in the system separately.

Inputs: Given a state i, a policy μ, and the trajectory index r.
Initialize TR, the total reward, to 0. Set k, the number of steps, to 0.
Loop until $k = k_{\max}$, where $k_{\max}$ denotes the number of state transitions in the trajectory:

- Start simulation in a given state i. Selection action $\mu(i)$ in state i. Let the next state be j. Observe the immediate reward, $r(i,\mu(i),j)$. Update TR as follows:

$$TR \leftarrow TR + \gamma^k r(i,\mu(i),j).$$

- Set $i \leftarrow j$ and $k \leftarrow k+1$.

Output: $\hat{V}_{\mu}^{r}(i) \leftarrow TR$.

Note that in the above, the value of $k_{\max}$ depends on γ. Clearly, when γ^k is very small, there is no point in continuing with the trajectory any further. Thus, in practice, $k_{\max}$ can be quite small; however, the above routine needs to be performed for numerous (R) trajectories and for each state separately. In what follows, we will assume that the above routine can be called whenever the value function of a state for a given policy is to be estimated.

Evolutionary Policy Iteration (EPI) Algorithm

Initialization: Set the iteration count, k, to 0. The population (collection) of policies in the kth iteration will be denoted by $\mathcal{L}^k$. Populate $\mathcal{L}^0$ with n arbitrarily selected policies. Set θ, the global/local selection probability, to an arbitrarily selected value between 0 and 1. Set a and b, the local and the global mutation probabilities, also to values in (0,1) such that $a < b$. Initialize action–selection probabilities, $P(i,a)$ for all $i \in \mathcal{S}$ and all $a \in \mathcal{A}(i)$, to some values such that $\sum_b P(i,b) = 1$ for every $i \in \mathcal{S}$. One example is to set $P(i,a) = 1/|\mathcal{A}(i)|$ for every $i \in \mathcal{S}$.

Loop until a termination criterion is met:

- Select an elite policy, denoted by π_*, from $\mathcal{L}^k$ via the following computation. For every state $i \in \mathcal{S}$, select an action a_i as follows:

$$a_i \in \left\{ \arg\max_{\pi \in \mathcal{L}^k} \tilde{V}_{\pi}(i) \right\} \text{ and set } \pi_*(i) = a_i.$$

- **Offspring generation:**

1. Generate $(n-1)$ subsets of $\mathcal{L}^k$, denoted by $\mathcal{Y}(t)$ for $t = 1,2,\ldots,n-1$. Each of these subsets will contain m policies where m itself is a random number from the discrete uniform distribution, $DU(2,n-1)$. The m policies that will be selected for $\mathcal{Y}(\cdot)$ will be selected with equal probability from the set $\mathcal{L}^k$.

2. Generate $(n-1)$ offspring (policies), denoted by π^t for $t = 1, 2, \ldots, n-1$, as follows: For each state $i \in \mathscr{S}$ and each value of t, select an action u_i^t as follows:

$$u_i^t \in \left\{ \arg\max_{\pi \in \mathscr{Y}^k(t)} \tilde{V}_\pi(i) \right\} \text{ and set } \pi^t(i) = u_i^t.$$

3. For each π^t for $t = 1, 2, \ldots, n-1$, generate a mutated policy, denoted by $\hat{\pi}^t$, from π^t as follows. With probability θ, use a as the mutation probability and with the remaining probability $(1-\theta)$, use b as the mutation probability. The mutation probability is the probability with which each state's action in the policy undergoes a change (mutation). If a state's action is to be changed, it is changed to an action selected via the action–selection probabilities, $P(\cdot,\cdot)$, defined above.

- Generating the new population: The new population $\mathscr{Q}^{k+1}$ is now defined to be the following set:

$$\{\pi_*, \hat{\pi}^1, \hat{\pi}^2, \ldots, \hat{\pi}^{n-1}\}.$$

- Set $k \leftarrow k+1$.

Output: The policy π_* is the best policy generated by EPI.

The idea of mutating policies can be explained via a simple example. Assume $a = 0.2$ and $b = 0.9$. Further assume that $\theta = 0.4$. Then, a random number from the distribution $U(0,1)$ is first generated. If this number is below $\theta = 0.4$, then the mutation probability will be a; otherwise, the mutation probability will be b. When it comes to mutating a given policy, consider the following example. Assume that there are three states and four actions in each state. Furthermore, for the sake of exposition, the policy, μ, will now be expressed as $(\mu(1), \mu(2), \mu(3))$. Consider the following policy which is to be mutated w.p. 0.9:

$$(3, 4, 1).$$

A random number, y, from the distribution $U(0,1)$ will be generated for each state. We illustrate these ideas for state 2 in which the action prescribed by the policy is 4. If for state 2, $y < 0.9$, then the state's action will be altered to one dictated by $P(\cdot,\cdot)$, defined above. Then an action v is selected for state 3 w.p. $P(3, v)$. On the other hand, if $y \geq 0.9$, the state's action will be unaltered. These mutations are performed for every state in the system for every policy that is to be mutated.

13.4 Average Reward MDPs

Average reward MDPs (infinite horizon) are commonly associated to problems where discounting does not appear to be appropriate. This is usually the case when the discount factor is very close to 1. Also, finite-horizon problems are often converted into infinite-horizon problems to make them tractable, and if the horizon is short, money does not lose value appreciably to factor that into the calculations. Under these conditions also, the average reward is a more appropriate performance metric. Finally, there are scenarios, e.g., in queueing networks, electrical systems and production systems, where the rewards/costs are often calculated in terms of non-monetary measures such as waiting time, utilization, inventory, etc. Again, in such instances, the average reward is a more suitable performance metric.

Surprisingly, simple extensions of the discounted MDP algorithms do not work in the average reward domain. Thus, for instance, setting $\gamma = 1$ in Q-Learning leads to an unstable algorithm in which the Q-factors become unbounded. For average reward, we cover two different algorithms: one based on relative value iteration and the other based on updating the value of average reward, alongside the Q-factors. Note that the Bellman optimality equation for average reward, Eq. (13.2), contains ρ^* as an unknown in addition to the Q-factors. Hence, a straightforward extension of Q-Learning is ruled out.

13.4.1 Relative Q-Learning

The algorithm we now describe employs the concept of relative value iteration in combination with Q-Learning and originates from [2]. The steps differ from those in Q-Learning as follows:

- In Step 0, any state-action pair in the system is selected, and henceforth termed the distinguished state-action pair. It will be denoted by (i^*, a^*).
- In Step 2, the update is performed as follows:

$$Q^{k+1}(i,a) = (1-\alpha^k)Q^k(i,a) + \alpha^k \left[r(i,a,j) + \max_{b \in \mathscr{A}(j)} Q^k(j,b) - Q^k(i^*,a^*) \right].$$

The algorithm can be shown to converge w.p.1 to the optimal solution [2], under conditions identical to those required for Q-Learning. Furthermore, it can be shown that w.p.1,

$$\lim_{k \to \infty} Q^k(i^*,a^*) = \rho^*,$$

which implies that in the limit the Bellman optimality equation for average reward is solved.

13.4.2 R-SMART

We now present another algorithm for average reward MDPs [38], which is a refined version of algorithms originally presented for semi-MDPs (a more general version of the MDP) in [26, 32]. We restrict the discussion here to the MDP, which is a special case of the semi-MDP. The algorithm is called R-SMART (short for Relaxed-Semi-Markov Average Reward Technique). Like SARSA, this algorithm relies on using ε-greedy action selection, and further it uses a separate update for the average reward term. There are two versions of this algorithm. One uses connection to the stochastic shortest path (SSP) problem [15], and the other uses an artificial contraction factor. We now present details.

SSP-Version

In this version, the average reward problem is essentially transformed into an SSP. Since we are not interested in the original SSP, we do not explain the SSP in detail, but refer the reader to [15]. What is important to note is that the SSP transformation creates a contractive mapping for the Q-factors enabling their convergence. The steps in the resulting algorithm will have the following differences with those of Q-Learning:

- In Step 0, select any state in the system, and call it the distinguished state i^*.
- In this algorithm, some parameters related to the average reward calculation will be needed. They are TR, the total reward, TT, the total time, and ρ^k, the estimate of the average reward in the kth iteration. In Step 0, initialize each of TR, TT, and ρ^1, to 0.
- In Step 1, select action a using an ε-greedy form of action selection (as discussed in the context of SARSA).
- In Step 2, update $Q(i,a)$ as follows:

$$Q^{k+1}(i,a) = (1-\alpha^k)Q^k(i,a) + \alpha^k \left[r(i,a,j) + \mathbf{1}\{j \neq i^*\} \max_{b \in \mathscr{A}(j)} Q^k(j,b) - \rho^k \right], \tag{13.16}$$

  where $\mathbf{1}\{\cdot\}$ is the indicator function that returns a 1 if the condition inside the brackets is satisfied and zero otherwise. This is followed by an update of ρ. If a greedy action was selected in Step 1, i.e., if $a \in \arg\max_{b \in \mathscr{A}(i)} Q(i,b)$, then update TR, TT and ρ^k in the order shown below:

$$TR \leftarrow TR + r(i,a,j);$$

$$TT \leftarrow TT + 1;$$

$$\rho^{k+1} \leftarrow (1-\beta^k)\rho^k + \beta^k (TR/TT).$$

  If a greedy action is not selected, set $\rho^{k+1} \leftarrow \rho^k$.

In the update for ρ^k, the step size β^k is chosen in a manner discussed in the actor-critic algorithm. The indicator function within the update in Eq. (13.16) ensures that the problem is essentially transformed into an SSP. Further note that the steps above ensure that Q-factors for state-action pair (i^*, a) are updated like every other Q-factor. Under some conditions shown in [38], the algorithm is shown to converge to the optimal solution and ρ^k to ρ^* w.p.1. It is also worthwhile pointing out that the state i^* plays the role of the (fictitious) absorbing state in the SSP when it is encountered as the next state (j) in a transition.

Contraction-Factor Version

In this version of R-SMART, an artificial contraction factor, $\bar{\gamma}$, will be used. This artificial factor will make the Q-factor transformation contractive, thereby enabling convergence. The average reward will be updated as done above in the SSP-version. The steps in the contraction-factor version of R-SMART will have the following differences with those of Q-Learning:

- In Step 0, select a suitable value for $\bar{\gamma} \in (0,1)$.
- As in the SSP-version above, in Step 0, initialize each of TR, TT, and ρ^1, to 0.
- In Step 1, select action a using an ε-greedy form of action selection.
- In Step 2, update $Q(i,a)$ as follows:

$$Q^{k+1}(i,a) = (1-\alpha^k)Q^k(i,a) + \alpha^k \left[r(i,a,j) + \bar{\gamma} \max_{b \in \mathscr{A}(j)} Q^k(j,b) - \rho^k \right].$$

  Thereafter, update ρ^k as shown in the steps of the SSP-version.

In the above algorithm, the value of $\bar{\gamma}$ must be guesstimated. Convergence can be assured under certain conditions that require the knowledge of the transition probabilities. In practice, a value close to 1 often generates optimal solutions. Although, it is not possible to guess the correct value of $\bar{\gamma}$ and one must use trial and error, on large-scale problems, this version often outperforms the SSP-version.

13.5 Finite Horizon MDPs

As noted earlier, the finite-horizon problem includes stages, which further adds to the curse of dimensionality posed by the state-action space of infinite-horizon problems. Nonetheless, these problems are important in their own right. They find many applications in operations research. Many inventory control problems and problems in revenue management belong to the finite time horizon. In machine learning, these problems are called *episodic tasks*. Sometimes finite-horizon problems can be converted into infinite-horizon problems by introducing an artificial transition from the absorbing state to a starting state; the reason for this transformation is that it can make the problem tractable [40].

In this section, we will first discuss an SSP algorithm for solving the finite-horizon problem. Thereafter, we will consider a learning automata algorithm that has been developed more recently.

13.5.1 Special Case of Stochastic Shortest Path (SSP)

The finite-horizon problem can be studied as a special case of the stochastic shortest path (SSP) problem in which the state-action pair is replaced by the state-stage-action triple [15]. Notation for this problem has been introduced earlier, and the reader should review it at this time.

We will assume that there are T stages in which decision-making is to be performed. When the system reaches the $(T+1)$th stage, there is no decision making to be done, and the simulator will return to the starting state, which is assumed to be known. The following description is based on [37].

Simulation-Based Algorithm for Finite-Horizon MDPs

Step 0. Input $k_{\max}$= total number of iterations. Initialize iteration count $k=0$, stage $s=1$, and all the Q-factors, $Q^k(i,s,a)$ for all $i \in \mathscr{S}$, all $s \in \mathscr{T}$ and all $a \in \mathscr{A}(i,s)$, to 0. Also set $Q^k(j,T+1,b)=0$ for all k, all $j \in \mathscr{S}$ and $b \in \mathscr{A}(j,T+1)$. Start system simulation at the starting state, which is assumed to be known with certainty.

Step 1. For current state i and current stage s, select action a w.p. $1/|\mathscr{A}(i,s)|$, and simulate to reach next state j in stage $(s+1)$, receiving reward $r(i,s,a,j,s+1)$.

Step 2. Update the Q-value of (i,s,a) as follows:

$$Q^{k+1}(i,s,a) \leftarrow (1-\alpha^k)Q^k(i,s,a) + \alpha\left[r(i,s,a,j,s+1) + \max_{b \in \mathscr{A}(j,s+1)} Q^k(j,s+1,b)\right].$$

Increment k by 1 and s by 1, and if the new value of s equals $(T+1)$, set $s=1$. If $k < k_{max}$, then set $i \leftarrow j$ and return to Step 1; otherwise, go to Step 3.

Step 3. For each $l \in \mathscr{S}$ and each $s \in \mathscr{T}$, select $d(l,s) \in \arg\max_{b \in \mathscr{A}(l,s)} Q^k(l,s,b)$. The policy (solution) generated by the algorithm is d.
Stop.

13.5.2 Pursuit Learning Automata (PLA) Sampling

We now present the pursuit learning automata (PLA) sampling algorithm due to [23]. The theory of learning automata for MDPs has been covered extensively in [60]. This particular algorithm is based on the concept of PLA [67, 85]. The idea here is to apply a simulation-based sampling algorithm at each state-stage pair in the system. It is possible to start the simulation at any given state-stage pair, and simulate different actions. The algorithm is remarkably robust, and can converge quickly to the optimal solution. In algorithms considered previously, one simulates a long trajectory in which states are visited in an asynchronous manner. This algorithm is also simulation-based, thereby avoiding the transition probability model, but requires that the system be simulated in each state-stage pair separately; also, a long simulation trajectory is not needed here. In many ways, it uses the power of simulation and at the same time does not leave the convergence to count on visiting each state-stage pair infinitely often. Thus, it appears to combine the graceful and systematic synchronous updating of dynamic programming with the power of simulation.

The algorithm will be presented for a given state-stage pair (i,s). Since the value function of the next state-stage pair will be required, a recursive call will be necessary to that pair if a forward pass is done. Alternatively, one could start at the last decision-making stage (T) in any state, update all the states in that stage, and then move backwards, one stage at a time, as is done in backward dynamic programming. Furthermore, in a backward pass style of updating, all states for that stage must be updated before moving to the previous stage. Remember that for stage T, the value function of the next stage is zero. Thus, if $V(l,s)$ denotes the value function for state l and stage s, then $V(l,T+1)=0$ for all l.

The steps below present a routine (function) that must be performed for each state-stage pair separately. A step-size α, which could be state-dependent, will be used in updating the probabilities of the learning automaton (LA). In each state, the LA will store an action–selection probability of $P(i,a)$ such that $\sum_a P(i,a) = 1$. Some other counters will be needed: $TR(l,s,u)$ will measure the total reward accumulated thus far within the routine when action u is tried in state l when encountered in stage s, and $N(l,s,u)$ will measure the number of times action u has been tried in the state-stage pair, (l,s).

The initialization step must be performed each time this routine is called. For each call, we have a given state, i, and a given stage, s.

Pursuit Learning Automata (PLA) Sampling Algorithm

Initialization: Input $k_{\max}(i,s)$ = number of iterations (simulations) allowed for each state-stage pair (i,s) (simulation budget), $\alpha \in (0,1)$. Initialize the LA probabilities as follows: For any $l \in \mathscr{S}$ and all $a \in \mathscr{A}(l)$, $P(l,a) = 1/|\mathscr{A}(l)|$. Initialize the counters, $TR(l,s,u)$ and $N(l,s,u)$, for all $l \in \mathscr{S}$, all $s \in \mathscr{T}$, and all $u \in \mathscr{A}(l,s)$, to 0; the iteration count $k = 0$; and $V(l,T+1) = 0$ for all l.

Step 1. For current state i and current stage s, select action a w.p. $P(i,a)$, and simulate to reach next state j and next stage $(s+1)$, receiving reward $r(i,s,a,j,s+1)$.
Step 2. Update the following quantities:

$$TR(i,s,a) \leftarrow TR(i,s,a) + r(i,s,a,j,s+1) + V(j,s+1); \tag{13.17}$$

$$N(i,s,a) \leftarrow N(i,s,a) + 1.$$

Then compute

$$Q(i,s,a) \leftarrow \frac{TR(i,s,a)}{N(i,s,a)}. \tag{13.18}$$

Step 3. Determine the greedy action as follows:

$$a^* \in \arg\max_{b \in \mathscr{A}(l,s)} Q(l,s,b).$$

Step 4. Update the action–selection probabilities: For all $u \in \mathscr{A}(i,s)$,

$$P(i,u) \leftarrow (1-\alpha)P(i,u) + \alpha \mathbf{1}\{a^* = u\}.$$

Step 5. Increment k by 1. If $k < k_{\max}(i,s)$, then return to Step 1; otherwise, go to Step 6.
Step 6. Set

$$V(i,s) = Q(i,s,a^*).$$

Stop.

Note that the update in Eq. (13.17) requires the value function of the next state-stage pair. In a backward pass application of the above routine, this value will already have been computed and will be available for use in the update. In case one does not use a backward pass, estimating this value will have to be done via a recursive call to the next state-stage combination $(j,s+1)$. It is also interesting to note that the computation of the Q-factor in Eq. (13.18) is based on *direct* averaging, a concept used in model-building (also called model-based) RL [79].

13.6 Numerical Results

In this section, we present numerical results on some small problems in order to illustrate the use of RL algorithms. The case studies covered here are for problems whose transition probabilities can be estimated; thus, it is possible to determine whether the optimal solution was reached. Furthermore, these problems can also be used as testbeds by other researchers for empirical investigation of their own algorithms. In practice, a simulation analyst is generally also interested in testing an algorithm on large-scale problems, whose transition probabilities are difficult to estimate and the only benchmarks available are problem-specific heuristics. Large-scale versions of both case studies that we cover here are also candidates for such tests. We begin with the infinite-horizon problem, and then discuss the finite horizon case.

13.6.1 Infinite Horizon

The case study we cover here is on preventive maintenance of machines and is primarily drawn from [34]. It is well-known that systems whose probabilities of failures increase as they age can benefit from preventive maintenance. Examples of such systems include production lines, bridges, roads, and electric power plants.

We consider the case of a production line which deteriorates with time and the deterioration can be captured by a function. After a preventive maintenance, generally, the system has a lower probability of failure than when it failed. Since it generally costs much lower to preventively maintain a line than to repair it after a failure, a significant volume of literature has appeared in the area of preventive maintenance. Toyota Motors have popularized the use of preventive maintenance in many automobile firms. A large chunk of the literature studies the problem of determining the time interval after which maintenance should be performed.

We make the following assumptions about the system:

- The production line is needed every day.
- If the line fails during the day, the repair takes the remainder of the day, and the line is available only the next morning. After a repair, the line is as good as new.
- When a line is shut down for preventive maintenance, it is down for the entire day. After a preventive maintenance, the line is as good as new.
- If σ denotes the number of days elapsed since the last preventive maintenance or repair (subsequent to a failure), the probability of failure during the σth day can be modeled as $(1-\xi\psi^{\sigma+2})$, where ξ and ψ are scalars in the interval $(0,1)$, whose values can be estimated from the data for time between successive failures of the system.
- For any given positive value of $\varepsilon \in \mathbb{R}$, we define $\bar{\sigma}_\varepsilon$ as the minimum integer value of σ such that the probability of failure on the $\bar{\sigma}$th day is less than or equal to $(1-\varepsilon)$. Since a fixed value of ε will be used, we will drop ε from the

notation. The definition of $\bar{\sigma}$ will allow us to truncate the countably infinite state space to a finite one. The resulting state space of the system will be assumed to be $\mathscr{S} = \{0,1,2,\ldots,\bar{\sigma}\}$, i.e., the probability of failure on the $\bar{\sigma}$th day will be assumed to equal 1. Furthermore, note that rounding of this nature is necessary to ensure that the probabilities in the last row of the finite Markov chain for one of the actions (the production action in particular) add up to 1.

- The costs of maintenance and repair are known with certainty, and will equal C_m and C_r respectively.

The underlying system transitions can be modeled via Markov chains as follows. Let the state of the system be defined by σ, the number of days elapsed since the last preventive maintenance or repair. Clearly, when a maintenance or repair is performed, σ is set to 0. If a successful day of production occurs, i.e., no failure occurs during the day, the state of the system is increased by 1. Each morning, the manager has to choose from two actions: $\{produce, maintain\}$. Then, we have the following transition probabilities for the system. We first consider the action *produce*. For $\sigma = 0,1,2,\ldots,\bar{\sigma}-1$,

$$p(\sigma, produce, \sigma+1) = \xi\psi^{\sigma+2}; \quad p(\sigma, produce, 0) = 1 - \xi\psi^{\sigma+2}.$$

For $\sigma = \bar{\sigma}$, $p(\sigma, produce, 0) = 1$. For all other cases not specified above,

$$p(\cdot, produce, \cdot) = 0.$$

For the action *produce* and all values of σ,

$$r(\sigma, produce, 0) = -C_r; \quad r(\sigma, produce, \sigma') = 0 \text{ when } \sigma' \neq 0.$$

For the action *maintain*, the mathematical dynamics will be defined as follows. For all values of σ, $p(\sigma, maintain, 0) = 1$ and $r(\sigma, maintain, 0) = -C_m$. For all other cases not specified above, $p(., maintain, .) = 0$ and $r(., maintain, .) = 0$.

We set $\xi = 0.99$, $\psi = 0.96$, $C_m = 4$, $C_r = 2$, and $\bar{\sigma} = 30$. Thus, we have 31 states and 2 actions. Our objective function is average reward, and the optimal policy, which is determined via policy iteration, is of a threshold nature in which the action is to produce for $\sigma = 0,1,\ldots,5$ and to maintain from $\sigma = 6$ onwards. The contracting factor version of the algorithm R-SMART was used in a simulator with the following specifications:

- The artificial contraction factor, $\bar{\gamma}$, was set to 0.99.
- The learning rates used were:

$$\alpha^k = \frac{1000}{5000+k}; \quad \beta^k = \frac{1000}{k(5000+k)} \text{ where } k \geq 1.$$

- The exploration probability was defined as shown below:

$$P^k = 0.5 \frac{log(k+1)}{k+1} \text{ where } k \geq 1.$$

- The system was simulated for 200,000 days.

The algorithm always generated the optimal solution in every replication. At least 20 different replications were performed.

The above problem can be studied for a much larger state space. If the time between failures is significantly higher and as a result $\bar{\sigma}$ is closer to 1,000, the transition probability matrix contains a million elements, which is not easy to handle. In other words, dynamic programming breaks down. However, it is not difficult to simulate this system, allowing us to use the simulation-based algorithm—if necessary in conjunction with some function approximation. It is also very critical to note here that unlike the problem considered above, usually the transition probability structure is not available, and the system can still be simulated as long as the distributions of the input random variables are available. Thus, for example, in an $M/G/1$ queue, if one observes the system at the instants when arrivals occur, one can formulate a Markov chain. However, the transition probabilities of this Markov chain are not necessary to simulate the system; rather one can simulate the queue using the distributions of the inter-arrival time and the service time. In preventive maintenance problems also, simulation of the system is often possible without generating the transition probabilities (see e.g., [26]).

In general, look-up tables work with up to maybe 2,000 Q-factors in regular computers, but for a state-action space larger than that, some sort of function approximation becomes necessary. Thus, for large-scale MDPs, it is imperative that the analyst either seeks to reduce the state-action space to a reasonable number or alternatively uses a function approximation scheme. Using function approximation introduces additional computational issues with regards to how to capture the state-action space in terms of the basis functions (architectures). It is not uncommon in practice to experiment with a large number of candidate architectures before the algorithm starts outperforming a heuristic or a set of heuristics known to generate reasonable results. It is this aspect of the large-scale problem that makes it necessary to study, or generate if necessary, some problem-specific heuristics. In summary, one can conclude that a successful implementation of RL on a large-scale problem is a significant computational exercise that requires patience.

13.6.2 Finite Horizon

We now consider a finite-horizon MDP where the objective is to maximize the expected value of the total undiscounted reward over the time horizon. We use an example, drawn from [24], on inventory control. Inventory control problems are ubiquitous in supply chain management, and indeed, most modern supply-chain software seek to solve problems of the nature considered here. Although the transition probabilities will be computed to determine the optimal solution, as in the previous subsection, these problems can be solved for any given distribution of the input random variable—the size of the demand (in one period/stage) in this case.

In many inventory control problems, the manager seeks to order raw material in a fashion so as to minimize the fixed costs of ordering (also called set-up costs), the costs of lost sales, and the costs of holding inventory. In general, large order quantities increase holding costs, but reduce the costs of lost sales and the ordering costs; on the other hand, small order quantities increase the risk of lost sales, the ordering costs, but minimize the inventory holding costs. Clearly, without holding costs, the problem has a trivial solution, which is: order the maximum possible quantity. A goal of modern inventory control [5], however, is to maximize inventory turns and minimize inventory. As such, the holding costs play a critical role in this problem. Moreover, this is a multiple-period problem in which decisions for ordering quantities have to be made in every period (stage) separately and unused inventory from a previous time period is carried over into the next time period.

Let D_s denote the demand during the sth stage (period), x_s denote the inventory at the start of the sth stage, and u_s denote the action chosen in the sth stage. The action u_s will equal the amount ordered at the start of the sth stage. The following assumptions will be made about the problem:

- There is no backordering, and a demand lost is lost forever.
- The size of demand in a given time period is a discrete random variable whose distribution is known.
- The holding costs per unit per unit time, h, the cost per order, A, and the lost sales cost per unit (opportunity cost), p, are known with certainty.
- We will assume that the demand will be realized at the end of the stage, and the order placed at the start of a stage will arrive at the end of the stage. This will simplify the computation of the inventory holding costs.
- There is an upper limit, M, on the amount of inventory that can be held (dictated by storage requirements). This implies that an ordering amount that causes inventory to exceed M will not be allowed.
- The number of stages, T, is deterministic and known.
- The starting inventory, x_1, is known with certainty.

Under these conditions, the inventory levels will then change from one stage to the next as follows:

$$x_{s+1} = (x_s + u_s - D_s)^+, \quad x_1 \geq 0,$$

where $x^+ \equiv \max(x, 0)$, which reflects the condition that inventory cannot be negative due to the no backordering assumption. The goal is to minimize the expected total costs of operating the system, where costs can arise out of holding inventory, lost sales, and the set-up (fixed) cost per order. The total cost in one trajectory can be computed as follows:

$$\sum_{s=1}^{T} \left[A \cdot \mathbf{1}\{u_s > 0\} + hx_s + py_s^- \right],$$

where $x^- \equiv \max(-x,0)$ and $y_{s+1} = x_s + u_s - D_s$ for $s = 1,2,\ldots,T$, and $y_1 = 0$.

The following values were used for the inputs in the experiments reported in [24]: $p = 10$, $A = 5$, $h = 1$, $M = 20$, $T = 3$, $x_1 = 5$, and $k_{\max}(i,s) = 4$ for all state-stage pairs. The demand was assumed to have a discrete uniform distribution, $DU(0,9)$. Backward dynamic programming was used to determine the optimal solution, which was also delivered by the PLA sampling technique. Then, via simulation, the value function for the starting state, $V_{PLA}(x_1,1)$, was computed for the optimal policy, using 30 replications and a large number of samples. This value was found to be 24.48 with a standard error of 0.51, which compares well with the true value, $V_*(x_1,1) = 27.322$, determined via backward dynamic programming (which needed the exact transition probabilities). Numerous results with other values for the inputs can be found in [24].

13.7 Extensions

The ideas underlying MDPs can be extended to at least three other domains: semi-MDPs (SMDPs), stochastic games, also called Competitive Markov decision processes (CMDPs), and Partially Observable MDPs (POMDPs).

In an SMDP, the time of transition from one state to another is not the same for every transition. In the most general case, this time is a generally distributed random variable. Although the SMDP is sometimes loosely referred to as a continuous time MDP (CTMDP), the latter name is usually reserved for the SMDP in which the transition times have the exponential distribution. The theory of SMDPs can be studied for both discounted and average reward objective functions [15]. RL for SMDPs has been studied in [19] (discounted reward) and [31,38] (average reward).

In a CMDP, there are multiple decision-makers, and the transition probabilities and rewards depend on the actions of all or a subset of all decision-makers. The problem becomes significantly more complex and has been studied in detail in [29]. These problems are of considerable interest to economists, and both of the early contributors [61,73] have been awarded Nobel prizes in economics. The work in [61] provides a critical idea, called Nash equilibrium, needed for solving a CMDP, while the work of Shapley [73] provides the first attempt at value iteration and solving the CMDP computationally.

In a POMDP, the underlying state is only partially observable to the decision maker via signals, and the decision-maker is required to choose the best possible action. The POMDP has been used in robotics and pattern recognition problems.

Not surprisingly, attempts have been made to use simulation to solve POMDPs and CMDPs via simulation. Some noteworthy works in CMDPs include the algorithms in [46,69]. Some special forms of sequential games in which the actions of the different decision-makers are not concurrently taken but are sequential are studied in [12]. For POMDPs, the body of literature is somewhat evolved. See [49] for a survey of early algorithms and [24] for some more recent simulation-based algorithms.

A large number of computational extensions can be found in computer science, e.g., hierarchical RL, although convergence guarantees for the underlying theory are somewhat sparse. Another topic that we did not cover is that of policy gradients [9, 77], which is rooted in the idea of using simulation and derivatives of the objective function to generate an optimal policy in MDPs. These algorithms suffer from large variance.

13.8 Convergence Theory

The convergence theory for RL algorithms has become quite rigorous. Here, we provide a brief account. At least three different lines of convergence arguments have been worked out for classical Q-Learning-type algorithms:

1. theory developed in [12], which is based primarily on the idea of "reducing cube sizes" (see Prop. 4.5 of [12]);
2. theory developed in [48, 75, 83], which is based on some remarkably simple arguments and draws on some basic results in stochastic approximation (see [48] in particular), and
3. theory based on ordinary differential equations, for which the reader is referred to [16].

Although one finds three distinct strands of convergence arguments, many of the proofs rest on showing some basic properties, e.g., the underlying transformation is contractive (shown for Q-Learning and the SSP-version of Q-Learning in [12]) and the iterates remain bounded (shown for the SSP-version of Q-Learning in [12] and for classical Q-Learning in [33]). In other words, by exploiting these fundamental convergence arguments, generally showing convergence boils down to stability and contraction arguments, which are much simpler.

More recently, convergence arguments when the algorithm is combined with function approximation have been developed; [15] provides an up-to-date account and presents some of the open problems. The algorithms based on evolutionary search and learning automata have an independent convergence theory, which has been developed in depth in texts such as [24, 60].

13.9 Concluding Remarks

This chapter presented selected topics on RL relevant to simulation optimization. We began with Q-Learning, along with a discussion on how to combine it with a simple linear-basis function approximator, which is critical for large-scale problems that simulation-based optimization seeks to solve. We also covered actor-critics, but only the most modern version. SARSA(λ) was covered in some detail, because although a heuristic, it remains popular and can be implemented in simulators. Thereafter, we presented algorithms belonging to the class called API. Our discussion included average reward and the finite-horizon problems, which are not covered in most texts in much detail. Extensions and convergence theory were briefly described. Here, we summarize some applications and some of the open problems that should be exciting topics for future research.

- *Applications*: Since this area has origins in machine learning, some of the initial applications were naturally in the area of robotics and computer science. Even today, these algorithms find applications in exciting areas in machine learning, e.g., autonomous helicopter control [1, 63] and fMRI studies [97]. However, the body of literature that applies these algorithms to industrial tasks is expanding. Some of the early applications in the area of operations management include jobshop scheduling [99], AGV routing [84], preventive maintenance [26], and airline revenue management [31, 40]. Some more recent work includes the beer game of supply chain management [22], wireless communication [3, 96], irrigation control [72], and reentrant semiconductor manufacturing [68].
- *Function approximation and convergence*: While many advances have been made in combining RL with function approximators, this is an area that needs significant additional research without which function approximation will remain the Achilles' heel of RL. Not surprisingly, recent coverage of RL seems to stress the function approximation aspects, many of which are heuristically applied without convergence analysis. The Bellman error development is also rather heuristic, and apart from the traditional techniques used in conjunction with Bellman error [6, 17, 18, 27, 89], recently other techniques have been sought to be used: see [64] (kernel-based function approximation), [92] (evolutionary function approximation), and [56] (Laplacian methods). Regardless of the nature of the function approximation technique used, a preliminary step in this process involves transforming the state space into the feature space, and some important references in this context include [43] (coarse coding), [4] (CMAC coding), and [50] (Kanerva coding). Another promising line of investigation includes the LSTD (least-squares temporal differences) algorithm [62, 98]. Much of this work has occurred along the lines of API, which also happens to be an active area of research [14]. See also [55] where a Q-function-based LSTD algorithm, which is suitable for a simulation-based setting, is presented, although the algorithm's convergence has never been proved.
- *TD(λ) in combination with Q-Learning*: Although the concept of TD(λ) has been around for a long time in this field, its convergence properties are known well for the case when it is combined with policy evaluation within an API algorithm. In API, whether $\lambda > 0$ provides any advantages over $\lambda = 0$ has not been tested empirically in a comprehensive manner; one clear disadvantage is that it requires eligibility traces that are not easily combined with function approximation. When combined with value iteration algorithms, such as Q-Learning (see Q(λ) learning

in [65]), the worth of TD(λ) becomes even more doubtful, since no convergence guarantees to optimality are available. However, in practice, Q(λ) appears to converge significantly faster than Q-Learning [65]. See [83] for an interesting result which shows that Q(λ) does converge but not necessarily to the optimal solution. It is clear that issues with respect to combining TD(λ) with algorithms such as Q-Learning and SARSA remain important open problems that require further investigation.

- *Optimistic API*: Although much research has occurred in classical versions of API, it is well-known that it is slow, since it requires numerous simulations to evaluate just one policy. As such, there is great interest in optimistic API in which the policy is evaluated via just one (or a few) state transition. The interest in API also stems from the empirical evidence suggesting that function approximation works better in conjunction with API rather than with value-iteration-based methods. Interestingly, the actor-critic, one of the earliest algorithms in RL [8,95], sought to attain the same objective. One recent algorithm in this direction is [13].
- *Model-building algorithms*: Model-building (also called model-based in the RL literature, but not to be confused with the same term in optimization) algorithms do not need the transition probability model, but build it within the simulator while simultaneously solving the Bellman equation. The earliest model-building algorithms for discounted and average reward are [7, 84], respectively. Other works on model building include [20,28,51,59,78,81,87] While most of these algorithms seek to generate the value function and are rooted in DP, some recent algorithms are based on Q-Learning and actor-critic frameworks [36,41]. These algorithms have attracted significant interest recently in applications: robotic soccer [94], helicopter control [1,54,63], function magnetic imaging resonance (fMRI) studies of brain [47,97], and vision [57].

References

1. P. Abbeel, A. Coates, T. Hunter, and A. Y. Ng. Autonomous autorotation of an RC helicopter. In *International Symposium on Robotics*, 2008.
2. J. Abounadi, D. Bertsekas, and V. S. Borkar. Learning algorithms for Markov decision processes with average cost. *SIAM Journal of Control and Optimization*, 40(3):681–698, 2001.
3. N. Akar and S. Sahin. Reinforcement learning as a means of dynamic aggregate qos provisioning. In *Lecture Notes in Computer Science, Volume 2698/2003*. Springer, Berlin/Heidelberg, 2003.
4. J. S. Albus. *Brain, Behavior and Robotics*. Byte Books, Peterborough, NH, 1981.
5. R. Askin and J. Goldberg. *Design and Analysis of Lean Production Systems*. Wiley, NY, 2002.
6. L. Baird. Residual algorithms: Reinforcement learning with function approximation. In *Proceedings of the Twelfth International Conference on Machine Learning*, pages 30–37. Morgan Kaufmann, 1995.
7. A. G. Barto, S. J. Bradtke, and S. P. Singh. Learning to act using real-time dynamic programming. *Artificial Intelligence*, 72:81–138, 1995.

8. A. G. Barto, R. S. Sutton, and C. W. Anderson. Neuronlike elements that can solve difficult learning control problems. *IEEE Transactions on Systems, Man, and Cybernetics*, 13:835–846, 1983.
9. J. Baxter and P. Bartlett. Infinite-horizon policy-gradient estimation. *Journal of Artificial Intelligence*, 15:319–350, 2001.
10. R. E. Bellman. The theory of dynamic programming. *Bull. Amer. Math. Soc*, 60:503–516, 1954.
11. R. E. Bellman. *Dynamic Programming*. Princeton University Press, Princeton, NJ, 1957.
12. D. Bertsekas and J. Tsitsiklis. *Neuro-Dynamic Programming*. Athena Scientific, Belmont, MA, USA, 1996.
13. D. Bertsekas and H. Yu. Q-learning and enhanced policy iteration in discounted dynamic programming. In *Proceedings of the 49th IEEE Conference on Decision and Control*, pages 1409–1416, 2010.
14. D. P. Bertsekas. Approximate policy iteration: A survey and some new methods. *Journal of Control Theory and Applications*, 9(3):310–335, 2011.
15. D. P. Bertsekas. *Dynamic Programming and Optimal Control*. Athena Scientific, Belmont, MA, USA, 4th edition, 2012.
16. V. S. Borkar. *Stochastic Approximation: A Dynamical Systems Viewpoint*. Hindusthan Book Agency, New Delhi, India, 2008.
17. J. A. Boyan and A. W. Moore. Generalization in Reinforcement Learning: Safely Approximating the Value Function. *Advances in Neural Information Processing Systems*, pages 369–376, 1995.
18. S. Bradtke and A. G. Barto. Linear least squares learning for temporal differences learning. *Machine Learning*, 22:33–57, 1996.
19. S. Bradtke and M. Duff. Reinforcement learning methods for continuous-time Markov decision problems. In *Advances in Neural Information Processing Systems 7*. MIT Press, Cambridge, MA, USA, 1995.
20. R. I. Brafman and M. Tennenholtz. R-max: A general polynomial time algorithm for near-optimal reinforcement learning. *Journal of Machine Learning Research*, 3:213–231, 2002.
21. X. R. Cao. *Stochastic Learning and Optimization: A Sensitivity-Based View*. Springer, New York, 2007.
22. S. K. Chaharsooghi, J. Heydari, and S. H. Zegordi. A reinforcement learning model for supply chain ordering management: An application to the beer game. *Decision Support Systems*, 45, 2008.
23. H. S. Chang, M. C. Fu, J. Hu, and S. I. Marcus. Recursive learning automata approach to Markov decision processes. *IEEE Transactions on Automatic Control*, 52(7):1349–1355, 2007.
24. H. S. Chang, M. C. Fu, J. Hu, and S. I. Marcus. *Simulation-based Algorithms for Markov Decision Processes*. Springer, 2007.
25. H. S. Chang, H. G. Lee, M. C. Fu, and S. I. Marcus. Evolutionary policy iteration for solving Markov decision processes. *IEEE Transactions on Automatic Control*, 50(11):1804–1808, 2005.
26. T. K. Das, A. Gosavi, S. Mahadevan, and N. Marchalleck. Solving semi-Markov decision problems using average reward reinforcement learning. *Management Science*, 45(4):560–574, 1999.
27. S. Davies. Multi-dimensional interpolation and triangulation for reinforcement learning. *Advances in Neural Information and Processing Systems*, 1996.
28. C. Diuk, L. Li, and B. Leffler. The adaptive k-meteorologists problem and its application to structure learning and feature selection in reinforcement learning. In *Proceedings of the 26th Annual International Conference on Machine Learning*, 2009.
29. J. Filar and K. Vrieze. *Competitive Markov Decision Processes*. Springer-Verlag, New York, NY, USA, 1997.
30. A. Gosavi. *Simulation-Based Optimization: Parametric Optimization Techniques and Reinforcement Learning*. Springer, Boston, 2003.

31. A. Gosavi. A reinforcement learning algorithm based on policy iteration for average reward: Empirical results with yield management and convergence analysis. *Machine Learning*, 55:5–29, 2004.
32. A. Gosavi. Reinforcement learning for long-run average cost. *European Journal of Operational Research*, 155:654–674, 2004.
33. A. Gosavi. Boundedness of iterates in Q-learning. *Systems and Control Letters*, 55:347–349, 2006.
34. A. Gosavi. A risk-sensitive approach to total productive maintenance. *Automatica*, 42:1321–1330, 2006.
35. A. Gosavi. Reinforcement learning: A tutorial survey and recent advances. *INFORMS Journal on Computing*, 21(2):178–192, 2009.
36. A. Gosavi. Reinforcement learning for model building and variance-penalized control. In *Proceedings of the 2009 Winter Simulation Conference*. IEEE, Piscataway, NJ, 2009.
37. A. Gosavi. Finite horizon Markov control with one-step variance penalties. In *Conference Proceedings of the Allerton Conference*. University of Illinois, USA, 2010.
38. A. Gosavi. Target-sensitive control of Markov and semi-Markov processes. *International Journal of Control, Automation, and Systems*, 9(5):1–11, 2011.
39. A. Gosavi. Approximate policy iteration for Markov control revisited. In *Procedia Computer Science, Complex Adaptive Systems, Chicago*. Elsevier, 2012.
40. A. Gosavi, N. Bandla, and T. K. Das. A reinforcement learning approach to a single leg airline revenue management problem with multiple fare classes and overbooking. *IIE Transactions*, 34(9):729–742, 2002.
41. A. Gosavi, S. Murray, J. Hu, and S. Ghosh. Model-building adaptive critics for semi-Markov control. *Journal of Artificial Intelligence and Soft Computing Research*, 2(1), 2012.
42. C. M. Grinstead and J. L. Snell. *Introduction to Probability*. American Mathematical Society, Providence, RI, 1997.
43. G. E. Hinton. Distributed representations. Technical Report, CMU-CS-84-157, Carnegie Mellon University, Pittsburgh, PA, USA, 1984.
44. J. H. Holland. *Adaptation in Natural and Artificial Systems*. University of Michigan Press, Ann Arbor, MI, USA, 1975.
45. R. Howard. *Dynamic Programming and Markov Processes*. MIT Press, Cambridge, MA, 1960.
46. J. Hu and M. P. Wellman. Nash Q-Learning for general-sum stochastic games. *Journal of Machine Learning Research*, 4:1039–1069, 2003.
47. S. Ishii, W. Yoshida, and J. Yoshimoto. Control of exploitation-exploration meta-parameter in reinforcement learning. *Neural Networks*, 15:665–687, 2002.
48. T. Jaakkola, M. Jordan, and S. Singh. On the convergence of stochastic iterative dynamic programming algorithms. *Neural Computation*, 6(6):1185–1201, 1994.
49. L. P. Kaelbling, M. L. Littman, and A. W. Moore. Reinforcement learning: A survey. *Journal of Artificial Intelligence Research*, 4:237–285, 1996.
50. P. Kanerva. *Sparse Distributed Memory*. MIT Press, Cambridge, MA, USA, 1988.
51. M. Kearns and S. P. Singh. Near-optimal reinforcement learning in polynomial time. *Machine Learning*, 49(2):209–232, 2002.
52. S. S. Keerthi and B. Ravindran. A tutorial survey of reinforcement learning. *Sadhana*, 19(6):851–889, 1994.
53. V. Konda and V. S. Borkar. Actor-critic type learning algorithms for Markov decision processes. *SIAM Journal on Control and Optimization*, 38(1):94–123, 1999.
54. R. Koppejan and S. Whiteson. Neuroevolutionary reinforcement learning for generalized helicopter control. In *GECCO: Proceedings of the Genetic and Evolutionary Computation Conference*, pages 145–152, 2009.
55. M. Lagoudakis and R. Parr. Least-squares policy iteration. *Journal of Machine Learning Research*, 4:1107–1149, 2003.
56. S. Mahadevan. Learning representation and control in Markov decision processes: New frontiers. In *Foundations and Trends in Machine Learning, Vol I(4)*, pages 403–565. Now Publishers, 2009.

57. J. Michels, A. Saxena, and A. Y. Ng. High speed obstacle avoidance using monocular vision and reinforcement learning. In *Proceedings of the 22nd International Conference on Machine Learning, Bonn, Germany*, 2005.
58. T. M. Mitchell. *Machine Learning*. McGraw Hill, Boston, MA, USA, 1997.
59. A. Moore and C. Atkeson. Prioritized sweeping: Reinforcement learning with less data and less real time. *Machine Learning*, 13:103–130, 1993.
60. K. Narendra and M. Thathachar. *Learning Automata: An Introduction*. Prentice Hall, Englewood Cliffs, NJ, USA, 1989.
61. J. F. Nash. Equilibrium points in n-person games. *Proceedings, Nat. Acad. of Science, USA*, 36:48–49, 1950.
62. A. Nedić and D. P. Bertsekas. Least-squares policy evaluation with linear function approximation. *Discret-event Dynamic Systems: Theory and Applications*, 13:79–110, 2003.
63. A. Y. Ng, H. J. Kim, M. I. Jordan, and S. Sastry. Autonomous helicopter flight via reinforcement learning. In *Advances in Neural Information Processing Systems 17*. MIT Press, 2004.
64. D. Ormoneit and S. Sen. Kernel-based reinforcement learning. *Machine Learning*, 49(2–3):161–178, 2002.
65. J. Peng and R. J. Williams. Incremental multi-step Q-learning. In *Machine Learning*, pages 226–232. Morgan Kaufmann, 1996.
66. W. Powell. *Approximate Dynamic Programming: Solving the curses of dimensionality*. Wiley-Interscience, NJ, USA, 2007.
67. K. Rajaraman and P. Sastry. Finite time analysis of the pursuit algorithm for learning automata. *IEEE Transactions on Systems, Man, and Cybernetics: Part B*, 26(4):590–598, 1996.
68. J. A. Ramirez-Hernandez and E. Fernandez. A case study in scheduling re-entrant manufacturing lines: Optimal and simulation-based approaches. In *Proceedings of 44th IEEE Conference on Decision and Control*, pages 2158–2163. IEEE, 2005.
69. K. Ravulapati, J. Rao, and T. Das. A reinforcement learning approach to stochastic business games. *IIE Transactions*, 36:373–385, 2004.
70. H. Robbins and S. Monro. A stochastic approximation method. *Annals of Mathematical Statistics*, 22:400–407, 1951.
71. G. Rummery and M. Niranjan. On-line Q-learning using connectionist systems. Technical Report CUED/F-INFENG/TR 166. Engineering Department, Cambridge University, 1994.
72. N. Schutze and G.H.Schmitz. Neuro-dynamic programming as a new framework for decision support for deficit irrigation sytems. In *International Congress on Modelling and Simulation, Christchurch, New Zealand*, pages 2271–2277, 2007.
73. L. Shapley. Stochastic games. *Proceedings of the National Academy of Sciences*, 39:1095–1100, 1953.
74. J. Si, A. G. Barto, W. B. Powell, and D. Wunsch, editors. *Handbook of Learning and Approximate Dynamic Programming*. IEEE Press, Wiley, Hoboken, NJ, 2004.
75. S. Singh, T. Jaakkola, M. Littman, and C. Szepesvári. Convergence results for single-step on-policy reinforcement-learning algorithms. *Machine Learning*, 39:287–308, 2000.
76. S. Singh and R. Sutton. Reinforcement learning with replacing eligibility traces. *Machine Learning*, 22:123–158, 1996.
77. S. Singh, V. Tadic, and A. Doucet. A policy-gradient method for semi-Markov decision processes with application to call admission control. *European Journal of Operational Research*, 178(3):808–818, 2007.
78. A. Strehl and M. Littman. A theoretical analysis of model-based interval estimation. In *Proceedings of the 22th International Conference on Machine Learning*, pages 856–863, 2005.
79. R. Sutton and A. G. Barto. *Reinforcement Learning: An Introduction*. The MIT Press, Cambridge, MA, USA, 1998.
80. R. S. Sutton. Learning to predict by the method of temporal differences. *Machine Learning*, 3:9–44, 1988.
81. R. S. Sutton. Integrated architectures for learning, planning, and reacting based on approximating dynamic programming. In *Proceedings of the 7th International Workshop on Machine Learning*, pages 216–224. Morgan Kaufmann, San Mateo, CA, 1990.

82. C. Szepesvári. *Algorithms for Reinforcement Learning*. Synthesis Lectures on Artificial Intelligence and Machine Learning. Morgan & Claypool Publishers, 2010.
83. C. Szepesvári and M. Littman. A unified analysis of value-function-based reinforcement-learning algorithms. *Neural Computation*, 11(8):2017–2060, 1999.
84. P. Tadepalli and D. Ok. Model-based average reward reinforcement learning algorithms. *Artificial Intelligence*, 100:177–224, 1998.
85. M. Thathachar and P. Sastry. A class of rapidly converging algorithms for learning automata. *IEEE Transactions on Systems, Man, and Cybernetics*, 15:168–175, 1985.
86. J. E. E. van Nunen. A set of successive approximation methods for discounted Markovian decision problems. *Z. Operations Research*, 20:203–208, 1976.
87. H. van Seijen, S. Whiteson, H. van Hasselt, and M. Wiering. Exploiting best-match equations for efficient reinforcement learning. *Journal of Machine Learning Research*, 12:2045–2094, 2011.
88. C. J. Watkins. *Learning from Delayed Rewards*. PhD thesis, Kings College, Cambridge, England, 1989.
89. P. J. Werbös. Building and understanding adaptive systems: A statistical/numerical approach to factory automation and brain research. *IEEE Transactions on Systems, Man., and Cybernetics*, 17:7–20, 1987.
90. P. J. Werbös. Consistency of HDP applied to a simple reinforcement learning problem. *Neural Networks*, 3:179–189, 1990.
91. R. M. Wheeler and K. S. Narenda. Decentralized learning in finite Markov chains. *IEEE Transactions on Automatic Control*, 31(6):373–376, 1986.
92. S. Whiteson and P. Stone. Evolutionary function approximation for reinforcement learning. *Journal of Machine Learning Research*, 7:877–917, 2006.
93. B. Widrow and M. E. Hoff. Adaptive Switching Circuits. In *Institute of Radio Engineers, Western Electronic Show and Convention, Convention Record, Part 4*, pages 96–104. 1960.
94. M. A. Wiering, R. P. Salustowicz, and J. Schmidhuber. Model-based reinforcement learning for evolving soccer strategies. In *Computational Intelligence in Games*. Springer Verlag, 2001.
95. I. H. Witten. An adaptive optimal controller for discrete time Markov environments. *Information and Control*, 34:286–295, 1977.
96. W. Yeow, C. Tham, and W. Wong. Energy efficient multiple target tracking in wireless sensor networks. *IEEE Transactions on Vehicular Technology*, 56(2):918–928, 2007.
97. W. Yoshida and S. Ishii. Model-based reinforcement learning: A computational model and an fMRI study. *Neurocomputing*, 63:253–269, 2005.
98. H. Yu and D. P. Bertsekas. Convergence results on some temporal difference methods based on least squares. *IEEE Transactions on Automatic Control*, 54(7):1515–1531, 2009.
99. W. Zhang and T. G. Dietterich. A reinforcement learning approach to job-shop scheduling. In *Proceedings of the Fourteenth International Joint Conference on Artificial Intelligence*, pages 1114–1120. Morgan Kaufmann, 1995.

Index

M.C. Fu (ed.), *Handbook of Simulation Optimization*, International Series in Operations Research & Management Science 216, DOI 10.1007/978-1-4939-1384-8

MIX
Papier aus verantwortungsvollen Quellen
Paper from responsible sources
FSC® C105338

If you have any concerns about our products,
you can contact us on
ProductSafety@springernature.com

In case Publisher is established outside the EU,
the EU authorized representative is:
Springer Nature Customer Service Center GmbH
Europaplatz 3, 69115 Heidelberg, Germany

Printed by Libri Plureos GmbH
in Hamburg, Germany